INTERNATIONAL

REVIEW OF CYTOLOGY

VOLUME 62

INTERNATIONAL

Review of Cytology

EDITED BY

G. H. BOURNE
St. George's University School of Medicine
St. George's, Grenada
West Indies

J. F. DANIELLI
Worcester Polytechnic Institute
Worcester, Massachusetts

ASSISTANT EDITOR
K. W. JEON
Department of Zoology
University of Tennessee
Knoxville, Tennessee

VOLUME 62

1980

ACADEMIC PRESS *A Subsidiary of Harcourt Brace Jovanovich, Publishers*
New York London Toronto Sydney San Francisco

ACADEMIC PRESS, INC.
111 Fifth Avenue, New York, New York 10003

United Kingdom Edition published by
ACADEMIC PRESS, INC. (LONDON) LTD.
24/28 Oval Road, London NW1 7DX

LIBRARY OF CONGRESS CATALOG CARD NUMBER: 52–5203

ISBN 0–12–364462–3

PRINTED IN THE UNITED STATES OF AMERICA

79 80 81 82 9 8 7 6 5 4 3 2 1

Contents

Immunocytochemical Localization of the Vertebrate Cyclic Nonapeptide Neurohypophyseal Hormones and Neurophysins

K. Dierickx

Recent Progress in the Morphology, Histochemistry, Biochemistry, and Physiology of Developing and Maturing Mammalian Testis

Sardul S. Guraya

Transitional Cells of Hemopoietic Tissues: Origin, Structure, and Development Potential

Joseph M. Yoffey

Human Chromosomal Heteromorphisms: Nature and Clinical Significance

Ram S. Verma and Harvey Dosik

List of Contributors

Numbers in parentheses indicate the pages on which the authors' contributions begin.

K. DIERICKX (119), *Department of Embryology and Comparative Histology, Rijksuniversiteit, B-9900 Gent, Belgium*

HARVEY DOSIK (361), *Division of Hematology and Cytogenetics, The Jewish Hospital and Medical Center, Brooklyn, New York 11238*

MITSURU FURUSAWA (29), *Laboratory of Embryology, Department of Biology, Faculty of Science, Osaka City University, Osaka, Japan*

SARDUL S. GURAYA (187), *Department of Zoology, College of Basic Sciences and Humanities, Punjab Agricultural University, Ludhiana, Punjab, India*

ALLAN PENTECOST (1), *Department of Applied Biology, Chelsea College, University of London, London SW10 OQX, England*

V. RAGHAVAN (69), *Department of Botany, The Ohio State University, Columbus, Ohio 43210*

RAM S. VERMA (361), *Division of Hematology and Cytogenetics, The Jewish Hospital and Medical Center, Brooklyn, New York 11238*

JOSEPH M. YOFFEY (311), *Department of Anatomy and Embryology, Hebrew University–Hadassah Medical School, Jerusalem, Israel*

INTERNATIONAL REVIEW OF CYTOLOGY, VOL. 62

Calcification in Plants

ALLAN PENTECOST

Department of Applied Biology, Chelsea College, University of London, London, England

I. Introduction

Calcification is a rather loose term applied to the process of deposition of sparingly soluble calcium salts within or adjacent to the cells of animals and plants. Plants have far less capacity to calcify than animals, and while it is often easy to point to a calcified structure of the latter (e.g., a bone or shell) and understand its structural and physiological significance, this is rarely possible with plants.

Early this century plant biologists sought relationships between calcification and photosynthetic carbon dioxide metabolism (Pia, 1933). This is understandable as calcium carbonate is the most common salt deposited by plants and its precipitation is influenced by the thermodynamic state of the aqueous carbon dioxide system. Recent studies have shown that in many of the more sophisticated calcifying systems there is no direct link with photosynthesis although the energy provided by it is undoubtedly utilized in the calcification mechanism. Another problem not encountered in many animals concerns the extent to which some plants are actually involved in depositing these salts. There is evidence that many species become "calcified" indirectly as a result of living in an environment where carbonates may be spontaneously precipitated and tbere are others which utilize or excrete substances which can effect local deposition. There might be some basis, on these grounds, to distinguish between those plants exhibiting a purposeful and orderly internal control over the process and those

ISBN 0-12-364462-3

which appear to be associated with the deposition but have no means of regulating it. Unfortunately we still lack the information requisite for such a distinction as the next section will show.

Some members of all plant phyla show a capacity for carbonate deposition and the means by which it is achieved are almost as diverse as those of the individuals involved. There are some parallels between the mechanism in the plant and animal kingdoms but the process probably evolved on many occasions. Recent hypotheses advanced to explain the mechanisms of biomineralization have been much influenced by studies on the structure and chemistry of the organic matrices which are intimately associated with most animal deposits, together with what is currently known of ion movement in cells. These hypotheses, originally applied to animals, may be equally applied to some plants and will be discussed in Section V. Meanwhile, a general overview of calcification in the various plant phyla will be presented.

II. Calcification in the Plant Phyla

A. Prokaryotes

1. *Bacteria*

a. *Calcium Carbonate.* Marine and freshwater bacteria are frequently associated with carbonate deposits and numerous ''calcifying'' bacteria have been described (Pia, 1933; Pautard, 1970) but few have been isolated and studied *in vitro*. The pioneers (Drew, 1914; Kellerman and Smith, 1916) implicated ammonia production [Eq. (1)] (Kellerman and Smith, 1914). This was considered a satisfactory explanation for the precipitation but in the absence of ammonia alternative schemes involving acetate oxidation have been proposed [Eq. (2)] (Berkely, 1919).

$$2NH_4OH + Ca(HCO_3)_2 \rightarrow CaCO_3 + 2H_2O + (NH_4)_2CO_3 \quad (1)$$

$$Ca(CH_3COO)_2 + O_2 \rightarrow CaCO_3 + 3H_2O + 3C \quad (2)$$

Those which oxidize acetic acid can cause a substantial pH rise during growth and this will favor calcium carbonate precipitation by increasing the carbonate ion activity. Monaghan and Lytle (1956) noted pH values of 11 in a medium containing an unidentified sulfate-reducing marine bacterium which precipitated calcium carbonate. Unfortunately few of these studies provide empirically useful information although it is believed that bacteria play only a minor role in marine carbonate sedimentation (Berner, 1970). They may, however, play an important part in the nucleation of carbonates without affecting the pH of their environment significantly. Intracellular deposits of calcite have been described in several freshwater organisms (Bersa, 1926; Pia, 1933) but they require confirmation.

b. *Calcium Hydroxyapatite*. Most of our knowledge of the microorganisms associated with this mineral comes from studies of the dental calculus flora (see review by Ennever and Takazoe, 1973). The oral actinomycete *Bacterionema matruchotii,* when grown in metastable solutions of calcium phosphate (Ennever *et al.,* 1971), deposits both intra- and extracellular calcium hydroxyapatite. Studies with cell extracts suggest that a phospholipid initiates deposition (Ennever and Takazoe, 1973). When mesosome development is inhibited (Ennever *et al.,* 1970) calcification is reduced, suggesting membrane participation, and this has been verified by work on isolated *Bacterionema* membranes (Vogel and Smith, 1976). Similar studies with oral *Streptococcus* spp. (Streckfuss *et al.,* 1974) and *Escherischia coli* (Ennever *et al.,* 1974), which also calcify in metastable solutions, have not yet been demonstrated as membrane dependent. Oral *Streptoccus salivarius* (Streckfuss *et al.,* 1974) and a *Bacterionema* mutant (Streckfuss and Ennever, 1972) do not calcify *in vitro,* suggesting some degree of genetic control.

Calcification occurs randomly within the cells of these bacteria and this is usually followed by the growth of radiating clusters of mineral from several nucleation sites (see Ennever and Takazoe, 1973; Streckfuss *et al.,* 1974). Intracellular calcification may proceed through the cell wall to form extracellular deposits (Streckfuss *et al.,* 1974) although Lustmann *et al.* (1976) suggest that the bacteria become passively engulfed with calcification proceeding from the outside in. It is likely that both processes occur. Often the cells become completely calcified but the stage at which death occurs is unknown. Postmortem calcification has been reported in *E.coli* (Ennever *et al.,* 1974) and *Bacterionema* (Takazoe *et al.,* 1963). It is important to know whether these organisms calcify *in vivo* before some of the mechanisms propounded by Pautard (1970) can be properly tested.

2. *Cyanobacteria*

Many blue-green algae occur associated with marine and freshwater carbonate deposits (Golubic, 1973) and this is an association which extends back to the Precambrian (Walter, 1976). Contemporary freshwater species form roughly spherical or flattened concretions (oncolites) and more massive accumulations of tufa in well-aerated hardwater streams and lakes. The deposits consist of a core of consolidated and sometimes recrystallized calcite and discarded Cyanophyte sheaths with a superficial skin of viable cyanobacteria. Deposition occurs by the trapping and subsequent accretion of calcite microcrystals upon the surfaces of the polysaccharide sheaths (Pentecost, 1978). Those species so far examined appear to be obligate photoautotrophs and often grow up toward light as filaments perpendicular to the deposit surface (Golubic, 1973; Pentecost, 1978). Members of the Oscillatoriaceae predominate and it is likely that the filaments, which have little capacity for growth but show considerable motility (Castenholz,

1973), keep pace with the carbonate accretion by phototaxy. The precipitation of calcium carbonate is caused primarily by the loss of aqueous carbon dioxide to the atmosphere through the equilibration of carbonate-rich groundwaters (Barnes, 1965; Cole and Bachelder, 1969; Pentecost, 1975). In a medium whose major dissolved ions are Ca^{2+} and HCO_3^- the low levels of CO_2 uptake observed by cyanobacterial photosynthesis were considered by Pentecost (1978) to result in the formation of 10% or less of the precipitated carbonate. This facility to form carbonate deposits is probably related to the sheath composition since some species of *Scytonema* fail to encrust under favorable conditions. It is nevertheless a general phenomenon in which many species may participate (Fjerdingstad, 1951).

Marine cyanobacterial mats and multilayered stromatolites are important sediment binders and trappers along tropical shores (Golubic, 1973; Walter, 1976). Oscillatoriaceae are again the most important trappers of calcium carbonate which results primarily from seawater evaporation. Aragonite and gypsum predominate but magnesium calcite sometimes occurs (Monty, 1967) and may be due to the preferential binding of Mg^{2+} to the sheaths of *Schizothrix calcicola* (Gebelein and Hoffman, 1970).

B. Eukaryotes

1. *Fungi*

a. *Myxomycota.* Carbonate deposits are associated with the sporangium wall (peridium), the stalk (Gray and Alexopoulos, 1968; Schoknecht and Small, 1972), and specialized filamentous structures (capillita) associated with the spores in the sporangial stage of many of these slime molds. The presence of calcite and possibly aragonite (Schoknecht, 1975) upon the fructifications together with their degree of crystallinity is an important taxonomic criterion. The extent of mineralization is also dependent, however, upon environmental factors such as substrate pH and humidity (Gray and Alexopoulos, 1968) and some species fail to calcify when grown *in vitro* (Schoknecht and Small, 1972).

During sporangium development, Gustafson and Thurston (1974) noted in *Didymium squamulosum* the accumulation of Ca- and P-rich granules in the mitochondria of the precleavage spore cells which disappeared after spore formation. They present a model whereby calcium salts are discharged onto the peridium surface to react with atmospheric CO_2 forming $CaCO_3$. Schoknecht (1975) also noted the secretion of salts through pores in the peridium of *Physarum*. Intracellular calcite has been recorded in the plasmodial stages of some species (Mangenot, 1932). Calcium phosphates (Schoknecht, 1975) and calcium oxalate (Schoknecht and Keller, 1977) may also occur upon the sporangia.

b. *Eumycota.* Yeast calcification has been described by Buck and Greenfield (1964). Cultures of marine and terrestrial forms acted as nucleation centers for calcium carbonate deposits when grown in a seawater medium. Ennever and Summers (1975) found evidence for intracellular nucleation of calcium hydroxyapatite in *Candida albicans* when the plant was grown in metastable phosphate solutions (cf. Section II,A,1). There are also isolated reports of carbonate deposits in some Ascomycetes and lichens (Pobeguin, 1954).

2. *Chlorophyta*

This phylum includes numerous important calcified genera which will be dealt with under family headings.

a. *Bryopsidophyceae.* The possession of multinucleate siphonaceous cells is the characteristic feature of this important family which has many calcified fossil representatives. These genera belong to two cohorts.

b. *Eusiphonidae–Caulerpales.* This group, characterized by its multiaxial structure and complex anatomy, includes *Penicillus, Udotea,* and the "reef builder" *Halimeda.* Calcification is extracellular and occurs in the intercellular spaces between the cortical utricles. These spaces possess a large surface area (Borowitzka and Larkum, 1976a) and present a diffusion barrier which facilitates calcification. On the basis of the uptake kinetics of ^{45}Ca, Borowitzka and Larkum (1976b) have shown that Ca^{2+} uptake into the intercellular spaces from seawater is slow because of the large diffusion barrier between seawater and these cells. Carbon dioxide may be removed from the intercellular spaces by the adjacent photosynthesizing cortex resulting in a pH rise in the space. This would increase the CO_3^{2-} activity thus favoring calcification. Borowitzka and Larkum (1976b) present a model of the process which includes fluxes of HCO_3^-, OH^-, and H^+ for which they present some evidence based upon isotope uptake (Borowitzka and Larkum, 1976b) and inhibitor studies (Borowitzka and Larkum, 1976c). A positive correlation between light-stimulated calcification and photosynthesis was found although no calcification occurred at low light intensities (Borowitzka and Larkum, 1976?). A threshold photosynthesis rate may be necessary to trigger aragonite formation and this would be consistent with their model. Respiration had an inhibitory effect on calcification and experiments with ^{14}C-labeled glucose showed that 50% of the respiratory CO_2 could return to the intercellular spaces in the dark (Borowitzka and Larkum, 1976b). The thallus of *Halimeda* consists of a series of flattened articulated disks. The apical segments are responsible for plant growth and an "age gradient" of calcification and photosynthesis occurs from the tip to the base of these plants (Borowitzka and Larkum, 1976a). Several calcium-binding polysaccharides have been isolated from the cell walls of *Halimeda* (Böhm and Goreau, 1973) but their role in calcification (if any) is obscure (Section III,B).

Rhipocephalus, Penicillus, and *Udotea* do not possess articulated thalli but are also calcified. In *Penicillus,* calcification is mostly extracellular, between the capitular filaments (Marszalek, 1971), although intracellular aragonite has also been reported (Perkins *et al.,* 1972).

c. *Cystosiphonidae–Dasycladales.* The Dasycladales contain more fossil than extent genera and all are calcified except *Batophora.* Calcification occurs in at least two regions, the spaces between the siphonaceous filaments between the axis and thallus surface, similar in many respects to that of *Halimeda* (Fritsch, 1935), and also within the gametangia in some genera. In the gametangia, calcification may be intracellular and result in the formation of a fused mass of cysts known as spicules (Fritsch, 1935). Excepting this, calcification appears to be extracellular (Borowitzka *et al.,* 1974). In *Acetabularia* there are no intercellular spaces but the outer walls of the stalk and umbrella are usually calcified (Puiseux-Dao, 1970).

d. *Charophyceae.* This class of predominantly freshwater algae is renowned for its remarkable cells (Hope and Walker, 1975). Calcification has been observed in several species of *Chara* (Wood and Imahori, 1965), *Lamprothamnium* (Feist-Castel, 1973), *Lychothamnus* (Horn af Rantzien, 1959), *Tolypella* (Daily, 1969) and *Nitellopsis* (Daily, 1973). This process is, nevertheless, a capricious phenomenon within the group, although a detailed study of the deposits surrounding the oogonia of some genera (Daily, 1975) suggested that the precise location and morphology of the deposit might be species-specific in some cases. Deposition is mostly extracellular and occurs upon and between the cortical cells (Groves and Bullock-Webster, 1924), upon the internodal cells, and surrounding the spiral oogonial threads (Fritsch, 1935; Feist-Castel, 1973; Daily, 1975). In *Nitella translucens,* Spear *et al.* (1969) have shown that the internodal cells possess alternating acidic and alkaline regions during photosynthesis, resulting in the deposition of band-shaped layers of calcite over the alkaline zones. In *Chara* these bands frequently coalesce and cover the entire plant. The zones have been shown to represent areas of HCO_3^- influx and OH^- efflux (Lucas and Smith, 1973; Lucas and Dainty, 1977) stimulated by photosynthesis. *Chara* species normally grow in calcium-rich, basic waters where local pH rises are likely to result in calcite precipitation. Pentecost (unpublished observations) found a strong positive correlation between light-stimulated deposition and photosynthesis rate in *C. globularis* and evidence that the degree of encrustation was related to the saturation state of the surrounding water with respect to calcite, similar to the findings of Boyd and Lawrence (1966). Unlike most other algae, charophytes frequently grow up, fruit, and die within 1 year, resulting in the formation of extensive deposits of marl (Davis, 1900).

Calcification of the oogonia of some charophytes begins, according to Daily (1975), after the fertilization and during the maturation of the oospore. Deposition often begins along the adaxial walls of the spiral cells and may eventually fill these cells with calcite. Intracellular calcite has been reported by this author in the spiral cells and sterile oogonial threads of *Chara*.

Calcification has also been reported in other families of the Chlorophyta. Members of the Oedogoniophyceae, Conjugatophyceae, and Chlorophyceae may become encrusted with calcite, particularly in calcareous streams and lakes, (Pia, 1933) but no species specificity is apparent. Deposition is probably dependent upon the rates of carbon dioxide exchange between the water and the atmosphere and cell wall interfaces (cf. Section II,A). Two flagellate genera, *Phacotus* and *Pteromonas,* possess loricas which may be calcified (Fritsch, 1935; Pobeguin, 1954; Bourrelly, 1966) but recent electron microscope studies on *Pteromonas* (Belcher and Swale, 1967) and *Phacotus lenticularis* (Pentecost, unpublished) have not substantiated this.

3. *Chrysophyta—The Coccolithophorids*

These unicellular plants belong to the Haptophyceae, a class distinguished by the presence of a haptonema, an organ with a flagellum-like structure, probably used for attachment. All forms are covered with layers of organic scales and in the Coccolithophorids these may be calcified and are then known as coccoliths. There are basically two kinds of coccoliths, those consisting of an array of rhombohedral or hexagonal microcyrstals which rest upon the scale surface (holococcoliths) and those composed of calcified elements forming a composite structure in which the crystallite shape is highly modified (heterococcoliths). The two forms do not occur upon the same individual but they may occur in different phases of the life cycle (Parke and Adams, 1960). The ontogeny of coccoliths is complex and varies with the species.

The formation of heterococcoliths is known in some detail. Those of *Hymenomonas carterae* are composed of two different elements, assembled into an oval ring around the perimeter of a scale (Pienaar, 1969). The scales first appear within a Golgi cisterna surrounded by vesicles containing granular elements termed coccolithosomes (Outka and Williams, 1971). The coccolithosomes appear to be released into a ''base-plate vesicle'' where they may contribute toward the organic matrix upon, or within which, calcification is initiated. Calcification presumably begins upon the surface of the scale, assisted by the surrounding matrix, but the details require clarification. Scale formation in a related but uncalcified species results from the polymerization and assembly of the component microfibrils within the endomembrane system (Brown, 1974). After calcification, the coccoliths are released onto the cell surface by exocytosis. Unlike calcification, the formation of the coccolith base-plate scale is light-independent (Pienaar, 1975). Coccolith formation in *Hymenomonas roseola* (freshwater species) is similar (Manton and Peterfi, 1969).

In contrast, the coccoliths of *Emiliania hyxleyi* are not associated with scales and a structure termed the "primary coccolith vesicle" develops at the nuclear membrane (Klaveness, 1976). Calcification of the coccolith elements begins near the perimeter of the vesicle and proceeds mainly toward the center. Calcite may sometimes be seen in contact with a membrane surface and it is apparent that the vesicles, possibly supported by fibrillar structures (Klaveness, 1972), may control the final coccolith shape (Klaveness, 1976). No coccolithosomes have been observed in this species but the endomembrane system is intimately involved with coccolithogenesis in all of the heterococcoliths so far investigated. Three morphotypes of *E. huxleyi* have been described: the "normal" coccolith-bearing cell without scales, a "naked" cell, and a scaly cell with uncalcified scales and flagella (Klaveness and Paasche, 1971). This ability for scale formation poses some intriguing questions for the geneticist interested in the control of mineralization. The "naked" cells appear to differ only in the absence of coccoliths, their microanatomy being otherwise the same as that of "normal" cells (Klaveness and Paasche, 1971).

Intracellular calcification has never been observed in forms bearing holococcoliths (Manton and Leedale, 1963; Paasche, 1968) although coccolithosome-like bodies have been noted in *Calyptrosphaera* (Klaveness, 1973) together with an abundance of peripheral vesicles which also occur in the motile holococcolith stage of *Coccolithus pelagicus*. Their function is uncertain and intracellular activity of some kind cannot be entirely dismissed (Paasche, 1969). *Emiliania huxleyi* occasionally forms unfinished coccoliths which closely resemble some holococcoliths (Klaveness, 1976). Some remarkable cold-water marine genera bearing holococcoliths have recently been described. *Pappomonas* and *Papposphaera* possess three different crystallites, two forming body scales and the third supported upon calcified tubular shafts perpendicular to the cell wall (Manton and Oates, 1975; Manton *et al.*, 1976a). *Turrisphaera* has perforated hexagonal crystallites forming tower-like protuberances (Manton *et al.*, 1976b) constructed upon an organic matrix, in common with all other coccoliths. Scales are invariably associated with holococcoliths where they form a base plate for the crystallites although in the more elaborate forms, e.g., *Pappomonas,* there is no direct contact between the scale and the terminal crystals. Since the scales of a related species appear to be composed of several polymers (Brown, 1974), it may be the presence of one of these, rather than the scale microfibrils, which initiates calcification.

The physiological studies of Paasche and his contemporaries (see reviews by Paasche, 1968; Isenberg and Lavine, 1973) have suggested that coccolith formation is light dependent but experiments employing inhibitors of photosynthesis (Paasche, 1965) and the action spectrum of coccolith formation (Paasche, 1966) have failed to produce results showing a direct link between photosynthetic carbon dioxide metabolism and calcification. Coccolith formation may also occur, at a reduced rate, when some species are grown heterotrophically on glycerol (Blankley, 1971). Studies on *Hymenomonas carterae* employing inhibitors of protein synthesis (Dorigan and Wilbur, 1973) and of transmembrane ion transport (Dorigan and Wilbur, 1973; Blankenship and Wilbur, 1975) have added little to our knowledge of the process.

When strontium is added to coccolithophorid cultures, it does not become deposited within the coccoliths (Paasche, 1968) but has a stimulatory effect upon calcification under conditions of low calcium concentration (Blackwelder *et al.*, 1976). The effect of magnesium is similar to that of strontium but less pronounced. How this affects calcification is uncertain but both ions influence the process of cell division in coccolithophorids under certain conditions (Weiss *et al.*, 1976).

Many Chrysophytes deposit silica in their cell walls (Darley, 1974) but there are few, except for the Coccolithophorids, which calcify. Perhaps one example is the coccoid *Chrysonebula holmesii* (Lund, 1953) whose mucilaginous cell walls are packed with calcite rhombohedra. However, this may result from environmental conditions rather than cellular metabolism as is the case with many calcicolous diatoms, which inhabit similar freshwaters, supersaturated with calcite (Pentecost, 1975).

4. *Dinophyta*

Motile dinoflagellates are uncalcified but the cysts of some fossilized (Déflandre, 1933) and extant (Wall *et al.*, 1970) species possess calcified spines external to the cytoplasm and contained between a pair of "membranes" (Wall *et al.*, 1970). Nothing is known of their origin.

Symbiotic Dinophyta have been implicated in the calcification of marine invertebrates, particularly the hermatypic corals and other reef-building Cnidaria (Goreau, 1963; Hartmans and Goreau, 1970). In corals, the phycobiont is contained within carrier cells of the gastrodermal epithelium, at some distance from the sites of mineralization. The stimulatory effects of light on hermatypic coral calcification have been intensively studied (Goreau, 1963; Muscatine, 1971, 1973; Chalker and Taylor, 1975) and the results suggest that the algae liberate components which are transferred to the calcification site where they may participate in matrix formation (Muscatine and Cernichiari, 1969; Pearse and Muscatine, 1971; Barnes and Taylor, 1973; Vandermuelen and Muscatine, 1974). In *Acropora,* calcification is reduced when it is incubated with inhibitors of photosynthesis and oxidative phosphorylation (Barnes and Taylor, 1973) although calcification in the dark remained unaffected. The calcification mechanism in corals is far from clear (Muscatine, 1973; Johnston, 1975) and may involve decarboxylation and nitrogen metabolism (Crossland and Barnes, 1974).

These algae are also associated with some foraminifera (Hedley and Adams, 1976) and one phase of test calcification is closely associated with these endosymbiotic cells (Angell, 1967).

5. *Phaeophyta*

Padina (Dictyotales) is the only genus of brown algae which is commonly calcified, and even here, the extracellular deposit is variable and occasionally absent (Borowitzka, 1977). Uptake of ^{45}Ca is light stimulated in *P. japonica* and a calcification gradient occurs from the margin of the frond to the plant base (Ikemori, 1970). The fan-shaped thallus grows from an apical meristem and the deposits, which develop on both surfaces (Miyata *et al.*, 1977), occur in zones which may have a seasonal origin. It would be of interest to know whether these bands correlate with polysaccharide synthesis, as this is also often seasonal in the Phaeophyta (Percival and McDowell, 1967).

6. *Rhodophyta*

Calcified species are known in two families of the order Cryptonemiales, Squamariaceae, and Corallinaceae, and in the Nemalionales.

a. *Cryptonemiales*. The Squamariaceae comprise a small family and the calcified species consist of crustaceous thalli with a basal layer of prostate radiating filaments. In *Peysonnelia,* aragonite is associated with the basal cells, the cortical cells, and the nemathecia (Denizot, 1968). Cystoliths (Section II,B,7) occur in *P. rubra*. Calcification is erratic but generally slight. Species of *Petrocelis,* a related, uncalcified genus, have been shown to be stages in the life cycle of Gigartinales (West, 1972).

The Corallinaceae is a more important group of calcified algae including many reef builders. In the crustaceous subfamily Melobesioideae all of the cell walls, except some involved with reproduction, become calcified to some degree, whereas in the subfamily Corallinoideae, comprising the articulated forms, the flexible genicular segments are also uncalcified (Johansen, 1974). Calcification in both groups begins soon after spore germination (Cabioch, 1972) and is most intense in the meristematic regions which are marginal in the Melobesioideae and apical in the Corallinoideae. Mineralization begins within these regions with the appearance of tabular, rhombohedral, or acicular crystals of calcite within the cell wall and orientated with their long axes perpendicular or parallel to it (Lind, 1970; Peel, 1974). According to Lind (1970), calcification begins close to the plasma membrane in *Porolithon gardineri,* whereas Peel (1974) observed initiation in the middle lamella of *Corallina officinalis*. In both cases, deposition is primarily extracellular and in *Corallina,* confined to the outer regions of the cell wall (Bailey and Bisalputra, 1970; Peel, 1974). Intracellular crystals have also been noted but attributed to artifacts of preparation or postmortem changes.

The apical cortical cells and the cover cells (whose function is uncertain) contain an extensive vesicular endomembrane system showing evidence of exocytotic activity (Bailey and Bisalputra, 1970; Peel, 1974). Isotopic studies have shown that calcification is light dependent in both subfamilies (Okazaki *et al.,* 1970; Lind, 1970; Ikemori, 1970; Pearse, 1972; Pentecost, 1979) and a strong positive correlation between light-stimulated calcification and photosynthetic rate has been obtained for *Corallina officinalis* (Pentecost, 1979). Calcification in the dark is appreciable but declines steadily after a period of illumination in *Corallina* and *Bossiella* (Pearse, 1972; Pentecost, 1979). Using ^{14}C-labeled pyruvate, Okazaki *et al.* (1970) suggested that respiratory carbon dioxide contributed little toward the deposition. Experiments with killed or dark-treated plants indicate that dark calcification is due, at least partially, to a physical absorbtion process (Pearse, 1972) combined with a delayed metabolic reaction (Pentecost, 1979).

Studies of the Corallinoideae have shown a rapid decrease in calcification and photosynthetic rate down the articulated stems (Pearse, 1972; Pentecost, 1979), whereas the degree of mineralization is approximately constant. This is consistent with apical growth and recalls the calcification "age gradients" of *Halimeda* (Borowitzka and Larkum, 1976b) and the coral *Acropora cervicornis* (Chalker and Taylor, 1975). *Corallina* can probably utilize HCO_3^- during photosynthesis (Blinks, 1963) and the cells contain abnormally high concentrations of organic acids (Furuya, 1965) but the significance of this awaits further investigation.

Kirkpatrick (1912) noted an association between Rhodophyte spores and the calcisponge *Astrosclera* but no symbiosis as exists within hermatypic corals has been propounded.

b. *Nemalionales.* *Galaxaura* and *Liagora* are the only common members. The former is articulated and reminiscent of some Corallinaceae but extracellular aragonite is deposited in the cortex (Borowitzka *et al.*, 1974). *Liagora* is unarticulated but otherwise similar.

7. *Bryophyta and Tracheophyta*

Plants in many families of these phyla possess calcium carbonate deposits of some kind. The deposits may be divided into those which are extracellular upon aquatic species and those which are intracellular—the cystoliths of terrestrial and some aquatic plants. The two forms of deposits are entirely unrelated. The amorphous and crystalline deposits associated with the pericarps of fruits and other plant structures have been reviewed elsewhere (Pobeguin, 1954; Arnott and Pautard, 1970; Arnott, 1975).

a. *Extracellular Deposits.* Aquatic macrophytes such as *Vallisneria, Elodea,* and *Potamogeton* often acquire an encrustation of calcite when growing in hardwaters. These plants can utilize HCO_3^- for photosynthesis (Ruttner, 1953; Steeman Nielsen, 1960) and in *Potamogeton* these ions are taken up at the lower leaf surface and OH^- ions are released from the upper surface (Helder, 1975). The result is a pH increment at the upper surface leading to calcite deposition. Bryophytes often assist in tufa formation (Section II,A) in hardwater streams but the extent to which this occurs as a result of these plants' photosynthetic activity, by pushing Eq. (6) (Section VI) to the right, is uncertain as far greater losses of carbon dioxide occur to the atmosphere (Pentecost, 1975). This is because these waters are enriched with carbon dioxide due to their equilibration with a soil atmosphere (Golubic, 1973; Pentecost, 1975), prior to their emergence. The encrustation of such species as *Eucladium verticillatum* and *Cratoneuron commutatum* can often be attributed to the presence of epiphytic cyanobacteria which trap the precipitate (Pentecost, 1975).

b. *Intracellular Deposits.* Cystoliths occur in numerous vascular plants including bryophytes and pteridophytes (Arnott and Pautard, 1970; Arnott, 1975). They consist of amorphous or microcrystalline calcium carbonate clumped around a clavate extension of the cell wall. Their fine structure requires study and their function is uncertain. Intracellular calcium oxalate deposits often occur as stacks of membrane-bound crystallites within the vacuole of a modified cell (Arnott, 1975). The development of these structures, termed idioblasts, show some features akin to the coccolith vesicle of *Emiliania huxleyi*.

III. Calcification Matrices in Plants

A. Matrix Concept and Composition

Organic matrices are commonly associated with biogenic calcium carbonate and phosphate deposits where they remain as a residue following decalcification, usually comprising less than 10% of the deposit weight. Matrices are believed to initiate and control the calcification process. Much of the original research and speculation on the way these substances function derives from vertebrate calcification (Schiffman *et al.*, 1970). Bone calcification, for instance, occurs among collagen fibrils in a sulfated mucopolysaccharide "ground substance" with calcium-binding properties (Eastoe, 1968; Isenberg and Lavine, 1973). Two postulates have been proposed for the function of these substances (Towe, 1968): The *epitaxy hypothesis* assumes epitaxial growth of crystals upon an organic template, thereby explaining the anisotropy of many biogenic deposits (Bachra, 1973), and the *compartment theory* visualizes nucleation and crystal growth within a discrete region bounded by a matrix. The latter hypothesis constrains the crystal within a fixed volume, whereas the former may permit further growth with no physical contact between the crystal surface and the matrix. These postulates may apply to certain calcification mechanisms but there is a serious lack of direct experimental evidence for them.

There are few examples in the plant kingdom where the development and participation of a matrix have been demonstrated. The best examples are seen in the Coccolithophorids. Klaveness (1976) noted the simultaneous deposition of a matrix with calcite in the coccolith vesicles of *Emiliania huxleyi* and suggested that it might function by suppressing homogeneous calcite nucleation. Calcite is nucleated at specific distal sites within the vesicle but the relationship between the inner matrix membrane (Klaveness, 1972) and the matrix itself, which appears to contain a glucoprotein (Klaveness, 1976), is unclear. Other attempts to analyze the material (Westbroek *et al.*, 1973; de Jong *et al.*, 1975) indicate the presence of a protein-free polysaccharide. Calcification in other heterococcoliths differs from that of *Emiliania* (Section II,B,3). Matrix formation in *Hymenomonas carterae* was considered to precede calcification by Outka and Williams (1971) and is presumably associated with the coccolithosomes and an organic scale template. Calcification only occurs at the perimeter of the distal scale surface which may be composed of several polymers (Brown, 1974). The composition of the matrix of *Hymenomonas carterae* is known in some detail (Isenberg *et al.*, 1965; Wilbur and Watabe, 1967; Isenberg and Lavine, 1973) and consists, after hydrolysis, of hexoses, pentoses, and some amino acids, including hydroxyproline. The latter is also found in collagen but it is not invariably associated with mineralization (Eastoe, 1968), casting some doubt over the significance of this discovery (Isenberg *et al.*, 1965).

When holococcolith crystallites are decalcified, an organic matrix usually remains (Klaveness, 1973; Manton *et al.*, 1976a). In addition, the crystals are associated with scales (Section II,B,3) and a delicate "skin" (Klaveness, 1973). The terminal crystallites of *Pappomonas* possess a faintly regular rugose surface which is probably matrix material (Manton and Oates, 1975; Manton *et al.*, 1976a).

The other group of algae for which there is ample evidence for matrix participation is the Corallinaceae (Rhodophyta). In the encrusting *Porolithon gardineri,* Lind (1970) observed a fibrous matrix system composed of sheets of electron-dense material sometimes showing a periodicity of 5 to 70 nm in the cell wall and associated with the calcite crystals. Sulfhydryl groups were detected within the cell walls which is noteworthy because of their association with calcium transport in the sarcoplasmic reticulum (Wasserman and Kallfelz, 1970). In *Corallina,* Peel (1974) found that the Golgi system in the apical cells secreted fibrous and other electron-dense components into the cell wall region, where a fibrous network could be observed on decalcification, but no periodicity was detected. The mineralization of higher plant idioblasts also occurs in the presence of a matrix which is probably deposited simultaneously with the calcium oxalate crystals (Arnott, 1975).

Some difficulties are encountered concerning the precise role of matrices and the demonstration of their active participation in mineralization. Since matrices may be defined as substances which assist in the initiation, and control the direction, of crystal growth, it is often difficult to decide to what extent such materials are active and also their site of activity. Heterococcolith matrices, for instance, appear to consist of an organic polymer situated within the crystal lattice. Although its distribution within the crystal is not known in detail, this is not a unique phenomenon since calcite crystals can be readily grown in gels, incorporating the polymer molecules within the crystal lattice (Nicl and Hemish, 1969), but in this case, there is no crystallographic orientation as in coccoliths (Watabe, 1967).

The problem is well illustrated in the carbonate deposits formed by the cyanobacteria *Rivularia haematites* and *Schizothrix calcicola*. Thin sections reveal regions where the calcite crystals show a definite orientation with respect to the sheath surfaces, whereas other regions show a random orientation (Pentecost, 1978). It may be argued that a matrix is present inasmuch as the sheath surfaces may influence the direction of crystal growth although control over the ultimate size and shape of the crystals is not apparent. Similar cases occur in the calcification of *Halimeda* and dinoflagellate cysts in which crystal formation develops at specific sites, but there is no apparent control over crystal shape or orientation (Wilbur *et al.*, 1969; Wall *et al.*, 1970). Evidence for matrix participation therefore requires (1) the presence of the material in contact with the solid phase, and (2) control over crystal orientation and morphology. These criteria only apply currently to calcification in the Coccolithophorids and Corallinaceae, although many genera remain to be examined in detail (see Table II).

B. Cell Wall Polysaccharides

These consist of simple or branched chains composed mainly of (1-4)- and (1-3)-linked pentoses and hexoses (Percival and McDowell, 1967; Mackie and Preston, 1974). Many algal polysaccharides have acidic properties because of the presence of carboxyl and ester sulfate groups (Table I) but their tertiary structures in most cases are unknown. These polymers are of interest as they are often components of plant and animal matrices and many have calcium-binding properties (Böhm and Goreau, 1973; de Jong *et al.*, 1975). Although these acidic polymers occur in most (perhaps all) of the calcifying algae (Table I) they also occur in most noncalcifying species (Borowitzka, 1977). These materials show such variation from species to species that it is not possible to ascribe any significance to the differences between those of calcifying and noncalcifying species. Probably all of the acidic polysaccharides act as ion exchangers and the polyuronic acids have a strong affinity for calcium ions with which they form insoluble salts. Kohn and Larson (1972) have shown that the interaction between calcium ions and carboxyl groups of polyuronide sols is partially dependent upon the linear charge density of the polymer. In polysaccharide gels it is believed that calcium ions form bridges between adjacent molecules via carboxyl groups (Katchalsky *et al.*, 1961) or through coordinate links with hydroxyl groups (see Percival and McDowell, 1967). A study of polyuronide ion-exchange properties led Smidsrød and Haug (1972) to the conclusion that sequences of calcium ions could bind the polymer chains in the gel state. The distance between carboxyl groups of polyuronide chains is about the same as the interatomic distances of

TABLE I
POLYSACCHARIDES ASSOCIATED WITH PLANT CALCIFICATION

Phylum/genus	Site	Uronic acid	Ester sulfate	Notes	Reference
Prokaryota					
Bacteria	Cell wall	√	—	No calcified spp. examined	
Cyanobacteria	Cell wall	√	—	No calcified spp. examined	
Eukaryota					
Chlorophyta					
Halimeda	Cell wall	√	√	Ca binding	Böhm and Goreau (1973)
Acetabularia	Cell wall	Glucuronic	Galactose -4 and -6		Percival and Smestad (1972)
Chrysophyta					
Emiliania	Matrix	?Galacturonic	√		de Jong *et al.* (1975)
Hymenomonas	Matrix	√	√		Isenberg and Lavine (1973)
Phaeophyta					
Padina	Cell wall	Glucuronic	√		Jabbar Mian and Percival (1973)
Rhodophyta					
Corallina	Cell wall		Galactose-6 ?Gal-2		Turvey and Simpson (1966)
Porolithon			√		Lind (1970)

calcium atoms in the calcite unit pseudocell and it is tempting to speculate upon the possibilities of epitaxial growth. However, this is subject to the same criticisms as those raised against the collagen–bone salt hypothesis (Towe, 1968). Furthermore, calcium binding can lead to the distortion of the polymer chains leading to a reduction in the crystallinity (Percival, personal communication) thus reducing the opportunities for epitaxy.

IV. The Solid Phase—Mineralogy and Crystal Orientation

This discussion is limited to the polymorphs and hydrates of calcium carbonate. For the mineralogy of calcium hydroxyapatite see McLean and Urist (1955) and Zipkin (1973). The three polymorphs of calcium carbonate, in order of stability at NTP, are calcite (trigonal), aragonite (orthorhombic-pseudohexagonal) and vaterite (orthorhombic). Vaterite is unstable and is rarely the result of plant calcification (Wilbur and Watabe, 1963). There are also some unstable crystalline hydrates of calcium carbonate (Brooks *et al.,* 1950), one of which has been tentatively identified in an infected rhodophyte *Nemalion helminthoides* (Dixon, 1973). These unusual minerals have only been observed under adverse or abnormal growth conditions and will not be considered further.

There are many factors influencing which form is precipitated; these include temperature (Wray and Daniels, 1957), rate of crystallization, pH, and presence of interfering ions and organic solutes (Kitano *et al.,* 1975). When magnesium ions are present in excess of ca.10 m*M,* aragonite is precipitated, so that most marine extracellular deposits are deposited in this form (seawater [Mg] is 54 m*M*). The presence of certain organic solutes can counteract the effect of magnesium (Kitano *et al.,* 1975) resulting in calcite formation, but the mechanism is obscure. It is noteworthy, in this respect, that in the marine calcite-depositing algae, the mineral is always associated with the cell wall or matrix material (Gebelein and Hoffman, 1970; Lind, 1970; Wall *et al.,* 1970; Peel, 1974; Klaveness, 1976).

Marine calcites frequently contain substantial amounts of magnesium carbonate (Borowitzka, 1977; Levy and Strauss, 1960), resulting in magnesium calcite. The degree of substitution ranges up to 20 mole% and varies according to the species, the growth rate, the water temperature, and the plant age (Chave and Suess, 1970; Vinogradov, 1953; Moberly, 1968). Which of these factors are most important is difficult to decide as few algae have been examined in detail. Growth of the encrusting Coralline *Phymatolithon* was found to be highest in the warmest months but the highest magnesium content was found in deposits which rapidly filled old conceptacles in late winter (Moberly, 1968). This led to the conclusion that the growth rate of the deposit was related to the degree of substitution. Since the results from other species were inconclusive and there is no evidence to indicate that the infilling was biogenic, the significance of these observations remains doubtful. If the infillings were, in fact, abiogenic, these findings support the observations of Baas-Becking and Galiher (1931) and Peel (1974) who found little magnesium in the apical segments of *Corallina* but increasing quantities in progressively older intergenicula, where there is evidence of cellular disorganization and abnormal deposition (Bailey and Bisalputra, 1970; Peel, 1974). No magnesium calcites have been reported from marine heterococcoliths (Paasche, 1968) and it would be of interest to know whether this also applies to the holococcoliths. In freshwaters, plants deposit calcite with little magnesium present (Pobeguin, 1954; Pentecost, 1975).

Aragonite may be formed either in the presence or absence of a matrix (Table II) and occurs, like calcite, both within or external to the cell wall. This polymorph is usually found in the absence of a matrix and outside the cell wall as in the marine Bryopsidophyceae although here it may nucleate at the cell wall surface (Wilbur *et al.,* 1969). The calcium atoms are held in 9-fold coordination with oxygen in aragonite, whereas in calcite, the coordination is 6-fold, although

TABLE II
THE MINERALOGY AND ORIENTATION OF CALCIUM CARBONATE DEPOSITS ASSOCIATED WITH PLANTS

Class/family	Presence of matrix	Marine/ freshwater[a]	Polymorph present	Crystal orientation	Notes	Reference
Prokaryota						
Cyanobacteria	x	m	Mg–C, A	Occasional		Gebelein and Hoffman (1970)
	x	f	C	Occasional		Pentecost (1978)
Bacteria	x	m	A	?	Deposits extracellular	Greenfield (1963)
	x	f	C	?	Oral spp. deposit hydroxyapatite	Pia (1933)
Eukaryota						
1. Fungi						
Myxomycota	?	Terrestrial	C, A?	x	Other minerals may be present	Schoknecht (1975)
Eumycota						
Endomycetales (yeasts)	x	m	?	x		Buck and Greenfield (1964)
Ascomycotina	?	Terrestrial	C	?		Pobeguin (1954)
2. Algae						
Chlorophyta						
Chlorophyceae	?	f	C	?		Pobeguin (1954)
Bryopsidophyceae						
Caulerpales	x	m	A	x		Wilbur *et al.* (1969)
Dasycladales	x	m	A	x		Borowitzka *et al.* (1974)
Charophyceae	x	f	C	Occasional	Deposits develop in bands	Borowitzka *et al.* (1974); Daily, 1975
Chrysophyta						
Haptophyceae	√	m,f	C	√		Manton *et al.* (1976a)
Dinophyta	?	m	?	√	Cysts only	Wall *et al.* (1970)
Phaeophyta	x	m	A	x		Miyata *et al.* (1977)
Rhodophyta						
Nemalionales	?	m	A	x		Borowitzka *et al.* (1974)
Corallinaceae	√	m	Mg–C	√		Baas-Becking and Galiher (1931)
3. Bryophytes	x	f	C	x		Pentecost (1975)
4. Tracheophytes						
Aquatic	x	f	C	x	Extracellular deposits	Pobeguin (1954)
Terrestrial spp.	√	f	C	x	Cystoliths	Arnott and Pautard (1970)

[a] m, Marine; f, freshwater; A, aragonite; C, calcite; Mg–C, magnesium calcite.

the internal arrangement of the molecules is similar. Because of the difference in coordination, the transformation of aragonite into calcite and vice versa is practically impossible at NTP without resolution, as it involves the breakage of large numbers of coordinate bonds.

In the Coccolithophorids and Corallinaceae, the calcite is deposited in crystals of a definite size and orientation. In *Corallina* the deposits consist of minute tabular or rhombohedral crystals with their trigonal axes perpendicular, and their long axes parallel, to the cell wall (Baas-Becking and Galiher, 1931; Bailey and Bisalputra, 1970; Peel, 1974). In older regions of the thallus orientation is less evident. Lind (1970) suggested that the cell walls of encrusting coralline algae may move as a response to growth stresses leading to a loss of orientation.

In the Coccolithophorid *Emiliania huxleyi,* electron diffraction studies have shown that the trigonal axis of the calcite is parallel to the radial axis of the coccolith and thus lies in the direction growth of the coccolith elements (Watabe, 1967). In the apparently unmodified crystals of holococcoliths the orientation is particularly striking but the position of the crystallographic *c*-axis is unknown. In many genera, the crystal form has not been confirmed by diffraction studies (e.g., *Pappomonas, Turrisphaera*) but in *Pappomonas* the tabular crystallites exhibit the characteristic angle of the calcite cleavage rhomb (see Fig. 9 in Manton and Sutherland, 1975). The large petaloid crystals of this species are of considerable interest since the blades appear to grow in a stepwise fashion from the distal edge (Manton *et al.,* 1976a). Coccolith microcrystals are generally associated with organic scales and a close structural relationship is apparent (Outka and Williams, 1971; Klaveness, 1973; Manton and Sutherland, 1975). In contrast to calcite, aragonite crystals rarely show any preferred orientation although few aragonite-depositing plants have been investigated with the electron microscope (Table II).

The type of crystal deposited is probably determined at the nucleation stage (Wray and Daniels, 1957) and this is difficult to investigate experimentally, especially in biological systems. Studies with mollusks have shown that the shell matrix can influence the polymorph deposited (Wilbur and Watabe, 1963), although it is known that trace quantities of catalysts or inhibitors can affect both the nucleation and the crystal growth rate (see Chave and Suess, 1967; Walton, 1967; Bachra, 1973; Simkiss, 1975, for discussions).

V. Calcification and the Transport of Ions

Before discussing the question of ion transport it will be necessary to distinguish between those plants in which calcification is intracellular, which probably necessitates the passage of ions across one or more membranes (if endocytosis, for which there is no evidence, is excluded), and those where it is extracellular and independent of membrane transport. Cellular calcium transport can only be considered relevant to systems where there is unequivocal evidence of intracellular calcification *in vivo,* which currently limits the discussion to the formation of heterococcoliths and idioblasts.

Calcium ion levels in animal cytoplasm are low (ca. 1 μM, see Simkiss, 1975), and evidence is accumulating for the participation of calcium pumps which transport the ion across animal membranes, particularly those of mitochondria and the sarcoplasmic reticulum (Wasserman and Kallfelz, 1970; Oxender, 1972). In the latter, calcium is transported reversibly by a membrane-bound ATPase whose structure and properties are known in some detail (Tada *et al.*, 1978). There is as yet no clear evidence for such a system operating in the calcification processes of plants or animals, although it has often been suggested (Clifford Jones, 1970; Simkiss, 1975; Granstrom and Linde, 1976). We have no knowledge of calcium ion activity within the cytoplasm of plant cells and no detailed calcium flux measurements for calcified species. For the Coccolithophorid *Emiliania huxleyi* it is possible to estimate the net inward flux of calcium from the data of Paasche (1964). Calculations indicate rates of 5 to 15 pmole of calcium cm^{-2} sec^{-1}, and although these are approximate figures only, they would appear to be too high for the passive influx of a divalent cation (cf. MacRobbie, 1974). Klaveness (1976) found evidence of ATPase activity at the surface of the coccolith vesicle membrane of *Emiliania* which would appear to support an active transport process.

Carbon dioxide may enter plant cells as an uncharged molecule (CO_2 or H_2CO_3) or as an ion (HCO_3^- or CO_3^{2-}). The distribution of these forms is governed by the pH of the medium, and in the region of pH 6–9, that of most natural waters, the bicarbonate ion predominates. Many aquatic plants can assimilate both CO_2 and HCO_3^- but there is no evidence for the uptake of CO_3^{2-}. Molecular carbon dioxide is thought to enter photosynthesizing cells by passive diffusion (Raven, 1970) but there are good reasons for assuming an active entry of HCO_3^- (Raven, 1968; Smith, 1968) with influx rates of around 10 pmole cm^{-2} sec^{-1} (Hope and Walker, 1975). Our knowledge of the activity and ionic distribution of carbon dioxide within the cytoplasm is limited but it is apparent that the enzyme carbonic anhydrase may play a significant role by catalyzing the reaction

$$CO_2 + H_2O \rightarrow HCO_3^- + H^+. \tag{3}$$

This enzyme has been noted in several calcified invertebrates (Simkiss, 1975) and is known in some plants where it is presumably active in photosynthetic carbon dioxide fixation (Raven, 1974) but it has not been conclusively demonstrated in any calcifying species. Despite this, several studies employing the apparently specific inhibitor of carbonic anhydrase, acetazolamide, have generally resulted in reduced rates of calcification but the effects are complex and may involve other processes. For instance, in *Hymenomonas* sp. Isenberg *et al.* (1963) obtained inhibition of calcification at 10^{-3} to 10^{-4} M but growth continued. The effect of the inhibitor was dependent upon culture conditions and the existence of two forms of the enzyme was postulated, one involved with calcification and another associated with photosynthesis and unaffected by acetazolamide. In *Emiliania* however, Paasche (1964) observed inhibition of both photosynthesis and calcification but only at high concentrations of inhibitor (10^{-2} to 10^{-3} *M*). Acetazolamide is toxic to *Balanus* at these levels (Costlow, 1973).

In *Halimeda,* the inhibitor decreased light-stimulated calcification and photosynthesis at concentrations of 2.5×10^{-4} *M* (Borowitzka and Larkum, 1976c; Borowitzka, 1977). Its effect upon photosynthetic oxygen evolution was pH dependent, consistent with the supposed role of the enzyme in carbon dioxide metabolism, but at higher levels (10^{-3} *M*), respiration was stimulated, further inhibiting calcification (Section II,B,2). Acetazolamide also inhibits light-stimulated calcification in *Corallina* and *Amphiroa* (Corallinaceae) and *Padina* (Phaeophyta) at these concentrations (Ikemori, 1970). No overall pattern emerges from these studies and more detailed investigations are required. Particular attention should be paid to the permeability of acetazolamide and its mode of action in the future.

Carbonic anhydrase can be regarded as a reversible acid–base generator [Eq. (3)] and, if situated within a membrane, could maintain a pH gradient, operating essentially as an ion pump. The accumulation of base (HCO_3^-) could lead to the precipitation of $CaCO_3$ (Eqs. (4) and (5)). Such a scheme has been suggested for mollusk calcification (Campbell and Boyan, 1975).

$$HCO_3^- + Ca^{2+} \rightarrow CaCO_3 + H^+, \tag{4}$$

$$HCO_3^- + OH^- + Ca^{2+} \rightarrow CaCO_3 + H_2O. \tag{5}$$

Equation (5) dispenses with acidosis (Simkiss, 1975) although there is ample evidence for the active transport of H^+ across plant membranes generated by a Mitchell-type scheme, ATP hydrolysis, or ion cotransport (Kitasato, 1968; MacRobbie, 1974; Slayman, 1974). Whether carbonic anyhydrase is involved is uncertain (Carter, 1972) but evidently any process involving H^+ movement affects pH directly thereby influencing the solubility of $CaCO_3$. There is indirect evidence for H^+ fluxes across the cell walls of *Chara corallina* (Lucas and Smith, 1973), *Halimeda* (Borowitzka and Larkum, 1976c), and *Acetabularia* (Saddler, 1970) although their origin is unknown. Carbonic anhydrase may catalyze reactions other than (3) (Carter, 1972); thus evidence of its existence does not demonstrate per se its participation in any of the above schemes.

Phosphatase activity is frequently associated with invertebrate calcification and several postulates involving Ca–ATP complexes have been advanced (Simkiss, 1975). There is no evidence for their involvement in plant calcification although *Corallina officinalis* has high alkaline phosphatase activity (Desruisseaux and Baudoin, 1949). This warrants further investigation.

VI. General Discussion

It should be clear that there is no singular mechanism responsible for plant calcification and this necessitates a consideration of the individual phyla or groups of species. Evidently, direct photosynthetic involvement cannot be used as a criterion to distinguish between plant and animal calcification as there are few well-substantiated examples of this. Light-stimulated deposition is, however, generally observed in the red, green, brown, and Chrysophyte algae and a 1:1 partition of carbon into organic compounds and $CaCO_3$ is often apparent, e.g., in *Emiliania* (Paasche, 1964), *Coccolithus* (Paasche, 1969), and *Corallina* (Pentecost, 1979). This is consistent with Eq. (6), but studies with photosynthetic inhibitors.

$$2HCO_3^- + Ca^{2+} \rightarrow CaCO_3 + CO_2\ (\mathrm{phs}) + H_2O \tag{6}$$

(Paasche, 1964, 1969), heterotrophic growth (Blankley, 1971), and action spectra (Paasche, 1966) indicate, as Paasche states, that these observations may be coincidental.

Calcification is an energy-demanding process and energy will be required, in intracellular deposition, for (i) transport of the reactants, Ca^{2+} and HCO_3^- or CO_3^{2-} and possibly the removal of inhibitors from the nucleation site, (ii) transport and synthesis of an organic matrix, if present, (iii) concentration of the reactants above the critical supersaturation and, perhaps, (iv) transport of the solid phase. In cases of extracellular calcification, the only energy-requiring process may be (iii) and even this can be provided in some cases by thermodynamically favorable events such as temperature fluctuations, evaporation, or loss of carbon dioxide to the atmosphere. Where access of the reactants is restricted, however, as in *Halimeda* (Borowitzka and Larkum, 1976a) there will be additional requirements. Future research should attempt to establish the origin of the energy sources leading to mineralization, to determine the extent of biological control. In many cases we still lack sufficient data to enable a clear distinction to be made between biogenic and abiogenic calcification in plants and further studies on cystolith ontogeny and calcification in the Squamariaceae, Nemalionales, and Siphonales would be valuable. In species with extracellular deposits, more attention should be paid to the chemistry of the medium in contact with the solid phase and there is a general need for more basic research on the nucleation and growth of the calcium carbonate polymorphs. Unfortunately the precipitation of these minerals cannot be predicted from a knowledge of their solubility products alone since they supersaturate to a high degree (Walton, 1967; Dragunova, 1970). The purpose of plant calcification is also uncertain, even in those species where its occurrence is regular and controlled. Braarud *et al.* (1952) suggested that coccoliths might protect cells from high light intensities but this now seems unlikely (Paasche, 1964). Outka and Williams (1971) suggest a cytoskeletal function which is plausible as the cells lack a rigid cell wall, and buoyancy control is also conceivable (Eppley *et al.,* 1967). Furthermore, the motile stage of *Coccolithus pelagicus* appears to be less heavily calcified than the sedentary stage (Paasche, 1969) which may result in greater mobility through a smaller density difference.

The development of calcified, articulated branches in the marine algae affords great flexibility to an otherwise brittle and easily damaged plant. *Corallina officinalis* is an example which inhabits temperate tide pools subject to high stresses due to wave action. This plant, of which 70–90% of the dry weight is calcium carbonate (Pentecost, 1979), is unlikely to be favored as a nutritious foodstuff by many organisms, since almost every cell is encrusted in mineral and these could not be ingested free of calcite. Indeed, this species does appear to be avoided by many grazers. It is also of interest to note that the potential grazers of *Corallina* in British rock pools are mostly gastropods which would have difficulty in gaining a purchase upon the articulated branches. However, during a study of the ecology of some mid-Atlantic coastal waters, Price (1978) found that *Amphiroa* (Corallinaceae) was heavily grazed by blackfish, and the encrusting corallines were readily eaten by echinoderms. The cropping of calcified algae by organisms which themselves require large quantities of calcium for their tests and shells can be readily understood, so that calcification is not necessarily a deterrant to grazers. A considerable disadvantage is also apparent for the heavily calcified algae, since the density of the deposit may occlude some of the incident radiation necessary for photosynthesis. A layer of 1- to 5-μm-diameter calcite crystals, 0.1 mm thick, for example, may attenuate light intensity by 70% (Pentecost, 1978). This suggests that for many species, optimal photosynthesis can only occur close to the thallus periphery, except at the high light intensities possible near the surface. Although such areas are often inhabited by calcified forms, they also occur in abundance in waters 10–20 m deep (Johansen, 1974). One means of obviating the problem is to concentrate all of the photosynthetic tissue at the thallus periphery and keep it clear of mineralization. This seems to be achieved to some extent by the Siphonales (e.g., *Halimeda, Dasycladus*), but it is not apparent in the calcified Rhodophytes whose growth rates are low (ca. 1–5 cm per year) in comparison to most other marine algae (see Johansen, 1974). Calcification could also retard the diffusion of nutrients into cells (cf. Spear *et al.,* 1969) and would be particularly serious for those prokaryotic organisms which become completely surrounded by mineral.

ACKNOWLEDGMENTS

I wish to thank Mrs. L. Irvine, Mr. W. Price, Dr. E. Percival, and Dr. M. C. Peel for their useful suggestions and advice.

References

Angell, R. W. (1967). *J. Protozool* **14,** 566.
Arnott, H. J. (1975). *In* "The Mechanisms of Mineralization in the Invertebrates and Plants" (N. Watabe and K. M. Wilbur, eds.), pp. 55–78. Univ. of S. Carolina Press, Columbia, South Carolina.
Arnott, H. J., and Pautard, F. G. E. (1970). *In* "Biological Calcification" (H. Schraer, ed.), pp. 375–446. North-Holland Publ., Amsterdam.
Baas-Becking, L. G. M., and Galiher, E. W. (1931). *J. Phys. Chem* **35,** 467–479.
Bachra, B. N. (1973). *In* "Biological Mineralization" (I. Zipkin, ed.), pp. 845–881. Wiley (Interscience), New York.
Bailey, A., and Bisalputra, T. (1970). *Phycologia* **9,** 83.
Barnes, D. J., and Taylor, D. L. (1973). *Helgol. Wiss. Meeresunters.* **24,** 284.
Barnes, I. (1965). *Geochim. Cosmochim. Acta* **29,** 85.
Belcher, J. H., and Swale, E. M. F. (1967). *Nova Hedwigia* **13,** 353.
Berkeley, C. (1919). *Trans. R. Soc. Can.* **13** (3), 15.
Berner, R. A. (1970). *In* "Carbonate Cements" (O. P. Bricker, ed.), *Studies of Geology,* **19,** pp. 247–251. Johns Hopkins Press, Baltimore, Maryland.
Bersa, E. (1926). *Planta* **2,** 373.
Blackwelder, P. L., Weiss, R. E., and Wilbur, K. M. (1976). *Mar. Biol.* **34,** 11.
Blankenship, M. L., and Wilbur, K. M. (1975). *J. Phycol.* **11,** 211.
Blankley, W. F. (1971). Auxotrophic and heterotrophic growth and calcification in Coccolithophorids. Ph.D. Thesis, University of California, San Diego, California.
Blinks, L. R. (1963). *Protoplasma* **57,** 126.
Böhm, E. L., and Goreau, T. F. (1973). *Int. Rev. Ges. Hydrobiol.* **58,** 117.
Borowitzka, M. A. (1977). *Oceanogr. Mar. Biol. Ann. Rev.* **15,** 189.
Borowitzka, M. A., and Larkum, A. W. D. (1976a). *J. Exp. Bot.* **27,** 864.
Borowitzka, M. A., and Larkum, A. W. D. (1976b). *J. Exp. Bot.* **27,** 879.
Borowitzka, M. A., and Larkum, A. W. D. (1976c). *J. Exp. Bot.* **27,** 894.
Borowitzka, M. A., Larkum, A. W. D., and Nockolds, C. E. (1974). *Phycologia* **13,** 195.
Bourrelley, P. (1966). "Les algues d'eau douce," Vol I. Boubée, Paris.
Boyd, C. E., and Lawrence, J. M. (1966). *Proc. Conf. Southeast. Ass. Game Fish Commnrs.* **20,** 413.
Braarud, T., Gaarder, K. R., Markali, J., and Nordii, E. (1952). *Nytt. Mag. Bot.* **1,** 129.
Brooks, R., Clark, L. M., and Thurston, E. F. (1950). *Philos. Trans. R. Soc. Lond. A* **243,** 145.
Brown, R. M., Jr. (1974). *In* "Proceedings, International Symposium on Plant Cell Differentiation" Preprint (M. S. Pais, ed.), pp. 1–55. Lisbon, Portugal.
Buck, J. D., and Greenfield, L. J. (1964). *Bull. Mar. Sci. Gulf Carib.* **14,** 239.
Cabioch, J. (1972). *Cah. Biol. Mar.* **13,** 137.
Campbell, J. W., and Boyan, B. D. (1975). *In* "The Mechanisms of Mineralization in the Invertebrates and Plants" (N. Watabe and K. M. Wilbur, eds.), pp. 109–134. Univ. of S. Carolina Press, Columbia, South Carolina.
Carter, M. J. (1972). *Biol. Rev.* **47,** 465.
Castenholz, R. W. (1973). *In* "The Biology of Blue-green Algae" (N. G. Carr and B. A. Whitton, eds.), pp. 320–339. Blackwell, London.
Chalker, B. E., and Taylor, D. E. (1975). *Proc. R. Soc. London B* **190,** 323.
Chave, K. E., and Suess, E. (1967). *U.S. Geol. Prof. Pap.* **350,** 138.
Chave, K. E., and Suess, E. (1970). *Limnol. Oceanogr.* **15,** 633.
Clifford Jones, W. (1970). *Symp. Zool. Soc. London* **25,** 91.
Cole, G. A., and Bachelder, G. L. (1969). *J. Ariz. Acad. Sci.* **5,** 271.

Costlow, J. D. (1973). *Physiol. Zool.* **32,** 177.

Crossland, C. J., and Barnes, D. J. (1974). *Mar. Biol.* **28,** 325.

Daily, F. K. (1969). *Proc. Indiana Acad. Sci.* **78,** 406.

Daily, F. K. (1973). *Bull. Torrey Bot. Club* **100,** 75.

Daily, F. K. (1975). *Phycologia* **14,** 331.

Darley, W. M. (1974). *In* "The Physiology and Biochemistry of Algae" (W. D. P. Stewart, ed.), pp. 655–675. Blackwell, London.

Davis, C. A. (1900). *J. Geol.* **8,** 485.

Déflandre, G. (1933). *Bull. Soc. Zool. Fr.* **58,** 265.

Denizot, M. (1968). "Les Algues Floridées Encroutants." Muséum National d'Histoire Naturelle, Paris.

Desruisseaux, G., and Baudoin, N. (1949). *C.R. Hebd. Séances Soc. Biol.* **144,** 519.

Dixon, P. S. (1973). "The Biology of the Rhodophyta." Oliver and Boyd, Edinburgh.

Dorigan, J. L., and Wilbur, K. M. (1973). *J. Phycol.* **9,** 450.

Dragunova, D. A. (1970). *Hydrobiol. J.* **6,** 50.

Drew, G. H. (1914). *Carnegie Inst. Washington Yearb.* **5,** 9.

Eastoe, J. E. (1968). *Calc. Tiss. Res.* **2,** 1.

Ennever, J., and Summers, F. E. (1975). *J. Bacteriol.* **122,** 1391.

Ennever, J., and Takazoe, I. (1973). *In* "Biological Mineralization" (I. Zipkin, ed.), pp. 1–33. Wiley (Interscience), New York.

Ennever, J., Takazoe, I., Vogel, J. J., and Summers, F. E. (1970). *Tex. Rep. Biol. Med.* **28,** 27.

Ennever, J., Vogel, J. J., and Streckfuss, J. L. (1971). *J. Dent. Res.* **50,** 1327.

Ennever, J., Vogel, J. J., and Streckfuss, J. L. (1974). *J. Bacteriol.* **119,** 1061.

Eppley, R. W., Holmes, R. W., and Stricklan, J. D. H. (1967). *J. Exp. Mar. Biol. Ecol.* **1,** 191.

Feist-Castel, M. (1973). *Geobios* **6,** 239.

Fjerdingstad, E. (1951). *Rev. Algol.* **4,** 246.

Fritsch, F. E. (1935). "The Structure and Reproduction of the Algae," Vol. 2. University Press, Cambridge, England.

Furuya, K. (1965). *Bot. Mag. (Tokyo)* **78,** 274.

Gebelein, C. D., and Hoffman, P. (1970). *In* "Carbonate Cements" (O. P. Bricker, ed.), *Studies of Geology,* **19,** pp. 319–326. John Hopkins Press, Baltimore, Maryland.

Golubic, S. (1973). *In* "The Biology of Blue-green Algae" (N. G. Carr and B. A. Whitton, eds.), Botanical Monographs, Vol. 9. Blackwell, London.

Goreau, T. F. (1963). *Ann. N. Y. Acad. Sci.* **109,** 127.

Granstrom, G., and Linde, A. (1976). *J. Histochem. Cytochem.* **24,** 1026.

Gray, W. D., and Alexopoulos, C. J. (1968). "The Biology of the Myxomycetes." Ronald Press, New York.

Greenfield, L. J. (1963). *Ann. N.Y. Acad. Sci.* **109,** 23.

Groves. J., and Bullock-Webster, G. R. (1924). "The British Charophyta," Vol. 2, *Charae.* Ray Society, London.

Gustafson, R. A., and Thurston, E. L. (1974). *Mycologia* **66,** 397.

Hartmans, W. D., and Goreau, T. F. (1970). *Symp. Zool. Soc. Lond.* **25,** 205.

Hedley, R. H., and Adams, C. G. (1976). "Foraminifera." Academic Press, New York.

Helder, R. J. (1975). *Proc. K. Ned. Akad. Wet. Ser. C.* **78,** 189.

Hope, A. B., and Walker, N. A. (1975). "The Physiology of Giant Algal Cells." University Press, Cambridge, England.

Horn af Rantzien, H. (1959). *K. Svenska Vetensk. Akad. Ark. Bot.* Ser. 2, **4,** 165.

Ikemori, M. (1970). *Bot. Mag. (Tokyo)* **83,** 152.

Isenberg, H. D., and Lavine, L. S. (1973). *In* "Biological Mineralization" (I. Zipkin, ed.), pp. 649–686. Wiley (Interscience), New York.

Isenberg, H. L., Lavine, L. S., Mandell, C., and Weissfellner, H. (1965). *Nature (London)* **206,** 1153.

Jabbar Mian, A., and Percival, E. (1973). *Carbohydrate Res.* **26,** 133.

Johansen, H. W. (1974). *Oceanogr. Mar. Biol. Rev.* **12,** 77.

Johnston, I. S. (1975). *In* "The Mechanisms of Mineralization in the Invertebrates and Plants" (N. Watabe and K. M. Wilbur, eds.), pp. 249–260. Univ. of S. Carolina Press, Columbia, South Carolina.

de Jong, L. W., Dam, W., Westbroek, P., and Crenshaw, M. A. (1975). *In* "The Mechanisms of Mineralization in the Invertebrates and Plants" (N. Watabe and K. M. Wilbur, eds.), pp. 135–153. Univ. of S. Carolina Press, Columbia, South Carolina.

Katchalsky, A., Cooper, R. E., Upadhyay, J., and Wassermann, A. (1961). *J. Chem. Soc.* 5198.

Kellerman, K. F., and Smith, N. R. (1914). *J. Washington Acad. Sci.* **4,** 400.

Kellerman, K. F., and Smith, N. R. (1916). *Zeitbl. Bakt. Parasitkde. Abt. 2* **45,** 371.

Kirkpatrick, R. (1912). *Proc. R. Soc. London B* **84,** 579.

Kitano, Y., Kanamori, N., and Yoshioka, S. (1975). *In* "The Mechanisms of Mineralization in Invertebrates and Plants" (N. Watabe and K. M. Wilbur, eds.), pp. 191–202. Univ. of S. Carolina Press, Columbia, South Carolina.

Kitasato, H. J. (1968). *J. Gen. Physiol.* **52,** 60.

Klaveness, D. (1972). *Protistologica* **8,** 335.

Klaveness, D. (1973). *Norw. J. Bot.* **20,** 151.

Klaveness, D. (1976). *Protistologica* **12,** 217.

Klaveness, D., and Paasche, E. (1971). *Arch. Mikrobiol.* **75,** 382.

Kohn, R., and Larson, B. (1972). *Acta Chem. Scand.* **26,** 2455.

Levy, L. W., and Strauss, M. R. (1960). *Colloques Int. Cent. Natn. Res. Scient.* **103,** 38.

Lind, J. W. (1970). Processes of Calcification in Coralline Algae. Ph.D. Thesis, University of Hawii, Honolulu.

Lucas, W. J., and Dainty, J. (1977). *J. Membr. Biol.* **32,** 75.

Lucas, W. J., and Smith, F. A. (1973). *J. Exp. Bot.* **24,** 1.

Lund, J. W. G. (1953). *New Phytol.* **52,** 114.

Lustmann, J., Lewin-Epstein, J., and Shieyer, A. (1976). *Calcif. Tissue Res.* **21,** 47.

Mackie, W., and Preston, R. D. (1974). *In* "Algal Physiology and Biochemistry" (W. D. P. Stewart, ed.), pp. 40–85. Blackwell, London.

McLean, F. C., and Urist, M. R. (1955). "Bone, an Introduction to the Physiology of Skeletal Tissue." Univ. of Chicago Press, Chicago, Illinois.

MacRobbie, E. A. C. (1974). *In* "Algae Physiology and Biochemistry" ('V. D. P. Stewart, ed.), pp. 676–713. Blackwell, London.

Mangenot, G. (1932). *C.R. Soc. Biol.* **111,** 936.

Manton, I., and Leedale, G. F. (1963). *Arch. Mikrobiol.* **47,** 115.

Manton, I., and Oates, K. (1975). *Br. Phycol. J.* **10,** 93.

Manton, I., and Peterfi, L. S. (1969). *Proc. R. Soc. London B* **172,** 1.

Manton, I., and Sutherland, J. (1975). *Br. Phycol. J.* **10,** 377.

Manton, I., Sutherland, J., and McCully, M. (1976a). *Br. Phycol. J.* **11,** 225.

Manton, I., Sutherland, J., and Oates, K. (1976b). *Proc. R. Soc. London B* **194,** 179.

Marszalek, D. S. (1971). *In* "Scanning Electron Microscopy" (O. Johari and T. Corvin, eds.), Pt. 1, pp. 273–280. I.T.T. Research Institute, Chicago, Illinois.

Miyata, M., Okazaki, M., and Furuya, K. (1977). *Bull. Jpn. Soc. Phycol.* **25,** 1.

Moberly, R. M. (1968). *Sedimentology* **11,** 61.

Monaghan, P. H., and Lytle, M. L. (1956). *J. Sedim. Petrol.* **26,** 111.

Monty, C. L. V. (1967). *Ann. Soc. Geol. Belg.* **90,** 55.

Muscatine, L. (1971). *In* "Experimental Coelenterate Biology" (H. M. Lenhoff, L. Muscatine, and L. V. Davis, eds.), pp. 227–238. Univ. of Hawaii Press, Honolulu.
Muscatine, L. (1973). *In* "The Biology and Geology of Coral Reefs" (R. Endean, ed.), Vol. 1, Pt. 2, pp. 81–112. Academic Press, New York.
Muscatine, L., and Cernichiari, E. (1969). *Biol. Bull.* **137,** 506.
Nicl, H. J., and Henisch, H. K. (1969). *J. Electrochem. Soc.* **116,** 1258.
Okazaki, M., Ikawa, T., Furuya, K., and Miwa, T. (1970). *Bot. Mag. (Tokyo)* **83,** 193.
Outka, D. E., and Williams, D. C. (1971). *J. Protozool.* **18,** 285.
Oxender, D. L. (1972). *Ann. Rev. Biochem.* **41,** 777.
Paasche, E. (1964). *Physiol. Plant. Suppl. 1* **3,** 82.
Paasche, E. (1965). *Physiol. Plant.* **18,** 138.
Paasche, E. (1966). *Physiol. Plant.* **19,** 770.
Paasche, E. (1968). *Ann. Rev. Microbiol.* **22,** 71.
Paasche, E. (1969). *Arch. Mikrobiol.* **67,** 199.
Parke, M., and Adams, I. (1960). *J. Mar. Biol. Ass. U.K.* **39,** 263.
Pautard, F. G. E. (1970). *In* "Biological Calcification" (H. Schraer, ed.), pp. 1–26. North-Holland Publ., Amsterdam.
Pearse, V. B. (1972). *J. Phycol.* **8,** 88.
Pearse, V. B., and Muscatine, L. (1971). *Biol. Bull.* **141,** 350.
Peel, M. C. (1974). Studies of Fine Structure in the Rhodophyta with Particular Reference to Spore Development. Ph.D. Thesis, University College of North Wales, Bangor.
Pentecost, A. (1975). Calcium Carbonate Deposition by Blue-green Algae. Ph.D. Thesis, University College of North Wales, Bangor, Gwynedd.
Pentecost, A. (1978). *Proc. R. Soc. London B* **200,** 43.
Pentecost, A. (1979). *Br. Phycol. J.* **13,** 383.
Percival, E., and McDowell, R. H. (1967). "Chemistry and Enzymology of Marine Algal Polysaccharides." Academic Press, New York.
Percival, E., and Smestad, B. (1972). *Carbohydr. Res.* **25,** 299.
Perkins, R. D., McKenzie, M. D., and Blackwelder, P. L. (1972). *Science* **175,** 624.
Pia, J. (1933). *Mineral. Petrogt. Mitt.* **1.**
Pienaar, R. N. (1969). *J. Cell Sci.* **4,** 561.
Pienaar, R. N. (1975). *In* "The Mechanisms of Mineralization in the Invertebrates and Plants" (N. Watabe and K. M. Wilbur, eds.), pp. 203–229. Univ. of S. Carolina Press, Columbia, South Carolina.
Pobeguin, T. (1954). *Ann. Sci. Natl. Bot.* **15,** 29.
Price, J. H. (1978). Progress in underwater Science, *Rep. Underw. Assoc., N.S.* **3,** 111.
Puiseux-dao, S. (1970). "*Acetabularia* and Cell Biology." Logos Press Ltd, London.
Raven, J. A. (1968). *J. Exp. Bot.* **19,** 193.
Raven, J. A. (1970). *Biol. Rev.* **45,** 167.
Raven, J. A. (1974). *In* "Algal Physiology and Biochemistry" (W. D. P. Stewart, ed.), pp. 434–455. Blackwell, London.
Ruttner, F. (1953). "Fundamentals of Limnology" (D. G. Frey and F. E. J. Fry, trans.). Univ. of Toronto Press, Toronto.
Saddler, H. D. W. (1970). *J. Gen. Physiol.* **55,** 802.
Schiffman, E., Martin, G. R., and Miller, E. J. (1970). *In* "Biological Calcification" (H. Schraer, ed.), pp. 27–67. North-Holland Publ., Amsterdam.
Schoknecht, J. D. (1975). *Trans. Am. Microsc. Soc.* **94,** 216.
Schoknecht, J. D., and Keller, H. W. (1977). *Can. J. Bot.* **55,** 1807.
Schoknecht, J. D., and Small, E. B. (1972). *Trans. Am. Microsc. Soc.* **91,** 380.
Simkiss, K. (1975). *In* "The Mechanisms of Mineralization in the Invertebrates and Plants" (N.

Watabe and K. M. Wilbur, eds.), pp. 135–160. Univ. of S. Carolina Press, Columbia, South Carolina.

Slayman, C. L. (1974). *In* "Membrane Transport in Plants" (U. Zimmerman and J. Dainty, eds.), pp. 107–119. Springer-Verlag, Berlin.

Smidsrød, O., and Haug, A. (1972). *Acta Chem. Scan.* **26,** 2063.

Smith, F. A. (1968). *J. Exp. Bot.* **19,** 207.

Spear, D. G., Barr, J. K., and Barr, C. E. (1969). *J. Gen. Physiol.* **54,** 397.

Steeman Nielsen, E. (1960). *In* "Handbuch der Pflanzenphysiologie" (W. Ruhland, ed.), Vol. 5 (1), pp. 70–84. Springer-Verlag, Berlin.

Streckfuss, J. L., and Ennever, J. (1972). *J. Dent. Res.* **52,** 1099.

Streckfuss, J. L., Smith, W. N., Brown, L. R., and Campbell, M. M. (1974). *J. Bacteriol.* **120,** 502.

Tada, M., Yamamoto, T., and Tonomura, Y. (1978). *Physiol. Rev.* **58,** 1.

Takazoe, I., Kurahashi, Y., and Takuma, S. (1963). *J. Dent. Res.* **42,** 681.

Towe, K. M. (1968). *Biomineral Res. Rep.* **4,** 1.

Turvey, J. R., and Simpson, P. R. (1966). *Proc. Internat. Seaweed Symp., 5th,* pp. 323–327.

Vandermuelen, J. H., and Muscatine, L. (1974). *In* "Symbiosis in the Sea" (W. B. Vernberg and F. J. Vernberg, eds.), pp. 1–19. Univ. of S. Carolina Press, Columbia.

Vinogradov, A. P. (1953). *Mem. Sears Fdn. Mar. Res.* **2**

Vogel, J. J., and Smith, W. N. (1976). *J. Dent. Res.* **55,** 1080.

Wall, D., Guillard, R. R. L., Dale, B., Swift, E., and Watabe, N. (1970). *Phycologia* **9,** 151.

Walter, M. R. (1976). "Stromatolites." Bureau of Mineral Resources, Geology and Geophysics, Canberra, A. C. T. Elsevier, Oxford, England.

Walton, A. G. (1967). "The Formation and Properties of Precipitates." Wiley (Interscience), New York.

Wasserman, R. H., and Kallfelz, F. A. (1970). *In* "Biological Calcification" (H. Schraer, ed.), pp. 313–345. North-Holland Publ., Amsterdam.

Watabe, N. (1967). *Calc. Tiss. Res.* **1,** 114.

Weiss, R. E., Blackwelder, P. L., and Wilbur, K. M. (1976). *Mar. Biol.* **34,** 17.

West, J. A. (1972). *Br. Phycol. J.* **7,** 299.

Westbroek, P., de Jong, E. W., Dam, W., and Bosch, L. (1973). *Calc. Tiss. Res.* **12,** 227.

Wilbur, K. M., and Watabe, N. (1963). *Ann. N.Y. Acad. Sci.* **109,** 82.

Wilbur, K. M., and Watabe, N. (1967). *Stud. Trop. Oceanogr. Miami* **5,** 133.

Wilbur, K. M., Colinvaux, L. H., and Watabe, N. (1969). *Phycologia* **8,** 27.

Wood, R. D., and Imahori, K. (1965). "A Revision of the Characeae," Vol. 1. Cramer, Weinheim, West Germany.

Wray, J. L., and Daniels, F. (1957). *Annu. Chem. Soc. J.* **79,** 2031.

Zipkin, I. (1973). "Biological Mineralization." Wiley (Interscience), New York.

INTERNATIONAL REVIEW OF CYTOLOGY, VOL. 62

Cellular Microinjection by Cell Fusion: Technique and Applications in Biology and Medicine

MITSURU FURUSAWA

Laboratory of Embryology, Department of Biology, Faculty of Science, Osaka City University, Osaka, Japan

I. Introduction

A reliable technique for introducing foreign substances into cells could be one of the most useful tools in cell biology. It not only enables the direct study of

ISBN 0-12-364462-3

some of the biological functions of molecules in living cells, but also may involve the possibility of developing a new field in cell biology. The major merits of using such a technique include the ability to directly examine biological phenomena occurring in a living cell, and to test biological functions of a given molecule in a living cell by introduction of the test substance into the cell, thereby evading the restrictions imposed by cell surface receptors or cell membrane barriers. It is well established that somatic cell hybridization experiments have brought us valuable information, especially with regard to the mechanism controlling gene expression in mammalian cells (Harris *et al.*, 1969; Ringertz *et al.*, 1971; Davidson and de la Cruz, 1974; Peterson *et al.*, 1972). However, the complexity of the constitution of the resulting hybrid cells which is inevitably brought about by cell fusion (usually a hybrid cell containing two sets of heterologous chromosomes and the mixed cytoplasm and cell membrane from two parent cells) makes it hard to perform unambiguous analysis. This difficulty might be overcome by the introduction of an intracellular injection technique.

Since 1914 when Barber (1914) succeeded in microinjecting fluids into cells, many methods of injecting or introducing materials into a living cell have been attempted. For example, Briggs and King (1952) established a method of nuclear transplantation in amphibian eggs. A decade ago, Gurdon (Gurdon *et al.*, 1969; Gurdon, 1974a) developed a bioassay system using oocytes or eggs of *Xenopus*. By introducing test molecules into them by means of a fine glass needle, their biological functions could be examined. More recently, Graessmann (1970) and Diacumakos (1973) have established a microinjection technique by which any foreign material can be injected effectively into individual culture cells. This was carried out by handling an extremely fine needle under a microscope, the field of which was amplified in a television picture. Such a technique makes it possible to inject any substance into single cells or even into their nuclei at a relatively high efficiency. There could be, however, two limitations to the use of this technique. First, it may be almost impossible to use it for small cells which are growing in a suspension form such as lymphocytes. Second, the injection requires a skillful technique and a special instrument and the number of cells which can receive injections within a certain time is limited. The latter limitation cannot be overlooked since a considerable amount of cells is needed if the sample is to be subjected to biochemical analysis, since the injected cells have to be subjected to culture for several days to obtain an adequate amount of cells. In other words, it is impossible to obtain biochemical data from the cells immediately after they are injected. In addition, when treating a phenomenon occurring at a very low efficiency, it is desirable to have a special injection technique by which a substance can be injected into a huge number of target cells in a short time.

In 1973 Yamaizumi and Okada, of the Institute for Microbial Disease at the University of Osaka, and I were performing an experiment involving the transplantation of erythrocyte membrane to mammalian cells. This was brought about by the fusion between human erythrocytes (HRBCs) and Ehrlich ascites tumor cells by Sendai virus (HVJ). Fusion efficiency was determined by Yamaizumi by background staining with Giemsa followed by benzidine. In these experiments, some cells showed a dark brown color in the cytoplasmic region, whereas the remaining cells were light blue, as indicated by light Giemsa staining. This means that the cell fusion of HRBCs and tumor cells resulted in the transfer of a detectable amount of hemoglobin to individual tumor cells. We then considered that, if erythrocyte ghosts containing a foreign substance instead of hemoglobin were used, introduction of that substance into cells would be feasible. Immediately, Nishimura, Aichi Medical College (who at that time was a student in my laboratory), and I prepared HRBC ghosts containing fluorescein isothiocyanate (FITC). These were obtained by performing hemolysis in FITC-containing hypotonic saline. Injection of the FITC molecules into cultured animal cells was attained by HVJ-mediated cell fusion between the ghosts and the cells, although the injection efficiency was very low (Furusawa *et al.*, 1974).

To raise the fusion efficiency Nishimura *et al.* (1976) improved the procedure for replacing the contents of erythrocyte ghosts by means of a gradual hemolysis. Using a dialyzing tube according to the method of Seeman (1967) with slight modifications, they obtained an efficiency of about 40%. Later, Loyter *et al.* (1975) and Schlegel and Rechsteiner (1975) reported independently similar injection methods. Thus, at the present time we have accomplished a refined technique, by which the injection of a definite amount of macromolecules into a huge number (e.g., 10^7 cells) of any culture cells in a short period of time is possible. In this review the detailed method of this injection technique and of its applications in cell biology, carried out primarily by our group, will be presented. Prospects of this method in biology and medicine will be also mentioned.

II. Method of Injection

The injection procedure consists of two steps: first, to load a foreign substance into HRBC ghosts, and second, to fuse the ghosts containing the substance with target cells by HVJ (Okada, 1962). Cell fusion is brought about by HVJ and is based on a method originated by Okada and Murayama (1966). Polyethyleneglycol can be used in the fusion of sheep erythrocyte ghosts (Kriegler and Livingston, 1977). In addition to human erythrocytes, dog, guinea pig, and cow erythrocytes are also suitable for this purpose. The fusion efficiency of these erythrocytes corresponds to 90, 50, and 42% of that of human erythrocytes, respectively. The outline of the injection method has been reported elsewhere (Furusawa *et al.*, 1976; Nishimura *et al.*, 1976).

A. Loading of Foreign Substances into Erythrocyte Ghosts

1. *Trapping of Foreign Substances in Erythrocyte Ghosts*

The standard procedure for loading is shown below, using bovine serum albumin (BSA) as the test substance to be injected. Human blood, irrespective of the blood type, was collected using a syringe with anticoagulant: isotonic sodium citrate, heparinized or EDTA-containing saline. The erythrocytes (HRBCs) were washed three times with phosphate-buffered saline (PBS: 137 m*M* NaCl, 2.7 m*M* KCl, 8.1 m*M* Na_2HPO_4, 1.5 m*M* KH_2PO_4, 4 m*M* $MgCl_2$, pH 7.2) to remove leukocytes and serum and with reverse phosphate-buffered saline (rPBS; 137 m*M* KCl, 2.7 m*M* NaCl, 8.1 m*M* Na_2HPO_4, 1.5 m*M* KH_2PO_4, 4 m*M* $MgCl_2$, pH 7.2) by centrifugation at 1500 rpm. PBS was usable in place of the rPBS. The final HRBC suspension was centrifuged at 1500 rpm for 10 minutes and packed RBCs were obtained. For the introduction of BSA into HRBC ghosts, Seeman's gradual hemolysis method (Seeman, 1967) was used with slight modifications.

A mixture of 1 vol (0.1 ml) of packed HRBCs (ca. 1×10^9 cells) and 9 vol (0.9 ml) of BSA (e.g., 1 mg/ml) dissolved in rPBS (or PBS) was dialyzed against

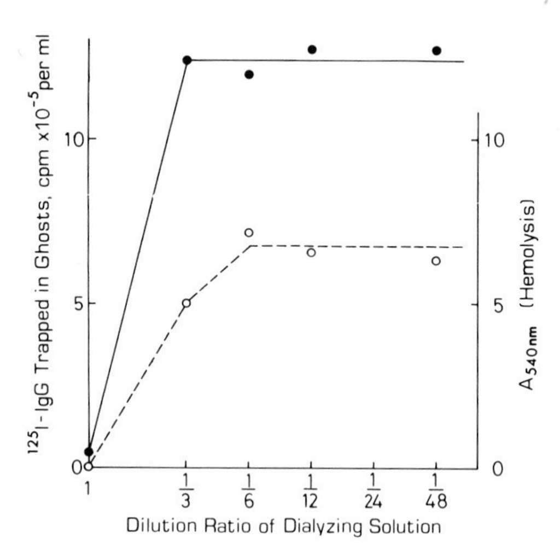

Fig. 1. Effect of the osmolarity of the dialyzing solution on the loading of ^{125}I-labeled immunoglobulin G (IgG) into human erythrocyte (HRBC) ghosts. A mixture of 1 vol of HRBCs and 9 vol of IgG (7×10^6 cpm) was dialyzed against reverse phosphate-buffered saline (rPBS) diluted with distilled water in various ratios for 30 minutes. After further dialysis against isotonic PBS for 30 minutes, the ghosts were washed five times with PBS. (●) Counts associated with the ghost fraction; (○) absorbance of the first supernatant of the centrifugation (for washing) at 540 nm of optical density of hemoglobin. The result shows that dialysis against 6-fold diluted rPBS causes complete hemolysis and that the amount of IgG molecules trapped in the ghosts reaches the maximum point in this dialyzing condition. (From Yamaizumi *et al.*, 1978a.)

a sufficient volume of 6-fold diluted rPBS (or PBS) with stirring for 15 to 30 minutes at room temperature, or, if necessary, at 0°C. In order to determine adequate conditions under which complete hemolysis takes place, hemolysis was performed in the stepwise diluted rPBS. Figure 1 shows that the minimum dilution sufficient to bring about complete hemolysis is 1/6 (in this case, immunoglobulin G, IgG, of 150,000 daltons was used). Since exposure of erythrocytes to a solution of low ionic strength causes a decrease in the fusion ability by HVJ, 1/6 diluted rPBS (or PBS) was used for the dialysis. During this dialyzing step hemolysis occurred gradually, and outflow of hemoglobin from the HRBCs and penetration of BSA into the ghosts took place. It is most important to check the accomplishment of hemolysis in all of the HRBCs under a phase-contrast microscope. In the case of BSA (1 mg/ml), dialysis for 25 minutes was required to attain the degree of hemolysis which causes the trapping of BSA. Longer dialysis did not result in an increase in uptake of BSA. The dialyzing tube was then transferred to another beaker containing isotonic PBS and dialyzed for 15 minutes as above. In this second dialyzing step, the ruptured membrane resealed and BSA was trapped in the HRBC ghosts. The ghosts were then washed with PBS to remove excess BSA and leaked hemoglobin by centrifugation at 2000

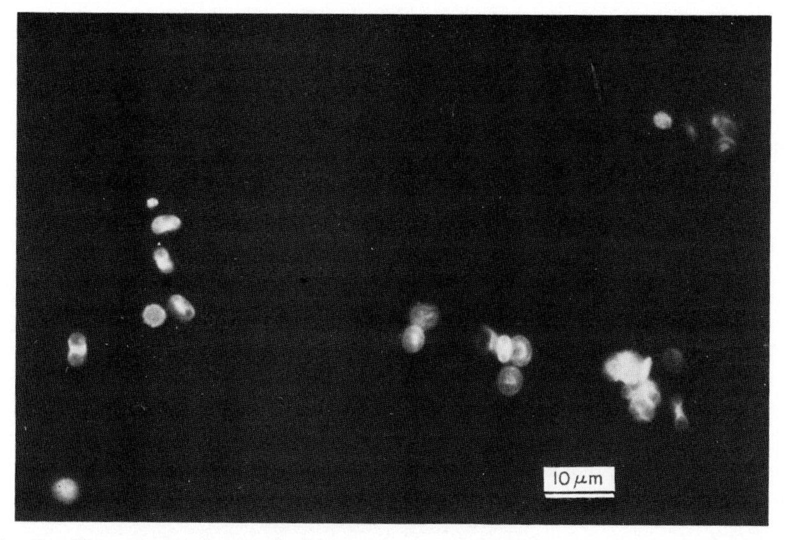

FIG. 2. Fluorescence micrograph of human erythrocyte (HRBC) ghosts containing fluorescein isothiocyanate (FITC)-conjugated bovine serum albumin (BSA). A mixture of 1 vol of packed HRBCs and 9 vol of PBS containing FITC-conjugated BSA (1 mg/ml) was dialyzed against 6-fold diluted rPBS for 30 minutes, followed by dialysis against isotonic PBS for 30 minutes. Each ghost shows fluorescence with FITC-conjugated BSA and retains a concave shape as in the intact erythrocyte. The analysis by cell sorter indicated that each ghost contained approximately an equal amount of BSA. ×850.

rpm for 10 minutes and finally suspended in balanced salt solution supplemented with $CaCl_2$ (BSA^+: 140 m*M* NaCl, 5.4 m*M* KCl, 0.34 m*M* Na_2HPO_4, 0.44 m*M* KH_2PO_4, 2 m*M* $CaCl_2$, buffered with 10 m*M* Tris–HCl at pH 7.6). The ghosts thus obtained retain a normal concave shape (Fig. 2) and still complete cell fusion activity.

An experiment was carried out on the amount of substance incorporated into ghosts using FITC-labeled BSA. The amount of BSA trapped in ghosts increased linearly with increases in the concentration of BSA in the dialyzing tube, and the concentration was about one-third of the original concentration in the tube (Fig. 3). This means that the amount of substance to be injected can be controlled. Trapping efficiency depends on the molecular weight of the protein. The concentration of proteins trapped in the ghosts is equal to that in the outside solution in a dialyzing tube in the case of 22,000-dalton proteins, one-half in 68,000-dalton proteins, and one-third in 150,000 dalton proteins, respectively (Yamaizumi *et al.*, 1978a,c,d). It should be emphasized that analysis with a cell sorter demonstrated that each ghost contained nearly equal amounts of the labeled BSA (Mekada *et al.*, 1978b). The experiment with ^{125}I-labeled BSA showed that 80% of the trapped BSA was in the stroma of the HRBC ghosts and the remaining 20% was associated with the membrane fraction (Yamaizumi *et al.*, 1978a,d).

The trapping conditions can be varied according to requirements. For example, if very small amounts of precious material are involved the following procedure is recommended. A mixture of 1 vol (e.g., 0.1 ml) of packed HRBCs and 2 vol (0.2 ml) of Hanks' solution containing the substance is dialyzed against cold pure water for 15 minutes using a specially devised dialyzing instrument (Fig. 4).

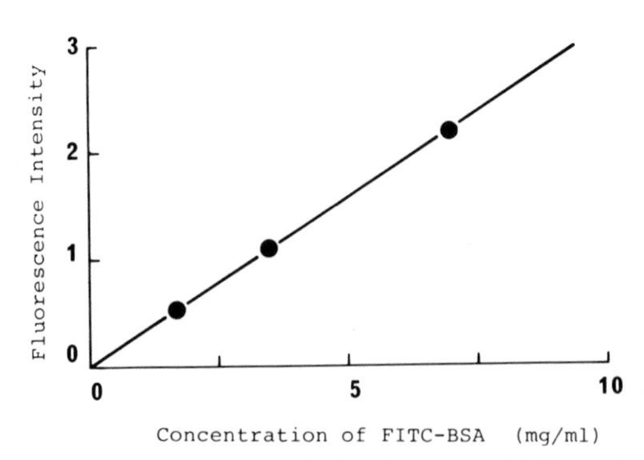

FIG. 3. Rate of incorporation of fluorescein isothiocyanate (FITC)-conjugated bovine serum albumin (BSA) into human erythrocyte (HRBC) ghosts. A mixture of 0.3 ml of packed HRBCs and 1.7 ml of various concentrations of FITC-conjugated BSA resolved in reverse phosphate-buffered saline (rPBS) was dialyzed against 6-fold diluted rPBS for 30 minutes and then dialyzed against PBS for 30 minutes. Fluorescence distribution of ghosts was measured with a cell sorter (FACS II). The mean intensity of the fluorescence of the ghosts increases linearly with increases in the concentration of FITC-conjugated BSA added to the dialyzing tube. (From Mekada *et al.*, 1978b.)

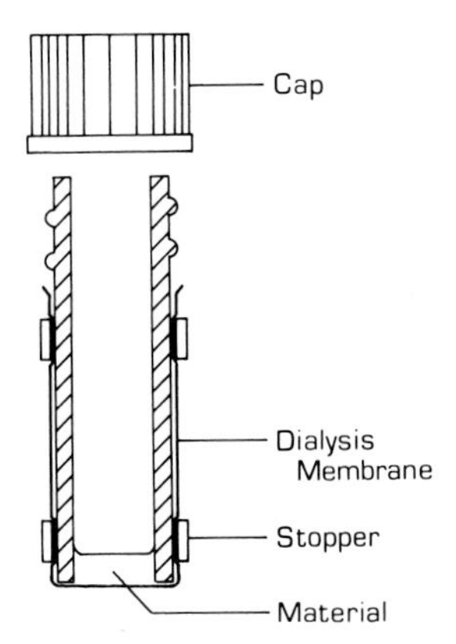

FIG. 4. Schematic figure of a micro-dialyzing instrument. The instrument consists of a glass tube (1 cm in diameter) with a cap; the bottom opening is covered with a dialyzing membrane. Before use, this instrument should be autoclaved. Materials (200 to 250 μl) to be dialyzed are poured on the membrane from the top opening of the tube with a pipet, and the cap is closed to avoid bacterial contamination. For dialysis, the bottom of the tube is dipped in a dialyzing solution and the tube is gently shaken occasionally to mix the contents. This is a slight modification of the method of Auer and Brandner (1976).

This can then be followed by dialysis against isotonic Hanks' solution for 15 minutes at 0°C and further incubation at 37°C for 90 minutes. With a similar procedure, T3 phage DNA of 2.5×10^7 daltons can be effectively trapped in the ghosts (Auer and Brandner, 1976). In addition, Wasserman *et al.* (1976) found that the presence of cytochrome *c* at a concentration of 0.5 mg/ml in the dialysis bag increases the ability of the ghosts to fuse to recipient cells.

2. *Stability of the Substance in HRBC Ghosts*

The trapped materials in the ghosts are quite stable, since virtually no proteolytic enzymatic activity (Yamaizumi *et al.*, 1978d), DNase, nor RNase are present in the ghosts. After the ghosts containing the foreign substances are stored for a few days in the refrigerator, they can be used for the injection experiment. Stability of the proteins trapped in HRBC ghosts was examined with ^{125}I-labeled BSA. When HRBC ghosts containing ^{125}I-labeled BSA were kept at 4°C for 4 days, 10% of the radioactivity leaked from the ghosts and no degradation of BSA molecules occurred during storage. In contrast, when kept at 37°C for 1 day, 90% of the radioactivity was lost from the ghosts. Again no degradation was observed in the leaked BSA molecules, so that this diminution of radioactivity in the ghosts is not due to the enzymatic digestion by protease (Morrison and Neurath, 1953) but merely to the leakage of BSA through the membrane. The leakage could be inhibited to some degree by using PBS in place of rPBS when performing a dialysis under hypotonic conditions (Yamaizumi *et al.* 1978d).

B. Intracellular Injection of BSA by Cell Fusion of the HRBC Ghosts and Target Cells

1. *Suspension Cells*

a. *Injection Procedures.* The injection procedure is schematically illustrated in Fig. 5. Injection of BSA is achieved by cell fusion between the target cells and the HRBC ghosts containing BSA. Cell fusion was performed according to the method of Okada and Murayama (1966). Target cells were washed with cold PBS to remove culture medium by centrifuging three times, suspended in cold BSS^+ at a density of 2.5×10^6 cells/ml, and kept in ice until used. Trypsinized cells can also be used as target cells for this purpose.

Ultraviolet-inactivated HVJ of 1000 hemagglutinating units (HAU) (Salk, 1944), suspended in BSS ($CaCl_2$-free BSS^+), was prepared and kept in ice. Depending on the kind of target cell, 200 to 2000 HAU of HVJ was used. One

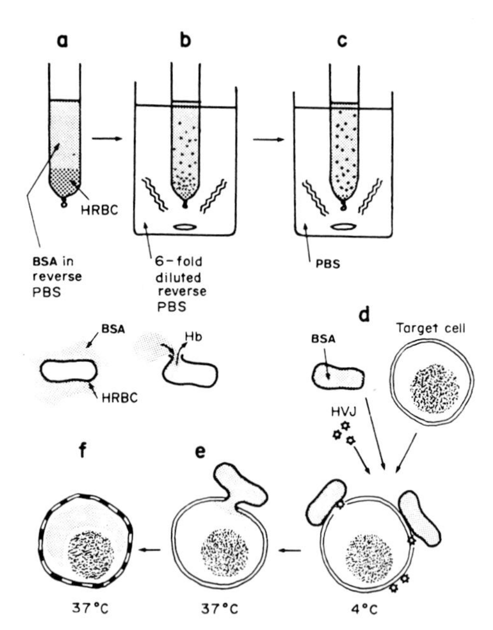

Fig. 5. Diagramatic representation of the steps for replacement of hemoglobin of human erythrocytes (HRBCs) with bovine serum albumin (BSA) and injection of BSA into target cells by cell fusion. (a) BSA and HRBCs are mixed in a dialyzing tube under isotonic conditions; HRBCs (shown in the middle row) are intact. (b) During dialysis against hypotonic reverse phosphate-buffered saline (rPBS), the outflow of hemoglobin (Hb) and the penetration of BSA into HRBC ghosts occur. (c) The ruptured HRBC membrane is resealed and BSA is trapped in the ghost during dialysis against isotonic PBS. (d) Mixing of ghosts containing BSA, target cells, and Sendai virus (HVJ) causes agglutination. (e,f) HRBC ghosts and target cells fuse at 37°C, BSA is introduced into the cytoplasm of the cells, and molecules of HRBC membrane and target cells are intermixed. (From Furusawa *et al.*, 1976.)

volume (0.5 ml) of the cell (5×10^6 to 1×10^7) suspension, 1 vol of the ghosts containing BSA suspended in cold BSS$^+$ (20% v/v), and 2 vol of the HVJ solution were mixed together in a round-bottom test tube. The mixture was kept in ice for 15 minutes to allow cell agglutination to take place. The mixture was incubated at 37°C for 5 minutes and then shaken gently in the bath for a further 15 minutes to allow cell fusion of the ghosts and the cells. The ratio of cell number of target cells to that of HRBC ghosts is approximately 1 to 100. This excess HRBC ghost proportion is important for preventing cell fusion between target cells and for increasing the efficiency of fusion between target cells and HRBC ghosts. As the cell fusion reaction needs an oxygen supply, it is better to use a relatively large test tube. During cell fusion a little leakage of trapped materials into the medium took place. The higher titer of HVJ causes more leakage of materials. After the last incubation, the cells were collected by centrifugation and resuspended in a culture medium for subsequent cultivation. If necessary, nonfused free HRBC ghosts can be removed by repeated centrifugations. The viability of the injected cells is usually quite high (more than 95%). Usually, almost a 90% injection efficiency is obtained, although this percentage varies in individual cases. Wasserman *et al.* (1976) showed that the presence of La^{3+} ion (0.2 to 0.35 m*M*) raised the fusion ability of the ghosts and minimized the leakage of materials from the ghosts. When cells with low fusion ability such as mouse lymphoma cells are used, the fusion efficiency can be raised by replacing the last incubation by centrifugation at 800 rpm for 20 minutes at 37°C (unpublished data). In the case of Ehrlich ascites tumor cells, one can perform a second injection to the same population 14 hours after the first injection, since the fusion ability of the cells is completely recovered by this time. Thus, repeated injections are possible to the same cells.

After fusion of the intact HRBC and target cell, hemoglobin diffuses throughout the cell body including the nuclear domain within 5 minutes. In the case of BSA injection, however, penetration into nucleus does not take place, although rapid diffusion is observed into the cytoplasm as well as in the case of hemoglobin (Yamaizumi *et al.*, 1978d).

b. *Estimation of Injection Efficiency.* There are three ways to estimate the injection efficiency: (1) Staining of residual hemoglobin with benzidine dye (Cooper *et al.*, 1974); HRBC ghosts contain a small amount of hemoglobin after hemolysis; consequently cells fused with the ghosts also carry a residual hemoglobin but nonfused ones do not, benzidine dye selectively staining the former. A small aliquot of the cell suspension is added in 4 ml of PBS in a plastic dish and several drops of benzidine solution are added. Benzidine staining is observed under a inverted microscope using a blue filter (Tabuse *et al.*, 1976). (2) Staining with the FITC-labeled anti-HRBC membrane protein serum: After cell fusion, HRBC membrane is inevitably transplanted to the cell membrane of target cells. One can detect the injected cells because the antibody stains the cell surface without killing the cells (Mekada *et al.*, 1978a). Anti-HRBC membrane protein serum is prepared in rabbits according to the method of Adachi and Furusawa (1968; Furusawa and Adachi, 1968). An indirect method is used: For staining, the cells are treated with the antibody at 0°C and then with FITC-labeled anti-rabbit IgG goat serum. The cells receiving the injection can be observed under a fluorescence microscope. (3) Trapping of marker molecules in the ghosts: Biological inactive substance such as BSA conjugated with FITC is added to the dialyzing tube with the substance to be injected. Hemolysis results in loading of the marker molecules together with the injecting substance into the ghosts. Cell fusion of the ghosts and target cells causes the introduction of marker molecules which can be detected under a fluorescence microscope.

2. *Plating Cells*

The adherent cell can receive the injection in an adherent form as it is. This method has the advantage of avoiding cell fusion between target cells. The outline of this method is shown in Fig. 6. Cells, 1×10^5 or fewer, were inoculated in a plastic dish 30 mm in diameter. The dish was incubated at 37°C until the cells settled onto the substratum, but before the onset of the first cell division. After removal of the culture medium, the dish was washed with cold BSS+ three times by gentle pipetting. One milliliter of UV-inactivated HVJ of

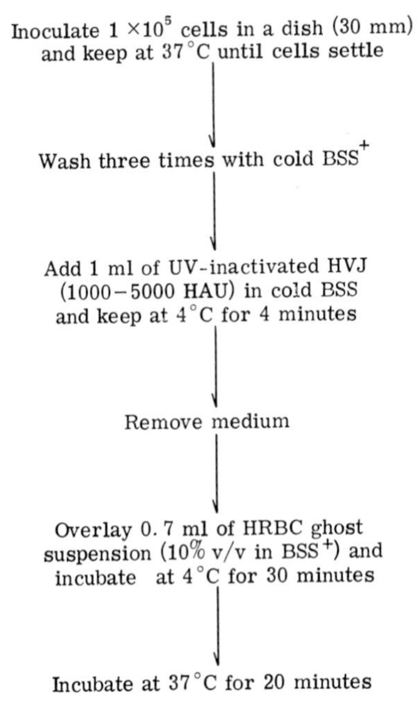

FIG. 6. Procedures for microinjection of cells growing in attached form.

1000 to 5000 HAU which was suspended in BSS was poured into the dish and kept at 4°C for 4 minutes to allow HVJ to adhere to the surface of the plating cells. Longer incubation at 4°C resulted in a decrease in fusion efficiency. The medium was then carefully removed, 0.7 ml of BSA-containing HRBC ghost suspension (10% v/v) was overlaid, and the dish was incubated at 4°C for 20 minutes. Fusion efficiency increased as the number of HRBC ghosts overlaid on the culture was increased. This incubation resulted in the adsorption of the ghosts onto the surface of the rosette-forming cells. The cultures were then incubated at 37°C for 20 minutes to allow fusion of the ghosts and the cells.

During this incubation most of the ghosts which had formed rosettes detached from the cell surface, and only a small number of them, usually one ghost to one cell, could fuse with a target cell. In general, the injection efficiency of BSA was lower than in the case of cells in suspension. The medium and the free ghosts were carefully removed by pipet, a culture medium was added, and the dish was incubated for further cultivation. The following is an experiment in which the fusion (injection) efficiency was 84%. Namely, 5×10^4 BudR-resistant mouse L cells were seeded in a 30-mm dish; 1 ml of chilled HVJ of 1000 HAU was added and the dish was kept at 4°C for 4 minutes; 0.7 ml of chilled HRBC ghost suspension (10% v/v) was added and the dish was kept at 4°C for 20 minutes and then incubated at 37°C for 20 minutes.

C. How to Collect the Cells Receiving the Injection of a Definite Amount of Substances

Collection of individual cells receiving an injection of a definite amount of substance can be attained by using a cell sorter with computer system. Two methods are used: One employs FITC-labeled anti-HRBC membrane protein serum and the other FITC-labeled molecules as the marker.

1. *Method 1*

As mentioned above, HRBC membranes transplanted to target cells during the cell fusion reaction could be a good marker for identifying cells receiving the injections, since HRBC surface antigens can be recognized by the staining of the FITC-labeled antibody. An indirect fluorescent antibody staining technique should be used. Heat-inactivated anti-HRBC serum was diluted to an appropriate concentration with PBS. The cell surface was treated with the antiserum for 10 minutes and washed three times with PBS by centrifugation. Then the FITC-conjugated anti-rabbit IgG goat serum dissolved in PBS was added to the cells. After 10 minutes, the cells were washed three times with PBS to remove excess antibodies. All treatments should be performed at 4°C. Fluorescence of injected cells (cells fused with the ghosts) but not of noninjected cells, could be observed under a fluorescence microscope. To subject the cells to cell sorter analysis, they were suspended in Hepes-buffered saline (pH 7.2) supplemented with 2% calf serum. Fluorescence intensity should be proportional to the number of HRBC ghosts fused to a single target cell. As already mentioned, the dose of trapped materials in each ghost can be controlled by adjusting the amount of material added to the dialyzing tube. Therefore, it is possible to select cells injected with a given amount of substance, using the cell sorter which can sort out and collect cells according to the intensity of fluorescence. This machine can also select cells according to their volume, meaning that fused target cells carrying plural nuclei can be recognized and selectively removed. The maximum scanning speed of this machine is 1 to 5 $\times$ 10^3 cells per second.

2. *Method 2*

An example of cell separation using a cell sorter (FACS II, Beck–Dickinson Co., Fluorescence Activated Cell Sorter-II) is shown below. This was done by Mekada *et al.* (1978b) and Yamaizumi *et al.* (1978c) and is the easiest technique for cell separation presently available. In this experiment FITC-labeled BSA was used as the marker molecule to be injected and mouse L cells as the target. FITC-labeled BSA (7 mg/ml) was added to a dialyzing tube together with HRBCs and the dialysis produced the BSA-containing ghosts according to the method described above. The ghosts and L cells were then fused to introduce FITC-labeled BSA into the L cells. Figure 7 indicates the light scattering pattern of the injected cells, which reflects relative cell size. It is clear that the machine has separated L cells from free ghosts. The sizes of almost all of the L cells do not change before and after cell fusion, which means that fusion between target L

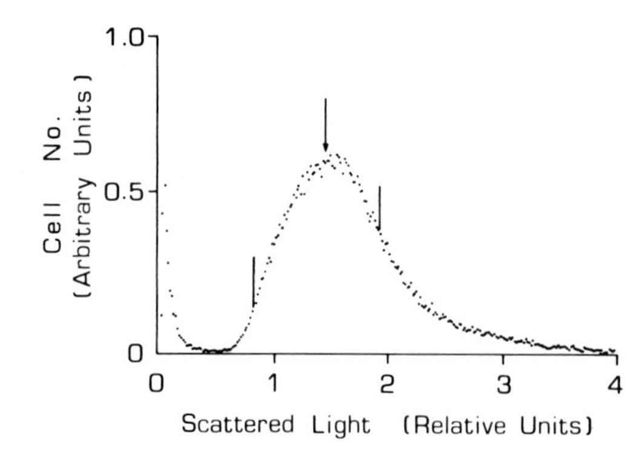

FIG. 7. Light scattering pattern of the bovine serum albumin (BSA)-injected cell sample. After cell fusion of human erythrocyte (HRBC) ghosts containing fluorescein isothiocyanate (FITC)-conjugated BSA and L cells, the sample was analyzed with a cell sorter. The cell sorter can separate L cells (the main peak) from free ghosts (left small peak). The intensity of scattered light reflects the relative cell size. Arrow indicates the mean value of the light scattering intensity of intact L cells. All of the cells distributing in the area between the two bars consist of mononucleated nonfused L cells. The cells in this area are a mixed population of injected (fused with HRBC ghosts) and noninjected cells. (From Mekada *et al.*, 1978b.)

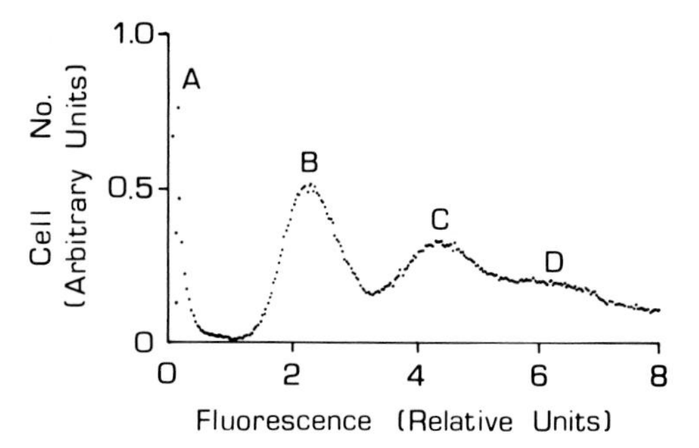

FIG. 8. Fluorescence profile of L cells fused with human erythrocyte (HRBC) ghosts containing fluorescein isothiocyanate (FITC)-conjugated bovine serum albumin (BSA). The HRBC ghosts were prepared by dialysis with 3.5 mg/ml FITC-conjugated BSA. The fluorescence pattern was obtained from the cell population consisting of mononucleated cells by light scattering analysis (see Fig. 7). Peak A corresponds to nonfused L cells, peak B to L cells fused with a single ghost, peak C to two ghosts, and peak D to three ghosts, respectively. (From Mekada *et al.*, 1978b.)

cells is rare under these experimental conditions (arrow indicates the mean value of the light scattering intensity of untreated L cells). The cell population distributing between the two bars shown in the figure was collected and scanned according to flourescence intensity. Giemsa staining demonstrated that this population consisted entirely of mononucleate cells. The result of the scanning (Fig. 8) shows four peaks. The first one from the left (A), without fluorescence, corresponds to nonfused L cells. The remaining three peaks correspond to L-cell populations injected with FITC-labeled BSA. Fusion of a single L cell with a single ghost results in the second peak (B), fusion with two ghosts results in the third (C), and with three ghosts in the fourth (D), respectively. In proportion to the increasing number of HRBC ghosts fused to a single L cell, the average intensity of the fluorescence is discretely strengthened twice and three times compared to the second peak where the cell fusion of one L cell to one ghost took place. Thus it can be said that individual cells in a given peak have received an injection of a given amount of BSA. The cells belonging to the first peak can be used as the noninjected control. When one collects samples from different peaks separately, one can examine the dose effect of the substance in one experiment. As far as we tested with various cell types, the proportion of the occurrence of 1:1 cell fusion was considerably high. Figure 9 shows the results of cell separation by the FACS II. Although the use of fluorescein-labeled molecules is indispensable for cell sorting with the cell sorter, it is not necessary to label the substance to be tested with fluorescein. To determine the relative amount of the injected test substance, a certain amount of FITC-labeled molecules which are biologically inactive, such as BSA, must be added to a dialyzing tube as a marker. Dialysis might result in penetration of the marker molecules into the ghosts at a certain ratio in proportion to the amount of test substance. Consequently, the machine can sort out the target cells which have been injected with

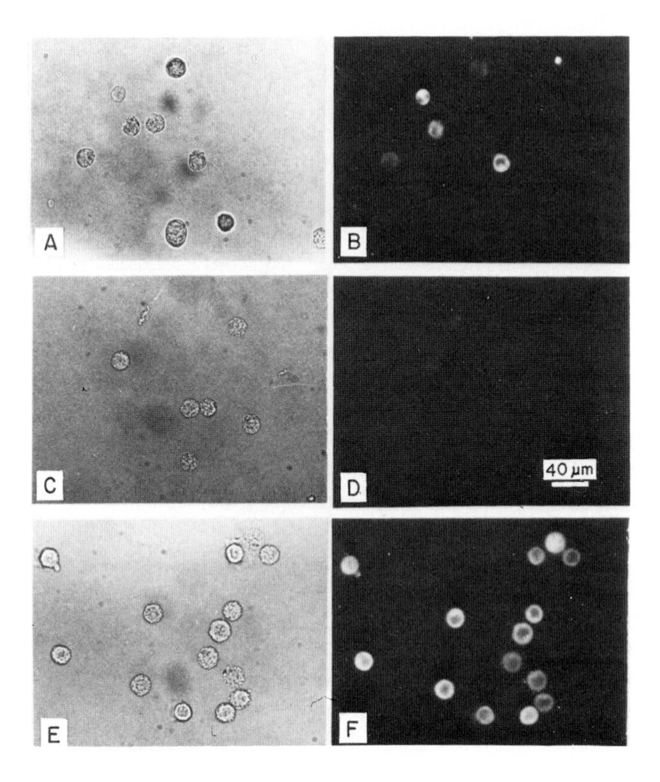

FIG. 9. Light (A,C,E) and fluorescent (B,D,F) micrographs of cells injected with fluorescein isothiocyanate (FITC)-conjugated bovine serum albumin (BSA). (A,B) Original sample before sorting. The population is not homogeneous in both cell size and fluorescent intensity. Five out of eight cells are injected. Two small fluorescing corpuscles are free human erythrocyte (HRBC) ghosts. (C,D) Cells obtained from the population of peak A in Fig. 8. Cell size is homogeneous and each has a single nucleus, but no injection of BSA molecules took place. (E,F) Cells obtained from peak B. Cell size is homogeneous. Each cell has fused with a single ghost, and therefore shows equal fluorescent intensity. In other words, cells receive injections containing equal amounts of BSA molecules. Two pictures in each row are from the same field of the microscope. ×100. (From Mekada *et al.*, 1978b.)

a given amount of the test substance. After cell sorting, the viability of the cells is quite high and they can grow in a culture medium as do nontreated cells (Yamaizumi *et al.*, 1978c; Mekada *et al.*, 1978a,b).

III. Advantages and Limitations of This Injection Method

The viability of mononucleated target cells receiving injections was quite high (more than 90%) when HVJ of around 1000 HAU was used. The cells fused with the ghosts can divide in a normal manner, but sometimes they show a delay of 1 day in the onset of cell division in comparison with nonfused cells (Yamaizumi *et al.,* 1978d). It should be emphasized that our injection method is relatively simple, and one can inject, within a short time, a given amount of foreign material into a considerably large number of cells. For example, when 1 mg of double-stranded DNA (5×10^6 daltons) was used as an original sample, under adequate experimental conditions introduction of about 1000 molecules into each target cell (total cell number: 10 million) is possible (Furusawa and Tonegawa, unpublished data). Almost all kinds of cells from any species of mammals and birds which carry HVJ receptors can be used for this method. The upper limit of the molecular size that can be injected by this method is in the region of 10^7 daltons. For instance, Auer and Brandner (1976) demonstrated that 2.5×10^7 daltons of T3-phage DNA could be loaded in HRBCs by a slightly modified method of dialysis. tRNA (Kaltoft *et al.,* 1976), SV40 cRNA (Auer and Brandner, 1976), and variously charged proteins (Yamaizumi *et al.,* 1978d) can also be injected. When small molecules which pass through a dialyzing membrane are desired to trap the ghosts, the molecules should be dissolved in solutions of the inside and outside of the dialyzing tube at the same concentration, during dialysis. Many points of the procedure for loading substances into HRBC ghosts remain to be improved. One should pay attention to loading substances into HRBC ghosts in order to avoid the exposure of HRBCs to a hypotonic condition for a long period, since this brings about a decrease in the fusion ability of HRBCs. The difference in electric charge of the trapped proteins has no effect on the fusion efficiency of the HRBC ghosts and the target cells: For example, HRBC ghosts loaded with BSA (isoelectric point being 4.6, 67,000 daltons), rabbit IgG (7.0, 150,000 daltons), or chick lysozyme (11.0, 14,600 daltons) all showed the same fusion efficiency of about 85%.

Disadvantages of this method are: (1) the unwanted introduction of human erythrocyte membrane, residual hemoglobin, HVJ envelope and its inactivated RNA, which could have undesirable effects on cells (however, these troubles can be practically avoidable using vacant HRBC ghosts as a control); (2) the possibility of denaturation of fragile materials due to exposure to hypotonic conditions during dialysis; (3) the difficulty of injection of some of the particular molecules which precipitate and form large aggregates under the ionic conditions in the dialyzing steps; (4) wastfulness of the materials to be injected (this also could be minimized to some degrees by improving the dialyzing procedures); (5) the impossibility of injecting substance into the nuclear domain.

IV. Comparison with Other Methods

In addition to the injecting method described in this article, other methods of introducing substances into a living cell have been attempted. We now have a syringe method in which a fine glass needle is used for introducing materials into cells. It has a long history since 1914 when Barber first tried to inject foreign materials into living cells with a micromanipulator (Barber, 1914). The second is a liposome method which uses the fusion ability of artificial lipid vesicles containing substances. The third is the so-called HVJ method in which the interaction between HVJ and cell membranes is used. Each method has merits and disadvantages. In this section a brief explanation of each method is given and comparisons among these four methods are attempted.

A. Syringe Method

Gurdon *et al.* utilized *Xenopus* eggs as a test system of the biological function of macromolecules in a living cell (Gurdon *et al.,* 1969). The introduction of test substances into the egg or oocyte can be easily attained with a fine glass needle, since the egg is extremely large in size compared with mammalian culture cells. Much valuable information has been elucidated using this system. Various kinds of mRNAs such as globin, lens crystalline, L chain of immunoglobulin G from myeloma, collagen, protamine, interferon, etc. were introduced into the eggs. The results showed that all of the mRNAs tested were translated in the *Xenopus* egg. Moreover, DNA from various sources, such as ribosomal genes, SV40, colicin El, adenovirus, ϕX174, etc., was also injected and the synthesis of RNA molecules which can be specifically hybridized against corresponding DNA was demonstrated. For details, see the monographs by Gurdon (1974a,b) and Lane and Knowland (1975). However, since the oocyte can be considered to be a specialized cell in that it is multipotent, there might naturally be some limitations in the application of this system for studies in cell biology. To overcome this limitation, Graessmann (1970) and Diacumakos (1973) established a microinjection technique for introducing substances into mammalian or avian culture cells. This was done with a very fine needle and a special instrument. As already mentioned, the main disadvantage of this technique is that it is impossible to perform injections into more than 500 cells in an hour.

Recently Stacey *et al.* (1977) performed a viral rescue experiment with the syringe method. A mutant strain of Rous sarcoma virus (RSV) with the ability to evoke transformation of infected cells but not to produce virions was used. That is to say, it is defective in *env* gene (gene of envelope protein) but has *src* gene (gene responsible for evoking the transformation of the infected cell). When they injected poly A containing RNA extracted from wild-type RSV-2-infected cells into cells infected with the mutant virus, production of RSV virions took place; thus the injected RNA fraction contained the envelope (*env*) mRNA. In this experiment the maximum volume of material to be injected into a single cell was 5% of a single cell volume. The injection efficiency was about 500 cells per 1 hour.

B. Liposome Method

Magee (1972) and Papahadjopoulos *et al.* (1974) established a novel method of injection using artificial lipid particles or liposomes. When phospholipids are suspended in a salt or sucrose solution at a temperature higher than the phase-transition temperature, closed particles which consist of unilamellar or multilamellar membranes with bilayer liquid molecules are formed. When liposomes containing foreign substances are added to a cell suspension, introduction of the substances into the cells takes place by the fusion between liposomes and cells or by endocytosis of the cells. This method, however, also has some disadvantages which cannot be overlooked. Trapping of substances in liposomes must be performed by sonication, which involves the risk of denaturation of fragile materials such as nucleic acids. Furthermore, liposomes tend to adhere onto the cell surface, so that the quantitative analysis of experimental results becomes difficult. Various kinds of lipids have been tested as to the surface charge of liposomes which might influence the ability to fuse to cells. As a matter of fact, these improved liposomes frequently showed cytotoxic effects. Another attempt was made to raise the fusion efficiency using unilamellar liposomes. Unilamellar liposomes can be obtained by vigorous sonication, which results in a decrease in liposome size, but the decrease in size causes a decrease in the amount of substances to be injected. To avoid this problem, Uchida *et al.* (1978) developed new liposomes which associate with the envelope spikes of HVJ. They succeeded in preparing rebuilt liposomes with a high fusion ability and low cytotoxicity. Large unilamellar liposomes prepared by treatment with Ca^{2+} and EDTA have been made (Wilson *et al.*, 1977). Using this liposome they showed that the injection of poliovirus into nonsusceptible CHO cells causes cell infection. It was estimated that the efficiency of injections was no more than 3% of the total cell number judging from the cell number producing the virus.

C. HVJ Method

There are three injection methods using the interreaction of HVJ and the cell membrane. (1) When a mixture of materials to be injected and UV-inactivated HVJ is poured on culture cells, the materials are inserted effectively into the cells, presumably through the small pores which are made by the interaction of HVJ and the cell membrane (Tanaka *et al.*, 1975). An example using this method is shown in Section VI. (2) When a mixture of materials and HVJ is sonicated, the materials are trapped in the viral envelope. Addition of the sonicated virus to cells caused introduction of the materials into cells by the reaction of the virus with the cell membrane. The amount of materials injected into a single cell was 10 times that in the former method using HVJ alone. Using this method, Uchida *et al.* (1979) demonstrated that introduction of fragment A of diphtheria toxin into cells resulted in the death of the cells. Fragment A cannot kill a cell when added in a culture medium, since it is lacking the specific fragment necessary to bind to the cell surface receptor (see Section V,A). (3) Reassembled HVJ envelope was prepared from HVJ virions solubilized with a nonionic detergent. If the reassembled envelope containing a foreign material is layered on culture cells, the material is introduced into the cells, because the envelope still retains the neuraminidase, hemogglutinating, cell fusion, and hemolytic activities present in intact HVJ. With this technique Uchida *et al.* (1977) succeeded in killing mouse L cells by introduction of CRM45, a nontoxic mutant protein related to diphtheria toxin. With the normal method of administration by intravenous injection the CRM45 is nontoxic for animal cells because it lacks the C-terminal sequence of amino acids necessary to bind to the cell surface receptors.

D. Conclusions

The principal drawback to the syringe method is the limited amount of cells that can be injected in a given period. However, this is the one and only method which can inject substances into the nucleus within the cell. On the other hand, the liposome method is disadvantageous in that the amount of injected substance per target cell and the efficiency of the injection are low. In contrast, injection with the HVJ method is highly efficient: Introduction of substance into all of the treated cells is possible. This method also surpasses others with regard to the cell number to be injected and ease of handling. But, the principle disadvantage of this method is the restriction in the amount of substance to be injected and in the molecular size of injectable substances.

In conclusion, it can be said that the present HRBC ghost method is very useful, since it almost satisfies the above-mentioned conditions necessary for microinjection. The amount of substance that can be injected depends on the volume ratio of ghost to target cells and on the number of ghosts fused to a target cell. The volume of material to be injected into a target cell is estimated to be in the range of 1 to 3% of the volume of cell body. For instance, when 10 mg/ml BSA was added to a dialysis tube (the ratio of packed HRBCs to rPBS being 1/9), the average number of BSA molecules injected per cell was 4×10^6. The most striking advantage of our method is that one can inject a definite amount of materials into a huge number of target cells in a short time.

V. Example of the Use of This Microinjection Method

Several experiments using the method of microinjection, carried out by our group, are presented below.

A. Analysis of the Mode of Action of Diphtheria Toxin in Living Cells

The aim of this experiment was to analyze the mode of action of diphtheria toxin, by introducing antibodies against the toxin into a living susceptible cell. This was the first trial to introduce antibodies into a living cell and examine whether the antibodies could function in the cytoplasm. Diphtheria toxin is a single polypeptide which inhibits protein synthesis in susceptible mammalian cells by catalyzing the transfer of the ADP-ribosyl moiety from NAD to covalent linkage with, and inactivation of, soluble polypeptide chain elongation factor 2 (Honjo *et al.,* 1968; Gill *et al.,* 1969). The toxin molecule consists of two fragments, A and B. The inhibitory activity lies in N-terminal fragment A of 22,000 daltons (Honjo *et al.,* 1968; Gill *et al.,* 1969), while C-terminal fragment B of 40,000 daltons has no toxic activity at all but carries a site for attachment to the specific receptors on the surface of a susceptible cell (Pappenheimer *et al.,* 1972; Uchida *et al.,* 1972). Fragment A was isolated from the 62,000-dalton toxin molecule following the proteolytic and reductive steps (Gill and Pappenheimer, 1971; Drazin *et al.,* 1971). Antibody against fragment A was raised in rabbits. Anti-fragment A neutralizes the enzymatic activity of fragment A, but will not neutralize the complete toxin molecule (toxin) consisting of fragments A and B. Accordingly, addition of an excess amount of anti-fragment A to culture medium did not protect susceptible human FL cells (amnion cell) from the attack of toxin added to the medium. In fact, 1 ''AU''/ml of anti-fragment A (1 ''AU'' can precipitate fragment A from 1 flocculating unit (Fl); 1 Lf is equivalent to 2 μg of the toxin) could not netralize 0.01 Lf of the toxin when added to the culture medium. Four hours after toxin administration, the culture medium was washed out and cultured in fresh medium for 1 week. The survival ratio of the FL cells after administration of the toxin was lower than 3% (217 colonies from 10,000 inoculated cells). These conditions were used in the following experiments.

To determine the mode of action of the toxin in the cell, we tested whether the susceptible cells became resistant to the toxin when anti-fragment A (which is not effective when added to the medium) had been previously introduced into the cells. The injection of anti-fragment A into FL cells was carried out according to the method described in Section II. The FL-cell population, 15 to 20% of which had received injections of anti-fragment A, were exposed to 0.01 Lf/ml toxin for 4 hours and cultured in a toxin-free medium for 1 week to allow formation of colonies. As a control, antibody against a ricin was used and the anti-ricin-injected cells were subjected to the same treatment. The results are shown in Fig. 10. The colony number obtained increased with increasing doses of injected anti-fragment A. At the highest dose of anti-fragment A (36 mg/ml in a dialyzing tube), it can be estimated that almost all of the cells receiving injections were rescued. Based on a plating efficiency of 60%, inoculated cell number of 10^4, and injection efficiency of 15–20%, the number of cells receiving the anti-

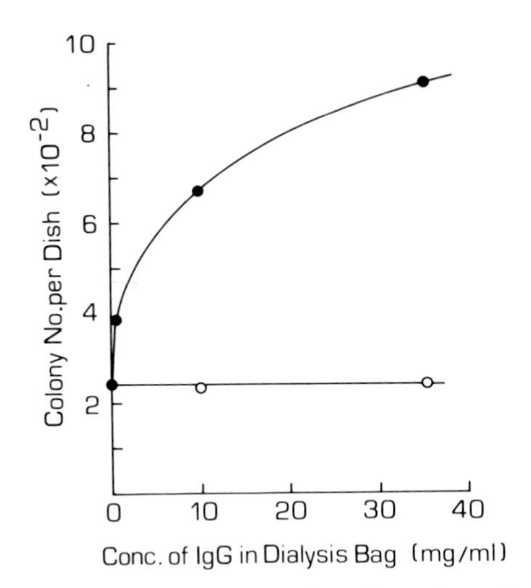

FIG. 10. Effects of diphtheria toxin treatment on the cells trapping antibodies. Ten thousand FL cells (human amnion cells) were fused with human erythrocyte (HRBC) ghosts containing various amounts of antibodies (IgG) and exposed to 0.01 Lf/ml diphtheria toxin for 4 hours. They were then cultured in a fresh toxin-free medium for 7 days. The colony number was counted: (●) anti-fragment A; (○) anti-ricin agglutinin (IgG). The graph shows that the injected anti-fragment A specifically protects the cells from the attack of the toxin. (From Yamaizumi *et al.*, 1978a.)

fragment A injection is calculated as follows: $10^4 \times 0.6 \times 0.15$–$0.20 = 900$–$1200$. This number coincides with the actual colony number of 911. Injection of anti-ricin did not have any effect on the ratio of colony formation. These results strongly suggest that the antibody retains its function even in a living cell cytoplasm, at least for 4 hours after the injection, and neutralizes the enzymatic action of fragment A. Concerning the mode of action of diphtheria toxin, it is speculated from our experiment that the toxin molecule which is trapped by the receptor on the cell surface is inserted into the cell membrane or cytoplasm in such a form that fragment A is exposed and can be neutralized with anti-fragment A (Yamaizumi *et al.*, 1978a). A plausible model for the mode of action of diphtheria toxin is illustrated in Fig. 11. The possibility of antibodies leaking out during cell fusion and neutralizing the toxin in the culture medium can be eliminated, because the antibody cannot react with the complete toxin molecule. From these data estimation of the number of fragment A molecules indispensable for killing a single cell is difficult, since it is demonstrated that injected immunoglobulin G is rapidly located in the nuclear domain (Wassermann *et al.*, 1977) and that degradation of immunoglobulin G molecules might occur.

Irrespective of the cell type, susceptible or resistant, fragment A cannot kill cells when added to the culture medium, because the fragment has no binding site to the cell surface receptors. *In vitro* experiments with a cell-free system (Honjo *et al.*, 1968; Gill *et al.*, 1969) and the above-mentioned experiment indicate that a naked form of fragment A existing in the cytoplasm can only exhibit its enzymatic activity and kill the cell.

The minimal dose of diphtheria toxin required to kill a single cell was determined by introducing various amounts of fragment A into the diphtheria toxin-resistant mouse L cell. To control the number of fragment A molecules injected, HRBC ghosts loaded with a known number of fragment A molecules with a constant amount of FITC-labeled BSA were prepared by a gradual hemolysis according to the method in Section II. The mixture of fragment A and FITC-labeled BSA was injected into L cells by HVJ-mediated cell fusion of the L cell

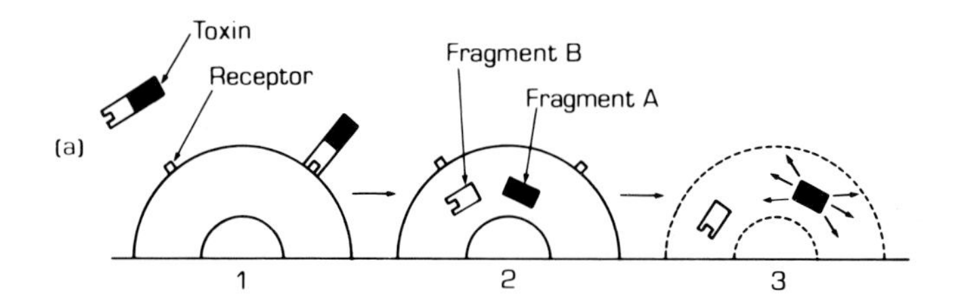

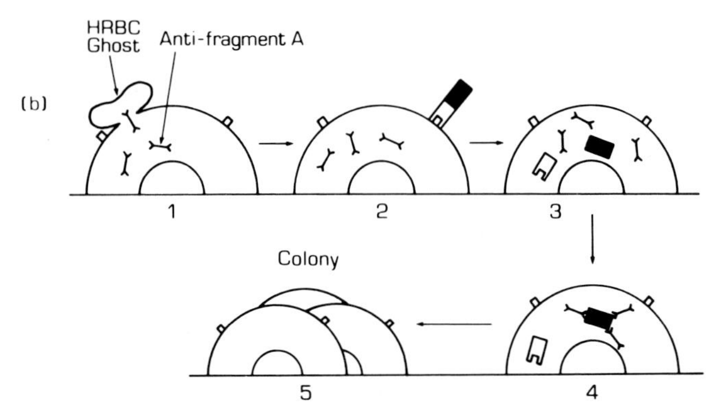

FIG. 11. A plausible mechanism for the mode of action of diphtheria toxin. (a) Under normal culture conditions, the toxin molecule which is added to the culture medium adheres to the receptor on the cell surface of a susceptible cell (1). Then, the toxin molecule enters the cytoplasm where the molecule is nicked and produces fragments A and B (2). The liberated fragment A kills the cell by blocking protein-synthesizing machinery (3). (b) Before administration of toxin, anti-fragment A is introduced into the cell by the fusion of a human erythrocyte (HRBC) ghost containing the antibody (anti-fragment A) and the cell (1). Thereafter, toxin is added to the medium. The toxin molecule penetrates into the cytoplasm and is divided into two fragments as described above (2,3). In this case, however, the preexisting anti-fragment A in the cytoplasm neutralizes the fragment A [remember that anti-fragment A cannot neutralize a complete toxin; therefore, the toxin should be divided into two fragments, A and B (4)]. Consequently, the antibody-injected cell can proliferate and form a colony even in the presence of the toxin (5).

and the ghosts. Then, mononuclear recipient L cells fusing with a single ghost were collected, using the cell sorter, on the basis of cell size and fluorescence intensity. After separation, the viability of cells injecting a known number of fragment A molecules was examined by measuring colony-forming ability. The molecular number of fragment A trapped per ghost was determined from the ghost number and the enzymatic activity of fragment A after lysis of the ghosts by treatment with 0.5% NP-40. The concentration of fragment A trapped in ghosts was equal to that in the mixture in the tube before dialysis. Therefore, the number of molecules of fragment A trapped per ghost can be calculated from the initial concentration of fragment A in the dialyzing tube. As shown in Fig. 11, the proportions of killed recipient L cells fused with ghosts containing one and two fragment A molecules were 60 and 85%, respectively. Provided that the introduction of fragment A into ghosts depends on a Poisson distribution, the result demonstrates that one molecule of fragment A can kill a cell.

This conclusion was further confirmed using a mutant toxin, CRM 176. In this mutant toxin, fragment B is completely functional, but the enzymatic activity of fragment A, which was measured on the inhibitory activity of protein synthesis of culture cells, is about 200-fold less than that of wild-type toxin (Uchida *et al.*, 1973). Fragment A of CRM 176 was injected into L cells by the same method as used above. Figure 12 shows that the minimal number of molecules of fragment A CRM 176 necessary for killing a single cell was between 100 and 200. This experiment directly proved the hypothesis that one molecule of fragment A may kill a cell (Uchida *et al.*, 1973; Duncan and Groman, 1969).

In short, the results presented here indicate that this new method for injection of foreign substances into cells is potentially very useful for quantitative investigations of the biological function of macromolecules in a living cell.

B. Translation of Rabbit Globin mRNA in Mouse Friend Erythroleukemia Cells

It is a well-known fact that animal mRNA can be translated in the cell-free system. Little information, however, has been obtained with regard to the behavior of a specific mRNA in a living animal cell. In addition to our knowledge there is no experimental data available on the translating ability of mRNA introduced into a living animal cell of different species, with the exception of Gurdon's experiments with *Xenopus* egg (Gurdon *et al.*, 1971, Gurdon, 1973). In our experiments, because of the ease of preparation, rabbit globin mRNA (g-mRNA) was used and Friend erythroleukemia cells which are known to be inducible to erythroid differentiation were used as a target cell. The aim of this experiment was to check whether single-stranded mRNA can be loaded into HRBC ghosts without being subjected to digestion by RNase, and whether the injected g-mRNA is translated and the α and β chains of rabbit hemoglobin are

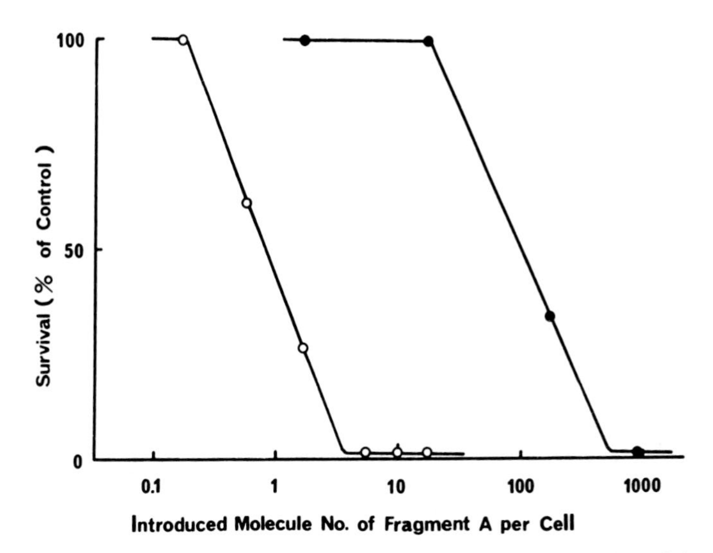

FIG. 12. Colony-forming ability of L cells injected with a known number of fragment A molecules. Mononuclear recipient L cells, fused with one ghost containing a known number of fragment A molecules, were collected with a cell sorter (FACS II). Three hundred cells thus collected were inoculated onto a plate. Seven days after inoculation, the colony-forming ability of recipient cells with introduced fragment A was compared to that of unfused L cells collected from the same samples. Colony-forming ability is expressed as the percentage of the number of colonies formed by unfused L cells. From these data it is concluded that one fragment A can kill one L cell. (○) Fragment A from wild-type toxin; (●) fragment A from CRM 176. (From Yamaizumi *et al.*, 1978c.)

synthesized in a normal way in Friend erythroleukemia cells. Polysomes from reticulocytes of anemic rabbits were kindly given by Dr. Staehelin, Basel Institute for Immunology. In collaboration with Drs. Ostertag and Volloch at the Max-Planck-Institute for Experimental Medicine in Göttingen, we prepared purified g-mRNA by oligo (dT) column affinity chromatography. Loading of g-mRNA into HRBC ghosts was performed according to the method of Auer and Brandner (1976) with slight modification. A mixture of 100 μl of packed HRBCs (1×10^9 cells) and 150 μl of PBS containing 1 mg of g-mRNA was placed in a small dialyzing chamber and dialyzed against 80-fold diluted PBS for 45 minutes with vigourous stirring at room temperature, followed by dialysis against isotonic PBS for 60 minutes. The trapping efficiency of g-mRNA into the ghosts was 30% in concentration, so it was calculated that each ghost contained, on the average, 3000 molecules of g-mRNA. Injection of g-mRNA was carried out by cell fusion between Friend erythroleukemia cells (3×10^7) and HRBC ghosts 1×10^9) containing g-mRNA. As a control, HRBC ghosts containing no g-mRNA were used. The cells fused with HRBC ghosts (with or without g-mRNA) were incubated with [^{3}H]leucine for various periods to detect the synthesis of rabbit globin molecules. The α and β chains of rabbit globin were separated by Sephadex G-100 column chromatography for size separation and then by a CM-52 column chromatography in pyridine–acetic acid. Approximately in the same positions of the two carrier rabbit α- and β-globin peaks, there were two peaks of radioactivity for [^{3}H]leucine. This indicates that the synthesis of rabbit α and β chains had occurred in mouse Friend erythroleukemia cells injected with rabbit globin mRNA. Six hours after injection, a detectable amount of the α and β chains of rabbit hemoglobin had been synthesized in a proportion of about 1:1. A similar result was obtained when induced Friend erythroleukemia cells pretreated for 3 days with dimethylsulfoxide were used. Experiments with the bioassay with *Xenopus* oocytes (Gurdon *et al.,* 1973) demonstrated that the speed of translation of α chain was one-fifth of that of β chain in the absence of heme, even if equal amounts of α-chain mRNA and β-chain mRNA were present (Giglioni *et al.,* 1973). In the present experiment, our sample of g-mRNA must contain equal amounts of both types of mRNA, since it was prepared from normal reticulocytes where the balanced synthesis of α and β chains necessarily takes place. Friend erythroleukemia cells are malignant transformed cells originating from erythroid committed cells or proerythroblastic cells in the spleen of Friend virus-infected mice. Therefore, it is likely that injected α- and β-chain messengers were translated at the same rate. It would, of course, be of interest to determine whether the g-mRNA introduced into nonerythroid cells is also translated in a similar manner.

Although the present experiments are preliminary, the following points are of interest. (1) Poly-(A)-containing mRNA is stable in the HRBC ghost. The trapping procedure into the ghosts and the process of cell fusion do not injure the translatability of the mRNA. (2) A lifetime of a given specific mRNA could be compared by injecting it into different types of cells. Depending on the species of mRNA, its lifetime might vary (Kafatos, 1972); e.g., the lifetime of an mRNA coding a specific protein which is needed for a very restricted period in a developing embryo should be very short, while that of the g-mRNA could be relatively long. Clearly it is important to determine the exact lifetimes of individual mRNAs and whether the difference in lifetime is due to the molecular characteristics of mRNA itself or to the specific RNase located in different cell types. These questions would be resolved by introducing microinjection methods. (3) By injecting an appropriate mRNA into a given cell, one could modify the cell in that the injected foreign mRNA is translated in the host cell. Thus the cell can acquire foreign specific proteins or enzymes in addition to its original constituents.

Recently, using the liposome method, the translation of rabbit globin mRNA introduced into a human cell line (Osttro *et al.,* 1978) and mouse lymphocytes (Dimitriadis, 1978) was demonstrated.

C. Injection of Rat Liver Nonhistone Chromosomal Proteins into Mouse Ehrlich Ascites Tumor Cells

One of the main aims of modern developmental biology is to elucidate the mechanism of regulation of gene expression in higher organisms. From the begining of this century, much data have been accumulated concerning cell differentiation and gene expression. Embryonic tissues in different states can be induced at will into differentiated states by the function of an "inducer," which, in some cases, is a tissue from another part of the embryo, or in other cases, purified molecules. In animal cells, it was believed that once determined and differentiated a cell never changes its direction of differentiation, except in the well-known phenomenon of metaplasia of transdetermination (Hadorn, 1968; Wolff, 1894). On the other hand, Gurdon demonstrated with a nuclear transplantation technique that nuclei from definitely differentiated cells such as the gut (Gurdon, 1962) or the skin of the hindleg (Gurdon *et al.,* 1975) retained pluripotency producing a complete tadpole or frog. These classical experiments indicate that these cells are stably determined and never change the direction of differentiation (transdetermination) *in situ,* but the gene constitutions have not altered during ontogenic development. Judging from these situations, the following can be said concerning the general profile of gene regulation. In an early stage of development, gene expression is easily influenced by stimulation from outside of the cell. However in the adult, specific genes once expressed in a given differentiated cell are considerably stable and continue the expressed state until cell death occurs. In addition, it seems likely that gene constitutions do not change during ontogenic development.

Recent experiments, however, suggest that these conceptions of gene regulation are not necessarily correct. In a heterokaryon consisting of double nuclei of hen's erythrocyte and human HeLa cell, the former, which had no function at all in a normal erythrocyte, is reactivated and the dormant genes are expressed (Harris *et al.,* 1969). Another example of the expression of dormant genes was proposed by Peterson *et al.* (1972) in which hybrid cells of rat hepatoma and mouse fibroblast produced mouse albumin, which normally is produced only by mouse liver *in vivo*. These two experimental facts indicate that even in the fully differentiated cell, dormant genes can be activated if the nucleus is placed in appropriate circumstances, such that the cell begins to synthesize a specific protein that is never synthesized under normal conditions. Furthermore, there is a possibility of the occurrence of gene translocation during ontogenic development in immunoglobulin genes (Tonegawa *et al.,* 1977a). Anyway, it should be remembered here that at present there is no definitive evidence of the presence of regulator gene or regulatory substances, or even of a promoter site, in the chromatin of higher organisms.

In our search for regulatory factors, we focused first on the experimental system of Weiss (Peterson *et al.*, 1972). Her data clearly show that in the hybrid cell, some unknown substance, which results in the expression of mouse albumin gene, activates the nucleus of mouse fibloblast. Concerning the source of the regulating substance there are two possibilities; one is the cytoplasm of rat hepatoma and the other is its nucleus. The former possibility is not unlikely. Cybrid experiments (Gopalakrishnan *et al.*, 1977) suggest that some gene-regulating substance is located in the cytoplasm though its function is inhibitory. That is, a cybrid cell consisting of a nucleus of Friend erythroleukemia cell and a cytoplasm of mouse fibroblast has lost the ability to perform erythroid differentiation in the presence of dimethylsulfoxide (DMSO). This characteristic is retained after repeated cell divisions. However, the latter possibility is more probable. There is considerable evidence suggesting that chromatin proteins, histone or nonhistone, carry gene-regulating functions. *In vitro* studies show that the gene-regulatory function of histone is nonspecific and rather inhibitory (Stein *et al.*, 1975a). In contrast, many data obtained from studies with a cell-free system strongly suggest that that of nonhistone chromosomal proteins is specific (Gilmour, 1974; Stein *et al.*, 1975a). It is possible that regulators are in the fraction of so-called "nonhistone chromosomal proteins (NHPs)." It might be possible that the identification of the regulator could be attained by introduction of the NHP fraction directly into living cells. Furthermore, intracellular injection techniques should enable us to locate it by repeated injection of stepwise purified samples.

We used rat liver cells which produce rat albumin or other liver-specific proteins as a source of NHPs.

Liver NHPs were injected into mouse Ehrlich tumor cell of mammary tumor origin. The final aim of this experiment was to induce the production of mouse serum albumin or liver-specific proteins in the Ehrlich tumor cells. In this preliminary experiment, we had to know whether injected NHPs are stable in the target cell and can attach to the DNA in a normal way. The detailed procedure for preparing NHPs is reported elsewhere (Yamaizumi *et al.*, 1978b). The following is a brief summary of the procedure. Adult male Wistar rat livers were irrigated with cold physiological saline to wash out blood. The liver cells were homogenized in 0.25 *M* sucrose with 3.3 m*M* $CaCl_2$ at 4°C with a Waring blender, without using detergent for disrupting cell membranes. The crude nuclei fraction was obtained by low-speed centrifugation. The nuclei thus obtained were crushed by homogenization in 0.01 *M* Tris–HCl (pH 8.0) and centrifuged at 10,000 rpm for 15 minutes. The precipitate was dissolved in 3 *M* NaCl, 5 *M* urea, 10 m*M* Tris–HCl (pH 8.3) and stirred at 0°C for 1 hour. The solubilized protein fraction was obtained by centrifugation at 64,000 rpm for 40 hours according to the method of Stein *et al.* (1975b). NHPs were separated from the histone on a Bio-Rex 70 column.

Rat liver NHPs were injected into Ehrlich tumor cells by the method described above. In order to determine the stability of the NHPs and to pursue their distribution in target cells, the NHPs were labeled with ^{125}I before injection. First, the stability of NHPs in HRBC ghosts was examined. SDS–polyacrylamide gel column chromatographic profiles of the original sample of NHPs and of NHPs trapped in the ghosts kept in ice for 24 hours were compared. No significant difference was observed in both profiles, indicating that NHPs are quite stable in the ghosts. Second, the distribution of injected labeled NHPs in the target cell was examined by autoradiography. Twenty minutes after injection, grains were equally distributed in number in the cytoplasmic and nuclear domain, while after 26 hours most of grains were found only in the nucleus. This result suggests that the injected proteins penetrated into the nucleus and were stabilized there, but in the cytoplasm they were quickly destroyed and eliminated from the cell body, or moved into the nucleus during incubation. To confirm this, the nuclear and cytoplasmic fractions were prepared (at 37°C) from the cells, at 30 minutes, 4 hours, and 24 hours after the injection, respectively. Then, the radioactivity of each fraction was compared. As shown in Table I, the NHPs injected into the cytoplasm were transferred at least within 30 minutes to the nuclei of recipient cells. The maximal amount of radioactivity was already found at 0 hour incubation (practically 30 minutes had passed from the start of injection). The amount of radioactivity in the nuclei did not decrease up to 24 hours of incubation, while the amount of radioactivity in the cytoplasm decreased quickly to about 20% of the initial amount. To check whether the radioactivity in the nuclear fraction was due to the NHPs bound to recipient chromatin, the chromatin was isolated from the nuclei. Sixty percent of the radioactivity in the nuclei

TABLE I

CHANGES IN RADIOACTIVITY OF ^{125}I-LABELED NONHISTONE CHROMOSOMAL PROTEINS (NHPs) AND BOVINE SERUM ALBUMIN (BSA) IN DIFFERENT FRACTIONS OF RECIPIENT CELLS DURING INCUBATION[a]

	Incubation (hours)	Radioactivity (cpm)			
		Cytoplasm	Nucleus	Chromatin	Nucleoplasm
NHPs	0	145,000	36,200	19,000	14,300
	4	51,800	33,100	19,200	10,500
	24	29,100	28,200	16,300	9,100
BSA	0	73,200	5,400	ND	ND
	26	13,600	1,300		

[a]Specific activities of the NHPs and BSA were 4×10^6 and 5×10^6 cpm/μg, respectively. Cells, 2×10^7, were collected, and radioactivity was counted for each sample. The efficiency of fusion between HRBC ghosts containing ^{125}I-labeled sample and target Ehrlich ascites tumor cells was approximately 80%. ND, Nondetectable. (From Yamaizumi *et al.*, 1978b.)

was found in the chromatin fraction irrespective of incubation time. A control experiment excluded the possibility that NHPs had been absorbed adventitiously to the nucler fraction. When BSA was injected into the cells, it remained mainly in the cytoplasm and was rapidly degraded (Table I), whereas the small amount of BSA shifted to the nuclei was degraded at the same rate as that in the cytoplasm.

Figure 13 shows the molecular profiles of nonhistone chromosomal proteins extracted from chromatin and the cytoplasm of recipient cells, which were analyzed by SDS–polyacrylamide gel electrophoresis. As shown in Figs. 13a–c, the profiles of the NHPs extracted from the chromatin of recipient cells at different times resembled that of the original sample of NHPs. In contrast, the profiles of the NHPs extracted from the cytoplasmic fraction markedly differed from those of the chromatin fraction. From the start of incubation, a single prominent peak corresponding to an approximately 40,000-dalton protein was observed as seen in Figs. 13d–f. It is uncertain whether this protein is a product of degradation of the original sample or a specific protein which is a component of the original NHPs that remained in the cytoplasm without being transferred into the nucleus. Judging from the changes in the profiles of the NHPs it is suggested that there was no exchange of injected NHPs between the cytoplasm and nucleus of recipient cells during incubation. In the case of BSA, at all times, the profiles of the ^{125}I-labeled molecules of the cytoplasm of recipient cells consistently showed a single sharp peak corresponding to a protein of 68,000 daltons, although the height of this peak decreased with time (Figs. 13g,h). There was no detectable radioactivity in the nucleus in these cases. The rapid degradation of NHPs and BSA in the cytoplasm of recipient cells may reflect the natural turnover of these proteins (Yamaizumi *et al.*, 1978b).

Our results clearly show that nonhistone chromosomal proteins were transferred within 30 minutes to the nucleus where they were stabilized, whereas BSA was scarcely transferred to the nucleus, remaining in the cytoplasm and being degraded. With regard to the migration of nuclear proteins, several investigators have reported that histones or other nuclear proteins are transferred to the nucleus (Gurdon, 1974a; Bonner, 1975; Gurdon *et al.*, 1976; Weintraub, 1972; Steplewski *et al.*, 1976). These findings suggest that there is some machinery in the cytoplasm by which nuclear proteins synthesized in the cytoplasm are transferred selectively to the nucleus.

D. Studies on Intracellular Injection Using the Present Method as Carried Out in Other Laboratories

The isolation and characterization of the nonsense mutation in a structural gene in mammalian cells should provide a valuable new tool for genetic analysis of mammalian cells. Using the same microinjection method introduced in this arti-

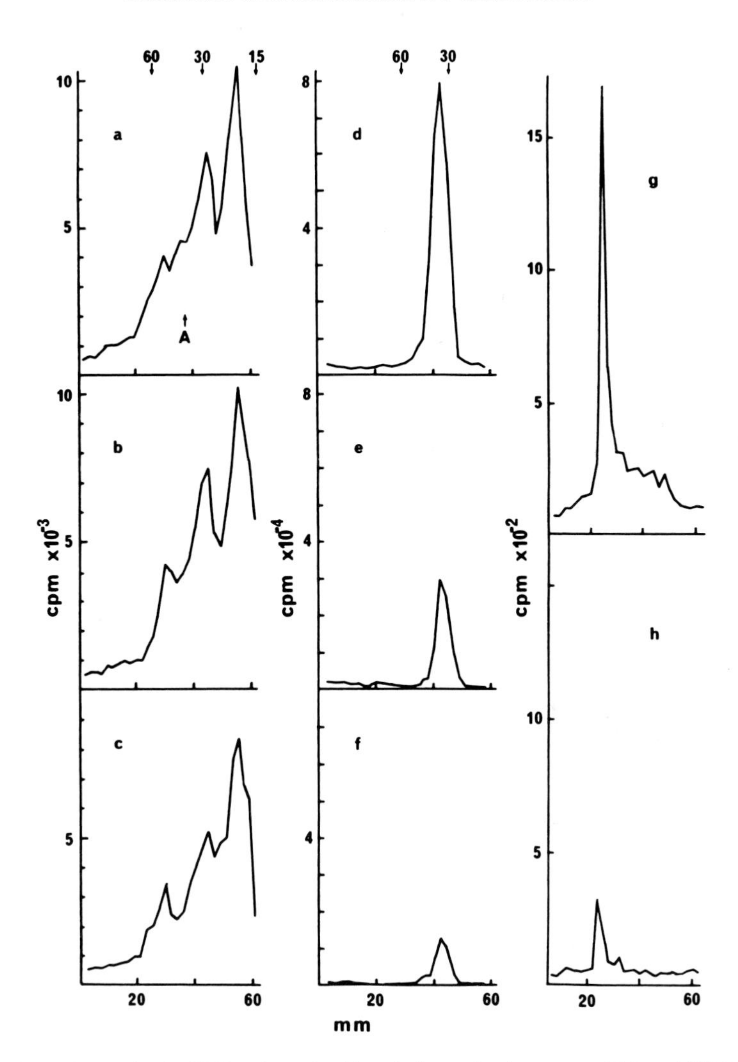

FIG. 13. Changes in the SDS–polyacrylamide gel electrophretic pattern of injected ^{125}I-labeled nonhistone chromosomal proteins (NHPs) and bovine serum albumin (BSA) in recipient cells during incubation. The ^{125}I radioactivity in 2-mm slices of the gels was counted. The arrows indicate the positions of indicator proteins of molecular weights 15,000, 30,000, and 60,000, respectively. Arrow A indicates the position of the peak protein which appeared in the cytoplasmic fraction (d–f) when tested by coelectrophoresis of chromatin and the cytoplasmic fraction. The result clearly shows that a major part of the introduced NHPs penetrates quickly into nuclei and adheres to chromatin in a stable form. Nonhistone chromosomal proteins: (a) 0 hour, in chromatin; (b) 4 hours, in chromatin; (c) 24 hours, in chromatin; (d) 0 hour, in cytoplasm; (e) 4 hours, in cytoplasm; (f) 24 hours, in cytoplasm. BSA: (g) 0 hour, in cytoplasm; (h) 26 hours, in cytoplasm. For details see the text. (From Yamaizumi *et al.*, 1978b.)

cle, Capecchi *et al.* (1977) isolated a mutant cell line with a nonsense mutation in a structural gene. They used the potential conditional-lethal system for locating nonsense mutants. They started by isolating mouse L-cell lines deficient for the nonessential enzyme hypoxanthine-guanine phosphoribosyl transferase (HGPRT) by screening with purine analogs. Such mutants remain viable because purine nucleotides can be synthesized by *de novo* pathways in the absence of HGPRT. The mutants thus selected include those which have lost the enzymatic activity but can still synthesize incomplete proteins which exhibit smaller subunit molecules relative to the wild-type HGPRT molecules. To ascertain whether the alteration occurred at the carboxyl-terminal end of the polypeptide chain as predicted for the nonsense mutation, the mutant cell lines were tested for sensitivity to phenotypic suppression by injection of a suppressor tRNA. Three kinds of tRNA were prepared from different *Escherichia coli* strains: S26 which is the suppressor-minus parental strain, S26RIE which contains Sul amber-suppressor tRNA, and PS2 which contains an ochre-suppressor tRNA. The results clearly showed that microinjection of ochre-suppressor tRNA into the $HGPRT^-$ cell line specifically restored HGPRT activity. A similar effect was obtained when yeast suppressor tRNA was used. From these properties they concluded that the mutant contains an ochre nonsense mutation (UAA) in the HGPRT structural gene.

There are basic studies of microinjection using HRBC ghosts and HVJ. In 1975, several months after we had reported our microinjection method (Furusawa *et al.*, 1974), Loyter *et al.* (1975) independently developed the same method. FITC-conjugated BSA, ferritin 120 Å in diameter, Latex spheres, 0.1 nm in diameter, and bacteriophage T4 the head of which is 650 Å in diameter were injected into human hepatoma tissue culture cells (HTC). However, they did not show any biological function of the injected materials in the cells.

Using the same method, Schlegel and Rechsteiner (1975) performed the injection of active enzyme into a living cell. Thymidine kinase extracted from Ehrlich ascites tumor cells was used as a test enzyme. They injected thymidine kinase into thymidine kinase-deficient mouse 3T3-4E cells, which cannot use exogenous thymidine for DNA synthesis. The amount of thymidine kinase injected into a single target cell was less than 1% of that normally present in a wild-type 3T3 cell. To examine whether uptake of thymidine into DNA does occur in deficient cells which have been injected with enzyme, the injected cells were incubated with [^{3}H]thymidine. Autoradiographic analysis revealed that introduction of thymidine kinase resulted in the usage of exogenous thymidine in the deficient cells.

Kaltoft *et al.* (1976) injected tRNA from yeast ($tRNA^{Phe}$ and 4 S RNA) and *E. coli* ($SuIII^+tRNA_I^{Try}$) into thymidine kinase-deficient Cl-lD mouse cells. The trapping of tRNA in cells was four times more efficient with HRBCs as a vehicle for injection than with HVJ alone. The injected tRNA remained intact after 6 hours of incubation.

Wasserman *et al.* (1976) improved the fusion procedure and injected cytochrome *c* and BSA into HTC. The use of La^{3+} ions raised the efficiency of fusion between HRBC ghosts and cells. They concluded that about 10^6 BSA molecules were injected per fused cell. With the same technique, Wille and Willecke (1976) introduced human serum albumin of 64,000 daltons and *E. coli* β-galactosidase of 520,000 daltons into murine CM thymidine kinase-deficient cells.

Kriegler *et al.* (1978) showed that phenotypic complementation of the SV40 mutant defect in viral DNA synthesis was brought about by introducing SV40 T antigens by microinjection. African green monkey cells (CV-lp) were injected with SV40 T antigen using HRBC ghosts containing T antigen and polyethylenglycol as fusagen. Then, the injected cells were infected with a temperature-sensitive mutant of SV40(SV40-tsA) which is defective in the initiation of viral DNA synthesis. The injection of T antigen resulted in an increase in both the amount of viral DNA sequences in the monolayer and the production of SV40 DNA.

VI. Prospects

A. General Aspects

Much information on the function of macromolecules and the mechanism of physiological regulation in living cells can be obtained from studies using intracelullar microinjection techniques. The present microinjection method may be useful for studies in cell biology, especially in (1) molecular biology at the cellular level, e.g., the introduction of a "gene regulating substance" into cells to evoke the expression of dormant genes, which may offer a clue to the analysis of the gene-controlling mechanism of animal cells; (2) cellular genetics, e.g., the introduction of genes or chromosomal fragments into cells to alleviate genetic defects or to change the genetic constitution of cells, or the introduction of plasmids with a defined gene to analyze a mechanism controlling gene expression in higher organisms; (3) cellular physiology, e.g., the introduction of biologically active substances into cells to study the mechanism of intracellular regulation; (4) virology, e.g., the introduction of oncoviral genome or virions into a normal cell to evoke *in vitro* transformation and to study the host–virus interactions.

Furthermore, a new field of cell biology, called "cellular engineering," could be developed with the present microinjection technique. In the future it might become possible to rebuild a cell at will as a carpenter rebuilds a house combining ready-made parts from different makers. For example, it will become possible to introduce cloned genes having a specific genetic code, or a special chromosome or cell organelle, into a given cell.

B. Experiments on Gene Expression

In the field of developmental biology, the most urgent question to be answered is: How is gene expression regulated in animal cells? Perhaps, gene expression of animals or eukaryotes might be controlled differently than that of bacteria. In the former, the DNA is protected by bonding with basic proteins, histones; there are several copies of the same gene in a genome; they develop isozymes and the genetic information of the subunits of an isozyme are coded in different genes; nascent mRNA or heterogenous RNA (pre-mRNA) in the nuclear domain is much longer in size than the mRNA translated on ribosomes (Brandhorst and McConkey, 1974); in differentiated cells some special genes are selectively and irreversibly expressed, etc. The characteristics listed above cannot be found in a bacterial system. A recent finding on the mouse immunoglobulin G gene is an unprecedented suggestive one. It is well known that the L chain of IgG is a single polypeptide consisting of V_L and C_L regions. It has been considered that structural genes corresponding to the V_L and C_L regions are initially independent and located separately in different parts of a chromosome, and that during ontogeny coupling occurs between them and consequently a complete functional L-chain gene is set up in th IgG-producing plasma cells. Recently Tonegawa *et al.* (1977a,b, 1978) and Hozumi and Tonegawa (1976) clearly demonstrated with a gene cloning technique that the positions of the V_L and C_L genes, which are isolated from embryonic mouse or any adult organs other than the plasma cell, are completely separated from each other, whereas those from IgG-producing myeloma cells are located in an adjacent position forming a single L-chain gene as a result of gene translocation. However, strange to say, there still exists an insertion of nontranslated DNA of 1250 base pairs between them. Similar instances of the insertion of nontranslated DNA in a structural gene are reported in the hemoglobin β-chain gene (Brack and Tonegawa, 1977; Tilghman *et al.*, 1977, 1978) and avian ovalbumin gene (Jefferys and Flavell, 1977; Breathnach *et al.*, 1977; Doel *et al.*, 1977; Weinstock *et al.*, 1978), although in both cases the possibility of gene translocation during ontogeny might not be possible. Anyway, it can be said that gene translocation should be taken into consideration as a possible mechanism of gene regulation in mammalian cell differentiation (Dr. Tonegawa's personal communication). Another possibility is that jumping or transposal genes are concerned with the regulation of gene expression (Nevers and Saedler, 1977).

Our microinjection technique could serve as a good tool for analyzing the mechanism of gene expression in animal cells, particularly in connection with the following two aspects: identification of a gene-regulating substance, and introduction of cloned genes into cells. As for the gene-regulating substance, the role of nonhistone chromosomal proteins (NHPs) has attracted some attention. Much of the experimental data on isolated chromatins support the concept of a positive regulatory role of NHPs in animal cell differentiation (Gilmour, 1974; Stein *et al.*, 1975a). Whether such a function of NHPs, clarified by an *in vitro* cell-free system, occurs in a living cell should be ascertained. This can be attained by employing our injection method. As stated in Section V, the introduction of NHPs into cells may possibly result in the expression of dormant genes, even in an already differentiated cell. Weiss *et al.* (Peterson *et al.*, 1972; Malawista and Weiss, 1974; Brown and Weiss, 1975) demonstrated that dormant genes were expressed in the hybrid cell. Further analysis or identification of the exact molecule which takes part in the expression of the gene could be technically possible. After confirmation of the occurrence of the expression of a dormant gene in a cell which has received an injection of a NHP fraction extracted from the differentiated donor cell, fractionation of the original crude sample is then feasible. The inducing ability of gene expression should be tested on each fraction by a similar injection method. Repeating this procedure, the identification or purification of the gene-regulating substance might be attained. However, it should be kept in mind here that there is no certain knowledge about the gene-controlling mechanism in animal cells. Therefore, any other fractions from the donor cell other than NHPs should be tested.

C. Experiments on Gene Introduction

First, the introduction of cloned genes into living cells could also be a promising method of analyzing the gene-controlling mechanism. Since the so-called gene engineering technique was developed, many attempts at gene cloning have been tried. Now four "functional" genes are isolated from different animals: the β-chain gene of mouse and rabbit hemoglobin (Tilghman *et al.*, 1977, 1978; Jefferys and Flavell, 1977), the L-chain gene of IgG from mouse myeloma (Tonegawa *et al.*, 1977a,b; Hozumi and Tonegawa, 1977), and the ovalbumin gene from birds (Breathnach *et al.*, 1977; Doel *et al.*, 1977; Weinstock *et al.*, 1978). More genes will be cloned in the near future. It is anticipated that soon some special animal DNA fragment corresponding to the regulator gene or promoter site in bacteria might be isolated. Our microinjection method may serve as a good bioassay system for checking a given cloned gene while maintaining its normal function.

Second, we can introduce foreign genes into animal cells, and if a recombination occurs in the host cell DNA, the gene composition of the cells can be changed. This recombinant will offer basic information concerning the gene therapy of genetic diseases. Third, the "gene construction" experiment may be attained by combining the gene engineering technique and the microinjection method. By joining a given gene to a specific DNA sequence which seems to correspond to the promoter site or regulator gene, a set of a certain functional genes (like an operon) could be artificially constructed *in vitro*. Then, the function of the constructed gene could be tested by injecting it into appropriate target cells. This type of experiment would be one of the best ways to clarify the gene-regulating system in eukaryotic cells.

D. Experiments with Mutant Cells

Combined use of mutant cells and the present microinjection technique may offer a clue to the analysis of the mechanism of some of the critical biological phonomena, such as the initiation of DNA synthesis. Namely, if a given mutant cell can be rescued by introducing the normal counterpart from the wild-type cells, it can be expected that the key molecule involved in the mutation could be recognized. This kind of "rescue" experiment might be fruitful in the study of cell biology.

E. Applications in Virology

Our technique could also be useful for studies in virology. Viral DNA or RNA, or even virions, can be introduced into any animal cell. In general this technique can introduce viral geneome into cells with or without specific surface receptors to the virus. Thus, it will become possible for virus to infect beyond its host range restriction. In an extreme case, even plant viruses can be introduced into animal cells. Such compulsory viral infection which never takes place under natural conditions may bring about new data on the host–virus interrelationship. Of course, this type of experiment should be done with special caution in connection with the biohazards involved. Moreover, attempts at *in vitro* transformation can be performed by introducing oncovirus or its genome into target cells. In fact Auer and Brandner (1976) reported that T4 phage, or its DNA, and SV40 DNA can be trapped in HRBC ghosts.

F. Applications in Medicine

The application of this microinjection method as a practical medical treatment may not necessarily be as promising, although it may contribute in basic studies on genetic disease, metabolic defects, or cancer. We can illustrate a beautiful example of the medical application of the injection method, which is one of the HVJ methods (see Section IV,C). Tanaka *et al.* (1975) found that when a mixture of HVJ and materials was poured on culture cells, the materials could be introduced into the cells. Using this technique they introduced endonuclease V of T4 phages into human xeroderma pigmentosum cells which are defective in the specific enzyme for repairing the damage to DNA caused by UV irradiation. Namely, UV-induced unscheduled DNA synthesis of xeroderma pigmentosum cells was restored to the normal level. From this result they concluded that T4 endonuclease was functional in human chromosomal DNA which had been damaged by UV irradiation in viable cells, and that all groups of xeroderma pigmentosum studied were defective in the first step of excision repair (Tanaka *et al.*, 1975). In addition, they demonstrated that the half-life of the introduced T4 endonuclease was 3 hours (Tanaka *et al.*, 1977). Although this is still in a preliminary stage, special attention should be paid to the possibility of finding a therapy for human genetic diseases by the introduction of normal enzymes.

As a possible application of our injection method to medicine, the treatment of a patient who has some defect in circulating cells such as lymphocytes, leukocytes, or macrophages will be possible. For example, defective lymphocytes are collected from the patient. In order to alleviate or repair the defect, specific drugs, or genes or enzymes of normal counterparts, are introduced into the defective cells by the present ghost method. Next, the treated cells are put back intravenously into the patient. The treated cells would function normally in the body of the patient.

There is another possibility that HRBC ghosts loaded with drugs themselves may serve as a good tool for the treatment of some diseases, especially some of the hepatic diseases. Erythrocytes are collected from the patient and a specific drug for the disease is loaded into the ghosts by the gradual hemolysis method described in Section II. As far as tested, substances of 389 molecular weight (FITC) can be trapped in ghosts, and leakage from the ghosts is negligible during storage in a refrigerator. Then the ghosts containing the drug are transfused into the vein of the patient. In the human body, old erythrocytes are mostly destroyed in the liver and spleen. Accordingly, the transfused ghosts could also be selectively trapped and destroyed in the liver where the loaded drugs might be liberated. This may result in the concentrated introduction of the drug into the liver. In general, the administration of drugs by this method appears to bring about an effect similar to that of continuous administration of a small quantity of drugs (Kinoshita and Tsong, 1978). Of course, there could be other unexpected side effects caused by the treatment. The main advantage of these types of treatment would be that there would be no immune reaction caused by the transfused cells or erythrocyte ghosts. Liposomes are also emerging as vectors for delivery of drugs, enzymes, and hormones for therapeutic purposes (Gregoriadis and Ryman, 1971, 1972; Gregoriadis, 1973; Colley and Ryman, 1974, 1975; Rahman *et al.*, 1973).

It is also possible that HRBC ghosts loaded with foreign substance would be used in immunotherapy. Usually, sensitization or vaccination is performed by injection of antigens into the vein, muscle, skin, or underskin. Therefore, in most cases, subsidiary ill effects such as inflammation cannot be avoided in the sensitization process. By introducing HRBC ghosts which are loaded with antigens into the vein, it may become possible to transport the antigens directly to macrophages. This new method of sensitization could minimize the risk of the unwanted side effects of antigens. To attain this, probably, an adequate treatment of the surface of HRBC ghosts would be required, since macrophages are required to recognize and distinguish the ghosts from the intact erythrocytes in the blood vessel.

On the contrary, several trials have been carried out with liposomes loaded with antigens. While it was suggested that liposomes would prevent the development of antibodies to proteins trapped within them, this proved not to be the case; liposomes act as adjuvants in enhancing antibody production to entrapped proteins (Allison and Gregoriadis, 1974; Heath *et al.*, 1976). It is therefore most important to search for adequate conditions under which the leakage of antigens from the vectors can be diminished, before the vectors reach the antibody-forming system in a body. For this purpose, more basic data on sensitizations using both liposomes and erythrocytes as a vector are required.

VII. Concluding Remarks

As described in this review, the present microinjection method using human erythrocytes and Sendai virus (HVJ) is broadly usable in studies of biology at a cellular level. In addition, it might be possible to apply this technique to medical treatments. In short, the most outstanding points of this technique are: (1) intrinsically no limitation in the number of target cells subjected to microinjection, and (2) the capability of performing a quantitative introduction of substance into each single cell. Despite these advantages, the present technique has an inevitable shortcoming in that intranuclear injection is impossible. Recently, F. Yamamoto (a student of the masters course in our laboratory) and I have developed a new instrument with which intranuclear injection can be accomplished easily using a fine micropipet, without the help of a conventional micromanipulator. The microinjection is carried out using a modified condenser of a normal light microscope. The micropipet is inserted and fixed in a straight perpendicular hole which is made through the optical axis of the condenser lens. Delicate three-dimensional movement of the tip of the micopipet can be controlled by handling the adjusting screws for the movement of the condenser. This modified condenser can be used with a normal and an inverted microscope. Compared with conventional microinjection methods using a micromanipulator (Graessmann, 1970; Diacumakos, 1973), this method has the considerable advantages of simplicity of handling and speed in performing microinjection or microsurgery using this new instrument. With this technique, experiments on nuclear transplanation, chromosome transplantation, or introduction of genetic materials into the nucleus of mammalian cells should become more easy to carry out. A detailed explanation of the structure and handling of this instrument will soon be published soon elsewhere (Yamamoto and Furusawa, 1978).

Finally, the author would like to call the microinjection method introduced in this review a "fusion injection," named by Dr. H. Eisen.

ACKNOWLEDGMENTS

The author wishes to express his gratitude to Professor Y. Okada and Dr. H. Mitsui for their encouragement and criticism. I am also grateful to Drs. T. Uchida, M. Yamaizumi, and E. Mekada for their valuable discussion and permission to use their original data, and to Dr. I. Pragnell and Miss Jane Theotokatos for their criticism of the manuscript.

REFERENCES

Adachi, H., and Furusawa, M. (1968). *Exp. Cell Res.* **50,** 490.
Allison, A. C., and Gregoriadis, G. (1974). *Nature (London)* **252,** 252.
Auer, D., and Brandner, G. (1976). *Z. Naturforsch.* **31,** 149.
Barber, M. A. (1914). *Philipp. J. Sci.* Ser. B. **9,** 307.
Bonner, W. M. (1975). *J. Cell Biol.* **64,** 421.
Brack, C., and Tonegawa, S. (1977). *Proc. Natl. Acad. Sci. U.S.A.* **74,** 5652.
Brandhorst, B. P., and McConkey, E. H. (1974). *J. Mol. Biol.* **85,** 451.
Breathnach, R., Mandel, J. L., and Chambon, P. (1977). *Nature (London)* **270,** 314.
Briggs, R., and King, T. J. (1952). *Proc. Natl. Acad. Sci. U.S.A.* **38,** 455.
Brown, J. E., and Weiss, M. C. (1975). *Cell* **6,** 481.
Capecchi, M. R., Von der Haar, R. A., Capecchi, N. E., and Sveda, M. M. (1977). *Cell* **12,** 371.
Colley, C. M., and Ryman, B. E. (1974). *Biochem. Soc. Trans.* **2,** 871.
Colley, C. M., and Ryman, B. E. (1975). *Biochem. Soc. Trans.* **3,** 157.
Cooper, M., Levy, J., Cantor, L., Marks, P., and Rifkind, R. (1974). *Proc. Natl. Acad. Sci. U.S.A.* **71,** 1677.
Davidson, R. L., and de la Cruz, F. (1974). "Somatic Cell Hybridization" pp. 131–150. Raven, New York.
Diacumakos, E. C. (1973). *In* "Method in Cell Biology" (D. M. Prescott, ed.), Vol. 7, pp. 287–311. Academic Press, New York.
Dimitriadis, G. (1978). *Nature (London)* **274,** 923.
Doel, R. A., Houghton, M., Cook, E. A., and Carey, N. H. (1977). *Nucl. Acids Res.* **4,** 3701.
Drazin, R., Kamdel, J., and Collier, R. J. (1971). *J. Biol. Chem.* **246,** 1504.
Duncan, J. L., and Groman, R. (1969). *J. Bacteriol.* **98,** 963.
Furusawa, M., and Adachi, H. (1968). *Exp. Cell Res.* **50,** 497.
Furusawa, M., Nishimura, T., Yamaizumi, M., and Okada, Y. (1974). *Nature (London)* **249,** 449.
Furusawa, M., Yamaizumi, M., Nishimura, T., Uchida, T., and Okada, Y. (1976). *In* "Method in Cell Biology" (D. M. Prescott, ed.), Vol. 14, pp. 73–80. Academic Press, New York.
Giglioni, B., Gianni, A. M., Corni, P., Ottolenghi, S., and Rungger, D. (1973). *Nature (London)* **246,** 99.
Gill, D. M., and Pappenheimer, A. M., Jr. (1971). *J. Biol. Chem.* **246,** 1492.
Gill, D. M., Pappenheimer, A. M., Jr., Brown, R., and Kurnick, J. T. (1969). *J. Exp. Med.* **129,** 1.
Gilmour, R. S. (1974). *In* "Acidic Proteins of the Nucleus" (I. L. Cameron and J. R. Jeter, Jr., eds.), pp. 297–317. Academic Press, New York.

Gopalakrishnan, T., Thompson, E. B., and Anderson, W. F. (1977). *Proc. Natl. Acad. Sci. U.S.A.* **74,** 1642.
Graessmann, A. (1970). *Exp. Cell Res.* **60,** 373.
Gregoriadis, G. (1973). *FEBS Lett.* **36,** 292.
Gregoriadis, G., and Ryman, B. E. (1971). *Biochem. J.* **124,** 58.
Gregoriadis, G., and Ryman, B. E. (1972). *Biochem. J.* **129,** 123.
Gurdon, J. B. (1962). *J. Embryol. Exp. Morph.* **10,** 622–640.
Gurdon, J. B. (1973). *Karolinska Symp. Res. Meth. Reprod. Endocrinol.* **6,** 225.
Gurdon, J. B. (1974a). *Nature (London)* **248,** 772.
Gurdon, J. B. (1974b). "The Control of Gene Expression in Animal Development" pp. 121–126. Oxford Univ. Press, London.
Gurdon, J. B., Birnstiel, M. L., and Speight, V. A. (1969). *Biochim. Biophys. Acta* **174,** 614.
Gurdon, J. B., Lane, C. D., Woodland, H. R., and Marbaix, G. (1971). *Nature (London)* **233,** 177.
Gurdon, J. B., Lingrel, J. B., and Marbaix, G. (1973). *J. Mol. Biol.* **80,** 539.
Gurdon, J. B., Laskey, R. A., and Reeves, P. R. (1975). *J. Embryol. Exp. Morphol.* **34,** 93.
Gurdon, J. B., Partington, G. A., and De Robertis, E. M. (1976). *J. Embryol. Exp. Morphol.* **36,** 541.
Hadorn, E. (1968). *Sci. Am.* **219** (5), 110.
Harris, H., Sidebottom, E., Grace, D. M., and Bramwell, M. (1969). *J. Cell Sci.* **4,** 499.
Heath, T. D., Edwards, D. C., and Ryman, B. E. (1976). *Biochem. Soc. Trans.* **4,** 129.
Honjo, T., Nishizuka, Y., Hayaishi, O., and Kato, I. (1968). *J. Biol. Chem.* **243,** 3533.
Hozumi, N., and Tonegawa, S. (1976). *Proc. Natl. Acad. Sci. U.S.A.* **73,** 3628.
Jefferys, A. J., and Flavell, R. A. (1977). *Cell* **12,** 1097.
Kafatos, F. C. (1972). *Karolinska Symp. Res. Meth. Reprod. Endocrinol.* **5,** 319.
Kaltoft, K., Zeuthen, J., Engback, F., Piper, P. W., and Celis, J. E. (1976). *Proc. Natl. Acad. Sci. U.S.A.* **73,** 2793.
Kinoshita, K., Jr., and Tsong, T. Y. (1978). *Nature (London)* **272,** 258.
Kriegler, M. P., and Livingston, D. M. (1977). *Somat. Cell Genet.* **6,** 603.
Kriegler, M. P., Griffin, J. D., and Livingston, D. M. (1978). *Cell* **14,** 761.
Lane, C. D., and Knowland, J. (1975). *In* "The Biochemistry of Animal Development" (R. Weber, ed.), Vol. 3, pp. 145–182. Academic Press, New York.
Loyter, A., Zakai, J., and Kulda, R. G. (1975). *J. Cell Biol.* **66,** 292.
Magee, W. E. (1972). *Nature (London)* **235,** 339.
Malavists, S. E., and Weiss, M. C. (1974). *Proc. Natl. Acad. Sci. U.S.A.* **71,** 927.
Mekada, E., Yamaizumi, M., and Okada, Y. (1978a). *J. Histochem. Cytochem.* **26,** 62.
Mekada, E., Yamaizumi, M., Uchida, T., and Okada, Y. (1978b). *J. Histochem. Cytochem.* **26,** 1067.
Morrison, W. L., and Neurath, H. (1953). *J. Biol. Chem.* **200,** 39.
Nevers, P., and Saedler, H. (1977). *Nature (London)* **268,** 109.
Nishimura, T., Furusawa, M., Yamaizumi, M., and Okada, Y. (1976). *Cell Struct. Funct.* **1,** 197.
Okada, Y. (1962). *Exp. Cell Res.* **26,** 98.
Okada, Y., and Murayama, F. (1966). *Exp. Cell Res.* **44,** 527.
Osttro, N. J., Giacomoni, D., Lavelle, D., Paxton, W., and Dray, S. (1978). *Nature (London)* **273,** 921.
Papahadjopoulos, D., Poste, G., and Mayhew, E. (1974). *Biochim. Biophys. Acta* **363,** 404.
Pappenheimer, A. M., Jr., Uchida, T., and Harper, A. A. (1972). *Immunochemistry* **9,** 891.
Peterson, J. A., and Weiss, M. C. (1972). *Proc. Natl. Acad. Sci. U.S.A.* **69,** 571.
Rahman, Y. E., Rosenthal, M. W., and Cerny, F. A. (1973). *Science* **180,** 300.
Ringertz, R., Carlsson, S., Ege, T., and Bolund, L. (1971). *Proc. Natl. Acad. Sci. U.S.A.* **68,** 3228.
Salk, J. E. (1944). *J. Immunol.* **49,** 87.

Schlegel, R. A., and Rechsteiner, M. C. (1975). *Cell* **5,** 371.
Seeman, P. (1967). *J. Cell Biol.* **32,** 55.
Stacey, D. W., Allfrey, V. G., and Hanafusa, H. (1977). *Proc. Natl. Acad. Sci. U.S.A.* **74,** 1614.
Stein, G. S., Stein, J. L., and Kleinsmith, J. L. (1975a). *Sci. Am.* **232** (2), 47.
Stein, G. S., Mans, R. J., Gabbay, E. J., Stein, J. L., Davis, J., and Adawadkar, P. D. (1975b). *Biochemistry* **14,** 1859.
Steplewski, Z., Knowles, B. B., and Koprowski, H. (1976). *Proc. Natl. Acad. Sci. U.S.A.* **59,** 769.
Tabuse, Y., Kawamura, M., and Furusawa, M. (1976). *Differentiation* **7,** 1.
Tanaka, K., Sekiguchi, M., and Okada, Y. (1975). *Proc. Natl. Acad. Sci. U.S.A.* **72,** 4071.
Tanaka, K., Hayakawa, H., Sekiguchi, M., and Okada, Y. (1977). *Proc. Natl. Acad. Sci. U.S.A.* **74,** 2985.
Tilghman, S. M., Tiemeier, D. C., Polsky, F., Edgell, M. H., Seidman, J. G., Leder, A., Engrist, L. W., Norman, B., and Leder, P. (1977). *Proc. Natl. Acad. Sci. U.S.A.* **74,** 4406.
Tilghman, S. M., Tiemeier, D. C., Seidman, J. G., Peterlin, B. M., Sullivan, M., Maizel, J. V., and Leder, P. (1978). *Proc. Natl. Acad. Sci. U.S.A.* **75,** 725.
Tonegawa, S., Brack, C., Hozumi, N., Mattyssens, G., and Schuller, R. (1977a). *Immunol. Rev.* **36,** 73.
Tonegawa, S., Brack, C., Hozumi, N., and Schuller, R. (1977b). *Proc. Natl. Acad. Sci. U.S.A.* **74,** 3518.
Tonegawa, S., Maxam, A. M., Tizard, R., Bernard, O., and Gilbert, W. (1978). *Proc. Natl. Acad. Sci. U.S.A.* **75,** 1485.
Uchida, T., Pappenheimer, A. M., Jr., and Harper, A. A. (1972). *Science* **175,** 901.
Uchida, T., Pappenheimer, A. M., Jr., and Greany, R. J. (1973). *J. Biol. Chem.* **248,** 3838.
Uchida, T., Yamaizumi, M., and Okada, Y. (1977). *Nature (London)* **266,** 839.
Uchida, T., Kim, J., Yamaizumi, M., Miyake, Y., and Okada, Y. (1979). *J. Cell. Biol.* **80,** 10.
Wasserman, M., Zakai, N., Loyter, A., and Kulka, R. G. (1976). *Cell* **7,** 551.
Wasserman, M., Kulka, R. G., and Loyter, A. (1977). *FEBS Lett.* **83,** 48.
Weinstock, R., Sweet, R., Weiss, M. C., Cedar, H., and Axel, R. (1978). *Proc. Natl. Acad. Sci. U.S.A.* **75,** 1299.
Weintraub, H. (1972). *Nature (London)* **240,** 449.
Wille, W., and Willecke, K. (1976). *FEBS Lett.* **65,** 59.
Wilson, T., Papahadjopoulos, D., and Taber, R. (1977). *Proc. Natl. Acad. Sci. U.S.A.* **74,** 3471.
Wolff, G. (1894). *Biol. Zentr.* **14,** 609.
Yamaizumi, M., Uchida, T., Okada, Y., and Furusawa, M. (1978a). *Cell* **13,** 227.
Yamaizumi, M., Uchida, T., Okada, Y., Furusawa, M., and Mitsui, H. (1978b). *Nature (London)* **273,** 782.
Yamaizumi, M., Mekada, E., Uchida, T., and Okada, Y. (1978c). *Cell* **15,** 245.
Yamaizumi, M., Furusawa, M., Uchida, T., Nishimura, T., and Okada, Y. (1978d). *Cell Struct. Funct.* **3,** 293.
Yamamoto, F., and Furusawa, M. (1978). *Exp. Cell Res.* **117,** 441.

INTERNATIONAL REVIEW OF CYTOLOGY, VOL. 62

Cytology, Physiology, and Biochemistry of Germination of Fern Spores

V. RAGHAVAN

Department of Botany, The Ohio State University, Columbus, Ohio

I. Introduction

The germination of a seed, spore, or similar other system, in general biological usage, connotes the beginning of growth in a resting organ during which its quiescent or dormant phase is replaced by a dynamic or active phase. In recent years, a large body of literature on the germination of seeds of angiosperms using methods of biochemistry and molecular biology has accumulated. The typical angiosperm seed consists of several layers of cells of protective seed coats enclosing the germ or the embryo, which in many cases may have formed the primordial shoot and root and rudiments of the first pair of leaves. In addition, seeds of many plants consist of a tissue known as the endosperm constituting a rich endowment of stored food including protein, fat, oil, and starch for later use by the embryo. During germination of the seed, its constituent parts such as the growing embryo axis and the nongrowing endosperm interact in a complex way by exchange of metabolites. Moreover, the reaction to a germination-inducing stimulus may occur in one part of the seed, while the actual growth itself may

ISBN 0-12-364462-3

occur in a different part. These cell and tissue interactions and associated growth processes will obviously require much more complex regulatory mechanisms for their coordination than could be required in the germination of single-celled spores. Because the latter represent experimental systems which could be much more easily analyzed, recent studies on the biochemical and molecular aspects of germination of spores of several fungi (see Lovett, 1975; Van Etten *et al.,* 1976, for reviews) have attained great significance in interpreting the regulatory mechanisms that switch the developmental pattern of resting organs.

Ferns (Filicales) represent another group of plants which yield spores with considerable potential for physiological and biochemical analysis of germination. The majority of ferns produce spores of only one type which are all similar in size and shape, a condition known as homospory. In certain water ferns such as those included in Marsileales and Salviniales, the condition of heterospory is found. Here there are two kinds of spores which differ distinctly in size, the egg-bearing megaspore being considerably larger than the sperm-yielding microspore. Because of the homogeneity they display in their morphology, functional analysis of germination has been largely undertaken with spores of homosporous ferns (hereafter referred to as fern spores).

Fern spores are single-celled structures which contain a quiescent protoplast closely invested by several layers of sporoderm constituting the exine and a thin inner layer or intine. In some species there may be an additional envelope termed perine which is found as an external covering (Lugardon, 1974; Gastony, 1974; Gastony and Tryon, 1976). The spore is born out of a reduction division in the spore mother cell and consequently represents the first cell of the gametophytic generation. Spores of many ferns germinate readily upon contact with water or a simple mineral salt medium in complete darkness or in response to a particular light quality or a specific growth hormone, and all organic substances for germination are derived from the metabolism of stored reserves in the spores. Germination initiates a succession of temporally coordinated and spatially integrated events which lead to the development of a mature gametophyte. The first gross morphological change in the germinating spore, resulting from an asymmetric division of the protoplast, is the outgrowth of a papillate structure which differentiates into the rhizoid. This is followed by the appearance of another cell, the prothallial cell, at an angle to the rhizoid. The prothallial cell may directly function as the protonemal cell or go through another round of division to yield a protonemal cell. In some species, the order of appearance of the rhizoid and the prothallial cell may be reversed. This relatively simple, but well-defined, differentiation pattern during germination of the spore provides a convenient experimental system for the study of cell development without the complexities associated with multicellular systems such as seeds. Moreover, the asymmetrical division of the protoplast which yields daughter cells with entirely different potentials is of significant morphogenetic interest. Of great importance is the fact that cell morphogenesis occurs in the absence of any nutritional or hormonal gradients imposed by surrounding cells. Since spores of several ferns germinate in response to specific wavelengths of light or specific growth hormones, they may also conceivably serve as model systems for interpreting the mechanism of seed germination. Despite these advantages, surprisingly little is known about the regulatory mechanisms that govern cell differentiation during germination of fern spores. When one realizes that ferns occupy a central position in the evolutionary history of the vegetable kingdom and have therefore attracted the attention of botanists more than any other group of plants, the relative paucity of data on the germination of spores should be questioned.

The purpose of this article is to review, summarize, and evaluate recent studies on the physiology, biochemical cytology, and biochemistry of germination of fern spores, and to focus attention on some currently critical areas of investigation. A number of examples of spore germination will be described to illustrate the hypothesis that there is probably but one control pattern which may be variously triggered. Reference is made to the inescapable parallels that exist between seed germination and spore germination when it is considered necessary to explain the problem further. To keep this essay within reasonable bounds, no effort is made to give a comprehensive coverage of the literature; for this purpose reference is made to Sussman (1965) and Miller (1968). In order to present the physiological and biochemical data in a proper perspective, the account will, however, begin with a brief consideration of the structure of the spore and morphological aspects of its germination.

II. Morphological Considerations

A. Internal Structure of the Spore

Although the exine morphology of spores of several ferns had been studied first with the light microscope and later with the scanning electron microscope, details of internal structure have been largely unexplored. One reason for this is that fern spore coats are much more durable and permanent than ordinary cell walls and thus impede the penetration of fixatives generally used in plant microtechnique. Another difficulty that besets workers in this field is that because of its thickness and the presence of various types of markings and sculpturings, the exine acts as an effective optical barrier of the spore contents. It is little wonder then that observations of fern spores with thin walls in whole mounts or in paraffin sections have generated much of what we know about their internal morphology. In recent years, however, the use of improved aldehyde fixatives combined with plastic embedding techniques has led to an enhanced appreciation of the internal structure of dormant spores and the early division pattern during their germination.

The fern spore is an extremely dry and desiccated structure whose most prominent internal feature is the presence of a nucleus surrounded by a blend of organelles and storage granules. The more or less round nucleus consists predominantly of dehydrated chromatin which stains moderately with azure B or with Feulgen. It is one of the peculiarities of the nucleus of dry, nonchlorophyllous spores that a nuclear membrane or nucleolus is not apparent, although in chlorophyllous spores, such as those of *Osmunda regalis, Matteuccia struthiopteris,* and *Onoclea sensibilis,* one or both of these nuclear components may be visible after proper fixation (Figs. 1 and 2). In the latter spore types the chloroplasts are arranged in a ring around the nucleus whereas the more peripheral cytoplasm contains numerous proteinaceous granules and vacuoles (Marengo, 1956; Gantt and Arnott, 1965). In the electron microscope the protoplast of the dormant spore of *M. struthiopteris* appeared virtually naked within the spore coats, although a new cell wall was secreted soon after imbibition preparatory to germination. In addition to chloroplasts, abundant ribosomes and scattered mitochondria made up an essential portion of the developmental organization of the dormant spore of this species, but there was hardly any ER (Marengo, 1973).

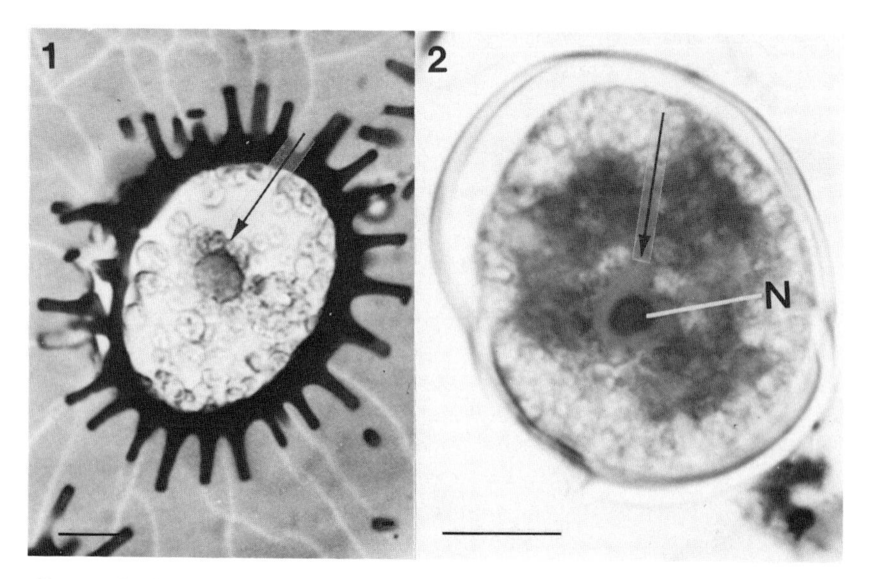

FIG. 1. Section of a dry spore of *Anemia phyllitidis*. Arrow indicates nucleus. Bar = 10 μm. (From Raghavan, 1976.)

FIG. 2. Section of a dry spore of *Onoclea sensibilis*. Arrow indicates nucleus. N, Nucleolus. Bar = 10 μm.

While the dormant, green spore has thus all the hallmarks of a typical plant cell, the paucity of ER reflects a reprograming of the cytoplasm for depressed metabolic activities. The cytoplasm of the newly formed spore, on the other hand, is in an active metabolic state, and as seen in spores of *O. sensibilis* just separated from the tapetum, a loosely defined ER is invariably present in them (Marengo and Badalamente, 1978).

Granules of various sizes and shapes which react positively to protein stains are found abundantly in nonchlorophyllous spores of *Anemia phyllitidis* (Raghavan, 1976) and *Pteris vittata* (Raghavan, 1977b), which are also characterized by a paucity of organelles (Fig. 3). Although appreciable amounts of lipids are also present in dry spores of *A. phyllitidis* (Gemmrich, 1977a) they could not be detected in sections processed for microscopic histochemistry through various organic solvents. On the other hand, histochemical examination of crushed spores of *Lygodium japonicum* has revealed the presence of lipids (Rutter and Raghavan, 1978), which, by analytical methods, have been found to constitute about 66% of the spore weight (Gemmrich, 1977b). While the majority of storage granules in dormant spores of *L. japonicum* were proteinaceous in nature, a few appeared to contain a centrally located periodic acid–Schiff (PAS) particle (Fig. 4). In nonchlorophyllous spores of *Thelypteris dentata* (Seilheimer, 1978) which contained 50% lipophilic materials, lipids generally found in association with chloroplasts such as glycolipids and phosphatidyl glycerol were absent. However, since chloroplasts appear rapidly in germinating nongreen spores, even in the absence of light, the presence of organelles answer-

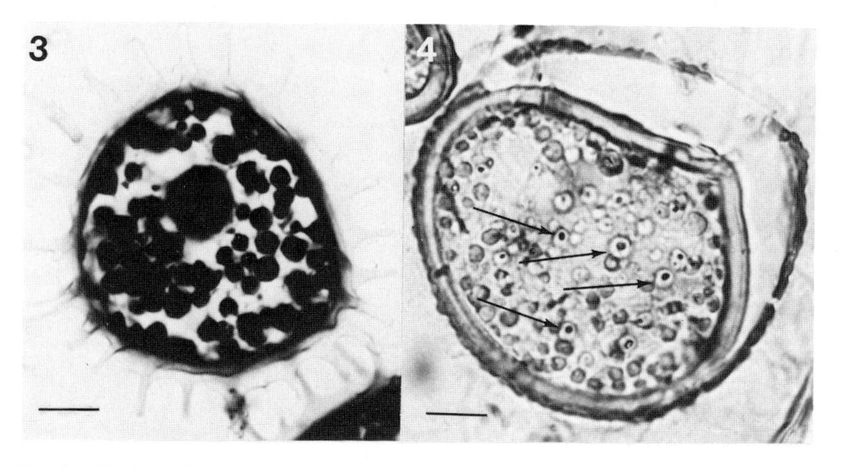

FIG. 3. Section of a dry spore of *Anemia phyllitidis* stained with aniline blue black showing protein granules. Bar = 10 μm. (From Raghavan, 1976.)

FIG. 4. Section of a dry spore of *Lygodium japonicum* showing storage granules, some of which contain a PAS-staining particle (arrows). Bar = 10 μm. (From Rutter and Raghavan, 1978.)

ing the description of proplastids as reported in dry spores (Gullvåg, 1971; Seilheimer, 1975) is of some interest. If proplastids are indeed present, lacking experimental evidence, it might be assumed that they are transformed into plastids by a nonphotochemical process. The possibility that primary storage products such as PAS-staining granules found in *L. japonicum* spores are precursors of chloroplasts or are used for membrane formation in plastids and ER also needs to be explored.

B. The Concept of Spore Germination

The problem which is encountered very early in any discussion on fern spore germination is the criterion used to define the process and detect the phenomenon. Although fern spores germinate within a morphological framework, the concept of germination itself has not been clearly defined or uniformly followed by earlier investigators. Some workers (Pietrykowska, 1962; Weinberg and Voeller, 1969a,b) associated the breakage of exine with germination, whereas others (Sugai and Furuya, 1967; Lloyd and Klekowski, 1970; Raghavan, 1970, 1973; Towill and Ikuma, 1973; Smith and Robinson, 1975) used the term to include protrusion of the rhizoid and/or prothallial cell followed by chloroplast formation in the latter. In analyzing the germination of spores of *Dryopteris filix-mas,* Mohr (1956) considered swelling, expansion, and greening of the spore contents following breakage of the exine as definitive signs of germination. In *Osmunda cinnamomea* and *O. claytoniana* (Mohr *et al.*, 1964), spore germination was separated into two stages, involving, respectively, breakage of the exine and protrusion of the rhizoid. This confusing terminology on spore germination is further complicated by the fact that spores of some species which have an obligate requirement for light for germination imbibe water in the dark and force open the exine, without actually germinating.

To ascertain the germination of spores of *O. sensibilis* in whole mounts as early as 50 hours after sowing, Edwards and Miller (1972) have developed a staining technique using a mixture of acetocarmine and chloral hydrate. The latter compound caused lysis of the chloroplasts and enhanced visibility of the nucleus stained with acetocarmine. Spores were considered to have germinated when the nucleus divided forming two daughter nuclei within the confines of the spore coats. This method, while eminently suitable for thin-walled spores, will hardly reveal the nuclei of thick-walled spores in whole mounts. In the absence of a clear operational definition, the term germination is used here loosely to include one or more of the processes involved such as breakage of the exine and protrusion of the rhizoid and the prothallial cell and does not necessarily indicate the stage of mitosis of the spore nucleus when the first irreversible developmental signal of germination is given.

C. Cell Division Pattern during Germination

The planes of segmentation of the spore protoplast which give rise to the rhizoid and protonemal cell during germination vary among the different groups of ferns, although genera included in any single family may exhibit a more or less uniform pattern. With the aid of whole mount preparations of spores collected at different stages of germination, a wide spectrum of cell division patterns ranging from the formation of a primary rhizoid and a protonemal cell by a simple asymmetric division to the generation of a mass of cells by irregular divisions has been reported in the different families of homosporous ferns. This made logical subsequent attempts to classify the known patterns of spore germination for use in the broad generic and familial grouping of ferns (Momose, 1942; Nishida, 1965; Nayar and Kaur, 1968). Recent studies on the cytology of germination of spores of certain ferns using sectioned materials have however placed some constraint on the significance of published descriptions of early ontogeny of gametophytes based principally on whole mount preparations and have helped to clear much of the confusion surrounding the conflicting accounts of earlier workers on one and the same species. A case in point is *Ceratopteris thalictroides,* whose gametophyte ontogeny was investigated successively by Javalgekar (1960), Nishida (1962), Pal and Pal (1963), and Nayar and Kaur (1969). Although Javalgekar (1960) and Pal and Pal (1963) made little mention of the actual events of spore germination in this species, their descriptions imply that the initial division of the spore is perpendicular to its polar axis. This division presumably gives rise to a proximal prothallial cell and perhaps to the rhizoid and to a distal basal cell forming a lateral rhizoid. According to Nishida (1962), the spore germinated to form a tripolar structure consisting of a pair of rhizoids and a protonemal cell, these probably arising by a series of divisions parallel to the polar axis of the spore. In the more detailed study of Nayar and Kaur (1969) the first division of the spore is said to give rise to a proximal rhizoid overlying a large distal prothallial cell. In yet another account, Hickok (quoted by Bierhorst, 1975) claimed that in a mutant form of *C. thalictroides,* two parallel divisions in the spore protoplast perpendicular to the polar axis give rise to rhizoid and prothallial cell. Recently, using thin sections of spores embedded in Spurr's resin, Endress (1974) attempted to resolve these contradictory observations and showed convincingly that the first division of the spore protoplast separated a proximal prothallial (protonemal) cell from a large basal cell and that the rhizoid was formed by an asymmetric division of the latter. As seen in the sequence illustrated in Figs. 5–10, in the germination of spores of *Anemia phyllitidis,* the first division is asymmetrical cutting off a large basal cell and a small protonemal cell, the latter appearing outside as the first sign of germination. The rhizoid is formed by another asymmetric division of the basal cell in a plane perpendicular

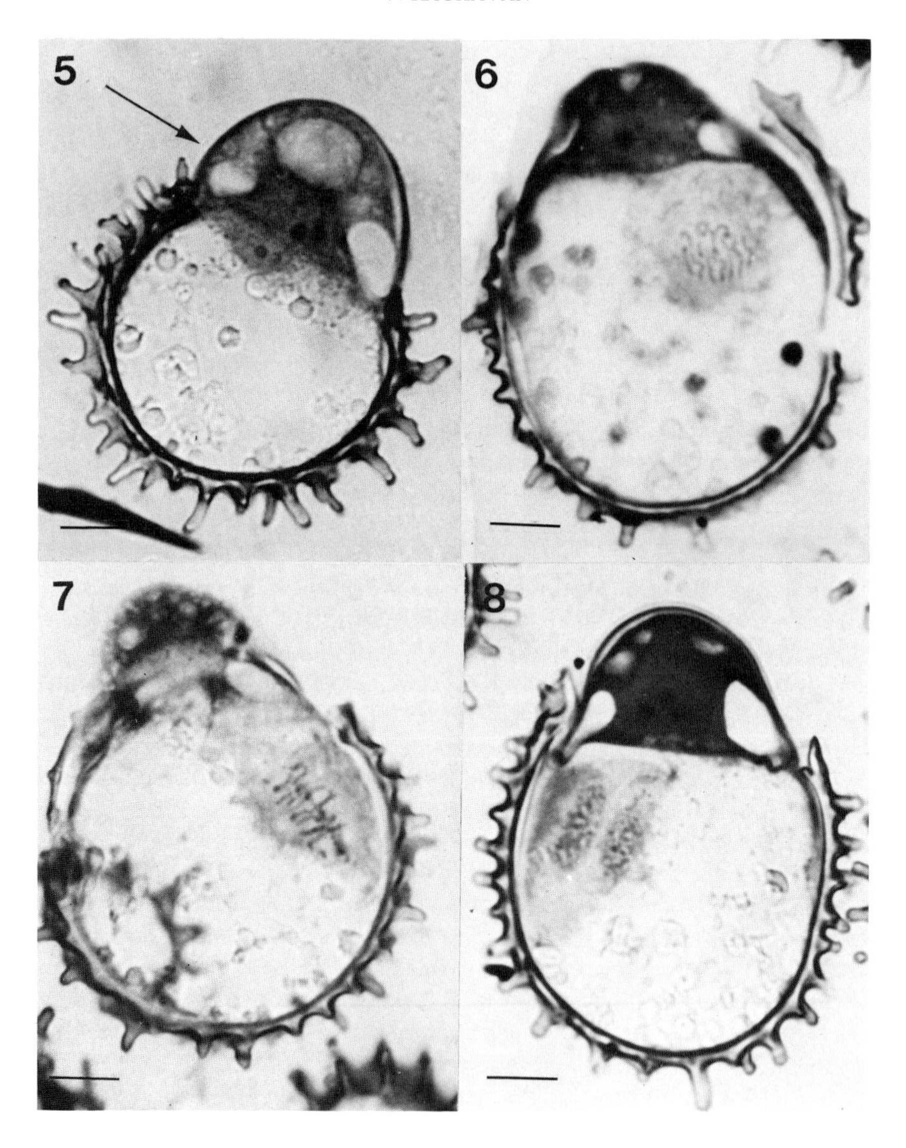

FIGS. 5–10. Sections of spores of *Anemia phyllitidis* showing origin of the protonemal cell and rhizoid. Bars = 10 μm. Fig. 5. Formation of protonemal cell (arrow). Fig. 6. Prophase stage of spore nucleus in preparation to form rhizoid. Fig. 7. Metaphase stage of dividing spore nucleus. Fig. 8. Division of the spore nucleus into two nuclei. Fig. 9. Formation of the rhizoid (arrow). P, Protonemal cell; SN, original spore nucleus. Fig. 10. Germinating spore with a two-celled protonemal filament (P) and rhizoid (R) with a nucleus (N). Arrow points to the original spore nucleus. (From Raghavan, 1976.)

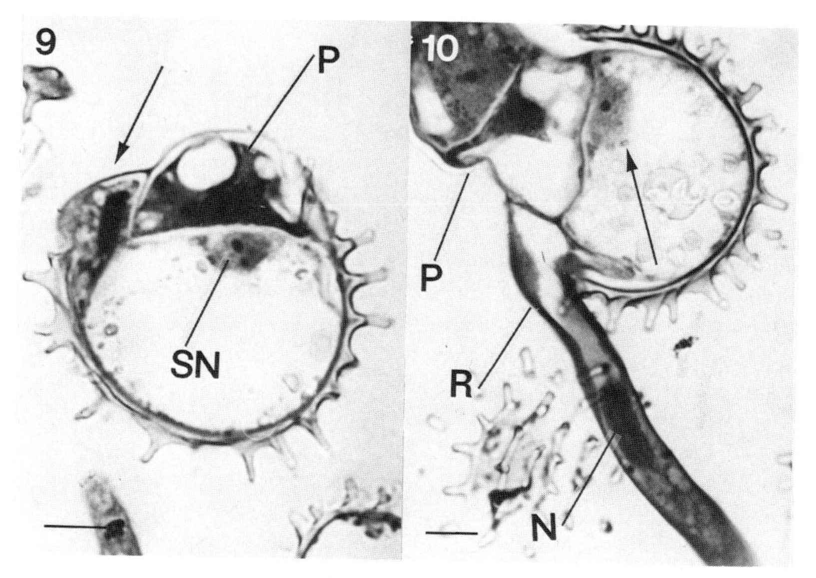

FIGS. 9 and 10.

to the first division (Raghavan, 1976). These observations do not agree with the early descriptions of germination of spores of *A. phyllitidis* and related species based on whole mounts (Twiss, 1910; Kaur 1961; Atkinson, 1962); the major points of difference are illustrated diagrammatically in Fig. 11.

According to Nayar and Kaur (1971) spore germination in Pteridaceae is of the "Vittaria" type in which the basal cell, after cutting off the rhizoid, divides by a wall perpendicular to the first into two equal cells. One of the daughter cells functions as the protonemal cell. However, as seen in sections of spores of *Pteris*

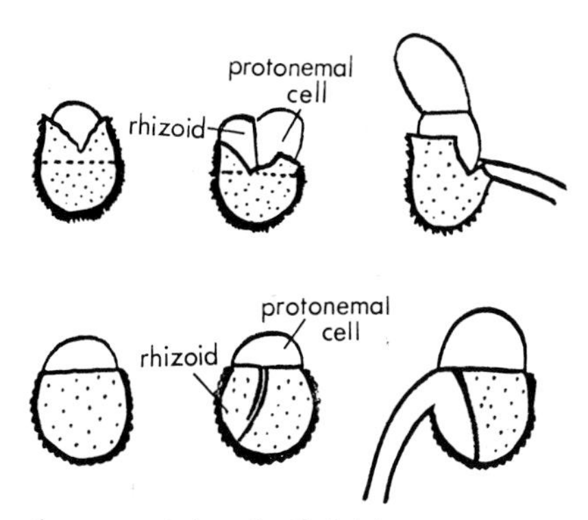

FIG. 11. A diagrammatic representation of cell division pattern during germination of spores of *Anemia phyllitidis* as seen in whole mounts (top: Atkinson, 1962) and in sections (bottom: Raghavan, 1976).

vittata, after the rhizoid is cut off, the basal cell does not divide further and functions directly as the protonemal cell (Raghavan, 1977b). Analysis of germination of spores of *Dryopteris borreri,* presumably showing Vittaria-type segmentation (Nayar and Kaur, 1971), has also given the impression that the large cell originating from an asymmetric division of the spore functions directly as the protonemal cell (Dyer and Cran, 1976). These studies would seem to provide good evidence that a single nuclear division is normally adequate to produce cells with divergent functions in a germinating fern spore.

Before leaving the subject, it is worth examining spores of *Lygodium japonicum* as yet another example of the level of information on cell division pattern during germination that can be obtained in sections. In successive investigations on the ontogeny of gametophytes of various species of *Lygodium,* the precise origin of the rhizoid has remained a point of contention between different workers (Twiss, 1910; Rogers, 1923; Clarke, 1936). Recently, Rutter and Raghavan (1978) showed that in *L. japonicum* neither the rhizoid nor the protonemal cell is delineated by the initial division of the spore protoplast. The first division of the spore gives rise to an apical cell which subsequently segments into the protonemal cell and an intermediate cell. This latter cell is the immediate precursor of the rhizoid (Fig. 12). In sum, results considered to this point have raised some skepticism on the value of the various systems of classification of early division sequences of germination based on whole mount preparations and have questioned the wisdom of their use in systematic and phylogenetic evaluation of ferns.

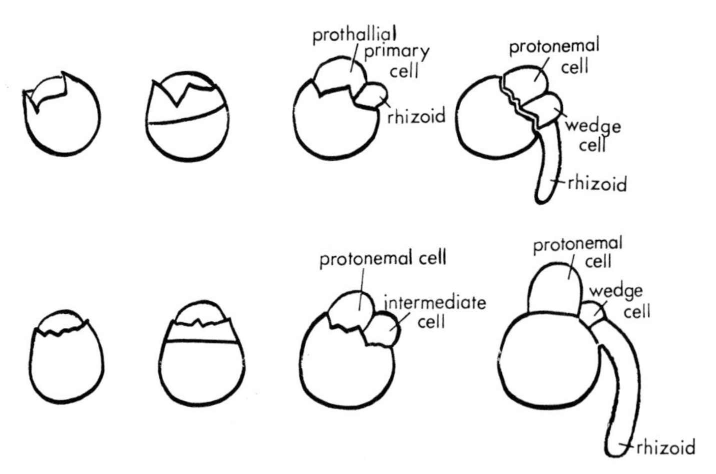

FIG. 12. A diagrammatic representation of cell division pattern during germination of spores of *Lygodium japonicum* as seen in whole mounts (top: Clarke, 1936) and in sections (bottom: Rutter and Raghavan, 1978).

D. Polarity of Spore Germination

Germination of a typical spore is intimately tied to an asymmetrical division heralding the appearance of two cells destined for different functions. In the simplest type of germination, one of these cells gives rise to the rhizoid while the other forms the protonemal cell. What factors determine the point of origin of the rhizoid and the protonemal cell in an ungerminated spore? In an early study, Kato (1957a,c) showed that in spores of *Dryopteris erythrosora* deprived of their exine by treatment with concentrated KOH or NaOH, the patch of cytoplasm at the point of origin of the future rhizoid was differentially colored, indicating that a unique cytoplasmic area destined to be the site of rhizoid initiation was fixed in the spore even before it began to germinate. Following nuclear division, each cell thus inherits a different cytoplasm in which specific developmental programs leading to the formation of rhizoid and protonema are later initiated. In *Matteuccia struthiopteris,* Gantt and Arnott (1965) found that irrespective of the orientation of spores on the medium, the nucleus always migrated to that end of the spore which originally faced the center of the tetrad. This occurred despite the fact that there were no cytological or biochemical signs of internal asymmetry in the spore suggestive of a specific morphogenetic event.

If determination of rhizoid by means of a cytoplasmic localization is a basic aspect of polarity of fern spores, one issue of importance is the extent to which polarity can be modified by external factors. Both unilateral as well as polarized illumination have been shown to alter the preexisting axis of spore polarity. According to Pietrykowska (1963) when spores of *M. struthiopteris* and *Athyrium filix-femina* were exposed to unilateral light, the rhizoid originated from the darkened side of the spore while the lighted side gave rise to the prothallial cell. In spores germinated in plane polarized white light, the rhizoid and the prothallial cell respectively appeared at points parallel to and perpendicular to the vibration planes. In spores of *Osmunda cinnamomea* illuminated with polarized blue light, the rhizoid appeared from that part of the spore which absorbed minimum light, this being circumscribed by the fact that the axes of maximum absorption of photoreceptor molecules were oriented parallel to the surface of the spore (Jaffe and Etzold, 1962). Several examples are also available in which the sharp differentiation between the rhizoid and the protonemal cell was lost when germinating spores were subjected to chemical treatments. Since widely different chemicals were used for these treatments, the extent to which the induced modifications represent functional manifestations of an altered cytoplasmic milieu of the spore is not known. Kato (1957a,b) reported that when spores of *Dryopteris erythrosora* were germinated in low concentrations of tryptophan or a high concentration of the auxin, naphthaleneacetic acid, instead of a rhizoid and a protonemal cell, two protonemal cells grew out of the same spore.

Similar results were also obtained upon treatment of spores of *Anemia phyllitidis* with *allo*-gibberic acid (Schraudolf, 1966). While these studies provided no evidence to show that formation of twin protonemata was coupled to cytokinetic symmetry in the germinating spore, Miller and Greany (1976) have shown that this was indeed the case in spores of *Onoclea sensibilis* allowed to germinate in a medium containing low concentrations of methanol. Apparently, an asymmetric division provokes differentiation of daughter cells into structurally and functionally different cell types, whereas a symmetric division invariably leads to cells with the same functional potential.

Formation of twin rhizoids in the absence of prothallial cells occurred when spores of *Pteris vittata* were germinated in distilled water under mineral nutrient starvation. The number of rhizoids formed per spore increased when germination occurred under submerged conditions in the presence of sucrose (Kato, 1973). Finally, treatment of spores with high concentrations of tryptophan or naphthaleneacetic acid (Kato, 1957a,c) resulted in the formation of enlarged, apolar cells which became committed to different kinds of division programs independent of the orientation of polar axis. These results clearly indicate that polarity is a labile feature of fern spore organization which may be reshuffled by cultural conditions.

III. Physiology of Spore Germination

The dry, dormant spore of a fern is an unlikely object, vastly different from the sporophyte from which it has evolved. Microscopic in size, it is produced in relative abundance, is resistant to extreme environmental conditions, and persists superficially unchanged for long periods of time. Since the very first division of the spore preparatory to germination leads to the formation of cells with different potentials, it is reasonable to assume that all physiological and biochemical activities that underlie cell diversification during germination operate within the limited territory of the spore. It is now necessary to examine the various physiological aspects of germination to obtain an insight into the processes that trigger spores in a quiescent or dormant state to initiate division. We start with observations concerning the phenomenal longevity of fern spores and end this section with an analysis of the various extracellular factors that control germination.

A. Viability of Spores

The length of time spores retain their ability to germinate, referred to as viability, and the rapidity of germination and subsequent growth of the gametophyte are basic to the successful competition of ferns in a new habitat. In general, the life span of spores is enhanced under conditions of storage where the metabolic activity is minimal. In addition, viability of spores is also determined by their genetic makeup and by the environmental factors to which they are exposed after harvest. The latter will determine whether spores will continue to live for the longest genetically possible period or will succumb at an earlier time.

Viability of spores varies enormously among ferns and periods ranging from a few days to a few years have been reported in the literature. What determines longevity of spores is a moot question, but it has become clear in recent years that a number of morphological and physiological attributes of spores have to be considered in formulating any general hypothesis to explain this. Lloyd and Klekowski (1970) have made comparisons between the viability of chlorophyllous and nonchlorophyllous spores of several families of Filicales. Their results show that under ordinary storage conditions green spores have relatively short periods of viability averaging about 48 days while nongreen spores live on an average up to 2.8 years. According to these workers, the chlorophyllous condition in spores might have evolved primarily as a safeguard against expenditure of energy required to lapse into dormancy and is probably a primitive state. Since chlorophyllous spores have a high respiratory activity it was naturally thought that they metabolized stored reserves in a shorter period of time than nongreen spores (Okada, 1929). In a few determinations made, it has also been established that short-lived green spores have a higher water content than long-lived nongreen spores (Okada, 1929) suggesting that uncontrolled loss of water may have some relevance to the life span of spores. Since the presence of an additional spore coat such as a perispore might ameliorate at least partially the desiccation of the cell contents, it is no wonder that out of 50 different species of chlorophyllous spores tested by Lloyd and Klekowski (1970) only spores of Onocleoideae which have a perispore appeared to have the longest viability. Notoriously short-lived spores such as those of different species of *Osmunda* can be kept in a viable condition for as long 3 years by storage at conditions which minimize desiccation (Stokey, 1951).

Even under favorable conditions of storage, increasing spore age also leads to a decline in viability. Such a loss of viability is not to be mistaken for a sudden or abrupt failure of all spores in a population to germinate, but is a slow decrease in the rate of germination. Although spores of *Pteridium aquilinum* (bracken fern) remain viable up to 10 years, germination percentage decreases by as much as 30% by the fourth month after collection and by 50% by the end of the first year (Conway, 1949). In a study of the viability of an aging population of spores of *Polypodium vulgare* conducted over a period of 7 years, Smith and Robinson (1975) obtained a sigmoid type of spore survival curve. The effects of age on viability were counteracted by a high spore density in the medium or by sowing spores in a conditioned medium. The decline in spore viability with age is similar to that caused by the breakdown of a store of essential metabolites or by the loss of ability to synthesize them during germination. Other observations on this species have indicated that increasing age of the spore delays the appearance of the first morphological sign of germination and leads to the formation of a disproportionate number of abnormal gametophytes in culture.

Closely related to the life span of spores is the length of time that elapses between sowing and actual germination and here again, comparisons between green and nongreen spores are germane. According to Lloyd and Klekowski (1970) the tendency is for chloroplast-studded spores to germinate more rapidly than nongreen spores, the former germinating on an average in 1.46 days after sowing compared to 9.5 days required for the latter. Much of the difference in the rapidity of germination between green and nongreen spores may result from the basic difference in the physiological state of the spore at the time of sowing. Green spores, endowed as they are with photosynthetically active pigments, pass into a phase of growth with relative ease when supplied with propitious conditions. Presumably, in nongreen spores the storage granules have to be hydrolyzed to simpler molecules to provide energy for the synthetic activities of germination, enlargement of the protoplast to break the exine and formation of chloroplasts from proplastids or other precursors.

B. Quiescence and Dormancy of Spores

As applied to fern spores, quiescence and dormancy are not sharply separated states and so come comments are in order concerning them. The assumption made here is that quiescent spores are those that germinate readily when confronted with water, oxygen, and a range of temperatures favorable for plant growth. On the other hand, dormancy may be visualized as a developmental block resulting in the failure of perfectly viable spores to germinate when supplied with favorable environmental conditions. This block may generally last for a variable period of time, although in some cases it may even continue almost indefinitely until a special condition is fulfilled. The enormous quantity of spores produced in a fern provides an effective method of dispersal in space, whereas a period of dormancy ensures their dispersal in time. Our account of spore dormancy in ferns has of necessity to be couched in less precise terms than can be applied to the vast majority of angiosperm seeds. The reason for this is that, unlike the angiosperm seed, dormancy mechanisms have not been systematically explored in fern spores. Spores of filmy ferns included in the Hymenophyllaceae and some members of Grammitidaceae are known to germinate within the sporangium and thus exhibit scarcely any dormancy, whereas at the other extreme are spores of genera with subterranean, saprophytic gametophytes, included in the Ophioglossaceae, which are known to remain dormant for long periods of time even when they are subjected to conditions normally favorable to germination. For example, Whittier (1973) has shown that in axenic cultures of *Botrychium dissectum* (Ophioglossaceae), a minimum period of 8 weeks was necessary for any spores to germinate. This is also probably true for spores of *B. multifidum* (Gifford and Brandon, 1978). Incomplete though our knowledge of the conditions required for germination of spores which yield subterranean gametophytes, it is obvious that dormancy mechanisms have evolved in such spores in response to competitive selective pressures. To support this statement, we must add yet another feature of germination of spores of these plants. When spores germinate in the soil, they invariably establish a symbiotic relationship with a mycorrhizal fungus. It is likely that the prolonged dormancy exhibited by spores could increase their chance to sift slowly down in the soil and become buried in close proximity to a fungal mycelium.

Spores of the large majority of ferns exhibit a form of dormancy which is overcome by light. A tabulation by Miller (1968) shows that out of 88 species tested, spores of only 7 species germinate in substantial numbers in complete darkness when given access to moisture, oxygen, and a favorable temperature range. Instances are also known where spores require storage in dry conditions for a period of time—a treatment known as after-ripening—before they will germinate (Hevly, 1963). There are probably no histochemical or morphological differences between after-ripened and non-after-ripened spores, yet some basic changes are induced in the former that govern their entire subsequent development. The literature on the germination of fern spores is also full of examples of dormant spores which progressively lose their readiness to germinate after application of a dormancy-breaking stimulus. This important observation continues to defy adequate explanation and it is not known whether this is due to loss of viability or to acquisition of secondary dormancy.

C. Factors Controlling Germination

The quiescent or dormant state of the spore is overtly terminated when it begins to germinate. Germination can take place only in environments where normal metabolic activity is possible. The requirements for germination of quiescent spores are surprisingly simple and essentially consist of the availability of moisture, a suitable temperature, and the normal composition of the atmosphere. Dormant spores will not germinate under conditions which favor the germination of quiescent spores, but as mentioned in passing earlier, they require some special treatments or a particular component of the environment to overcome the dormant state.

The requirements for germination of quiescent and dormant spores are described below. It should be emphasized that the conditions are applicable when spores are sown at a moderate density, which presumably varies for each species.

1. *Water and Medium Composition*

Imbibition of water by spores is the initial step in the germination process. Imbibition generally leads to the rehydration of the spore protoplast, in particular of the chromatin and the storage granules. Entry of water is however determined by the thickness of spore coats and the state of hydration of spore contents, and for this reason, variable periods of imbibition are necessary for spores of different species before they resume synthetic activities which can be identified with germination. In dormant spores which require an exposure to light for germination, photosensitivity increases with time after imbibition and remains stable thereafter. According to Mohr (1956), spores of *Dryopteris filix-mas* responded maximally to radiant energy after they were soaked on a solidified medium for 36 hours; corresponding times for spores of *Pteris vittata* (Sugai and Furuya, 1967), *Asplenium nidus* (Raghavan, 1971b), *Cheilanthes farinosa* (Raghavan, 1973), and *Matteuccia struthiopteris* (Jarvis and Wilkins, 1973) floated on liquid media were 96, 48, 72, and 96 hours, respectively. For chlorophyllous spores, the presoaking times required for photosensitivity are even shorter—for example—3 hours for *Osmunda cinnamomea* and 24 hours for *O. claytoniana* (Mohr *et al.*, 1964). Spores whose dormancy is broken by chemicals such as gibberellic acid are also known to require a period of imbibition before they respond optimally to the chemical signal (Raghavan, 1976).

The physical conditions of the medium are also of importance in the germination of spores because they determine the rate of imbibition and the survival rate of germinated spores. Although both liquid and solid media have been successfully employed, the former have been found to present some advantages for spore germination over the latter, but not for growth of the germinated spores into mature prothalli (Okada, 1929; Hurel-Py, 1950). It is likely that a solid medium affects germination by means of osmotic effects, preventing the breakage of exine and the appearance of rhizoid and prothallial cell. Some investigators have shown that artificially increasing the osmolarity of the medium by the application of sugars or mineral salts generally impedes the germination process (Gistl, 1928; Stephan, 1929; Raghavan, 1977a,b).

2. *Temperature*

The outcome of early investigations in which spores of a number of ferns germinated quite readily at a range of laboratory temperatures thwarted attempts to study germination as a function of temperature. In general, the tolerated range of temperatures at which spores germinate optimally is about 20–25°C whereas the minimal and maximal temperatures which permit germination are 15 and 50°C, respectively. Germination of spores is inhibited at temperatures higher than the optimum probably due to changes produced in the molecular organization of the proteins and other storage components. The precise sensitivity to temperature varies with the species and is invariably related to the temperature

requirements for the subsequent growth of the germinated spore. The ability of spores of *Notholaena sinuata, N. cochisensis, N. grayi, Pellaea limitanea, Cheilanthes eatoni,* and *C. lindheimeri* to withstand temperatures as high as 40–50°C in the laboratory has been attributed to their natural growth in a xeric environment (Hevly, 1963). The lowest temperature tolerances reported for germination of spores are 5°C for *Cyathea bonensimensis* (Kawasaki, 1954a) and 1–2°C for *Pteridium aquilinum* (Conway, 1949). Since both these species are not acclimatized to high altitudes, ecological extrapolations on their temperature tolerance must be made with some caution.

The effects of temperature on the germination of spores of certain species appear somewhat complex because of interactions with other environmental parameters, especially light or its lack thereof. The response of spores of *Dryopteris filix-mas,* which germinate optimally at 20°C, is quite typical of this kind of interaction between light and temperature. Mohr (1956) found that when spores were kept at temperature regimes up to 36°C for 24 hours following irradiation with a saturating dose of light, and then kept at 20°C for completion of germination, there was a progressive decrease in germination. However, as the interval between the end of the light treatment and beginning of high temperature treatment became longer, spores escaped from inhibition and in about 36 hours after the light exposure high temperature did not inhibit germination. Since spore germination in *D. filix-mas* is controlled by phytochrome, the simplest explanation of these results is that the action of the pigment in promoting germination is completed before the high temperature shock becomes effective. Somewhat similar results have also been recorded for spores of *Onoclea sensibilis* using 40°C as the inhibitory temperature and 25°C as the germination temperature (Chen and Ikuma, 1979). In spores of this species, a temperature of 30°C overcame the requirement for inductive red light and induced germination in complete darkness (Towill, 1978).

3. *pH Effects*

It is to be expected that germination of the spore, as a process related to living cells, is limited in its ability to occur within a narrow range of pH on either side of neutrality. On the other hand, the survival and growth of various species of ferns in ecological niches subjected to extremes of climate and consisting of unusual soil types make it likely that pH changes of the substratum have at best only a minor influence on spore germination. A survey of published accounts of pH ranges for optimum germination of spores of ferns shows that spores of most species germinate when the pH of the medium is in the slightly acidic or neutral range as shown, for example, by Conway (1948) for spores of *Pteridium aquilinum*. However, in some cases at least, pH near or above neutrality is less desirable for germination. This appears to be so for *Dryopteris filix-mas,* whose germination increased as a function of pH from 3.0 to 5.4 and decreased with further increase in pH. No germination was recorded at pH 9–10 (Mohr, 1956). Although spores of a number of ferns are reported to withstand pH as low as 3.2, the percentage of germination is reduced and rhizoid and protonema elongate very little (Courbet, 1955). The tolerance of spores of some Cheilanthoid ferns such as *Notholaena cochisensis* and *Pellaea limitanea* to pH in the range of 9–10 has been found to correlate well with pH of the soil in which these ferns grow (Hevly, 1963).

Weinberg and Voeller (1969b) have demonstrated a striking pH dependence of gibberellic acid-induced dark germination of spores of *Anemia phyllitidis*. Maximum dark germination occurred in a medium containing suboptimal concentrations of gibberellic acid at pH 3.5 to 4.5 and none germinated at pH above 6.0. By increasing the concentration of gibberellic acid, however, the inhibitory effect of high pH on germination was overcome. Interestingly enough, inhibitory effects of pH were not observed in spores germinating in light in the absence of gibberellic acid.

4. *Gases*

The composition of the bulk atmospheric gas is also an important feature of the environmental control of germination of fern spores, although published accounts on the effects of particular gaseous phases on germination are scanty. Respiratory processes in spores are probably stimulated soon after the protoplast becomes hydrated and so it is reasonable to assume that they should have access to oxygen during this period. This has been found to be the case. Towill and Ikuma (1975a) found that when spores of *Onoclea sensibilis* were allowed to imbibe in an atmosphere containing N_2, they gradually lost their sensitivity to germinate in response to a subsequent light stimulus. Recovery from anaerobiosis was quick as spores attained half-maximal sensitivity to irradiation after about 1 hour in air. The requirement for aerobic processes for the synthesis of metabolites initiating cell division during germination is underscored by the fact that administration of N_2 to spores of *O. sensibilis* following irradiation also inhibited their germination. Maximum inhibition of germination was obtained when N_2 was applied at 6 to 8 hours after irradiation. Spores subjected to anaerobiosis following irradiation recovered relatively slowly from inhibition suggesting that the characteristics of oxidative processes during the imbibition phase of germination and during the postirradiation period are quite different.

The presence of ethylene in the ambient atmosphere also inhibits both light-induced and dark germination of spores of *Onoclea sensibilis* (Edwards and Miller, 1972; Fisher and Miller, 1975). In photoinduced spores, application of the gas 1–3 hours subsequent to illumination was most effective in blocking germination in a large percentage of spores. Certainly the gas interacts with other environmental conditions of culture in its effects on germination. If ethylene-treated spores were exposed to high intensities of white light or to nonlethal doses of CO_2, the effect of the hydrocarbon was cancelled and spores germinated normally (Fisher and Miller, 1975; Edwards, 1977). In view of the extreme sensitivity of spores of *O. sensibilis* to ethylene and CO_2, conflicting reports in the literature on the dark germination of spores may be ascribed to the presence of varying amounts of these gases in the ambient atmosphere. The effect of ethylene on fern spore germination is the opposite of its effect on seed germination (Abeles and Lonski, 1969; Ketring and Morgan, 1969, 1971; Esashi and Leopold, 1969). The mechanism of ethylene effects in seeds and spores is in no way clear to account for these differential effects. It should be kept in mind that besides being a gas at ordinary temperature, ethylene is a normal plant metabolite that acts as a hormone in controlling a variety of plant growth processes such as hook formation in etiolated plants, flowering, abscission of leaves, and induction of climacteric period of fruit ripening. Thus it seems more likely that the primary action of ethylene on spore germination concerns a fundamental process common to a number of morphogenetic phenomena. Fisher and Miller (1978) have suggested that the target for ethylene action in germinating spores of *O. sensibilis* may be some fundamental process which occurs before the nucleus begins to synthesize DNA.

5. *Chemicals and Other Factors*

A number of unrelated chemicals have been tested for their effects on germination of spores of diverse ferns. In these trials, the test substances were generally incorporated in a range of concentrations into a basal medium and spores were cultured under conditions favorable for germination. No evidence was, however, sought to determine whether the test substance was essential to trigger germination or could substitute for the requirement for a critical factor such as light. These facts need to be kept in mind while considering the following data. Moreover, since many compounds inhibit spore germination and none of them exhibit any dramatic promoting effects, it is clear that the single-celled spore has an endogenous reserve of essential nutrients to germinate under favorable conditions.

Although supplementation of the medium with 2% sucrose was without any effect on the germination of spores of *Alsophila australis,* addition of both glucose and fructose at 1% appeared to inhibit germination (Hurel-Py, 1955). In contrast, these latter sugars were claimed to promote germination of spores of *Athyrium filix-femina* (Courbet, 1957). Glucose at a range of concentrations between 0.5 and 1.5% and fructose between 0.5 and 4.0% promoted germination of spores appreciably over the unsupplemented control medium. No comparable promotion of germination was observed with mannose, D-xylose and L-xylose, which were noninhibitory at low concentrations, whereas other sugars such as D-ribose, L-arabinose, L-rhamnose, and galactose were found to be totally inhibitory even at very low concentrations. From these results it is evident that spores of different ferns may have both different sugar requirements for germination and different programs to metabolize them.

Coumarin is a phenolic derivative of fairly widespread occurrence in plants and has been extensively used as an inhibitor of seed germination. This compound was found to inhibit germination of spores of *Gymnogramme calomelanos* (Sossountzov, 1961) and *Anemia phyllitidis* (Schraudolf, 1967). The inhibitory effect of coumarin on spores of *G. calomelanos* was partially offset upon withdrawal of the inhibitor from the medium whereas simultaneous addition of gibberellic acid reversed its effect on spores of *A. phyllitidis*. Germination of spores of the latter fern was also inhibited by the fluorene derivative, fluorene-9-carboxylic acid, but this inhibition was not counteracted by gibberellic acid. Unidentified substances inhibitory for spore germination are reported to be present in the centrifuged sediment of gametophytes of *Dryopteris filix-mas*; this is surprising since a water extract of the gametophyte had an opposite effect on spore germination (Bell, 1958). Coconut milk, the liquid endosperm of coconut, well known for its growth-promoting effects in plant tissue cultures has been shown to inhibit germination of spores of *Nephrolepis hirsutula* (Smith and Yee, 1975). However, the concentrations used (25–100%) are much higher than those routinely used for angiosperm and gymnosperm tissue cultures, which casts considerable doubt on the significance of the data.

Colchicine has been shown in a number of cases to delay germination of the spore. However, the effect of this chemical is much less on delaying or inhibiting germination than in inducing morphological and cytological abnormalities in the germinated spore (Rosendahl, 1940; Mehra, 1952; Yamasaki, 1954).

According to Pringle (1970), certain simple saturated fatty acids, starting with valeric acid and ascending the homologous series up to capric acid and the unsaturated oleic and linoleic acids, inhibited germination of spores of *Onoclea sensibilis*. The addition of a culture filtrate of gametophytes of *Pteridium aquilinum* containing antheridogen, a species-specific hormone that controls antheridia formation in fern gametophytes, however, overcame this inhibition and allowed complete germination of spores. Since these same fatty acids inhibiting spore germination are capable of enhancing the antheridium-inducing potency of the culture filtrate of *P. aquilinum* (Pringle, 1961), an interaction between antheridogen and fatty acids in the regulation of spore germination and induction of antheridia on gametophytes could be of advantage in the survival of bracken population in nature.

Some workers have described a general stimulation of germination of spores when a fungus or a fungal extract is present in the medium. Hutchinson and Fahim (1958) found that sterilized bracken fern spores sown over fungal colonies covered with a cellophane sheet germinated better and exhibited correspondingly better protonemal growth than spores sown in pure cultures. The presence of the fungus also seemed to halt, but not prevent completely, the loss of viability that accompanied storage of spores. According to Bell (1958), chance contamination of cultures of *Thelypteris palustris* by an unidentified fungus increased their germination. In her studies, Hurel-Py (1944, 1950) noted that not only did fern spores offer little resistance to attack by microorganisms, but they also germinated better. From these results we are led to believe that fungal association imparts to the spore some vital, volatile substances necessary to initiate germination. The nature of this substance is not known, but a compound isolated from a culture of *Fusarium oxysporum* causing change in the morphology of gametophytes of *Polypodium vulgare* has been identified as ethanol (Smith and Robinson, 1969).

IV. Overcoming Spore Dormancy

In the vast majority of fern spores, dormancy develops as a result of the action of subtle chemical processes within the spore protoplast during maturation such that spores will germinate only under conditions which reverse these changes. Rarely have the spore coats themselves been shown to influence dormancy or its breakage. The immediate relevance of our knowledge of the methods of overcoming dormancy in spores to their survival strategy is not obvious since there have as yet been no comparative studies on the evolutionary consequences of dispersal of ferns with dormant and nondormant spores.

In contrast to angiosperm seeds, only a small number of signals have been identified in the breakage of dormancy of fern spores, light being the principal one. Admittedly, in very many cases, aside from the primary need to hydrate air dry spores, the triggering mechanisms are not understood.

A. Light

Recognition of the importance of light in overcoming dormancy of fern spores came despite the fact that early studies (see Sussman, 1965) were somewhat bedeviled by conflicting reports from different laboratories on the light requirements for germination of spores of one and the same species. In contrast to the favorable effects of light on spore germination, inhibition of germination by light has been reported only rarely. In a recent work, Whittier (1973) showed that when axenically cultured spores of *Botrychium dissectum* were exposed to even 5 lux fluorescent light for 12 hours daily, no germination occurred; successful germination of the spores required culture in relatively long periods of darkness. The inhibitory effects of light on spore germination in *B. dissectum* are of some ecological advantage, since prothallia of this species are mycorrhizal in nature and subterranean in habitat. Presumably both these conditions are met by spores which lie buried in the soil where light would not normally permeate. Although spores of several other ferns are known to germinate in complete darkness, there is no unequivocal evidence to show that light inhibits their germination.

In discussing the effects of light on spore germination, we will consider the effects of light quality and light intensity and the photoperiodic effects of light.

1. *Effects of Light Quality*

Although our main concern here is with the effects of visible light, it is worth mentioning that in a few cases studied, ionizing radiations such as α rays (Zirkle, 1932), fast neutrons, X rays (Zirkle, 1932, 1936; Snyder, 1961; Palta and Mehra, 1973), and ultraviolet irradiation (Charlton, 1938) have been found to cause a decrease in spore germination, and delay in the display of germination symptoms in the surviving spores, accompanied by various abnormalities. Sensitivity of spores of *Osmunda regalis* to X irradiation was found to increase throughout the S phase of the mitotic cycle, and in spores which succumbed to radiation damage rhizoid production was suppressed (Howard and Haigh, 1970; Haigh and Howard, 1973). Spores of some ferns are also known to be tolerant to ionizing radiations and germinate normally or even at an accelerated rate following irradiation (Döpp, 1937; Rottmann, 1939; Knudson, 1940; Maly, 1951; Kawasaki, 1954b; Kato, 1964).

The last two decades have seen a remarkable broadening of our understanding of the control of fern spore germination by visible light. The primary stimulus for this has been the startling new discoveries in the field of seed germination and the impressive technological progress achieved in the manufacture of filter systems which permitted isolation of relatively narrow wavelengths of light. Mohr (1956; see also Bünning and Mohr, 1955) determined the action spectrum for promotion and inhibition of germination of fully imbibed spores of *Dryopteris filix-mas* under 24 hours of continuous irradiation with equal intensities of monochromatic light isolated with interference filters. Evaluation of the results 7 days after the beginning of irradiation showed that red light (650–670 nm) was the most effective region of the spectrum in inducing germination; subsequent exposure of red-induced spores to 12 hours of far-red light (733–750 nm) and to some extent blue light (400–450 nm) inhibited their germination. One other point is critical: Namely, the effects of red and far-red light were mutually photoreversible. In other words, germination of spores potentiated by red light was inhibited by far-red light and if further irradiated with red light, spores behaved as if they were not subjected to any far-red light at all, and germinated. These effects are very similar to, if not identical with, those known for germination of lettuce seed, control of flowering in photoperiodically sensitive plants, unfolding of the plumular hook of etiolated plants, and a variety of other photomorphogenetic processes in which the chromoprotein phytochrome participates, to suggest that germination of fern spores is also mediated by the same pigment.

From physiological studies on flowering and seed germination, it has been deduced that phytochrome exists in two photoeversible forms, Pr (red-absorbing form) and Pfr (far-red-absorbing form). Exposure of spores to red light presumably leads to the transformation of existing inactive Pr phytochrome to active Pfr form which is capable of absorbing far-red light. When the pigment in Pfr form, spore germination is favored, whereas in Pr form it is prevented.

With varying degrees of complexity, participation of phytochrome in the control of germination of spores of other ferns has also been documented. In spores of *Osmunda claytoniana* and *O. cinnamomea,* in which breakage of the exine occurs after exposure to brief periods of red light and emergence of the rhizoid requires a long period (4 days) of red or blue light in an atmosphere of CO_2, explanations along the lines of phytochrome control of the former and photosynthetic control of the latter have been sought (Mohr *et al.,* 1964). Other examples of phytochrome involvement in germination are found in spores of *Pteris vittata, Asplenium nidus, Cheilanthes farinosa,* and *Lygodium japonicum* but each of these species is peculiar in its own way and so requires, separate, but brief treatments.

According to Sugai and Furuya (1967, 1968) at least three photoreactions, two of which are phytochrome controlled, are involved in the germination of spores of *Pteris vittata.* The first phytochrome-mediated photoreaction is in the promotion of germination which occurs typically in response to low-energy red light (400 erg/cm^2/second for 5 seconds) and is reversible by far-red light (7.5 kerg/cm^2/second for 16 seconds). Since red-light-induced germination is also inhibited by blue light with peaks at 380 and 440 nm, there is reasonable evidence for the existence of a photoreaction in which a blue-light-absorbing pigment participates (Sugai, 1977). The other phytochrome-mediated reaction is in the recovery of germination from blue light inhibition which is accelerated by red light and annulled by far-red light (Sugai and Furuya, 1968). In further work, Sugai (1970) has shown that addition of a low concentration of ethanol to the medium protected spores from blue-light-induced inhibition of germination. While many details remain to be filled in, the picture which has emerged is that blue light is absorbed by carotenoids which lead to the production of an inhibitor of spore germination. Apparently, the synthesis of this compound is prevented by ethanol which acts as a reducing agent. How far this situation applies to spores of other species in which blue light inhibition of germination has been demonstrated remains to be seen. The ecological significance of the interaction between phytochrome and the blue-absorbing pigment in the germination of spores of *P. vittata* has been discussed by Furuya (1978).

While spore germination in some species is clearly dependent upon the action of phytochrome, there are also cases in which such action is not obvious or is masked by the presence of accessory pigments. For example, spores of some ferns germinate in response to a prolonged red light exposure. Is the ubiquitous pigment phytochrome seemingly active in the germination process? To answer this question, the photoresponses during germination of spores of *Asplenium nidus* may be considered (Raghavan, 1971b). When spores of *A. nidus* are exposed to continuous red light of moderate intensity for 72 hours and then returned to darkness for 8 days, they form both rhizoid and protonemal cell. If spores are held in darkness following 12-hour exposure to red light, they form only rhizoid and no protonemal cell. Irradiation of spores with far-red light immediately after red light reversed the red light effect and led to a significant decrease in the percentage of spores with rhizoid. Spores which initiated rhizoid in response to exposure to red light for 12 hours formed protenemal cell when given additional red light for 12 hours. The effect of the second red light was also annulled by a subsequent irradiation with far-red light. As shown in Fig. 13, promotion of rhizoid and protonema initiation by red light and its inhibition by far-red light are repeatedly reversible. These results are consistent with the hypothesis that outgrowth of rhizoid and initiation of protonemal cell during germination of spores of *A. nidus* are controlled by separate phytochrome reactions perhaps with different Pfr requirements. A requirement for separate Pfr reactions for initiation of rhizoid and protonema is also evident in the phytochrome-controlled germination of spores of *Polypodium aureum* (Spiess and Krouk, 1977). Finally, although spores of *A. nidus* require a 12-hour expo-

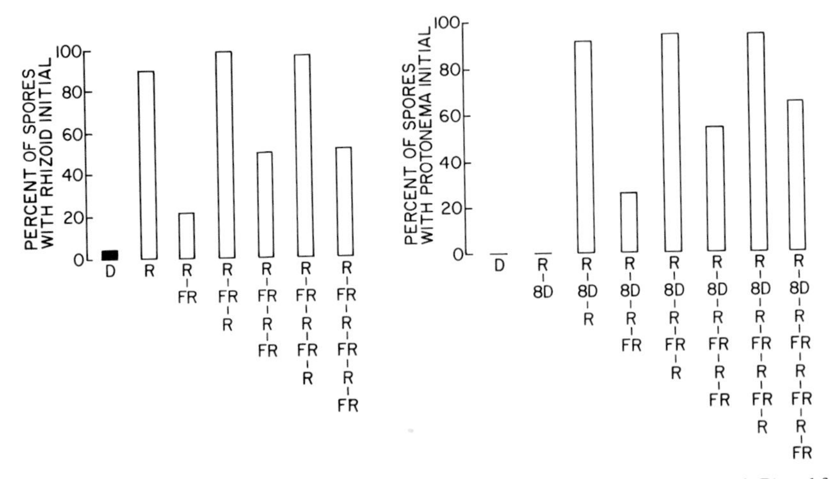

FIG. 13. Left: Initiation of rhizoid in spores of *Asplenium nidus* after exposure to red (R) and far red (FR) light in sequence. Initial exposure to R and FR were for 12 and 48 hours, respectively, and subsequent exposures for 10 minutes and 8 hours, respectively. Spores were examined 8 days after the last light exposure. D, Dark control. Right: Initiation of protonema in spores of *A. nidus*. Spores were exposed to 12 hours red light and then held in darkness for 8 days (8D) before light treatments described in the figure were given. Initial exposures to R and FR in the sequence were for 12 and 48 hours, respectively, and subsequent exposures for 10 minutes and 24 hours, respectively. Spores were examined 4 days after the last light exposure. D, Dark control. (From Raghavan, 1971b.)

sure to red light for rhizoid initiation, the process is potentiated by as little as 2 minutes of red light if they are preirradiated with far-red light for 48 hours. An explanation of these data based on analogous results obtained in the flowering of *Chenopodium rubrum* (Kasperbauer *et al.*, 1963) and germination of the seeds of some Bromeliaceae (Downs, 1964) is that far-red light acts like red light by maintaining a low level of Pfr sufficient to trigger germination in response to a subsequent brief exposure to red light. This example of control of germination of spores of *A. nidus* presents a picture different enough from that encountered in spores of other species to warrant further analysis.

An analysis of the germination of spores of *Cheilanthes farinosa* (Raghavan, 1973) will be discussed as an example in which phytochrome action is masked by the presence of another pigment. Spores of *C. farinosa* are insensitive to far-red light following an initial inductive exposure to red light. Germination, however, is inhibited by blue light given before or after red light. Yet, following blue light inhibition, if spores are repromoted to germinate by red light, far-red light nullifies this effect. Moreover, successive exposures of spores at this stage to far-red and red irradiances lead to mutual photoreversibility of germination. This fulfills the operational criterion for the existence in spores of phytochrome and control of germination may be ascribed to this pigment. Although there is no convincing explanation for the initial lack of photoreversibility of red light effect by far-red light, its importance leads us to some speculations. Probably this is due to the presence of a blue-light-absorbing pigment which prevents in some way the Pfr $\rightarrow$ Pr conversion. On the basis of this hypothesis one would expect that following exposure to blue light, this pigment was removed in the ensuing photochemical reaction and any additional Pfr formed during a subsequent saturating exposure to red light might be available for reversal by far-red light. Since the evidence is at present too tenuous to warrant associating transformations of unknown pigments with light quality, this explanation must await proof from critical action spectrum studies and analysis of pigments isolated from spores.

In spores of *Lygodium japonicum* the mode of action of phytochrome and the blue-absorbing pigment is evidently dependent upon the particular sample of spores used (Sugai *et al.*, 1977). Some lots of spores exhibited typical low-energy red-far–red reversibility and inhibition by blue light. Germination of other lots of spores required a prolonged period of red light or intermittent exposure to red light. The effect of the latter alone was negated by far-red light or by blue light.

According to Jarvis and Wilkins (1973), when irradiated with equal quantum flux density of light in the different spectral bands for 96 hours, spores of *Matteuccia struthiopteris* germinate in all wavelengths between 400 and 750 nm with two major peaks of promotion at 550 and 625 nm and a minor peak at 450 nm. However, promotion of germination by red light was not reversed by far-red light given for 48 hours. Although this may rule out phytochrome involvement in the control of germination, the long period of irradiation required and possible interference by other pigments may also appear to be stumbling blocks in the physiological demonstration of phytochrome action in this system. In fact, the action spectrum for germination is so unusual that it does not show even remote resemblance to the absorption spectrum of any known plant pigments.

Determination of the action spectrum for germination of spores of *Onoclea sensibilis* has revealed maximum quantum effectiveness in the red region and increasing insensitivity in both shorter and longer wavelengths (Towill and Ikuma, 1973). A vexing problem in establishing phytochrome involvement in germination of spores of this species was their apparent escape from inhibition by far-red light administered immediately after red light. Although photoreversibility by far-red light is probably prevented by the presence of other pigments, it might be argued that Pfr formed in response to red irradiation brings about almost immediately the first irreversible step of the germination process, thus allowing spores to escape from inhibition by far-red light.

Thus far only physiological evidence for phytochrome control of germination of fern spores has been documented and in no case has the presence of pigment or its transformation from one form to another been demonstrated in spores. This is surprising since such data are available for photosensitive seeds. The presence of a thick exine with its many layers of sporoderm material which characteristically insulates the spore from the external environment seems to pose the most formidable challenge in the demonstration of phytochrome and its phototransformation in spores by *in vivo* spectrophotometry.

Phytochrome should obviously function as the pigment in an ensuing photochemical reaction, but the manner in which the latter induces spore germination is not known. Extensive studies in other photomorphogenetic systems have led to the formulation of two hypotheses on the mode of action of phytochrome. The evidence, especially from some rapid-action responses, favors the idea that the pigment acts primarily on the membranes of plant cells (Hendricks and Borthwick, 1967), whereas in long-term responses, phytochrome-mediated photoreaction may serve to modulate gene activity leading to the inhibition or derepression of specific enzymes regulating certain morphogenetic processes (Hock and Mohr, 1964; Mohr, 1966). When cytological changes in spores of *Pteris vittata* collected at various times after photoinduction were followed in thin sections it was found that the first sign of germination visible as early as 12 hours after exposure to a saturating dose of red light was hydrolysis of storage protein granules (Raghavan, 1977b). Proteolysis was inhibited by addition of cycloheximide, an inhibitor of protein synthesis, but not by actinomycin D, presumed to inhibit the synthesis of mRNA. From these findings it has been suggested that intact, preexisting mRNA provides templates for the synthesis of enzyme proteins necessary for proteolysis and that activation of mRNA is the principal target of phytochrome action during fern spore germination.

2. *Effects of Light Intensity*

A different picture of light requirement is afforded by spores whose germination is dependent not so much upon the quality of light as on the intensity of light. Although no systematic investigations have been undertaken on the light intensity factor in the germination of fern spores, its importance has been mentioned in passing in the literature. When spores of *Microsorium ensatum* and *Pyrrosia lingua* were exposed to various photoperiods of light of three different intensities, germination increased with intensity of light and duration of exposure (Isikawa and Oohusa, 1954). Generally speaking, at lower intensities of light or when light is limiting, it takes much longer for a given lot of spores to reach maximum germination than at higher ones, as shown by Pietrykowska (1963) for spores of *Matteuccia struthiopteris*.

3. *Photoperiodic Effects*

A third category of light response is that of spores whose germination is influenced not by the quality or intensity of light, but by alternations between light and dark periods, i.e., by photoperiod. The existence of photoperiodism in the germination of fern spores was recognized by Isikawa (1954) who found that spores of *Athyrium niponicum* germinated maximally when they were exposed to continuous light. Interruption of the light regime by dark periods led to decreased germination. In another study (Isikawa and Oohusa, 1954), spores of a number of other ferns were also found to behave similarly in germinating in long photoperiods. A striking case of promotion of germination by a dark period in a photoperiodic regime is that of *Dryopteris crassirhizoma* which germinated best when exposed to a single long (18 to 24 hours) period of illumination. The total duration of light required for maximum germination was reduced when a dark period of 15 to 20 hours was inserted between light periods of 3 hours each (Isikawa and Oohusa, 1956).

Although there is a requirement for high-intensity light for germination of spores *Matteuccia struthiopteris,* the situation is complicated by the involvement of a photoperiodic effect. A short period of illumination at high intensity results in much less germination in spores of this species than when light is given for a longer period (Pietrykowska, 1962). Similar results have been reported in the germination of spores of *Acrostichum aureum* (Eakle, 1975).

B. Chemicals

Germination, whether initiated in a quiescent or dormant spore, triggers growth processes in a rapid burst of metabolic activity. We have seen earlier that one component of the external environment, namely, the quality of light, induces germination of dormant spores of many species of ferns. From the physiological point of view, interest in the environmental control of spore germination focuses on the pigments involved and in the resulting photochemical reaction. But it is clear that a number of partial processes must occur before the photochemical reaction is consummated in the final outcome of germination. Some understanding of these processes has been gained by the use of chemicals which substitute for light and promote spore germination in the dark.

Stimulation of dark germination of spores of *Anemia phyllitidis* with gibberellic acid reported by Schraudolf (1962) was the first clear case of a specific chemical circumvention of light requirement in spore germination. Whereas spores sown in the basal medium failed to germinate in the dark, addition of a low concentration of gibberellic acid induced complete dark germination. In spores of other ferns, addition of gibberellic acid does not completely substitute for light requirement, and as documented in *Lygodium japonicum* and *Mohria caffrorum* only 10–20% of the spores germinate in the dark in the presence of gibberellic acid (Näf, 1966; Weinberg and Voeller, 1969a). The spectacular effect of gibberellic acid on *A. phyllitidis* spores is quite different from the effects of this hormone in enhancing spore germination in light, as reported in *Athyrium filix-femina,* (von Witsch and Rintelen, 1962) or in increasing vegetative growth of gametophytes by pretreating spores with gibberellic acid, as reported in *Dryopteris filix-mas* (Knobloch, 1957).

Schraudolf's initial observations have been confirmed and extended by other investigators (Näf, 1966; Weinberg and Voeller, 1969a,b; Raghavan, 1976, 1977a). From their work, Weinberg and Voeller (1969b) have offered proof to show that normal germination of spores in light is moderated by the production of a gibberellin-like substance. When a concentrate from the culture medium used for germination of *Anemia phyllitidis* spores was tested in the well-known barley endosperm bioassay for gibberellic acid, it consistently exhibited gibberellin-like activity, whereas a similar concentrate from dark-imbibed spores did not. Light-induced germination of spores was also inhibited by AMO-1618, a selective inhibitor of gibberellin biosynthesis in some microorganisms and angiosperms. Finally, when fully imbibed spores were given suboptimal levels of light and gibberellic acid, an additive effect on germination was noted. This could happen only if small amounts of gibberellic acid are produced in light regimes insufficient to induce full germination.

How gibberellic acid induces germination of the spore is unknown. Since the first sign of gibberellin action seen in spores as early as 12 hours after application of the hormone is hydrolysis of storage protein granules one guess is that it might act by activating mRNAs coding for particular enzymes, especially those needed for proteolysis (Raghavan, 1977a). This of course is one of the several hypotheses that may be formulated to explain the mechanism of action of gibberellic acid in fern spore germination. Whatever hypothesis eventually turns out to be the valid one, the simplified system of the fern spore will make the hypothesis useful to explain the mode of action of gibberellin in other more complex systems.

Näf (1966) found that a liquid culture medium of 7-week-old gametophytes of *Anemia phyllitidis* (*Anemia* medium) completely canceled the light requirement for germination of spores of this species and to some extent of spores of *Lygodium japonicum*. The culture filtrates from gametophytes of *Anemia* and *Lygodium* are relatively species selective since *Lygodium* medium, which induces a weak germination response in spores of *Lygodium*, is not reciprocally active against *Anemia* spores. Recently, Endo *et al.* (1972) have shown that the substance with the dark germination-inducing ability in *Anemia* medium is identical to antheridogen. Antheridogen belongs to the class of diterpenes with a carbon skeleton structurally related to that of gibberellin (Nakanishi *et al.*, 1971). Since antheridogen acts like a weak gibberellin in several bioassays (Sharp *et al.*, 1975), the striking parallelism between the actions of these substances in circumventing the light requirement for spore germination is scarcely surprising. The antheridogen from *Lygodium* gametophytes has not been successfully isolated and characterized, and so its relationship to gibberellic acid remains to be elucidated.

No stimulatory effects have been attributed to other chemicals or hormones in the germination of fern spores. According to Spiess and Krouk (1977), the addition of gibberellic acid and kinetin together, or of cyclic AMP, sensitized spores of *Polypodium aureum* so that a smaller dose of red light was sufficient to induce their germination. These compounds were also effective in preventing far-red light reversal of red-light-induced germination of spores.

V. Biochemical Cytology and Biochemistry of Spore Germination

The potential of appropriately induced spores to germinate obviously depends upon the interplay of a complex pattern of metabolic processes and regulatory mechanisms that govern cell differentiation. On the one hand, storage food reserves present in the spore have to be degraded into simple compounds to provide energy and substrates for the protoplast to divide. Equally pressing is the need for synthesis of proteins and for biogenesis of organelles such as mitochondria for the catabolic activity associated with degradation of food reserves and of chloroplasts for initiating photosynthetic activity as soon as the first visible signs of germination occur. Not long after, cell division ensues with all the complex biochemical processes entailed, including protein and DNA synthesis and assembly of the mitotic spindle. Unlike in seeds of angiosperms, where major catabolic and anabolic activities during germination are segregated to morphologically different tissues, there is no conceivable division of labor in fern spores. Here, as will be seen later, both degradative and synthetic processes preparatory to germination take place in the same cell.

In this section an attempt is made to summarize what is known about the breakdown and metabolism of storage reserves, respiratory changes, and nucleic acid and protein metabolism during germination of spores of a selected number of species of ferns. This restriction in the number of species to be discussed is of necessity due to the fact that our knowledge of the biochemical aspects of germination of fern spores has been in general rather superficial and has lagged behind advances made in the study of germination of seeds of angiosperms and spores of certain lower cryptogams. The difficulty of obtaining sufficient material for experimentation is the primary cause of this situation, but other factors are also probably involved. In any case, hopefully, a few common principles that may apply generally to spores of a large number of ferns will emerge from this account. The initial events of germination, before division of the protoplast occurs, are of particular biochemical significance, because it is then that molecular events which determine the fate of the daughter cells presumably occur.

A. Metabolic Changes during Germination

Light microscopic histochemistry has made important contributions to the study of metabolic changes occurring during germination of spores and has thrown some light on the manner in which storage granules and their hydrolytic products are segregated into the daughter cells formed. These studies will be reviewed first because of the close relationship between structure and biochemical activities.

1. *Histochemical Changes*

The application of histochemical methods combined with enzyme digestions to the study of cell metabolism during germination was initiated by Gantt and Arnott (1965) in their work on spores of *Matteuccia struthiopteris*. In this work it was found that although starch granules were completely absent in dormant spores, they appeared in small numbers in spores imbibed in the dark and in great abundance in spores exposed to a dormancy-breaking light stimulus. Starch accumulation was later followed by extensive degradation of storage protein granules and by the time of the first mitosis, massive accumulation of starch and complete hydrolysis of protein granules were characteristically observed. The inverse relationship between starch synthesis and protein degradation might in theory suggest a connection between the two, but this remains to be demonstrated. Nonetheless, the abundance of protein granules in the dormant spore and their hydrolysis during germination support the contention that they are the primary storage products which provide raw materials for germination.

Another histochemical change associated with germination of spores of *Matteuccia struthiopteris* was an increase in the concentration of cytoplasmic RNA surrounding the nucleus at the site of the presumptive rhizoid. RNA content remained high in the rhizoid initial following division of the spore, but as rhizoid elongated, RNA content decreased with a concomitant increase in cytoplasmic proteins. Moreover, in contrast to the prothallial cell with its rich endowment of starch, protein, and lipid granules, the rhizoid was characterized by a sparsity of food reserves. These observations lead to the inference that differentiation of rhizoid and prothallial cell in the germinating spore is correlated with gradients in the distribution of food reserves, nucleic acids, and proteins.

During gibberellic acid-induced germination of spores of *Anemia phyllitidis,* it is possible to distinguish a period when hydrolysis of protein granules begins followed by a period of synthesis of starch (Raghavan, 1976). In fully imbibed spores, the first sign of germination seen as early as 12 hours after addition of the hormone was hydrolysis of storage proteins. Only after hydrolysis of protein granules had advanced considerably did PAS-staining granules begin to appear in a part of the spore cytoplasm destined to form the protonemal cell (Fig. 14). A sustained degradation of protein reserves beginning as early as 12 hours after exposure to a saturating dose of red light was also observed during phytochrome-controlled germination of spores of *Pteris vittata* (Raghavan, 1977b). Proteolysis was relatively rapid and during the next 12 to 24 hours most of the protein granules disappeared leaving the spores as empty shells criss-crossed by numerous cytoplasmic strands.

During later stages of germination of spores of *Anemia phyllitidis* hydrolysis of storage granules continued in the large basal cell whereas accumulation of PAS-staining granules was restricted to the apical protonemal cell (Fig. 15). In a

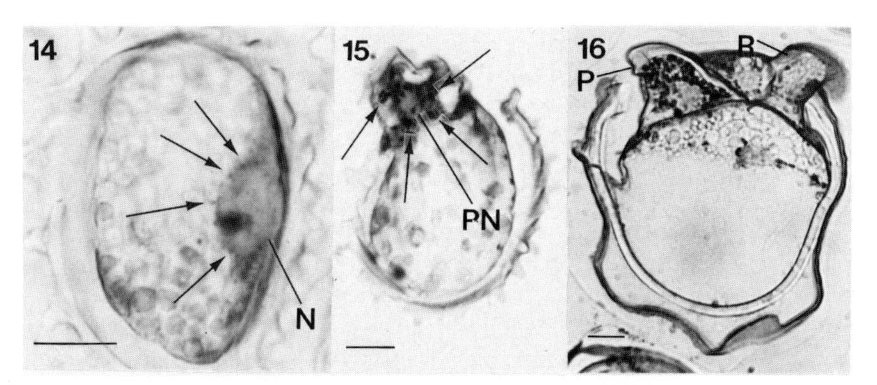

FIG. 14. Section of a germinating spore of *Anemia phyllitidis* showing the first appearance of PAS-positive granules (arrows) in the cytoplasm. N, Spore nucleus. Bar = 10 μm. (From Raghavan, 1976.)

FIG. 15. Section of later stage of a germinating spore of *A. phyllitidus* showing PAS-staining granules (arrows) surrounding the nucleus of the protonemal cell (PN). Bar = 10 μm. (From Raghavan, 1976.)

FIG. 16. Section of a germinating spore of *Lygodium japonicum* showing PAS-staining granules in the protonemal cell (P); none are present in the rhizoid (R). Bar = 10 μm. (From Rutter and Raghavan, 1978.)

similar manner, in spores of *Lygodium japonicum,* the capacity for accumulation of PAS-staining material was confined to the protonemal cell formed by a division of the apical cell (Fig. 16) whereas the rhizoid derived from the sister cell of the apical cell did not stain with PAS (Rutter and Raghavan, 1978). In spores of *Blechnum spicant,* membrane-bound lipid bodies which form the main storage reserves with proteins disintegrate into a large number of small granules during imbibition preparatory to germination (Gullvåg, 1969, 1971; Beisvåg, 1970). According to Fraser and Smith (1974), in spores of *Polypodium vulgare* which have their storage reserves exclusively as lipids, microbidies are found in close contact with disintegrating lipid bodies. This has given rise to the thought that microbidies are akin to glyoxysomes that contain enzymes for conversion of fatty acids to succinate. The association of microbodies with lipids during germination is well known in cotyledons and endosperm of seeds which store lipids. An important and general conclusion from these studies is that irrespective of the type of spores and the nature of storage reserves they contain, morphogenesis of germination is accompanied by visible changes in the relative proportions of certain cellular components which may reflect accompanying biochemical functions.

The metabolic activity of germinating spores must also be reflected in the presence of an array of enzymes, although the task of characterizing them has hardly begun. The only published work on enzyme histochemistry of germinating spores is that of Olsen and Gullvåg (1971) on acid phosphatase activity in *Matteuccia struthiopteris.* These investigators have shown that the enzyme activity resides primarily in the protein storage granules and within the ER which is in direct contact with the protein granules. Since high acid phosphatase activity coincides with the period of intense metabolic activity in the spore, it has been suggested that liberation of phosphate which is bound to the storage protein granules is one of the functions of the enzyme.

2. *Biochemical Changes*

Quantitative changes in the composition of cellular components during germination have been investigated in spores of *Pteridium aquilinum* (Raghavan, 1970) and *Onoclea sensibilis* (Towill and Ikuma, 1975b). As seen in Fig. 17, in dark-germinating spores of *P. aquilinum,* starting from about 12 hours after sowing, there was a decrease in protein content which dominated the pattern of protein changes and growth for the next 48 hours. During this period, initiation and growth of rhizoid were the primary morphogenetic manifestations of germination. During the period from 72 to 96 hours after sowing when initiation and growth of the protonemal cell occurred, a small increase in the protein content was registered. RNA content of the spore, on the other hand, showed a slow increase from sowing until initiation of the protonemal cell. Protein content of spores of *O. sensibilis* exposed to a germination-inducing light stimulus also decreased with progress of germination, whereas in dark-imbibed spores, the

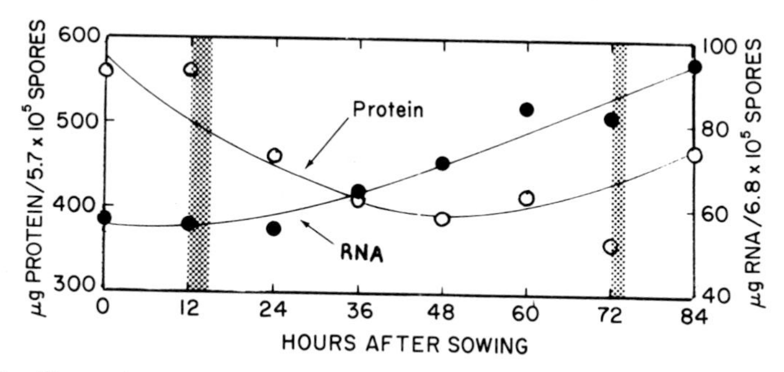

FIG. 17. Changes in protein and RNA contents of spores of *Pteridium aquilinum* during germination. Stippled portions indicate periods of red light treatment. (From Raghavan, 1970.)

protein content remained constant. Concomitant with the decrease in protein content, the concentration of free amino acids determined by the ninhydrin test increased. A similar relationship was seen between hydrolysis of sucrose and accumulation of starch in spores during germination.

In spores of *Gymnogramme sulphurea* (Guervin, 1972), *Polypodium vulgare* (Robinson *et al.*, 1973), *Anemia phyllitidis* (Gemmrich, 1977), and *Thelypteris dentata* (Seilheimer, 1978) which contain abundant lipophilic materials, germination is accompanied by the breakdown of endogenous triglycerides and a resynthesis of several new classes of di- and triglycerides. In *P. vulgare, de novo* synthesis of new triglycerides following the breakdown of reserve lipids has been established by incorporation of labeled acetate predominantly into the neutral lipid fractions. The neutral lipid fractions synthesized are the same ones found in the mature sporophytic fronds, suggesting the presence of enzyme systems for synthesis of the sporophytic type of lipids even in the very young gametophyte (Robinson *et al.*, 1973). These data indicate that the overall trend in quantitative changes in the cellular components of germinating spores is largely similar to what is predicted by histochemical methods. Generally, during germination, stored proteins, carbohydrates, and lipids are broken down and as the germinated spore becomes an autotrophic photosynthetic unit, a renewed accumulation of metabolites is initiated. Simultaneously, enzymes which catalyze interconversions of amino acids appear in the spore cytoplasm (Lever, 1971; Gemmrich, 1975). However, no data on the biosynthetic pathways of the different metabolites during germination of fern spores are available. Thus there is a wide open field for biochemical work on spores of ferns using modern analytical techniques to provide a picture interpretable in metabolic terms.

3. *Respiratory Changes*

Since germination is an energy-requiring process it is to be expected that oxygen will be indispensible for the normal germination of fern spores. The characteristics of the aerobic oxidation system during different phases of germination of spores of *Onoclea sensibilis* has received a particularly thorough study in recent years at the hands of Towill and Ikuma (1975a). These investigators showed that exposure of spores to a light stimulus increased the rate of respiration by about 30–50% over that prior to irradiation. It may be no coincidence that increased respiratory activity was noticed before accelerated carbohydrate metabolism, suggesting that energy from respiratory processes was channeled for carbohydrate breakdown and resynthesis.

In another study, Towill and Ikuma (1975c) observed that ATP content of spores increased steadily during dark imbibition, whereas, following irradiation, after a transient increase, it declined. This may be partly due to the provision for ATP utilization in starch synthesis during germination. Clearly, the succession of events during germination of spores is intimately controlled by a chain of oxidative reactions, but it is difficult to attribute the effects of light to any particular segment of the reaction chain. It is possible that the same activated pigment independently induces multiple processes such as germination, respiratory activity, and carbohydrate metabolism.

B. Nucleic Acid and Protein Synthesis

It is a widely held notion that cell differentiation to a considerable degree is determined by the transcription and translation of specific information-carrying molecules. To the extent this is true during fern spore germination, it must take place against a background of concurrent changes in the overall rates of DNA and protein biosynthesis. For this reason, an analysis of nucleic acid and protein synthesis during spore germination assumes great significance, insofar as with this information, one can determine whether morphogenesis is inextricably coupled to changing patterns of macromolecule synthesis. Since DNA is the well-established information-containing genetic material of the cell, this account will logically begin with a consideration of DNA synthesis during germination.

1. *DNA Synthesis*

When fully imbibed spores are exposed to a germination-inducing stimulus, in addition to biochemical reactions such as respiration and hydrolysis of storage reserves, cellular processes that culminate in the division of the spore nucleus are also initiated. Generally, after hydrolysis of storage granules is underway, the spore nucleus acquires granularity and comes to resemble a normal nucleus in morphology. Breakdown of storage granules probably provides the energy to prepare the nucleus for division. As studied by autoradiography of [^{3}H]thymidine incorporation (Raghavan, 1976, 1977b; Rutter and Raghavan, 1978) the essential precondition for actual division of the nucleus is DNA synthesis appropriate for the haploid genome.

One striking feature of DNA synthetic activity during fern spore germination is the occurrence of extranuclear DNA synthesis. This was first demonstrated in *Anemia phyllitidis* in which DNA appeared to be synthesized in both the nucleus and the immediately surrounding cytoplasm of the undivided spore (Raghavan, 1976). Cytoplasmic DNA synthesis continued for several hours in the protonemal cell after it was formed, but as this cell elongated, a strictly nuclear incorporation of [^{3}H]thymidine was observed (Figs. 18–20). In spores of *Pteris vittata,* cytoplasmic DNA synthesis was initiated after nuclear DNA synthesis was underway and remained quite vigorous even after the latter ceased following nuclear division (Figs. 21–23). With the initiation of the protonemal cell, [^{3}H]thymidine incorporation into the spore cytoplasm decreased and completely stopped after full emergence of the protonemal cell (Raghavan, 1977b). In a similar way, extranuclear DNA synthesis observed during germination of spores of *Lygodium japonicum* showed an increasing gradient from the basal spore cell to the protonemal cell (Rutter and Raghavan, 1978). From these results we must assume that each species can utilize local differences in cell metabolism to varying degrees during DNA synthesis preparatory to division. Since in all species investigated cytoplasmic DNA synthesis reaches its climax just before the germinating spore begins to acquire chloroplasts, there is probably some relationship between extranuclear DNA synthesis and chloroplast biogenesis in the spore.

Depending upon the species, the visible hallmark of germination of the spore is the appearance of the protonemal cell, rhizoid, or both outside the exine. Is DNA synthesis a prerequisite for activities connected with the visible sign of

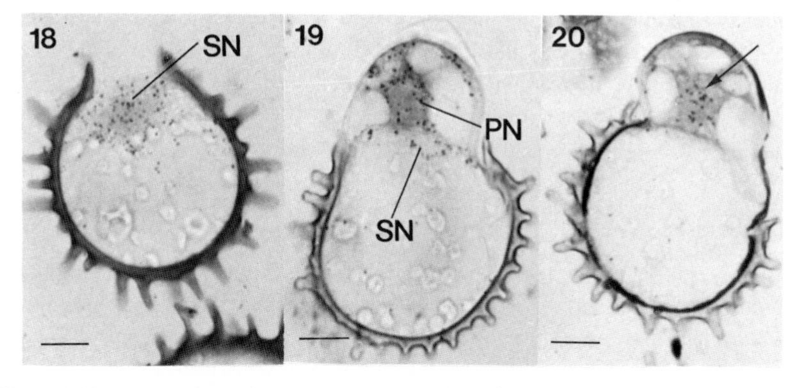

FIGS. 18–20. Autoradiographs showing incorporation of [^{3}H]thymidine into germinating spores of *Anemia phyllitidis*. Bars = 10 μm. Fig. 18. Incorporation of the label into both nucleus (SN) and cytoplasm of the spore before division. Fig. 19. Section of a spore, several hours after the first division, showing [^{3}H]thymidine incorporation into the cytoplasm of the protonemal cell. The spore nucleus (SN) as well as the nucleus of the protonemal cell (PN) have not started a second round of DNA synthesis. Fig. 20. Incorporation of [^{3}H]thymidine exclusively into the nucleus (arrow) of a fully formed protonemal cell. (From Raghavan, 1976.)

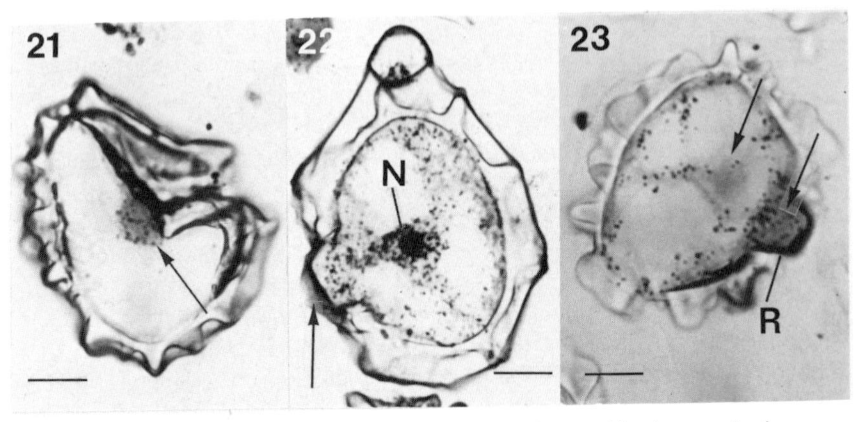

FIGS. 21–23. Autoradiographs showing incorporation of [^{3}H]thymidine into germinating spores of *Pteris vittata*. Bars = 10 μm. Fig. 21. Incorporation of the label into nucleus (arrow); cytoplasm unlabeled. Fig. 22. Heavy [^{3}H]thymidine incorporation into cytoplasm of a germinating spore; nucleus (N) is also labeled. The exine is open (arrow). Fig. 23. Absence of [^{3}H]thymidine incorporation into the newly formed daughter nuclei (arrows); cytoplasmic incorporation persists. R, Rhizoid. (From Raghavan, 1977b.)

germination? In *Anemia phyllitidis,* the only example in which this has been examined, spores were found to germinate in a medium containing hydroxyurea, an inhibitor of DNA synthesis and indirectly of mitosis. What was remarkable about the effect of hydroxyurea was that hydrolysis of storage reserves and enlargement of protoplast associated with germination occurred in spores incubated in a medium containing gibberellic acid and a dose of the drug sufficient to inhibit DNA synthesis in the nucleus or cytoplasm. From this one gains the impression that no unique DNA is made during germination which can thus occur even when chromatin is not replicated. In terms of the role of gibberellic acid in inducing germination, we might exclude the possible action of hormone at the level of DNA synthesis (Raghavan, 1977a).

2. *RNA Synthesis*

In spores of all species examined, the RNA synthetic system has been found to be totally inactive until several hours after administration of a germination-inducing stimulus. In autoradiographs of spores of *Anemia phyllitidis* fed with [^{3}H]uridine, a precursor of RNA synthesis, the first silver grains are seen in about 36 to 48 hours after addition of gibberellic acid, just at about the same time as DNA and protein synthesis is initiated (Raghavan, 1976). Thus, macromolecule synthesis during germination appears to be coupled closely in time and coincident with the appearance of granularity in the nucleus. Some differences were noted in the magnitude of incorporation of the isotope into the nucleolus, chromatin, and cytoplasm of spores at early and late stages of germination.

Generally, [^{3}H]uridine was incorporated almost exclusively into the nucleolus and chromatin of the undivided spore nucleus, whereas, coincident with the formation of the protonemal cell, increasing levels of incorporation were observed into the cytoplasm of the latter (Figs. 24, 25). Similarly, a study of RNA synthesis during germination of spores of *Pteris vittata* revealed a decreased nucleolar labeling with progress of germination (Raghavan, 1977b). These results are consistent in implying a changing pattern of RNA metabolism as the germinating spore becomes autotrophic. In *A. phyllitidis,* a low level of RNA synthesis is always found in the original spore cell, rhizoid, and basal cells of the protonemal filament even after they cease division (Raghavan, 1976). The ribosomes thus generated permit protein synthesis in these cells commensurate with their continued metabolic activity.

How RNA synthetic activity controls specific cytological events of germination has been studied in dark-germinating spores of *Pteridium aquilinum* (Raghavan, 1970, 1971a). Here, by administering doses of red light at two different times during a germination time course of 84 hours, a temporal separation of phases involving initiation of rhizoid (36–48 hours after sowing), elongation of rhizoid (48–60 hours after sowing), initiation of protonemal cell (74–77 hours after sowing), and elongation of protonemal cell (81–84 hours after sowing) was obtained. The synchrony in the display of germination symptoms seen

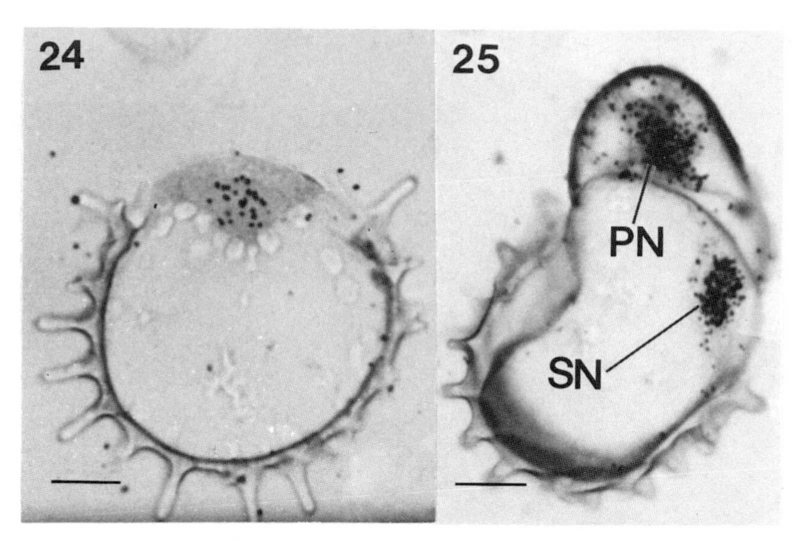

FIGS. 24 and 25. Autoradiographs showing [^{3}H]uridine incorporation into germinating spores of *Anemia phyllitidis*. Bar = 10 μm. Fig. 24. Strictly nuclear incorporation before division. Fig. 25. Spore after forming the protenmal cell. Incorporation of isotope into the nucleus of the spore cell (SN) and into the nucleus (PN) and cytoplasm of the protonemal cell is seen. (From Raghavan, 1976.)

in populations of spores also afforded a means of lining morphological and biochemical events in time. When spores were pulse-labeled with [^{3}H]uridine at different times after sowing, precursor accumulation into RNA increased slowly beginning at 36 hours until the end of the time course. From sucrose gradient centrifugation analysis of phenol-extracted RNA, it was clear that transcription products formed during initiation of rhizoid were confined almost exclusively to 4 S and lighter RNA species, and counts in the relatively heavier regions of the gradient were barely above background (Fig. 26). On the other hand, during rhizoid elongation, there was an accelerated incorporation of the label into all parts of the gradient including peaks roughly coincident with 23 S and 16 S RNA. The main differences thus pertain to types of RNA synthesized during stages of rhizoid initiation and elongation.

From experiments using actinomycin D, it appears that synthesis of new mRNA is not necessary for the initiation of rhizoid in spores of *Pteridium aquilinum*. When spores were incubated in concentrations of actinomycin D as high as 100 mg/liter, elongation of the rhizoid was completely blocked, but the exine had ruptured at the site of the presumptive rhizoid indicating that events preparatory to rhizoid initiation were unaffected. Moreover, for any specific period of time, or specified concentration of actinomycin D, the degree of inhibi-

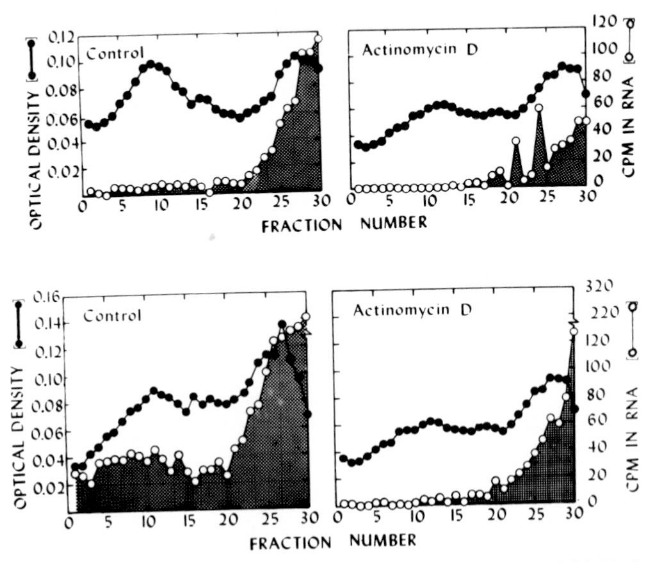

Fig. 26. Top: Sedimentation patterns of RNA synthesized during rhizoid initiation in spores of *Pteridium aquilinum* in the presence (right) or absence (left) of actinomycin D (100 mg/liter). The stippled areas indicate the extent of radioactivity. Bottom: Sedimentation patterns of RNA synthesized during rhizoid elongation in the presence (right) or absence (left) of actinomycin D (100 mg/liter). The stippled areas indicate the extent of radioactivity. (From Raghavan, 1970.)

tion of elongation of rhizoid was greater than the degree of inhibition of its initiation. This suggests a need for continued mRNA synthesis for rhizoid elongation in germinated spores. Evidently, while rhizoid initiation programmed without simultaneous RNA synthesis is taking place, new transcripts for the subsequent elongation of rhizoid are being synthesized. When spores were pulse-labeled with [^{3}H]uridine in the presence of actinomycin D during periods of rhizoid initiation and elongation, it became clear from sucrose gradient centrifugation profiles that the drug completely eliminated any heavy RNA whatsoever with messenger properties synthesized during these periods (Fig. 26). Since actinomycin D prevents rhizoid elongation without preventing its initiation, this reinforces the earlier conclusion that there is no commitment of newly synthesized mRNA for rhizoid initiation. Synthesis of actinomycin D-sensitive heavy RNA appears to be indispensible for rhizoid elongation although one is not certain to what extent the heavy new RNA synthesized is of the messenger type.

Differential effects of actinomycin D on the initiation and elongation of the protonemal cell in spores of *Pteridium aquilinum* were similar to the effects of the drug on rhizoid elongation and initiation. Sucrose gradient centrifugation showed that some heavy, probably nonribosomal RNA and light RNA were synthesized during initiation of the protonemal cell and actinomycin D virtually eliminated the synthesis of all heavy RNA. However, the drug was not nearly as effective in preventing synthesis of heavy RNA when applied during protonemal cell elongation. Since initiation and growth of the protonemal cell are associated with the appearance of chloroplasts, the possibility cannot be excluded that the fraction of heavy RNA insensitive to actinomycin D represents the product of an autonomous RNA replication mechanism of chloroplasts. However, this issue does not obscure the evidence that at least in *P. aquilinum* spores developmental events like rhizoid initiation and protonemal cell initiation can take place in the absence of simultaneous mRNA synthesis. When the protonemal cell begins to elongate, the germinated spore may be considered to have assumed complete genomic control of its own subsequent development.

Actinomycin D has been shown to prevent gibberellic acid-induced germination of spores of *Anemia phyllitidis* (Raghavan, 1977a). Histological examination showed that hydrolysis of storage granules and the first nuclear division delimiting the protonemal cell were insensitive to actinomycin D, although elongation of the protonemal cell was inhibited by the drug (Fig. 27). In contrast, division of the nucleus was inhibited in spores germinating to a stage of discernible protonemal initial in a medium containing 5-fluorouracil (Fig. 28), an analog of uracil known to block the synthesis of rRNA and tRNA in certain plants (Gressel and Galun, 1966; Key, 1966). From these results it seems that a fraction of RNA apparently synthesized in the presence of actinomycin D is sensitive to 5-fluorouracil and that the synthesis of 5-fluorouracil-sensitive RNA is necessary for initiating division delimiting the protonemal cell. Spores germinating in a

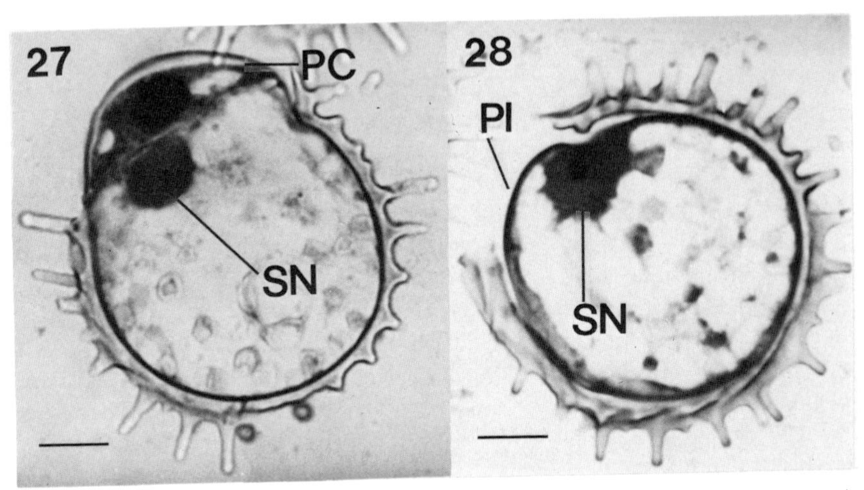

FIG. 27. Section of a spore of *Anemia phyllitidis* germinated in a medium containing 200 mg/liter actinomycin D. PC, Protonemal cell; SN, spore nucleus. Bar = 10 μm. (From Raghavan, 1977a.)

FIG. 28. Section of a spore of *Anemia phyllitidis* germinated in a medium containing 50.0 mg/liter 5-fluorouracil. The spore nucleus (SN) has not divided, although exine is open and protonemal initial (PI) is visible. Bar = 10 μm. (From Raghavan, 1977a.)

medium containing 5-flourouracil incorporated [^{3}H]thymidine into the nucleus indicating that the block to division occurred during the postsynthetic period of the cell cycle. Although photoinduced spores of *Pteris vittata* failed to show visible signs of germination in a medium containing 100–200 mg/liter actinomycin D, failure of the drug to prevent hydrolysis of storage granules and incipient rupture of the exine suggests that new RNA synthesis is not necessary only for a limited period in the early stage of germination (Raghavan, 1977b). It follows that these early events of germination, however limited they may be structurally, cytologically, or biochemically, must be largely programmed, operated, and controlled by developmental systems which were present in the dry spore at the time of sowing. A role for stored mRNA in coding for the first proteins of germination of fern spores will be discussed in a later part of this article.

3. *Requirement for Protein Synthesis*

Following imbibition of water for an appropriate length of time, the spore is poised on the threshold of growth that serves to induce division in the quiescent protoplast. Since at least the initial phase of growth takes place against a background of continuous loss of protein content, a direct evaluation of new protein synthesis for germination is difficult. However, histochemical and biochemical approaches combined with the use of specific inhibitors have made it possible to demonstrate the involvement of protein synthesis in the germination of spores of several ferns.

In spores of *Anemia phyllitidis,* autoradiographically detectable incorporation of [^{3}H]leucine, indicative of protein synthesis, occurred at about 36 to 48 hours after addition of gibberellic acid and coincident with the initiation of RNA synthesis (Raghavan, 1976). Incorporation of the isotope was noted mainly in the nucleus with little label in the immediately surrounding cytoplasm of the spore. Later, as the nucleus migrated toward one side of the spore establishing the future site of the protonemal cell, protein synthesis was particularly heavy in the cytoplasm in this region (Figs. 29, 30). It thus seems that the presumptive site of initiation of the protonemal cell in the single-celled spore is quite different from the rest of the spore with respect to its biochemical activities. The pattern of protein synthesis observed in spores of *Pteris vittata* was very similar to that in *A. phyllitidis.* [^{3}H]Leucine was incorporated into both nucleus and cytoplasm of the undivided spore; following nuclear division, an exclusive cytoplasmic incorporation occurred (Raghavan, 1977b). As stated earlier, hydrolysis of storage reserves was noticeable in fully imbibed spores of both *A. phyllitidis* and *P. vittata* as early as 12 hours after administration of a germination-inducing stimulus. The significance of the absence of any [^{3}H]leucine incorporation into spores before breakdown of storage granules is initiated remains uncertain, but it does seem apparent that the metabolism of the spore at this stage is channeled primarily into pathways not involving the use of exogenous amino acids.

In an alternate approach, the requirement for protein synthesis for hydrolysis of storage reserves during germination of spores of *Anemia phyllitidis* was inves-

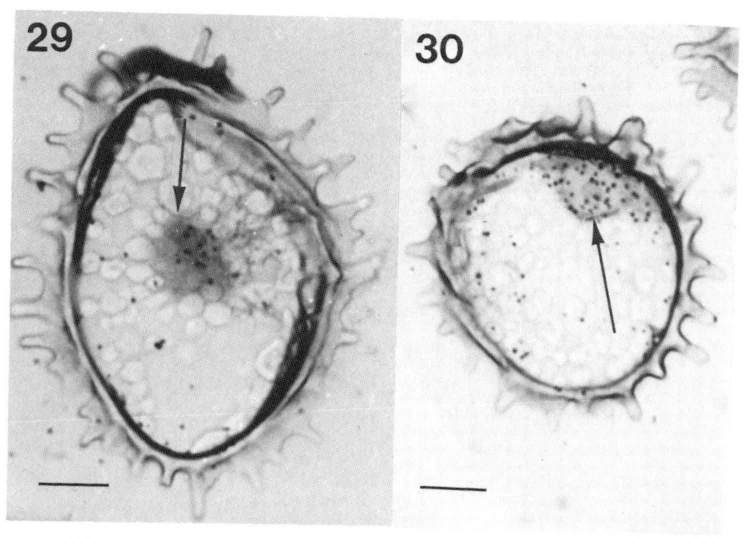

FIGS. 29 and 30. Autoradiographs showing incorporation of [^{3}H]leucine into germinating spores of *Anemia phyllitidis*. Bars = 10 μm. Fig. 29. Beginning of [^{3}H]leucine incorporation into spore nucleus (arrow). Fig. 30. Incorporation of the label into the nucleus (arrow) and cytoplasm of a spore at an early stage of germination. (From Raghavan, 1976.)

tigated using cycloheximide. Continuous incubation of fully imbibed spores of *A. phyllitidis* in gibberellic acid and cycloheximide (0.5 mg/liter) inhibited germination in the entire population of spores when examined 4 days later. This was also associated with complete inhibition of protein synthesis, as seen by autoradiography of [^{3}H]leucine incorporation. In sections, spores treated with the drug exhibited none of the usual symptoms of germination such as hydrolysis of protein granules and were very much similar in structure to the dry spores. It thus seems that hydrolysis of storage reserves and subsequent initiation of protonemal initial and rhizoid during germination of spores of *A. phyllitidis* are associated with the synthesis of cycloheximide-sensitive proteins (Raghavan, 1977a). The sustained degradation of protein granules that occurred during phytochrome-controlled germination of spores of *Pteris vittata* was also found to be dependent upon the synthesis of proteins since incubation of photoinduced spores in cycloheximide (0.1 mg/liter) not only inhibited germination but also did not result in any detectable hydrolysis of storage granules (Raghavan, 1977b). Very recently, Nagy *et al.* (1978) have shown that a protein with a molecular weight of 4.2×10^4 is synthesized in spores of *Dryopteris filix-mas* in response to red light exposure, even before any visible morphological signs of germination appear.

The ease with which the critical phases of germination can be separated in synchronously developing spores has been advantageously used to analyze the requirement for protein synthesis during germination of spores of *Pteridium aquilinum* (Raghavan, 1970). When spores were pulse-labeled with [^{14}C]leucine there was no detectable incorporation of isotope into proteins until about 36 hours when breakage of exine and protrusion of rhizoid ensued. The labeled amino acid, however, entered the spore readily throughout the early phase of germination, despite the absence of any incorporation during this period. By applying cycloheximide (0.07 mg/liter) at different times after sowing, it was concluded that proteins essential for rhizoid initiation were synthesized during a sensitive period beginning 26 hours after sowing. In this experiment, it was found that introduction of the drug any time up to 23 hours inhibited rhizoid initiation in the entire population of spores when examined at 72 hours after sowing. If the drug was applied between 26 and 48 hr after sowing about 15–40% of the spores had visible rhizoid initials, whereas addition of the drug at 48 hours or later after sowing had little effect on the percentage of spores with visible rhizoids. Results of addition of cycloheximide to 48-hour-old spores which had already formed rhizoid initials indicated that their elongation depended upon continued synthesis of proteins. Cycloheximide has also been used to treat spores at other stages of germination. Thus, in spores poised on the threshold of forming the protonemal cell, administration of the drug effectively prevented this morphogenetic event (Raghavan, 1971a).

According to Towill and Ikuma (1975b) in spores of *Onoclea sensibilis* which require exposure to red light for germination, protein synthesis during the preinduction phase was essential for maintenance of sensitivity to irradiation. Addition of cycloheximide to fully imbibed spores for 45 minutes was sufficient to induce 50% inhibition of germination whereas introduction of the drug at zero time after exposure of spores to red light induced 50% inhibition of germination in 30 minutes. These workers also showed that the kinetics of recovery of spores from preinduction and postinduction treatments with cycloheximide were very similar, suggesting that the same enzymes with half-lives of ca. 30 to 45 minutes were required for both phases of germination. When spores were treated with cycloheximide at different times after irradiation, escape from inhibition yielded a biphasic curve with a sharp deflection at 10 hours. Taken together, these interesting approaches to the study of protein synthesis during fern spore germination indicate that a short-lived enzyme probably initiates the germination process as early as 10 hours after photoinduction. These data are of course consistent with cytological studies of germinating spores of other species which have unequivocally shown that enzymatic breakdown of storage reserves occurs well before any detectable protein synthesis.

Finally, it may be mentioned at this point that fully functional enzymes are present in dry spores of certain ferns. Recently, Gemmrich (1979) showed that *de novo* synthesis of lipids occurred within 1 hour of dark imbibition of spores of *A. phyllitidis* in the absence of a germination inducing stimulus. Since lipid synthesis requires enzymes, it is presumed that they are preexisting in the dry spore and activated during imbibition. However, in view of the inability of dark imbibed spores to germinate, it is doubtful whether stored enzymes play a role in the germination process.

4. *The First Proteins of Germination*

In the wake of recent conceptual advances in molecular genetics and RNA and protein biosynthesis, thoughts about the biochemical aspects of germination of fern spores have been profoundly influenced by ideas of gene action, information transfer, and metabolic regulation. Sustained investigations on germination of seeds of angiosperms and spores of fungi and bacteria have convincingly demonstrated the existence of preformed, template-active mRNA which carry the codes for the first proteins of germination as these reproductive units are aroused from their dormant state. The similarities observed between angiosperm seeds and fern spores such as the requirement for a specific pretreatment to break dormancy and the presence of abundant storage materials suggest the existence of similar types of control mechanisms during their germination.

Here we will attempt to integrate our present knowledge about the molecular basis for change of gene expression as the developmental pattern of a single-celled fern spore is switched on at germination. According to the well-established tenets of molecular biology, gene action results from intracellular or extracellular signals that derepress particular segments of the genome. A tangible expression of gene action in a cell is the synthesis of new proteins. However, the derepressed segment of DNA of a cell nucleus does not itself direct the synthesis of proteins. Rather, once a portion of the genome is derepressed, information contained therein is transferred to mRNA molecules, and gene copies in the form of mRNA are then translated to form specific proteins. Proteins function as active enzymes or may require further activation. Against this background of established principles we may presume that at some early stage of germination of spores, the genes coding for enzymes which hydrolyze the storage granules are derepressed. Therefore, great interest lies in establishing how contact of spores with water or how administration of a germination-inducing stimulus to them regulates the availability of mRNAs which are translated into enzyme proteins. If germination trigger regulates the availability of mRNAs to code for enzyme proteins for the first events of germination, are such mRNA molecules already present in the spore, or are they formed anew from biosynthetic processes started after imbibition? If the essential ingredients for protein synthesis are already present in the dry spore, how are they kept in an inactive form until spores establish contact with water or are subjected to a germination-inducing stimulus? At this stage, definitive answers to these questions cannot be given. Yet, the questions represent a strategy in which a dormant spore is dissected and analyzed in detail at a crucial period when the first symptoms of change are initiated.

On the basis of discussion in the preceding pages, it is abundantly clear that there is a requirement for protein synthesis for initiation of the very early events of germination in fern spores. One of the most direct assays of gene activity for the synthesis of proteins would be to analyze mRNAs which code for them. This, however, presents a formidable task in view of the number of proteins involved and the difficulty of selecting specific mRNAs. A widely used indirect method is to inhibit mRNA synthesis by drugs which are relatively specific for the task and to determine the extent of morphogenesis and macromolecule synthesis occurring when gene activity is suppressed. It is in this context that actinomycin D has been used in the experiments described earlier. It is now known that this drug has many side effects aside from inhibiting mRNA synthesis and so it is impossible to treat the results of actinomycin D experiments as other than suggestive or corroborative of data obtained by other means.

With this due caution dictated by the possible and real shortcomings of actinomycin D use, it is nonetheless striking that in media supplemented with relatively high concentrations of this drug, spores are able to germinate up to the stage of complete hydrolysis of starch granules and incipient exine breakage as in *Pteris vittata* (Raghavan, 1977b), or up to the initiation of the rhizoid or protonemal cell as in *Pteridium aquilinum* (Raghavan, 1970) and *Anemia phyllitidis* (Raghavan, 1977a), respectively. Equally striking is the fact that in all three species studied, total protein synthesis seems not to be greatly affected by the drug even at doses where all mRNA synthesis is cut off. From these results it would seem that the ungerminated spore contains, as a superfluous holdover from sporogenesis, enough mRNA to code for proteins of the first events of germination.

Results on the effects of actinomycin D and 5-fluorouracil on protein synthesis during phases of rhizoid initiation and elongation in spores of *Pteridium aquilinum* also argue strongly for programming of early protein synthesis on preformed mRNA. DeMaggio and Raghavan (1972) electrophoretically separated ^{14}C-labeled proteins synthesized during early stages of germination of *P. aquilinum* spores into five peaks. When actinomycin D or 5-fluorouracil was added to the medium during rhizoid initiation and radioactively labeled proteins were separated by gel electrophoresis, the original peaks were still present with a somewhat lowered amplitude. This confirms that for the most part, gene expression during rhizoid initiation as reflected in the synthesis of proteins is not the result of control exerted at the level of transcription. If this were the case, some of the peaks should have been eliminated by actinomycin D treatment. Actinomycin D administered to spores during rhizoid elongation completely eliminated two peaks in the electrophoretic protein profile and nearly erased two others. This contrasts with the effect of 5-fluorouracil which, like actinomycin D, inhibited rhizoid elongation, but preserved all five peaks in the protein profile. Presumably, proteins synthesized at peaks insensitive to actinomycin D are formed on stored mRNA, whereas those at peaks eliminated by the drug are synthesized on newly formed mRNA. The persistence of radioactive peaks of proteins even after treatment of spores during rhizoid elongation with 5-fluorouracil suggests that the effect of the drug on this morphogenetic event is relatively nonspecific. These experiments thus provide additional proof of the independence of early protein synthesis from new transcription during spore germination.

Assuming that the dormant spore harbors a stockpile of mRNAs in an inactive or sequestered state, the means by which they are activated for use by the translational apparatus of the germinating spore remains unknown. It has been suggested that external signals for germination such as actinic radiation or a specific hormone activates stored mRNAs to provide information translatable into proteins (Raghavan, 1977a,b). An experiment reported by Raghavan (1977a) renders it likely that polyadenylation is the control mechanism that activates mRNA during gibberellic acid-induced germination of spores of *Anemia phyllitidis* and that the hormone specifically induces the synthesis of poly(A) RNA. The experiment involves the use of cordycepin (3′-deoxyadenosine), an analog of adenosine, believed to block post-transcriptional polyadenylation of mRNA. The rationale for this experiment is that processing of mRNA in *A. phyllitidis* spores which involves post-transcriptional addition of adenylic acid residues to the 3′-OH end of RNA molecules is probably delayed until spores are subjected to gibberellic acid action. While addition of cordycepin

(8.0 mg/liter) anytime up to 24 hours after transfer of spores to gibberellic acid inhibited their germination, later additions of the drug led to escape of virtually the entire population of spores from inhibition. Since a 24-hour pretreatment of spores in gibberellic acid allows their complete germination subsequently in the absence of the hormone, this minimum period of hormone action required to trigger germination lies will within the period of sensitivity to cordycepin when gibberellic acid-induced polyadenylation of mRNA is taking place. In cotton embryogenesis (reviewed by Walbot *et al.*, 1975) there is suggestive evidence to show that the hormone abscisic acid prevents processing of mRNA earmarked for storage until a certain period in the development of the ovule. In view of reports on the presence of abscisic acid in ripening and mature spores of *A. phyllitidis* (Cheng and Schraudolf, 1974; Bürcky, 1977), it is possible that this hormone interacts in a similar way with the mRNA processing system of spores. However they may operate, the mechanisms by which stored mRNA is activated must be partially specific for each species, since addition of cordycepin to photoinduced spores of *Pteris vittata* did not inhibit their germination (Raghavan, unpublished). The activation of mRNA is possibly related to the internal location of the molecule, and for this reason a complete explanation of the role of stored mRNA in fern spore germination will also require an understanding of the localization of the molecule in a cytological as well as a molecular sense.

VI. Concluding Comments

An attempt has been made in this review to cover the various aspects of cytology, physiology, and biochemistry of germination of spores of ferns. The subject of physiology of fern spore germination is especially vast, with several contributions having been made prior to the exciting period when photomorphogenetic studies in angiosperm seeds blossomed and general principles applicable to spore germination emerged. To do proper justice to the more recent studies, coverage of earlier investigations has been less than complete. As pointed out in this article, the overall response of spores to light quality and chemicals has many striking similarities to the behavior of seeds. This leads one to hope that further analysis of the regulatory aspects of germination of single-celled fern spores will provide significant information in understanding the mode of action of extracellular factors in the initiation of development in quiescent and dormant systems in the plant kingdom.

In a cytological and molecular sense, germination processes are triggered soon after spores are hydrated or after they have been subject to a dormancy-breaking treatment. At least in some cases, there is evidence that molecular mechanisms probably involve activation of stored mRNAs and that the complex processes of cell morphogenesis during germination are in part dependent upon this stored mRNA. This seems likely to be a reflection of the need for the germinated spore to insure its survival before synthetic machinery for the production of complex macromolecules becomes functional. Additional control mechanisms will no doubt be discovered as investigations on the germination of fern spores continue.

References

Abeles, F. B., and Lonski, J. (1969). *Plant Physiol.* **44,** 277.
Atkinson, L. R. (1962). *Phytomorphology* **12,** 264.
Beisvåg, T. (1970). *Grana* **10,** 121.
Bell, P. R. (1958). *Ann. Bot.* **22,** 503.
Bierhorst, D. W. (1975). *Am. J. Bot.* **62,** 319.
Bünning, E., and Mohr, H. (1955). *Naturwissenschaften* **42,** 212.
Bürcky, K. (1977). *Z. Pflanzenphysiol.* **85,** 181.
Charlton, F. B. (1938). *Am. J. Bot.* **25,** 431.
Chen, C.-Y., and Ikuma, H. (1979). *Plant Physiol.* **63,** 704.
Cheng, C.-Y., and Schraudolf, H. (1974). *Z. Pflanzenphysiol.* **71,** 366.
Clarke, H. M. (1936). *Am. J. Bot.* **23,** 405.
Conway, E. (1949). *Proc. R. Soc. Edinburgh* **B63,** 325.
Courbet, H. (1955). *C.R. Hebd. Acad. Sci.* **241,** 441.
Courbet, H. (1957). *C.R. Hebd. Acad. Sci.* **244,** 107.
DeMaggio, A. E., and Raghavan, V. (1972). *Exp. Cell Res.* **73,** 182.
Döpp, W. (1937). *Beitr. Biol. Pflanzen* **24,** 201.
Downs, R. J. (1964). *Phyton* **21,** 1.
Dyer, A. F., and Cran, D. G. (1976). *Ann. Bot.* **40,** 757.
Eakle, T. W. (1975). *Am. Fern J.* **65,** 94.
Edwards, M. E. (1977). *Plant Physiol.* **59,** 756.
Edwards, M. E., and Miller, J. H. (1972). *Am. J. Bot.* **59,** 458.
Endo, M., Nakanishi, K., Näf, U., McKeon, W., and Walker, R. (1972). *Physiol. Plant.* **26,** 183.
Endress, A. G. (1974). *Ann. Bot.* **38,** 877.
Esashi, E., and Leopold, A. C. (1969). *Plant Physiol.* **44,** 1470.
Fisher, R. W., and Miller, J. H. (1975). *Am. J. Bot.* **62,** 1104.
Fisher, R. W., and Miller, J. H. (1978). *Am. J. Bot.* **65,** 334.
Fraser, T. W., and Smith, D. L. (1974). *Protoplasma* **82,** 19.
Furuya, M. (1978). *Bot. Mag. (Tokyo) Spec. Issue* **1,** 219.
Gantt, E., and Arnott, H. J. (1965). *Am. J. Bot.* **52,** 82.
Gastony, G. J. (1974). *Am. J. Bot.* **61,** 672.
Gastony, G. J., and Tryon, R. M. (1976). *Am. J. Bot.* **63,** 738.
Gemmrich, A. R. (1975). *Phytochemistry* **14,** 353.
Gemmrich, A. R. (1977a). *Plant Sci. Lett.* **9,** 301.
Gemmrich, A. R. (1977b). *Phytochemistry* **16,** 1044.
Gemmrich, A. R. (1979). *Z. Pflanzenphysiol.* **91,** 317.
Gifford, E. M., Jr., and Brandon, D. D. (1978). *Am. Fern J.* **68,** 71.
Gistl, R. (1928). *Ber. Dtsch. Bot. Ges.* **46,** 254.
Gressel, J., and Galun, E. (1966). *Biochem. Biophys. Res. Commun.* **24,** 162.
Guervin, C. (1972). *Rev. Gen. Bot.* **79,** 109.
Gullvåg, B. M. (1969). *Phytomorphology* **19,** 82.

Gullvåg, B. M. (1971). *In* "Pollen and Spore Morphology. IV" (G. Erdtman and P. Sorsa, eds.), pp. 252–295. Almquist & Wiksell, Stockholm.
Haigh, M. V., and Howard, A. (1973). *Radiat. Bot.* **13,** 37.
Hendricks, S. B., and Borthwick, H. A. (1967). *Proc. Natl. Acad. Sci. U.S.A.* **58,** 2125.
Hevly, R. H. (1963). *J. Ariz. Acad. Sci.* **2,** 164.
Hock, B., and Mohr, H. (1964). *Planta* **61,** 209.
Howard, A., and Haigh, M. V. (1970). *Int. J. Radiat. Biol.* **18,** 147.
Hurel-Py, G. (1944). *C.R. Hebd. Acad. Sci.* **218,** 326.
Hurel-Py, G. (1950). *Rev. Gen. Bot.* **57,** 637.
Hurel-Py, G. (1955). *C.R. Hebd. Acad. Sci.* **241,** 1813.
Hutchinson, S. A., and Fahim, M. (1958). *Ann. Bot.* **22,** 117.
Isikawa, S. (1954). *Bot. Mag. (Tokyo)* **67,** 51.
Isikawa, S., and Oohusa, T. (1954). *Bot. Mag. (Tokyo)* **67,** 193.
Isikawa, S., and Oohusa, T. (1956). *Bot. Mag. (Tokyo)* **69,** 132.
Jaffe, L. F., and Etzold, H. (1962). *J. Cell Biol.* **13,** 13.
Jarvis, S. J., and Wilkins, M. B. (1973). *J. Exp. Bot.* **24,** 1149.
Javalgekar, S. R. (1960). *Bot. Gaz.* **122,** 45.
Kasperbauer, M. J., Borthwick, H. A., and Hendricks, S. B. (1963). *Bot. Gaz.* **124,** 444.
Kato, Y. (1957a). *Phyton* **9,** 25.
Kato, Y. (1957b). *Bot. Mag. (Tokyo)* **70,** 258.
Kato, Y. (1957c). *Cytologia* **22,** 328.
Kato, Y. (1964). *New Phytol.* **63,** 21.
Kato, Y. (1973). *Cytologia* **38,** 117.
Kaur, S. (1961). *Sci. Cult.* **27,** 347.
Kawasaki, T. (1954a). *J. Jpn. Bot.* **29,** 201.
Kawasaki, T. (1954b). *J. Jpn. Bot.* **29,** 294.
Ketring, D. L., and Morgan, P. W. (1969). *Plant Physiol.* **44,** 326.
Ketring, D. L., and Morgan, P. W. (1971). *Plant Physiol.* **47,** 488.
Key, J. L. (1966). *Plant Physiol.* **41,** 1257.
Knobloch, I. W. (1957). *Am. Fern J.* **47,** 134.
Knudson, L. (1940). *Bot. Gaz.* **101,** 721.
Lever, M. (1971). *Phytochemistry* **10,** 2995.
Lloyd, R. M., and Klekowski, E. J., Jr. (1970). *Biotropica* **2,** 129.
Lovett, J. S. (1975). *Bacteriol. Rev.* **39,** 345.
Lugardon, B. (1974). *Pollen Spores* **16,** 161.
Maly, R. (1951). *Z. Indukt. Abstamm. Vererbungsl.* **83,** 447.
Marengo, N. P. (1956). *Am. Fern J.* **46,** 97.
Marengo, N. P. (1973). *Bull. Torrey Bot. Cl* **100,** 147.
Marengo, N. P., and Badalamente, M. A. (1978). *Am. Fern J.* **68,** 52.
Mehra, P. N. (1952). *Ann. Bot.* **16,** 49.
Miller, J. H. (1968). *Bot. Rev.* **34,** 361.
Miller, J. H., and Greany, R. H. (1976). *Science* **193,** 687.
Mohr, H. (1956). *Planta* **46,** 534.
Mohr, H. (1966). *Photochem. Photobiol.* **5,** 469.
Mohr, H., Meyer, U., and Hartmann, K. (1964). *Planta* **60,** 483.
Momose, S. (1942). *J. Jpn. Bot.* **18,** 49, 139, 187.
Näf, U. (1966). *Physiol. Plant.* **19,** 1079.
Nagy, A. H., Paless, G., and Vida, G. (1978). *Biol. Plant.* **20,** 193.
Nakanishi, K., Endo, M., and Näf, U. (1971). *J. Am. Chem. Soc.* **93,** 5579.
Nayar, B. K., and Kaur, S. (1968). *J. Palynol.* **4,** 1.

Nayar, B. K., and Kaur, S. (1969). *Can. J. Bot.* **47,** 395.
Nayar, B. K., and Kaur, S. (1971). *Bot. Rev.* **37,** 295.
Nishida, M. (1962). *J. Jpn. Bot.* **37,** 193.
Nishida, M. (1965). *J. Jpn. Bot.* **40,** 161.
Okada, Y. (1929). *Sci. Rep. Tohoku Imp. Univ. Ser. 4. Biol.* **4,** 127.
Olsen, L. T., and Gullvåg, B. M. (1971). *Grana* **11,** 111.
Pal, N., and Pal, S. (1963). *Bot. Gaz.* **124,** 405.
Palta, H. K., and Mehra, P. N. (1973). *Radiat. Bot.* **13,** 155.
Pietrykowska, J. (1962). *Acta Soc. Bot. Pol.* **31,** 437.
Pietrykowska, J. (1963). *Acta Soc. Bot. Pol.* **32,** 677.
Pringle, R. B. (1961). *Science* **133,** 284.
Pringle, R. B. (1970). *Plant Physiol.* **45,** 315.
Raghavan, V. (1970). *Exp. Cell Res.* **63,** 341.
Raghavan, V. (1971a). *Exp. Cell Res.* **65,** 401.
Raghavan, V. (1971b). *Plant Physiol.* **48,** 100.
Raghavan, V. (1973). *Plant Physiol.* **51,** 306.
Raghavan, V. (1976). *Am. J. Bot.* **63,** 960.
Raghavan, V. (1977a). *J. Cell Sci.* **23,** 85.
Raghavan, V. (1977b). *J. Exp. Bot.* **28,** 439.
Robinson, P. M., Smith, D. L., Safford, R., and Nichols, B. W. (1973). *Phytochemistry* **12,** 1377.
Rogers, L. M. (1923). *Bot. Gaz.* **75,** 75.
Rosendahl, G. (1940). *Planta* **31,** 597.
Rottmann, W. (1939). *Beitr. Biol. Pflanzen* **26,** 1.
Rutter, M. R., and Raghavan, V. (1978). *Ann. Bot.* **42,** 957.
Schraudolf, H. (1962). *Biol. Zentralbl.* **81,** 731.
Schraudolf, H. (1966). *Plant Cell Physiol.* **7,** 277.
Schraudolf, H. (1967). *Z. Pflanzenphysiol.* **56,** 375.
Seilheimer, A. V. (1975). *Am. J. Bot.* **62** (Suppl.), 47 (abstract).
Seilheimer, A. V. (1978). *Am. Fern J.* **68,** 67.
Sharp, P. B., Keitt, G. W., Clum, H., and Näf, U. (1975). *Physiol. Plant.* **34,** 101.
Smith, C. W., and Yee, R. N. S. (1975). *Am. Fern J.* **65,** 13.
Smith, D. L., and Robinson, P. M. (1969). *New Phytol.* **68,** 113.
Smith, D. L., and Robinson, P. M. (1975). *New Phytol.* **74,** 101.
Snyder, C. R. (1961). *Am. Biol. Teach.* **23,** 506.
Sossountzov, I. (1961). *C.R. Hebd. Soc. Biol.* **155,** 1006.
Spiess, L. D., and Krouk, M. G. (1977). *Bot. Gaz.* **138,** 428.
Stephan, J. (1929). *Jahrb. Wiss. Bot.* **70,** 707.
Stokey, A. G. (1951). *Am. Fern J.* **41,** 111.
Sugai, M. (1970). *Dev. Growth Differ.* **12,** 13.
Sugai, M. (1971). *Plant Cell Physiol.* **12,** 103.
Sugai, M., and Furuya, M. (1967). *Plant Cell Physiol.* **8,** 737.
Sugai, M., and Furuya, M. (1968). *Plant Cell Physiol.* **9,** 671.
Sugai, M., Takeno, K., and Furuya, M. (1977). *Plant Sci. Lett.* **8,** 333.
Sussman, A. S. (1965). *In* "Encyclopedia of Plant Physiology" (W. Ruhland, ed.), Vol. 15/2, pp. 933–1025. Springer-Verlag, Berlin and New York.
Towill, L. R. (1978). *Plant Physiol.* **62,** 116.
Towill, L. R., and Ikuma, H. (1973). *Plant Physiol.* **51,** 973.
Towill, L. R., and Ikuma, H. (1975a). *Plant Physiol.* **55,** 150.
Towill, L. R., and Ikuma, H. (1975b). *Plant Physiol.* **55,** 803.
Towill, L. R., and Ikuma, H. (1975c). *Plant Physiol.* **56,** 468.

Twiss, E. M. (1910). *Bot Gaz.* **49,** 168.
Van Etten, J. L., Dunkle, L. D., and Knight, R. H. (1976). *In* "The Fungal Spore. Form and Function" (D. J. Weber and W. M. Hess, eds.), pp. 243–300. Wiley (Interscience), New York.
von Witsch, H., and Rintelen, J. (1962). *Planta* **59,** 115.
Walbot, V., Harris, B., and Dure III, L. S. (1975). *In* "Developmental Biology of Reproduction" (C. L. Markert, ed.), pp. 165–187. Academic Press, New York.
Weinberg, E. S., and Voeller, B. R. (1969a). *Am. Fern J.* **59,** 153.
Weinberg, E. S., and Voeller, B. R. (1969b). *Proc. Natl. Acad. Sci. U.S.A.* **64,** 835.
Whittier, D. P. (1973). *Can. J. Bot.* **51,** 1791.
Yamasaki, N. (1954). *Cytologia* **19,** 249.
Zirkle, R. E. (1932). *J. Cell. Comp. Physiol.* **2,** 251.
Zirkle, R. E. (1936). *Am. J. Roentgenol.* **35,** 230.

INTERNATIONAL REVIEW OF CYTOLOGY, VOL. 62

Immunocytochemical Localization of the Vertebrate Cyclic Nonapeptide Neurohypophyseal Hormones and Neurophysins

K. DIERICKX

Department of Embryology and Comparative Histology, Rijksuniversiteit, Gent, Belgium

ISBN 0-12-364462-3

I. Introduction

The vertebrate hypophysis consists of two main parts: the adenohypophysis and the neurohypophysis (Fig. 1). The adenohypophysis can, generally, be divided into the pars tuberalis, the pars distalis, and the pars intermedia. The neurohypophysis also consists of three parts: the median eminence, the neural stalk, and the neural lobe (lobus nervosus).

The *neural lobe* consists (1) of an accumulation of axon terminals ending on blood capillaries, and (2) of specialized neuroglial cells, the pituicytes. The pituicytes are diffusely spread in the interstitium of the network of axon terminals. The great majority[1] of the axon terminals of the neural lobe are the storage site of (1) two different cyclic nonapeptide hormones, commonly called the neurohypophyseal hormones (Table I), together with (2) their cysteine-rich carrier proteins, the neurophysins. In response to a variety of physiological stimuli, these axon terminals release neurohypophyseal hormone(s), together with neurophysin(s), into pericapillary spaces surrounding the blood capillaries of the neural lobe. Via these blood capillaries, the hormones and neurophysins reach the general blood circulation.

The *median eminence* constitutes (a part of) the wall of the infundibular cavity (infundibular recess) which, in higher vertebrates, is a funnel-shaped extension of the third ventricle. Like the neural lobe, it consists of nerve fibers, pituicytes, and blood vessels. The median eminence can be divided into an internal region and an external region. The *internal* region mainly consists of coarse nonmyelinated nerve fibers running, via the neural stalk, to the neural lobe of the hypophysis. Its inner, ventricular surface is covered with ependyma. The *external* region mainly consists of fine nonmyelinated nerve fibers ending on blood capillaries of the primary capillary plexus of the hypophyseal portal vascular system. In mammals, its external surface is covered with the pars tuberalis of the adenohypophysis.

As represented in Fig. 1, most nerve fibers of the neural lobe are the terminal

[1]Recently, immunocytological studies have shown that the neural lobe also contains a minority of other hormone-containing nerve fibers, e.g., somatostatin fibers (see Hökfelt *et al.*, 1975).

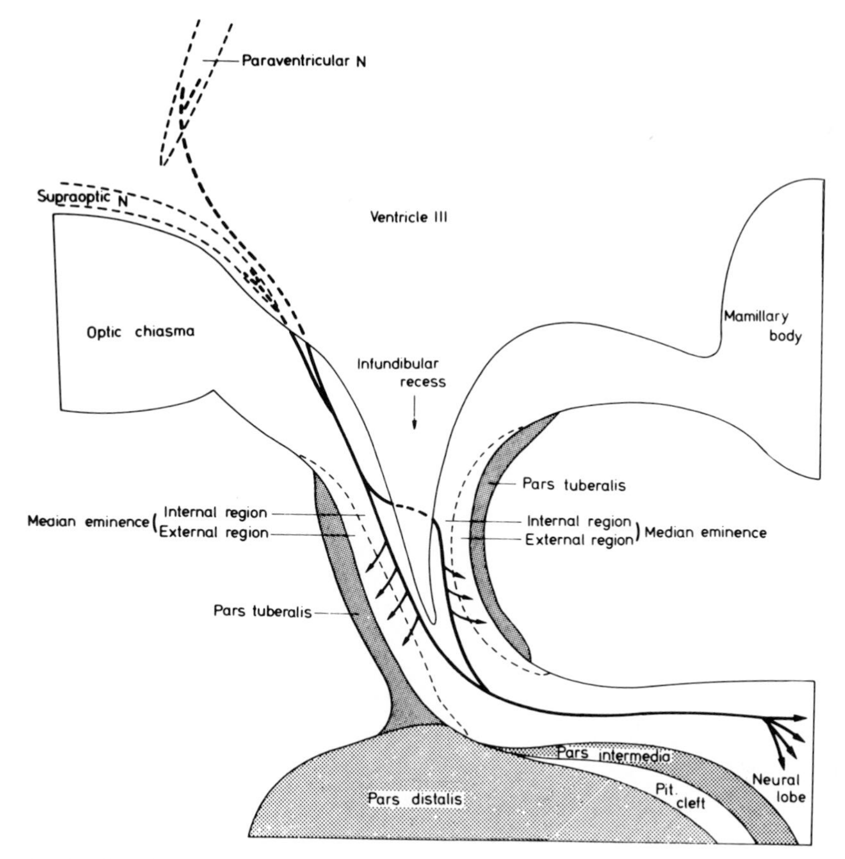

FIG. 1. Drawing representing a sagittal section of the mammalian hypothalamohypophyseal region, on which the paraventricular and supraoptic nuclei are projected. The heavy lines indicate the axonal pathways from the supraoptic and paraventricular nuclei, forming the supraopticohypophyseal tract running through the internal region of the median eminence and ending in the neural lobe. The arrows in the median eminence represent neurohypophyseal hormone-containing pathways to blood capillaries of the external region.

parts of the unmyelinated axons of large neurons located in the so-called magnocellular neurosecretory nuclei of the hypothalamus. In the anamniote vertebrates, these nuclei are represented by the two hypothalamic magnocellular *preoptic* nuclei, located bilaterally in the wall of the prechiasmatic part of the third ventricle (Fig. 25). In mammals, as well as in the other amniote vertebrates, each magnocellular preoptic nucleus of the anamniote vertebrates has evolved into two nuclei: the supraoptic nucleus and the paraventricular nucleus (Figs. 1, 20, 21, and 22).

In addition to a rich capillary network, nerve fibers, and neuroglial cells, the

TABLE I
SIMILARITIES AND DIFFERENCES IN AMINO ACID COMPOSITION OF SIX NEUROHYPOPHYSEAL HORMONES MENTIONED IN THE TEXT

Sequence	Hormone
1. Neutral oxytocin-like hormones (positions 3, 4, 8)	
Cys-Tyr-Ile-Gln-Asn-Cys-Pro-Leu-Gly(NH_2)	Oxytocin
Cys-Tyr-Ile-Gln-Asn-Cys-Pro-Ile-Gly(NH_2)	Mesotocin
Cys-Tyr-Ile-Ser-Asn-Cys-Pro-Ile-Gly(NH_2)	Isotocin
2. Basic vasopressin-like hormones	
Cys-Tyr-Ile-Gln-Asn-Cys-Pro-Arg-Gly(NH_2)	Arginine–vasotocin
Cys-Tyr-Phe-Gln-Asn-Cys-Pro-Arg-Gly(NH_2)	Arginine–vasopressin
Cys-Tyr-Phe-Gln-Asn-Cys-Pro-Lys-Gly(NH_2)	Lysine–vasopressin

hypothalamic magnocellular neurosecretory nuclei mainly consist of an accumulation of large cell bodies which show the combined cytological features of neurons and of peptide hormone-producing endocrine cells. The unmyelinated axons of these endocrine, or neurosecretory, neurons run toward the postchiasmatic floor of the third ventricle, where they form the supraopticohypophyseal tract (in anamnia, the preopticohypophyseal tract). This nerve tract traverses the internal region of the median eminence. Via the neural stalk, it ends in the neural lobe, where its swollen axon terminals form the main mass of this organ (Fig. 1).

It has been shown that the neurohypophyseal hormones, together with neurophysins, are packed in so-called elementary neurosecretory granules, which can be visualized in the cytoplasm of the cell bodies with the aid of transmission electron microscopy: They consist of a dense core enveloped by a membrane. From the cell body, the granules are transported in packets, along the axons, to the neural lobe, where they are stored in the swollen neurosecretory axon terminals, located around the blood capillaries. The neurosecretory granules are too small to be visible in the light microscope. But, their contents, the so-called neurosecretory material, can be visualized in the light microscope with the aid of staining methods depending upon the presence of high concentrations of protein-bound disulfide groups, which are present in both the neurohypophyseal hormones and the neurophysins. Moreover, as will be described later, specific staining of the hormones and neurophysins can be obtained with immunocytochemical methods. The *function* of the endocrine neurons of the hypothalamic magnocellular neurosecretory system is to synthesize, transport, and secrete neurohypophyseal hormones and neurophysins in response to a variety of physiological stimuli.

On the basis of results of isotope studies, a precursor model has been proposed for the neurohypophyseal hormone biosynthesis. According to this model, the biosynthesis of the *peptide bonds* of these cyclic nonapeptide hormones would occur solely in the cell body, on ribosomes, via pathways common to the biosynthesis of other peptide chains. Initially, the hormone would be constructed as part of a large precursor molecule, a protein, in which the hormone would be present in a bound, biologically inactive form. The liberation of the biologically active hormone from the precursor molecule would take place (1) during the formation of the neurosecretory granules in the Golgi apparatus and also (2) during the "maturation" of the neurosecretory granules while they are transported, in the axons, to the neural lobe.

Although analytic and isotope studies have indicated that hormone and neurophysin biosyntheses are intimately related events, so far, it is unknown whether or not neurophysin biosynthesis occurs via a similar mechanism. But, on the basis of results of several investigations, it has been proposed that neurophysin(s) and hormone(s) are made as part of the same precursor molecule(s). The activity of the hypothalamic magnocellular endocrine neurons is controlled by humoral (e.g., osmotic) and neurogenic stimuli. The influence of osmotic stimuli on the biosynthesis and release of the neurohypophyseal hormones was established long ago. But, so far, the nature and location of the receptors of these stimuli, the so-called osmoreceptors, remain under discussion. There is also a dearth of information about the afferent pathways and the afferent synapses transmitting the neurogenic stimuli to the magnocellular endocrine neurons. The incomplete knowledge about the afferent synapses is, in part, due to our very incomplete information on the dendrites of the hypothalamic magnocellular endocrine neurons. The nature of the cellular events which intervene between reception of physiological stimuli and hormone biosynthesis also remains a largely unsolved problem. On the other hand, progress has been made regarding the nature of the cellular processes involved in hormone and neurophysin release at the level of the neurosecretory axon terminals in the neural lobe. Electrophysiological studies support the view that nerve impulses, traveling down the supraopticohypophyseal tract, trigger the release of hormone. Other studies have indicated that hormone release occurs following a chain of events initiated by membrane depolarization. Calcium ions and contractile proteins appear to play an important role in the phenomenon of release.

Experiments carried out both *in vivo* and *in vitro* have demonstrated that the release of neurohypophyseal hormones is accompanied by the release of neurophysin. These results and electron microscopic studies strongly suggest that the neurosecretory granules fuse with the axonal membrane and empty their contents of hormone and neurophysin into the pericapillary space. After release of hormone and neurophysin contents, there appears to occur a re-uptake of the membrane of the granule into the interior of the axon terminal. Evidence has been obtained that the huge pool of the neural lobe hormones is metabolically heterogeneous and that about 10 to 20% of the total hormone content of the neural lobe is a "readily releasable" pool. After this readily releasable pool of hormone has been discharged, the neurohypophysis is still capable of releasing hormones in response to appropriate stimuli, but at a greatly reduced rate.

Finally, despite the intimate morphological contacts seen between the neurosecretory neurons and the interstitial neuroglial cells, their functional interrelationships remain incompletely known.

It is in the hypothalamic magnocellular neurosecretory system that, for the first time, the existence of hormone-producing neurons was clearly demonstrated. Its neurohormones were isolated from the neural lobe which, at that time, was the best known part of the neurohypophysis. Hence, these neurohormones were called "neurohypophyseal" hormones. Only later was it found that the neurohypophysis (median eminence and neural lobe) also contains a number of other peptide neurohormones, the so-called releasing and release-inhibiting hormones. These neurohormones play an important role in the control of the release of the hormones of the pars distalis of the hypophysis. Like the neurohypophyseal hormones, the releasing and release-inhibiting hormones present in the neurohypophysis are contained in nerve fibers. These nerve fibers are terminals of axons belonging to different parvocellular peptide hormone-producing neuron systems the cell bodies of which are located in the hypothalamus, and possibly in other parts of the brain.

The preceding short overview is intended for the reader not familiar with the subject. Detailed information can be found in the reviews by Diepen (1962), Sloper (1966), Bargmann (1968), Ginsburg (1968), Sachs (1970), Dodd *et al.* (1971), Acher (1974), Lederis (1974), Holmes and Ball (1974), Daniel and Prichard (1975), Cross *et al.* (1975), and Hayward (1977).

Although the hypothalamic magnocellular endocrine system is commonly used as a model system for the other, less known, peptide hormone-producing neuron systems (see Sachs, 1970; Ochs, 1977), from the preceding overview, it is clear that our knowledge about this magnocellular system is far from complete and that a number of problems remain to be solved. One of these problems concerns the localization of the neurohypophyseal hormones and of the neurophysins. This problem will be the major topic of this paper.

In the past, light microscopic studies of tissue sections stained with chrome–alum–hematoxylin or aldehyde–fuchsin (the so-called Gomori methods), with pseudoisocyanins, performic acid–Alcian blue, and other related methods (see Sloper, 1966) have played a considerable role in the localization of the "neurosecretory material" in the hypothalamic magnocellular endocrine neurons. Although, in these neurons, the neurosecretory material, stainable by these methods, mostly corresponds to accumulations of neurohypophyseal hormone–neurophysin complexes (see Sachs, 1970), these methods have only a limited value for the localization of neurohypophyseal hormones and (or) neurophysins. Indeed, as these methods depend upon the presence of high concentrations of protein-bound disulfide groups, besides neurohypophyseal hormone–neurophysin complexes, they also stain other nonrelated disulfide group-containing substances present in lysosomes, lipofuscin granules, the subcommissural organ, and some other structures of the brain (see Sloper, 1966; Stutinsky, 1970). Subsequent biochemical and biophysical investigations, combined or not with light or electron microscopy, have been of limited value for the localization of neurohypophyseal hormones and neurophysins (see Sachs, 1970). Recently developed immunological methods have proved to be more successful. Specific radioimmunoassay techniques have made it possible to identify and quantify neurohypophyseal hormones and neurophysins in small brain areas (see George and Jacobowitz, 1975; George, 1976, 1978; George and Forrest, 1976; Brownstein *et al.*, 1976). However, they give no information about the intracellular location of these substances. This aim may be reached with the aid of immunocytochemical methods, the results of which will be critically reviewed in the present paper.

II. Technical Remarks

A. Raising of Antisera

As is well known, for raising "specific" antisera to neurophysins and to neurohypophyseal hormones, the antigens used for immunizing purposes must be as pure as possible. Neurophysins are molecules large enough to have, by themselves, adequate antigenic properties. But, neurohypophyseal hormones, whose molecules are too small, have to be coupled with a large carrier molecule (e.g., thyroglobulin), in order to enhance their immunogenicity. Therefore, the antisera raised contain antibodies to these carrier molecules, as well as to the neurohypophyseal hormones. This has to be taken into consideration in immunostaining.

B. Tissue Preparation

As harsh fixation, dehydration, and embedding of tissues destroys and (or) removes the major part of tissue antigens, immunostaining of frozen sections of lightly fixed material appears to be the method of choice for maximal preservation of tissue antigens (see Goldsmith and Ganong, 1975; Bigbee *et al.*, 1977). Since for our purposes, maximal preservation of tissue antigen was not absolutely required, in our laboratory, postembedding immunoenzyme (PAP) staining was adopted. For light microscopic demonstration of neurophysins, quite satisfactory staining results were obtained after fixation of tissues in Bouin–Hollande fluid. We experienced, however, that for adequate preservation of neurohypophyseal hormones, sublimate should be added to the Bouin–Hollande fixative.

C. Immunocytochemical Methods

For the detection of neurophysins and of neurohypophyseal hormones in tissue sections, immunofluorescence methods and immunoenzyme methods have been used. In contrast to the immunofluorescence methods, the immunoenzyme techniques have the great advantage that the preparations are permanent, so that they do not need to be studied and photographed immediately after staining and can be compared at ease. Moreover, immunoenzyme cytochemistry is suited for both light and electron microscopy. The unlabeled antibody peroxidase–antiperoxidase complex (PAP) technique of Sternberger *et al.* (1970), which may be even more sensitive than radioimmunoassay, has proved to be by far the best immunocytochemical method. Owing to its great sensitivity, highly diluted primary antisera can be used, so that nonspecific staining is significantly reduced. For the same reason, even antisera of poor quality, and not suited for radioimmunoassay, may give excellent results when used in the PAP technique (see Moriarty, 1973; Moriarty *et al.,* 1973; Sternberger, 1974).

On the other hand, it has to be noted that, also partly due to the high sensitivity of the PAP technique, antisera with slight cross-reactivity in radioimmunoassay may cause important cross-reactive staining (see Vandesande and Dierickx, 1975, 1976a).

As pointed out by Bigbee *et al.* (1977), in the PAP technique, it is of the utmost importance that optimal primary antiserum dilutions are used, since too high antiserum concentration, as well as too low concentration, are causal factors of false negative results. Moreover, since in PAP staining many factors are involved (e.g., native amounts of antigen, accessibility of the antigen, regional fixation differences), serum dilutions appropriate for one tissue may not be appropriate for another. Therefore, in order to avoid false negative results, for *each* tissue, the optimal dilution of *each* primary antiserum should be determined on a broad serum dilution series.

D. Method Specificity and Antiserum Specificity

A distinction has to be made between method specificity and antiserum specificity. In the PAP procedure, method specificity can be tested (1) by omitting each reagent, one at a time, from the regular staining sequence, leaving all other staining factors unchanged, and (2) by replacing the primary antiserum, in the regular staining sequence, by the same optimally diluted antiserum from which all "specific" antibodies have been removed by (solid phase) adsorption with pure antigen against which the antiserum was raised.

These test procedures allow one to check whether or not positive staining, obtained in the complete PAP staining, is exclusively due (1) to the sequence of the successive immunological steps of the PAP procedure, and (2) more particularly to antibodies, present in the primary antiserum, directed against the antigen to be detected in the tissue sections. As has been mentioned above, false negative results can be avoided if, before PAP staining, for each tissue under study, the optimal dilution of the primary antiserum used is determined.

Antisera, raised against a given antigen, exhibit considerable variability with respect to specificity. Apart from the fact that highly specific antisera are rare, it is generally impossible to ascertain the "monospecificity" of an antiserum (see Cameron and Erlanger, 1977; Moore *et al.*, 1977).

The great majority of antisera show cross-reactivity with substances structurally related to the antigen against which the antiserum was raised. Tissues in which cross-reactive staining with substances structurally related to the antigen to be detected can be conclusively avoided are exceptions. The vertebrate hypothalamic magnocellular neurosecretory system is such an exception. Indeed, it has been shown that it contains only two, well-known neurohypophyseal hormones, together with their respective precursors and possible prohormones (see Sachs, 1970; Acher, 1974). Therefore, for the identification of a neurohypophyseal hormone in this system, a distinction has only to be made between the one neurohypophyseal hormone (together with possible precursors), on the one hand, and the other hormone (together with possible precursors), on the other. It has been shown (Swaab and Pool, 1975; Vandesande and Dierickx, 1975; Swaab *et al.*, 1975b; Vandesande *et al.*, 1975a,c) that such a distinction can be made by using anti-neurohypophyseal hormone sera which have been differentially adsorbed with the heterologous neurohypophyseal hormone bound to a solid phase (cyanogen bromide-activated Sepharose 4B). As, in addition, each of the two hormones is synthesized in a separate type of neuron (see Section III,B,1), the completeness of differential adsorption of the antisera can be directly and adequately tested on immunochemically double-stained hypothalamic tissue sections.

On the other hand, in some mammalian species, it has been shown that the hypothalamic magnocellular neurosecretory system contains two major neurophysins. As will be described (see Section III,B,2), with the aid of a procedure similar to that used for the neurohypophyseal hormones, it has been possible, in the cow and man, to identify two different neurosecretory cell types, each producing only one of the two major neurophysins.

Neurohypophyseal hormones and neurophysins have been described to also be present in other brain regions (e.g., the external region of the median eminence, the suprachiasmatic nuclei). So far, however, in these regions, it is impossible to ascertain whether positive immunoreactivity is due to these substances or to other structurally related cross-reacting antigens (see Sections IV and V).

For detailed information about immunocytochemical procedures the reader is referred to Sternberger (1974), Moriarty (1973), and Nairn (1969).

III. Immunocytochemical Localization of Neurohypophyseal Hormones and Neurophysins in the Hypothalamic Magnocellular Neurosecretory System

A. Introduction—Techniques

Evidence from studies on isolated neurosecretory nerve endings and from light and electron microscopic observations in hereditary diabetes insipidus rats has led to the hypothesis that, in mammals, vasopressin and oxytocin may be synthesized in two separate types of neurons, one type producing vasopressin, the other oxytocin (Orkand and Palay, 1966, 1967; Sokol and Valtin, 1967; Bindler *et al.*, 1967). Although this so-called one neuron–one neurohypophyseal hormone hypothesis has been the subject of numerous subsequent investigations, until recently, there was still disagreement on whether this concept is completely correct (Zimmerman, 1976a,b; Zimmerman and Anthunes, 1976; Zimmerman and Robinson, 1976; Sokol *et al.*, 1976). Therefore, in our laboratory, immunocytochemical studies were undertaken to identify specifically the neurons which produce the neurohypophyseal hormones, and their associated neurophysins, in mammals and in the other vertebrates. As our investigations led to conclusive results, we give now a brief survey of these studies and will discuss our results together with those obtained by other investigators.

The hypothalami from adult species of mammals, birds, reptiles, amphibians, teleosts, and cyclostomes were used. For the *light microscopic* studies, serial sections from fixed and paraffin-embedded hypothalami were made. Sections of the area containing the hypothalamic magnocellular neurosecretory system were stained with the unlabeled antibody peroxidase–antiperoxidase complex (PAP) method of Sternberger *et al.* (1970). In addition to the usual *single* PAP staining method, a *double* PAP staining method was used which enabled us to identify and localize *two* tissue antigens in single tissue sections. The details of this method have been published previously (Vandesande and Dierickx, 1976b; Vandesande *et al.*, 1977). In principle, it consists of two successive PAP stainings, performed on the same tissue section. In the first PAP staining, one tissue antigen (e.g., oxytocin) is selectively localized with the use of 3,3′-diaminobenzidine: yellow-brown amorphous precipitate. Then, the antibodies and the peroxidase, used in the first PAP staining, are selectively eluted from the tissue section, without removing the yellow-brown reaction product identifying the sites of the first antigen. As the second antigen (e.g., vasopressin), to be identified subsequently, is not denaturated by the elution procedure, in the next PAP staining, this second antigen is selectively localized with the use of 4-Cl-l-naphthol: gray-blue amorphous precipitate. The final result is a simultaneous selective localization, in the same section, of oxytocin (yellow-brown) and of vasopressin (gray-blue). In order to eliminate cross-reactive staining, in both staining methods, differentially adsorbed rabbit anti-neurohypophyseal hormone sera and differentially adsorbed rabbit anti-bovine neurophysin sera were used. Differential adsorption (i.e., complete removal of cross-reacting antibodies) of the anti-bovine neurophysin I or II sera was carried out by adding, respectively, either bovine neurophysin II or bovine neurophysin I, and subsequent elimination of the precipitated cross-reacting antibody–neurophysin complexes. Differentially adsorbed anti-neuropophyseal hormone sera were obtained by treatment of the antisera with the heterologous hormone, covalently bound to a *solid* phase (cyanogen bromide-activated Sepharose 4B beads).

For the *electron microscopic* studies, ultrathin serial sections were made from small tissue pieces fixed in glutaraldehyde and embedded in Araldite. These sections were stained with the unlabeled antibody peroxidase–antiperoxidase complex (PAP) method, using highly diluted differentially adsorbed rabbit anti-neurohypophyseal hormone sera and rabbit anti-bovine neurophysin sera. Successive small groups of serial sections were stained using alternately one of these antisera (see Aspeslagh *et al.*, 1976; Van Vossel *et al.*, 1976, 1977).

With the immunocytochemical stainings used, background staining was always minimal. Tests of method specificity, done by omitting each reagent, one at a time, from the regular staining sequence, resulted only in a diffuse minimal staining of the tissue sections. The same negative staining result was obtained when the differentially adsorbed primary antisera were replaced by primary antisera from which all specific antibodies had been removed.

In the double staining method, tests for selectivity and completeness of the elution of the antibodies and of the peroxidase from their antigens showed that the elution was complete and selective. Furthermore, it was also shown that the order in which the two primary antisera were used did not influence the final staining result.

It should be noted that the presence, in the magnocellular hypothalamic nuclei, of two different types of selectively stainable neurons provided us with a means to directly and adequately demonstrate the specificity of our *differentially adsorbed* antisera in the double-stained tissue sections. Similarly, it could also be shown that staining with a *non*adsorbed anti-hormone or anti-neurophysin serum always resulted in an additional, more or less pronounced, cross-reactive staining of the other neurohypophyseal hormone or neurophysin.

B. Localization in the Perikarya

1. *Immunocytochemical Localization of the Neurohypophyseal Hormones*

a. *Results in Mammals.* In the bovine hypothalamus, we (Vandesande *et*

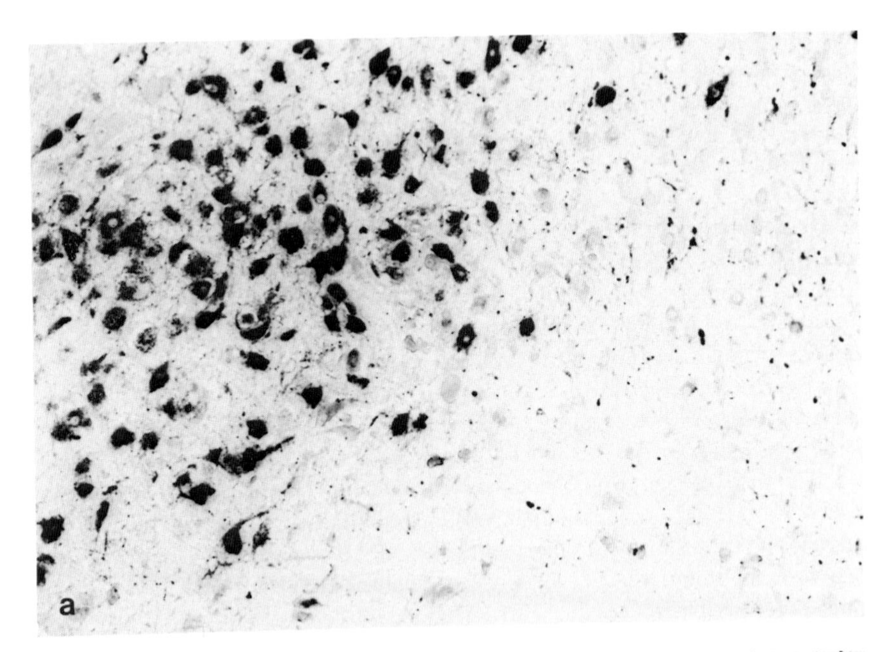
a

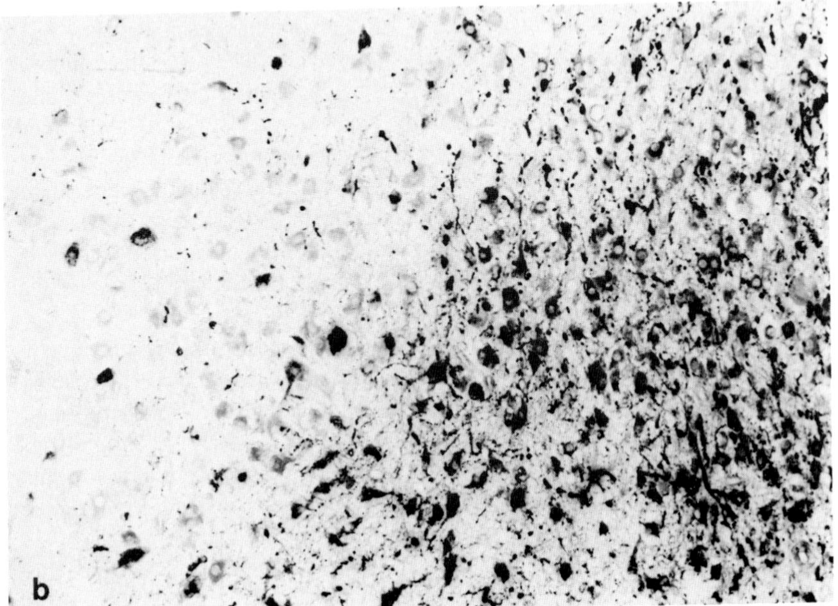
b

al., 1975a) demonstrated that vasopressin and oxytocin are synthesized in separate vasopressinergic and oxytocinergic neurons (Fig. 2). These two bovine neurosecretory cell types show distinct morphological differences and a preferential localization (De Mey *et al.*, 1974; Vandesande *et al.*, 1975a).

In the rat, we (Vandesande and Dierickx, 1975; Vandesande *et al.*, 1977) demonstrated that, in accordance with indications from the literature (Orkand and Palay, 1966, 1967; Sokol and Valtin, 1967; Bindler *et al.*, 1967; Burford *et al.*, 1971; Cross, 1974; Swaab *et al.*, 1975a,b; Van Leeuwen *et al.*, 1976; Choy and Watkins, 1977), in both the supraoptic and the paraventricular nuclei, vasopressin and oxytocin are synthesized in separate neurons, which show a tendency to preferential location. In their immunofluorescence studies on the distribution of oxytocin and vasopressin in the rat supraoptic and paraventricular nuclei, Swaab *et al.* (1975a,b) also used anti-neurohypophyseal hormone sera which had been differentially adsorbed with the heterologous hormone bound to cyanogen bromide-activated Sepharose 4B beads. However, their immunofluorescence technique did not conclusively demonstrate that oxytocin and vasopressin are contained in separate neurons, as has been admitted by the authors in the following: "Whether vasopressin and oxytocin are synthesized in separate cells cannot be proved definitively by the present technique. Our data are, however, consistent with this possibility" (Swaab *et al.*, 1975b), and "Whether vasopressin and oxytocin are synthesized in separate cells cannot be proved by the present data, although this possibility is supported" (Swaab *et al.*, 1975a).

Additional support for our light microscopic observation in the rat came from our electron microscopic studies (Aspeslagh *et al.*, 1976) which demonstrated that, in the rat neural lobe, vasopressin and oxytocin are located in separate nerve fibers. This observation has been confirmed by Van Leeuwen and Swaab (1977).

In the human hypothalamus, we (Dierickx and Vandesande, 1977b) demonstrated that the neurohypophyseal hormones are also synthesized in *separate* vasopressinergic and oxytocinergic neurons. In the supraoptic nuclei, oxytocinergic neurons are rare; their neurons are almost exclusively vasopressinergic. The paraventricular and accessory neurosecretory nuclei contain oxytocinergic as well as vasopressinergic neurons, but here the ratio of vasopressinergic to oxytocinergic neurons shows pronounced local differences. As in cattle, the two human neurosecretory cell types show distinct morphological differences (Dierickx and Vandesande, 1977b).

Studies on immunochemically double-stained tissue sections of the bat, the

---

FIG. 2. Same area, in two adjacent sagittal sections, of the supraoptic part of the bovine supraoptic nucleus: stained (a) with differentially adsorbed anti-oxytocin serum; (b) with differentially adsorbed anti-vasopressin serum (PAP staining). Note in (a) the selectively stained oxytocinergic neurons; in (b) the selectively stained vasopressinergic neurons. Note also the preferential location of the two cell types. (From Vandesande *et al.*, 1975a.)

cat, and the dog (to be published) showed that, in these species also, both neurohypophyseal hormones are synthesized in separate neurons of both the supraoptic and paraventricular nuclei. Recently, indications in favor of our findings have been described in the pig by Watkins and Choy (1976, 1977).

From all these observations, obtained from great numbers of immunochemically single- and double-stained tissue sections, it may be concluded that, in mammals, the neurohypophyseal hormones are exclusively synthesized in separate vasopressinergic and oxytocinergic neurons, and that both the supraoptic and the paraventricular nuclei contain both types of neurons. It has to be stressed that these results can be obtained only by means of carefully controlled immunocytochemical methods, and by using anti-hormone sera differentially adsorbed with the heterologous hormone bound to a *solid phase*. Indeed, staining with a nonadsorbed antiserum to one of the neurohypophyseal hormones always results in a more or less pronounced cross-reactive staining of the other neurohypophyseal hormone (Swaab and Pool, 1975; Swaab *et al.*, 1975b; Vandesande and Dierickx, 1975; Vandesande *et al.*, 1975a). On the other hand, differential adsorption, by adding the heterologous neurohypophyseal hormone to the antiserum, causes binding of the cross-reacting antibodies to the hormone added. But, as the resulting hormone–antibody complexes are soluble, these complexes remain dissolved in the antiserum. Staining with such a differentially adsorbed antiserum may greatly diminish cross-reactive staining. But, as the soluble hormone–antibody complexes present in the antiserum may dissociate during the staining, cross-reactive staining cannot be completely eliminated. Therefore, the immunocytochemical data (Zimmerman, 1976a, 1976b; Zimmerman and Anthunes, 1976; Zimmerman and Robinson, 1976; Sokol *et al.*, 1976) suggesting that at least some mammalian hypothalamic neurons may synthesize both neurohypophyseal hormones must be ascribed to cross-reactive staining (see also Swaab and Pool, 1975). Consequently, cross-reactive staining can only be avoided *completely* if, before staining, the cross-reacting antibodies are totally removed from the antisera, as is the case with antisera differentially adsorbed with a solid phase.

b. *Results in the Other Vertebrates.* In *birds* (Goossens *et al.*, 1977a) and *reptiles* (Goossens *et al.*, 1979), we found that vasotocin and mesotocin are synthesized in separate neurons of both the supraoptic and paraventricular nuclei.

In amphibians, we (Vandesande and Dierickx, 1976b) demonstrated that vasotocin and mesotocin are also synthesized in separate neurons of the preoptic magnocellular neurosecretory nuclei. This light microscopic observation has been confirmed by the electron microscopic immunocytochemical demonstration of separate vasotocinergic and mesotocinergic nerve fibers in the neural lobe of the amphibian hypophysis (Van Vossel *et al.*, 1976). Also in the preopticohypophyseal neurosecretory system of dipnoi, vasotocin and mesotocin are synthesized in separate vasotocinergic and mesotocinergic neurons (Goossens *et al.*, 1978). Similarly, in teleosts, vasotocin and isotocin are produced in separate hypothalamic neurosecretory neurons (Goossens *et al.*, 1977b). Finally, in cyclostomes, which produce little or no oxytocin-like peptide (Sawyer *et al.*, 1961; Follett and Heller, 1964; Sawyer, 1965, 1968; Rurak and Perks, 1974, 1976, 1977), we (Goossens *et al.*, 1977c) showed that the preopticohypophyseal neurosecretory system is entirely vasotocinergic.

c. *Conclusions.* From these extensive light and electron microscopic results, obtained by means of specific immunocytochemical methods and *specific* antisera, it may be concluded that, in all vertebrates, the neurohypophyseal hormones are synthesized in separate types of neurons. In mammals and in the other amniote vertebrates, both the supraoptic and the paraventricular nuclei contain vasopressinergic (vasotocinergic) and oxytocinergic (mesotocinergic) neurons. The validity of this one neuron–one neurohypophyseal hormone concept may be the more accepted since it has also been confirmed by recent electrophysiological studies (Brimble and Dyball, 1977; Poulain and Wakerley, 1977; Poulain *et al.*, 1977).

2. *Immunocytochemical Localization of the Neurophysins*

In a number of mammalian species (cow, pig, rat, dog, man) at least two major pituitary neurophysins have been isolated. In other species (guinea pig) only a single major neurophysin would be present (see reviews by Robinson, 1978; Pickering and Jones, 1978).

As described above, in mammals, the neurohypophyseal hormones are synthesized in separate vasopressinergic and oxytocinergic neurons. Since in the neurohypophyseal hormone-producing cells, the hormones are associated with neurophysins (see Sachs, 1970), it is evident that the vasopressinergic and oxytocinergic neurons each produce their own neurophysin(s). Theoretically, there are two possibilities. One possibility is that the neurophysins of the vasopressinergic and oxytocinergic neurons are different, as could be the case in animals in which, at least, two different neurophysins have been found. The other possibility is that, although the two endocrine cell types produce their own neurophysin(s), these neurophysins are identical, as could be the case in the guinea pig.

From a number of mammalian species, evidence has been obtained suggesting that vasopressin and oxytocin are each associated with a specific neurophysin. So far, indications for such association have been found in cattle (Dean *et al.*, 1968; Robinson *et al.*, 1971; Zimmerman *et al.*, 1974; Legros *et al.*, 1974, 1975, 1976), in the rat (Burford *et al.*, 1971; Sunde and Sokol, 1975; Choy and Watkins, 1977); in the pig (Pickup *et al.*, 1973; Watkins and Choy, 1976, 1977), and in man (Robinson 1975a,b; Husain *et al.*, 1975; Zimmerman *et al.*, 1975;

Robinson *et al.*, 1977). Consequently, these indications strongly suggest that, in these animals, the neurophysins produced by the vasopressinergic and oxytocinergic cells are different.

In our laboratory, immunocytochemical studies were undertaken to identify specifically the hypothalamic magnocellular endocrine neurons which, in the cow and in man, produce the different neurophysins. As described above, the bovine and human hypothalami were studied on serial sections which had been stained either with the single or with the double PAP method, using carefully differentially adsorbed anti-bovine neurophysin I, anti-bovine neurophysin II, and anti-neurohypophyseal hormone sera.

In the bovine hypothalamus, we (De Mey *et al.*, 1974; Vandesande *et al.*, 1975a) found that the vasopressinergic neurons produce only neurophysin II, while the oxytocinergic neurons synthesize only neurophysin I. Thus, both the bovine supraoptic and paraventricular nuclei contain separate neurophysin II–vasopressinergic neurons, and neurophysin I–oxytocinergic neurons. As already mentioned, the two bovine neurosecretory cell types show distinct morphological differences and a preferential localization.

The human hypothalamus was studied with the single and the double PAP method. Six different double stainings were performed, using in each double staining different pairs of differentially adsorbed antisera (see Table II). From the results, we (Dierickx and Vandesande, 1979) concluded that:

1. Both the human supraoptic and paraventricular nuclei contain, at least, two different neurophysins: one selectively stainable with differentially adsorbed anti-bovine neurophysin I serum, the other selectively stainable with differentially adsorbed anti-bovine neurophysin II serum.
2. These two human neurophysins are immunologically related to bovine neurophysin I and neurophysin II, respectively.
3. One human neurophysin is associated with vasopressin (vasopressin–neurophysin); the other with oxytocin (oxytocin–neurophysin).
4. The human vasopressin–neurophysin and oxytocin–neurophysin are separately located in two different types of neurons, which respectively correspond to the vasopressinergic and the oxytocinergic neurons of both the supraoptic and paraventricular nuclei.

C. Localization in the Axons to the Neural Lobe

As already mentioned, the great majority of the neurosecretory material-containing axons of the vertebrate hypothalamic magnocellular endocrine neurons converge, meet each other in the supraoptico (or preoptico-)-hypophyseal tract, and end in the neural lobe of the hypophysis.

TABLE II

PERCENTAGE AND MORPHOLOGICAL FEATURES OF SELECTIVELY STAINED NEUROSECRETORY CELL TYPES PRESENT IN *DOUBLE*-STAINED SECTIONS OF THE SUPRAOPTIC NUCLEUS[a]

	Pars dorsolateralis of supraoptic nucleus (percentage of neurosecretory cell types)	
Pairs of primary antisera used in double staining	Perinuclear accumulation of colored reaction product; large mean cell size	Colored reaction product diffusely spread; smaller mean cell size
1. Anti-vasopressin serum Anti-oxytocin serum	Vasopressinergic neurons >95%[b]	Oxytocinergic neurons <5%[b]
2. Anti-bovine NpII serum Anti-bovine NpI serum	Anti-NpII neurons >95%	Anti-NpI neurons <5%
3. Anti-vasopressin serum Anti-bovine NpI serum	Vasopressinergic neurons >95%	Anti-NpI neurons <5%
4. Anti-bovine NpII serum Anti-oxytocin serum	Anti-NpII neurons >95%	Oxytocinergic neurons <5%
5. Anti-vasopressin serum Anti-bovine NpII serum	Mixed-stained neurons >95%	Nonstained neurons <5%
6. Anti-oxytocin serum Anti-bovine NpI serum	Nonstained neurons >95%	Mixed-stained neurons <5%

[a]Anti-bovine NpI and anti-bovine NpII neurons: neurons selectively stained with anti-*bovine* neurophysin I serum and anti-*bovine* neurophysin II serum, respectively.

[b]Size and percentages are semiquantitative estimations made by two observers.

From the immunocytochemical identification, in the vertebrate hypothalamic magnocellular endocrine nuclei, of two different neurohypophyseal hormone-producing cell types, it could be inferred that there also exist two different types of neurohypophyseal hormone-containing axons running to the neural lobe of the hypophysis.

In the rat, the existence of these two different types of neurohypophyseal hormone-containing axons has been *simultaneously* demonstrated in immunochemically *double*-stained tissue sections. It has been shown that vasopressinergic and oxytocinergic perikarya, of both the supraoptic and paraventricular nuclei, are the origins of separate vasopressinergic and oxytocinergic axons, respectively. The two different types of nerve fibers can be followed, from their origin, through the internal region of the median eminence, and into the neural lobe of the hypophysis. Moreover, it has been found that, in the rat neural lobe, the distribution of the vasopressinergic and of the oxytocinergic axons is partly different (Vandesande and Dierickx, 1975). These light microscopic observa-

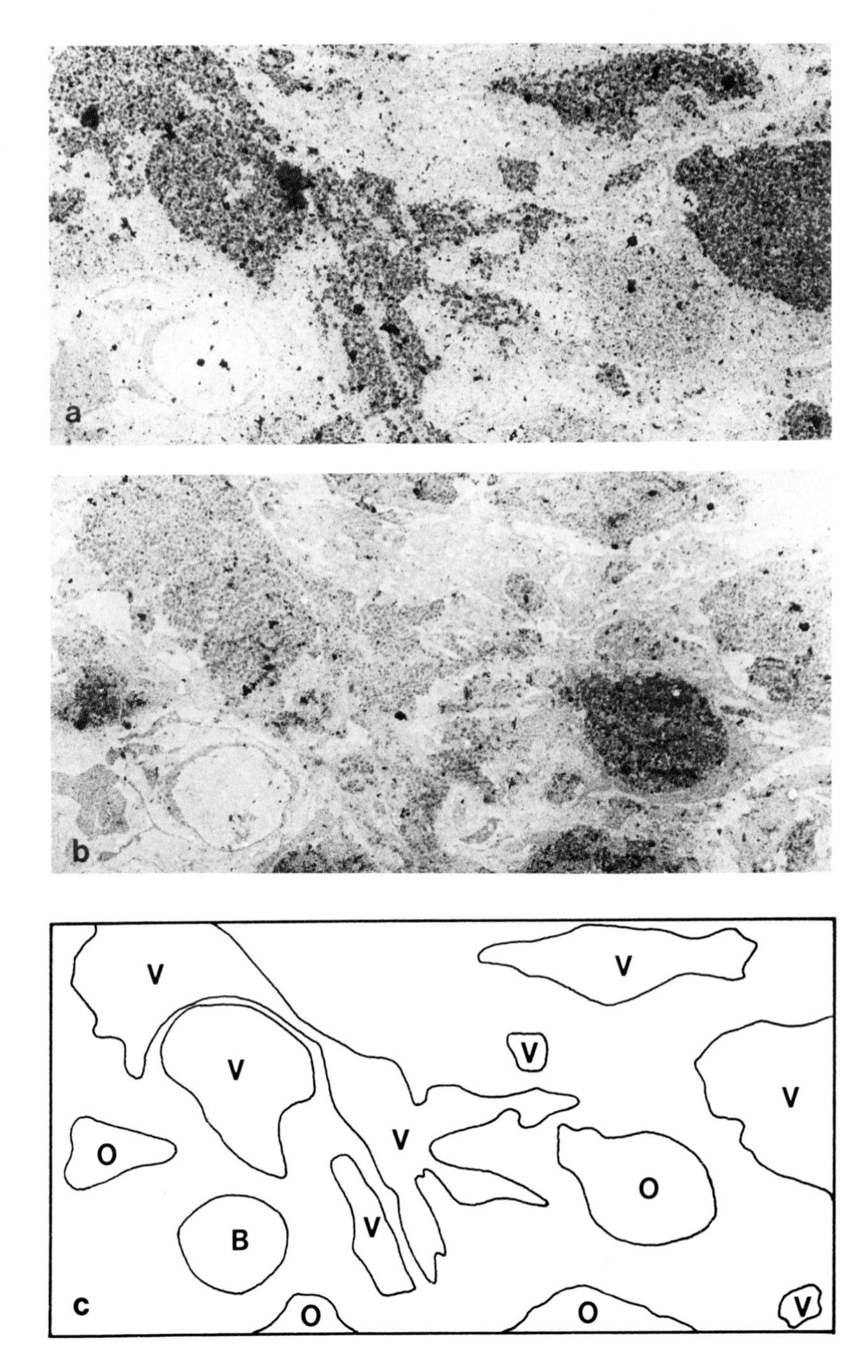
a
b
V
V
V
V
V
V
V
V
V
O
O
B
O
O
c

tions have been confirmed by subsequent immunocytochemical studies at the electron microscopic level (Fig. 3) which demonstrated that, in the rat neural lobe, vasopressin and oxytocin are located in separate nerve fibers (Aspeslagh *et al.*, 1976; Van Leeuwen and Swaab, 1977).

In amphibians, the existence of separate vasotocinergic and mesotocinergic axons, running from the magnocellular preoptic nuclei to the neural lobe, has been demonstrated by means of immunocytochemistry at the light microscopic level (Vandesande and Dierickx, 1976b) and at the electron microscopic level (Fig. 4; Van Vossel *et al.*, 1976).

In addition, studies on immunochemically double-stained tissue sections demonstrated the two separate types of hormone-containing axons in birds (Goossens *et al.*, 1977a) and in fishes (Goossens *et al.*, 1977b,c, 1978).

Vasopressin-containing and neurophysin-containing pathways to the neural lobe have also been described, at the light and electron microscopic levels, by other investigators. However, in these publications, convincing evidence of the existence of the two different types of hormone-containing axons is lacking (see reviews by Watkins, 1976; Zimmerman, 1976b; Zimmerman and Robinson, 1976).

D. Diabetes Insipidus and the One Neuron–One Neurohypophyseal Hormone Concept

Homozygous Brattleboro rats are characterized by their inability to synthesize vasopressin (see Moses and Miller, 1974; Valtin *et al.*, 1974, 1975; Van Wimersma Greidanus *et al.*, 1974; Rosenbloom and Fisher, 1975; George and Forrest, 1976; Valtin, 1977). There are indications that they are also unable to synthesize a vasopressin-associated neurophysin (Burford *et al.*, 1971; Sunde and Sokol, 1975). Evidence has been obtained suggesting that, in these animals, the inability to synthesize vasopressin and its associated neurophysin is due to a separate type of defective neuron, while oxytocin and its associated neurophysin are produced in another, different type of neuron (Orkand and Palay, 1967; Sokol and Valtin, 1967; Kalimo and Rinne, 1972; Sokol *et al.*, 1976; Tasso *et al.*, 1977).

On the other hand, immunocytochemistry has shown that, in the normal rat, vasopressin, together with neurophysin, is synthesized in separate neurophysin–vasopressinergic neurons, while oxytocin, together with neurophysin, is pro-

Fig. 3. Electron micrographs of two serial sections of the rat neural lobe: stained (a) with differentially adsorbed anti-vasopressin serum; (b) with differentially adsorbed anti-oxytocin serum. In (a), specific staining of the vasopressinergic axons and absence of specific staining of the oxytocinergic axons. In (b), the staining of the *same* axons is exactly the opposite. (c) Sketch for general orientation. V, Vasopressinergic axons; O, oxytocinergic axons; B, blood vessel. PAP staining. (From Aspeslagh *et al.*, 1976.)

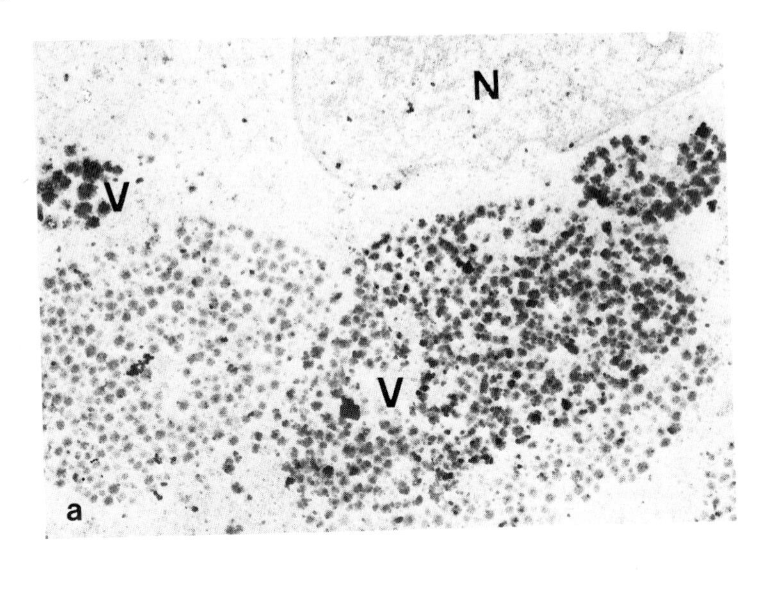

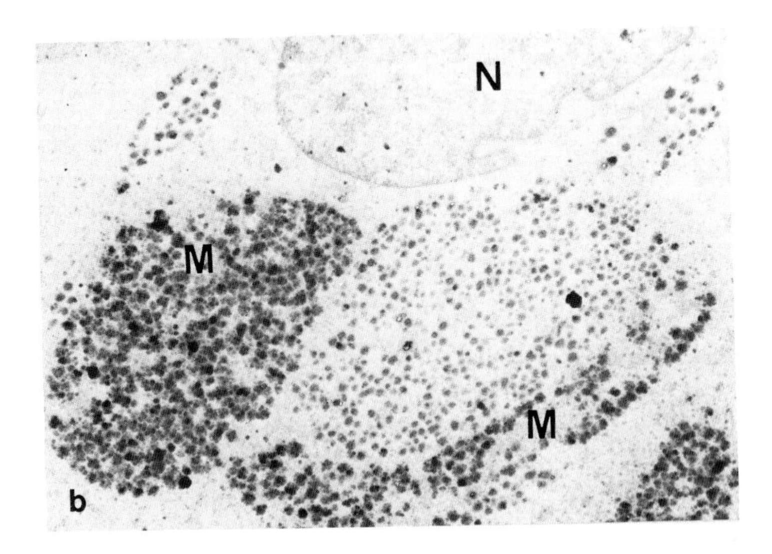

FIG. 4. Electron micrographs of two serial sections of the neural lobe of *Rana temporaria:* stained (a) with differentially adsorbed anti-vasotocin serum; (b) with differentially adsorbed anti-mesotocin serum. In (a), specific staining of the vasotocinergic axons and absence of specific staining of the mesotocinergic axons. In (b), the staining of the *same* axons is exactly the opposite. V, Vasotocinergic axon; M, mesotocinergic axon; N, nucleus. PAP staining. (From Van Vossel *et al.*, 1976.)

duced in another type of neuron: neurophysin–oxytocinergic neurons. In the supraoptic part of the supraoptic nucleus, and also in the paraventricular nucleus, both cell types show a distinct preferential location (Vandesande and Dierickx, 1975, 1976a; Vandesande *et al.*, 1977). Hence, since in the normal rat, vasopressin and its associated neurophysin are synthesized in separate neurophysin–vasopressinergic neurons, it may be expected that, in the homozygous Brattleboro rat, the neurons of this type are unable to synthesize vasopressin and its associated neurophysin. Studies on immunochemically single- and double-stained tissue sections of the homozygous Brattleboro rat have demonstrated that this is indeed the case (Vandesande and Dierickx, 1976a). Taking advantage of the preferential location of both cell types in the normal rat, it has been shown that, when during staining differentially adsorbed anti-oxytocin (anti-mesotocin) serum or anti-bovine neurophysin I serum was used, only those neurons of the supraoptic and paraventricular nuclei, the location of which corresponded to the preferential location of the oxytocinergic neurons of the normal rat, were stained. In contrast, in the same tissue sections, the other neurons of the supraoptic and paraventricular nuclei, the location of which corresponded to the preferential location of the vasopressinergic neurons of the normal rat, were not stained (Fig. 5). So, it was concluded that, like in the normal rat, the supraoptic and paraventricular nuclei of the homozygous Brattleboro rat contain separate neurons which produce oxytocin together with neurophysin.

Moreover, staining of the supraoptic and paraventricular nuclei with differentially adsorbed anti-vasopressin serum resulted in a complete absence of staining of all neurons of both nuclei. Consequently, the total absence of staining for oxytocin, for neurophysin, and for vasopressin in the neurons which correspond to the vasopressinergic neurons of the normal rat shows that, in the homozygous Brattleboro rat, this type of neuron is unable to synthesize either vasopressin or neurophysin. Thus, it may be concluded that the hereditary inability of the homozygous Brattleboro rat to synthesize vasopressin and its associated neurophysin is due to a biochemical defect in separate neurophysin–vasopressinergic neurons of the supraoptic and paraventricular nuclei (Vandesande and Dierickx, 1976a). These results provide additional evidence in favor of the one neuron–one neurohypophyseal hormone concept and of the one hormone–one neurophysin concept.

E. Increased Activity and the One Neuron–One Neurohypophyseal Hormone Concept

Prolonged osmotic stimuli lead to an enhanced release of both neurohypophyseal hormones (Jones and Pickering, 1969; George, 1976). They also activate neurohypophyseal hormone biosynthesis (see Sachs, 1970). There is evidence

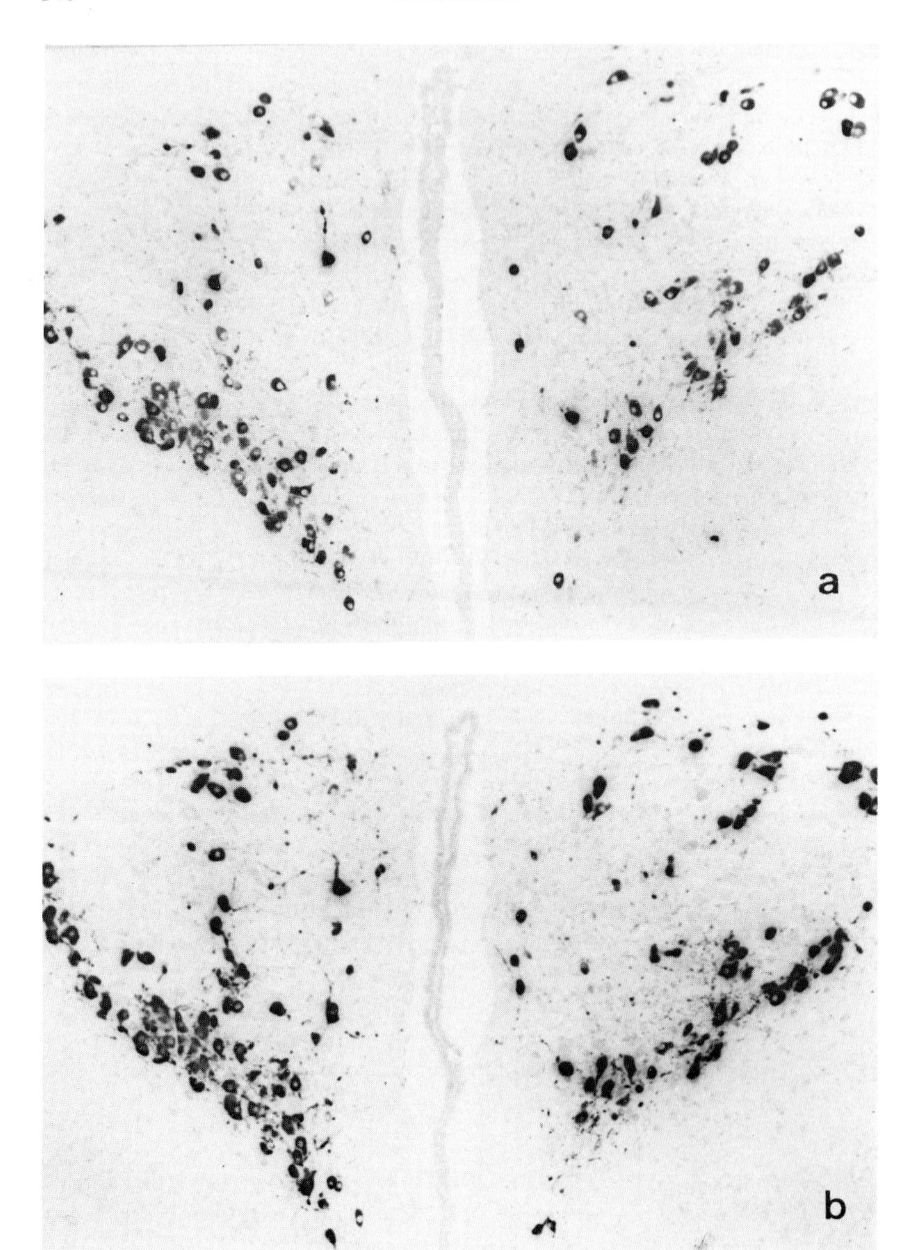
a
b

that estrogens influence neurohypophyseal hormone biosynthesis (Stutinsky, 1970) and stimulate the release of neurophysin(s) and neurohypophyseal hormones (see Robinson, 1978).

As has been described above, in *normal* vertebrates, the two neurohypophyseal hormones are synthesized exclusively in two different types of endocrine neurons. However, it may be questioned (see Paterson and Leblond, 1977) whether this is also the case in animals the hypothalamic magnocellular endocrine system of which has been subjected to intense and prolonged stress situations.

In our laboratory this problem is currently being studied in the rat. Two groups of 10 normal adult Wistar rats were dehydrated by deprivation of drinking water and feeding with dry standard food, for 4 days. During dehydration, one group received a daily intramuscular injection of 0.2 mg stilbestrol. After the fourth day of dehydration, both groups of animals were intracisternally injected with 50 μg colchicine dissolved in 50 μl water. The animals were killed 2 days after the injection of colchicine. The brains were removed and areas of the hypothalami were processed for immunocytochemistry, at the light and electron microscopic levels, as has been indicated above.

The light microscopic study of immunochemically double-stained sections of both groups of rats showed that both the supraoptic and paraventricular nuclei contained exclusively separate vasopressinergic and oxytocinergic neurons. Mixed-stained endocrine neurons were completely absent. The same result was found with immunocytochemistry at the electron microscopic level (Fig. 6). These results were the more convincing, as the neurosecretory perikarya contained large amounts of neurohypophyseal hormones, due to activation of neurohypophyseal hormone biosynthesis, on the one hand (see Sachs, 1970), and the effect of colchicine, on the other (see Boudier *et al.*, 1972). Moreover, the respective number of both cell types, present in the supraoptic and paraventricular nuclei of the treated animals, was compared with that of normal animals. From our estimations, no difference between the treated and the normal animals could be found (Vandesande and Dierickx, unpublished). Thus, so far as has been studied, switching of neurohypophyseal hormone synthesis during stress situations does not occur.

FIG. 5. Adjacent transverse sections of the paraventricular nuclei of a Brattleboro rat: stained (a) with differentially adsorbed anti-oxytocin serum; (b) with anti-bovine neurophysin I serum. The selective staining of the *same* neurons in both micrographs shows that these neurons contain oxytocin together with neurophysin. The preferential location of these oxytocinergic neurons at the periphery of the paraventricular nucleus corresponds to that of the oxytocinergic neurons of the normal rat. Note the complete absence of staining of the other neurons, whose preferential location in the center of the paraventricular nucleus corresponds to that of the vasopressinergic neurons of the normal rat. (From Vandesande and Dierickx, 1976a.)

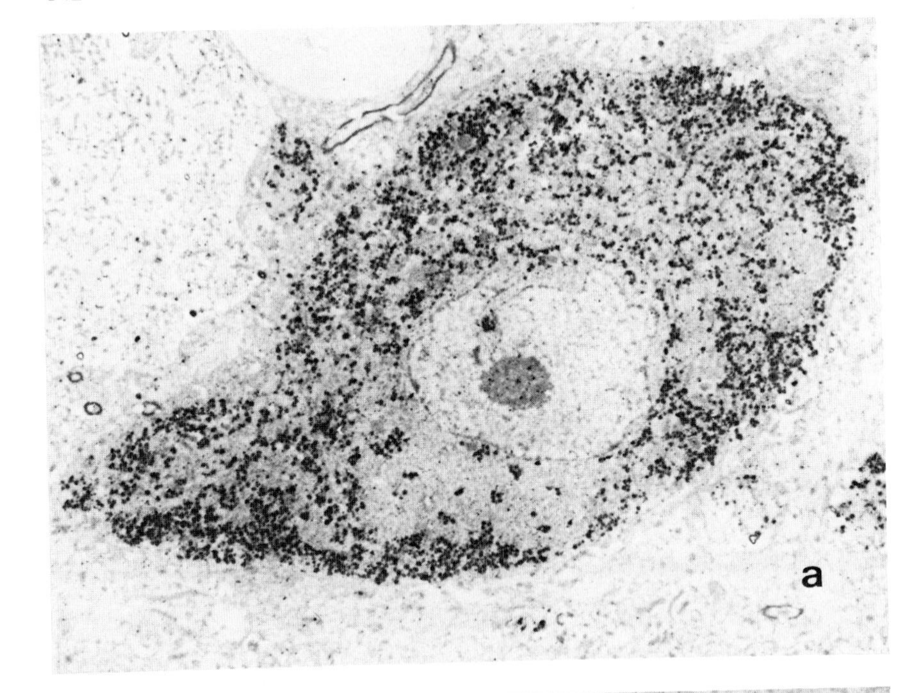
a

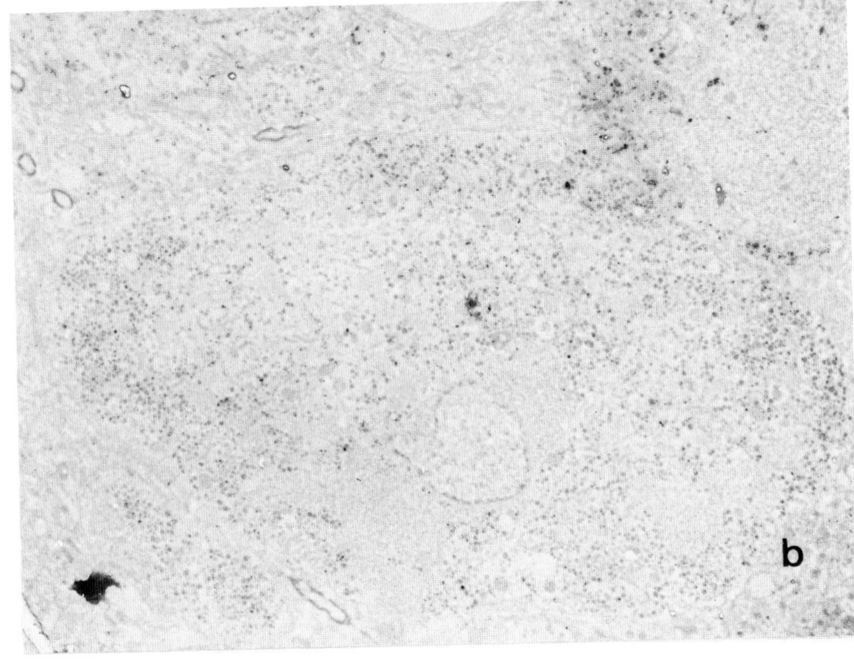
b

F. Nonimmunoreactive Magnocellular Neurons

Among the large, immunoreactive neurons of the vertebrate hypothalamic magnocellular neurosecretory nuclei, always a minority of interspersed, large neurons are found which does not stain either for neurohypophyseal hormones or for neurophysins. It may be questioned whether the nonimmunoreactivity of these neurons is due to transient neurohypophyseal hormone and neurophysin depletion, or if these neurons belong to other, additional types of endocrine neurons. Such additional types of endocrine neurons have been postulated in the amphibian (Goos, 1969; Van Oordt *et al.*, 1972; Notenboom *et al.*, 1976) and in the rat. So far, in the amphibian, these neurons have not been identified. In the rat supraoptic and paraventricular nuclei, magnocellular neurons, containing somatostatin together with neurophysin, have been described (Dubois and Kolodziejczyk, 1975). However, in our laboratory (Vandesande and Dierickx, unpublished), the existence, in the rat, of such magnocellular somatostatin-containing neurons could not be confirmed. On the contrary, in our study, all rat hypothalamic somatostatin-containing neurons were found to be parvocellular. They did not contain neurophysin and their localization corresponded mainly to that described by Hökfelt *et al.* (see reviews by Elde and Hökfelt, 1978; Hökfelt *et al.*, 1978).

G. The Parvocellular Neurons of the Magnocellular Neurosecretory Nuclei

In fishes, the hypothalamic magnocellular preoptic nucleus can be divided into a dorsal region, the neurons of which are very large (magnocellular part), and a ventral region, most neurons of which, as compared with those of the dorsal region, are distinctly smaller (''parvocellular'' part).

In amphibians also, the hypothalamic magnocellular preoptic nucleus can be divided into a dorsal, magnocellular part and a ventral, ''parvocellular'' part. Here, however, the difference between the cell size of the two regions is not so pronounced as in fishes. With the exception of a few interspersed nonimmunoreactive neurons, in the parvocellular part as well as in the magnocellular part of both the fishes and the amphibians, all neurons are immunoreactive for neurohypophyseal hormones (Vandesande and Dierickx, 1975; Goossens *et al.*, 1977b, 1978).

In birds, both the supraoptic and the paraventricular nuclei contain perikarya

---

Fig. 6. Electron micrographs of two serial sections of the same vasopressinergic cell of a rat supraoptic nucleus: stained (a) with differentially adsorbed anti-vasopressin serum, intense specific staining; (b) with differentially adsorbed anti-oxytocin serum, total absence of specific staining. PAP method. (From Vandesande and Dierickx, unpublished.)

belonging to different size classes. Of the small as well as of the large neurons, at least the great majority (if not all) were immunoreactive for neurohypophyseal hormones (Goossens *et al.*, 1977a).

In mammals, the supraoptic nucleus is entirely magnocellular. In addition to a magnocellular part, the paraventricular nucleus also contains a parvocellular part (see Diepen, 1962).

In the rat paraventricular nucleus, the parvocellular component lies close to the ependymal lining of the third ventricle and medial to the magnocellular component (see Bodian and Maren, 1951; Diepen, 1962). Morris (1974) has shown that this periventricular parvocellular component contains a distinctive population of small neurons characterized by relatively large numbers of small dense-cored granules, which can be discriminated from classical neurosecretory granules. Recent immunocytochemical studies, done in our laboratory (Vandesande and Dierickx, unpublished), have demonstrated that the neurons of this so-called parvocellular part of the rat paraventricular nucleus are, in fact, somatostatin cells, which do not contain neurophysin. The great majority of these somatostatin cells are small neurons. A number of them invade the medial part of the magnocellular component of the paraventricular nucleus (Fig. 7).

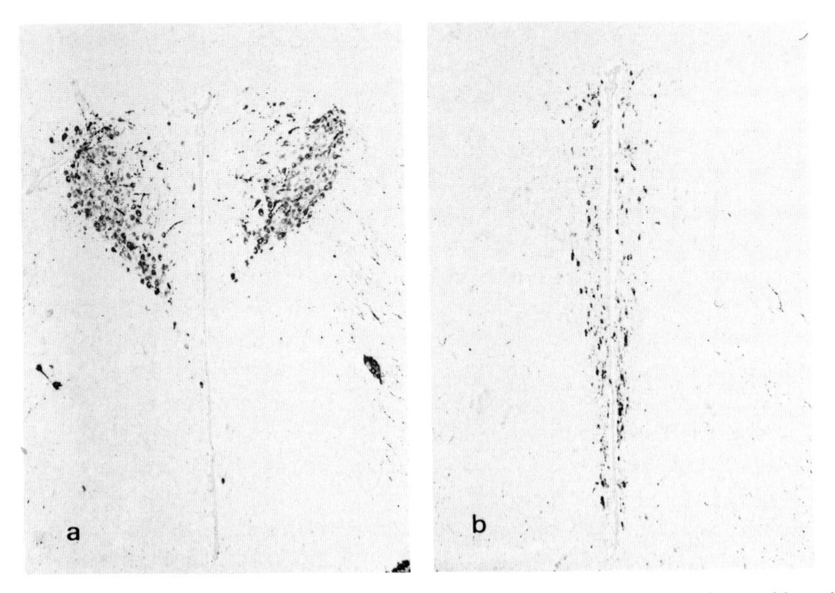

FIG. 7. Adjacent transverse sections of a rat paraventricular nucleus: stained (a) with anti-vasopressin and anti-oxytocin sera; (b) with anti-somatostatin serum. PAP method. In (a) only the magnocellular part, and in (b) only the parvocellular part, of the paraventricular nucleus are selectively stained. Staining with anti-bovine neurophysin I serum gave the same result as in (a). (From Vandesande and Dierickx, unpublished.)

The human paraventricular nucleus, unlike the supraoptic nucleus, is a heterogeneous structure consisting of magnocellular neurons and smaller neurons. A varying degree of interspersion of large and small neurons occurs throughout the entire nucleus. In contradiction with the results of Defendini and Zimmerman (1978), we have found that the small as well as the large neurons of the human paraventricular nucleus are either vasopressinergic or oxytocinergic neurons (Dierickx and Vandesande, 1977b).

In other mammals, additional studies on the parvocellular component of the paraventricular nucleus need to be performed.

IV. Immunocytochemical Localization of Neurohypophyseal Hormones and Neurophysins in the External Region of the Median Eminence

A. Introduction

In the external region of the median eminence of amphibians, birds, and mammals, Gomori-positive nerve fibers have been described, ending around blood capillaries of the hypophyseal portal system (see Benoit and Assenmacher, 1955; Dierickx and Van den Abeele, 1959; Oksche and Farner, 1974). In the rat and mouse, the number of the Gomori-positive nerve fibers of the external region of the median eminence strongly increases after bilateral adrenalectomy and also after hypophysectomy (see Bock, 1970; Rinne, 1970). This increase can be prevented by treatment with corticoids started immediately after adrenalectomy, or by treatment of the hypophysectomized rats with corticotropin (see Bock, 1970; Brinkmann and Bock, 1973). These findings indicate that the Gomori-positive fibers of the external region of the median eminence could influence the adrenocortical function and this via a neurohumoral control of the hypophyseal corticotropin secretion. Results, obtained in bilaterally adrenalectomized normal and diabetes insipidus (Brattleboro) rats, suggest that the majority of the Gomori-positive nerve fibers of the external region of the median eminence of the normal rat contain vasopressin–neurophysin complexes and that vasopressin could be involved in the control of corticotropin release (Schwabedal *et al.*, 1977). Moreover, observations in bilaterally adrenalectomized rats with hypothalamic lesions indicate that the Gomori-positive fibers of the external region of the median eminence could be terminals of neurons the perikarya of which are located in magnocellular hypothalamic nuclei (Bock and Jurna, 1977).

On the other hand, from electron microscopic studies it appeared that the secretory granules, present in the Gomori-positive fibers of the external region of the median eminence, are distinctly smaller than those of the classical hypothalamic magnocellular neurosecretory system (Wittkowski *et al.*, 1970; Wittkowski and Bock, 1972). This observation suggests that the neurons, from

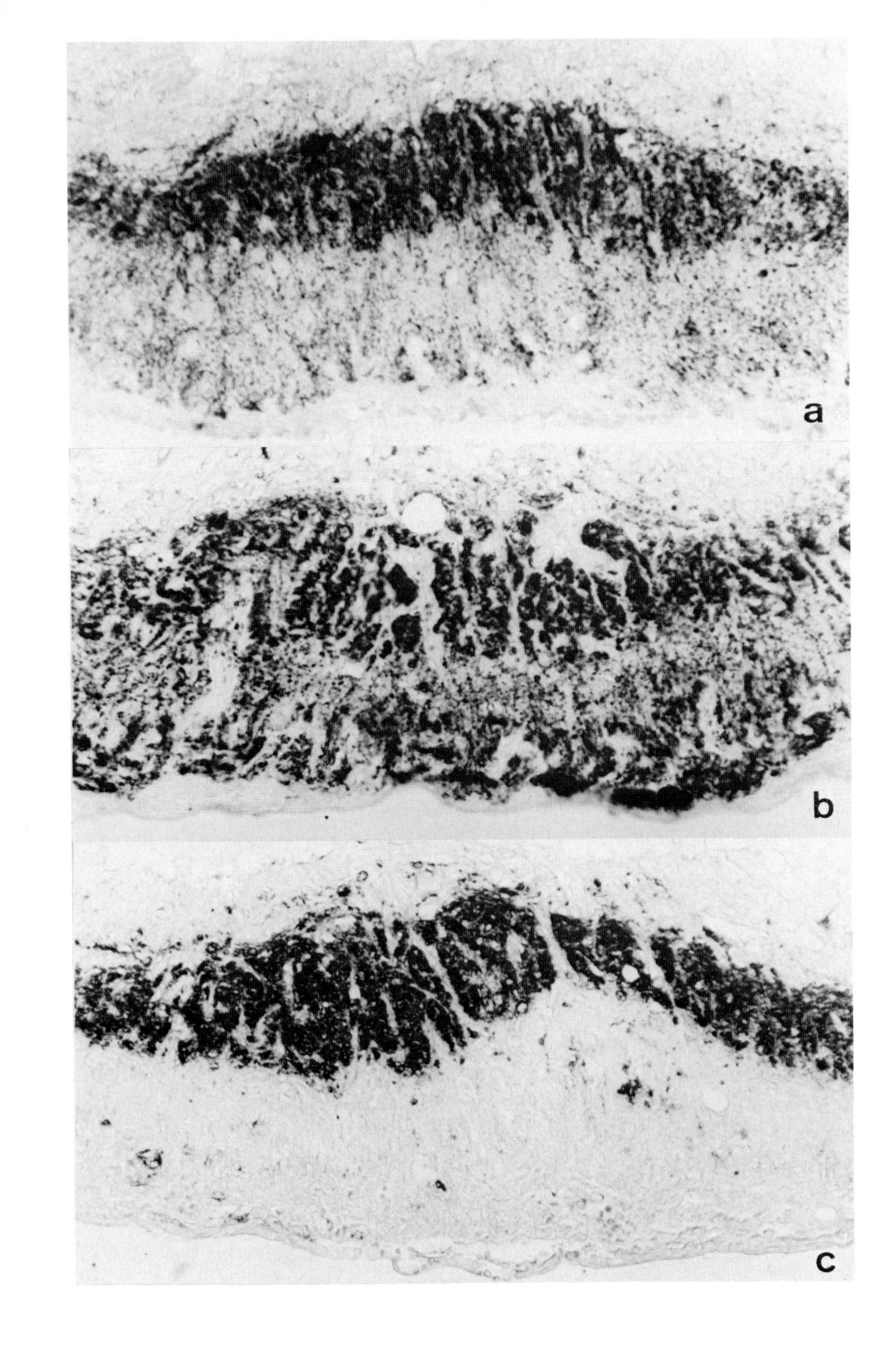
a
b
c

which the Gomori-positive fibers of the external region of the median eminence originate, differ from those of the classical hypothalamic magnocellular neurosecretory system. As Gomori methods lack specificity, immunocytochemistry has been used for the identification of the substances which are contained in the Gomori-positive granules of the external region of the median eminence.

B. Results of Immunocytochemistry

With the use of immunocytochemistry, at the light or electron microscopic level, ''vasopressin''-containing fibers have been described in the external region of the median eminence of the rat (Burlet *et al.*, 1973, 1974; Zimmerman *et al.*, 1975), mouse (Zimmerman *et al.*, 1975; Castel and Hochman, 1976), guinea pig (Silverman and Zimmerman, 1975; Silverman, 1976), and monkey (Zimmerman and Anthunes, 1976; Zimmerman *et al.*, 1976). ''Vasotocinergic'' fibers have been described in the external region of the median eminence of birds (Calas, 1975).

Recently, ''oxytocin''-containing fibers have been described in the external region of the median eminence of the guinea pig (Silverman, 1976) and monkey (Zimmerman and Anthunes, 1976). It has to be noted, however, that the reliability of these results can be questioned. Indeed, in all these studies, anti-hormone sera have been used which were not differentially adsorbed with the heterologous hormone bound to a solid phase. As has already been argued above, immunocytochemical stainings in which such antisera are used cannot (completely) differentiate one neurohypophyseal hormone from the other.

Also with the use of immunocytochemistry, ''neurophysin''-containing fibers have been found in the external region of the median eminence of several mammalian species (Watkins, 1973; Parry and Livett, 1973; Robinson and Zimmerman, 1973; Zimmerman *et al.*, 1973; Vandesande *et al.*, 1974), birds (McNeil *et al.*, 1975) and amphibians (Dierickx and Vandesande, 1977a). In the external region of the rat median eminence, the number of immunoreactive neurophysin-containing fibers greatly increases after bilateral adrenalectomy (Vandesande *et al.*, 1974; Watkins *et al.*, 1974; Stillman *et al.*, 1977). This increase of neurophysin-containing fibers shows a striking resemblance with the similar increase of Gomori-positive fibers in the external region of the median eminence of the bilaterally adrenalectomized rat (see Vandesande *et al.*, 1974; Watkins *et al.*, 1974). In contrast with the neurophysin-containing fibers of the internal region (Fig. 8), the immunoreactive neurophysin-containing fibers of the exter-

---

FIG. 8. Transverse sections of the median eminence: (a) normal rat; (b) bilaterally adrenalectomized rat; (c) rat treated with colchicine subsequent to bilateral adrenalectomy. PAP staining, using anti-bovine neurophysin serum. (From Vandesande *et al.*, 1974.)

nal region of both the normal rat and the bilaterally adrenalectomized rat dramatically disappear after intracisternal injection of colchicine (Vandesande *et al.*, 1974; Dierickx *et al.*, 1976).

In our laboratory, we hypothesized that—on the analogy of our findings in the classical hypothalamic magnocellular neurosecretory system—the Gomori-positive fibers of the external region of the median eminence represent two different types of neurophysin, plus neurohypophyseal hormone-containing fibers. We have tested this hypothesis on immunochemically single-stained and double-stained hypothalamic tissue sections. In all stainings, differentially adsorbed antisera were used. The results of our investigation follow.

In the external region of the bovine median eminence two different types of neurophysin plus neurohypophyseal hormone-containing fibers are present: (1) neurophysin II–vasopressinergic fibers, and (2) neurophysin I–oxytocinergic fibers (De Mey *et al.*, 1975a; Vandesande *et al.*, 1975c).

Similarly, the external region of the rat median eminence also contains separate neurophysin–vasopressinergic fibers and neurophysin–oxytocinergic fibers. Bilateral adrenalectomy always causes a great increase of the number of immunoreactive neurophysin–vasopressinergic fibers of the external region of the rat median eminence. On the contrary, the influence of adrenalectomy on the neurophysin–oxytocinergic fibers of the external region of the median eminence is variable and always much less pronounced than that exerted on the neurophysin–vasopressinergic fibers (Figs. 9 and 10). In contrast with the neurohypophyseal hormone-containing nerve fibers of the internal region, the immunoreactive neurohypophyseal hormone-containing fibers of the external region of the bilaterally adrenalectomized rat dramatically disappear after intracisternal injection of colchicine (Table III). The external region of the median eminence of the homozygous diabetes insipidus (Brattleboro) rat contains neurophysin–oxytocinergic fibers, but no immunoreactive neurophysin–vasopressinergic fibers (Dierickx *et al.*, 1976, see Table III). These observations have been confirmed by Stillman *et al.* (1977).

The external region of the amphibian median eminence contains separate vasotocinergic and mesotocinergic fibers (Dierickx and Vandesande, 1977a). Immunocytochemical studies, at the electron microscopic level (Fig. 11–14), have shown that both the vasotocinergic and mesotocinergic fibers of the external region of the amphibian median eminence are distinctly smaller than those of the classical hypothalamic magnocellular neurosecretory system (Van Vossel *et al.*, to be published). These results are in accordance with data obtained in the rat

---

FIG. 9. Adjacent transverse sections of the median eminence of a normal rat: stained (a) with differentially adsorbed anti-vasopressin serum; (b) with differentially adsorbed anti-oxytocin serum. PAP method. (From Dierickx *et al.*, 1976.)

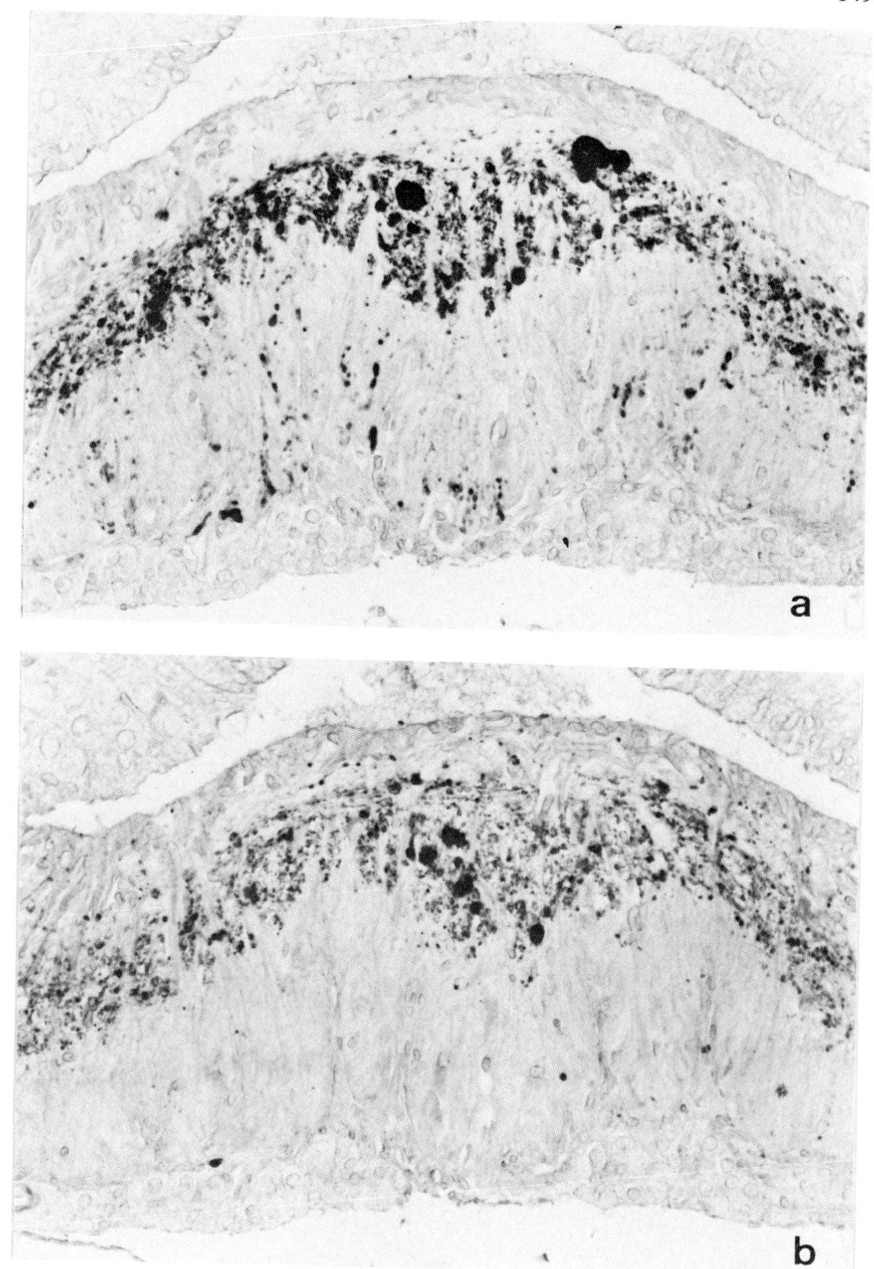
a
b

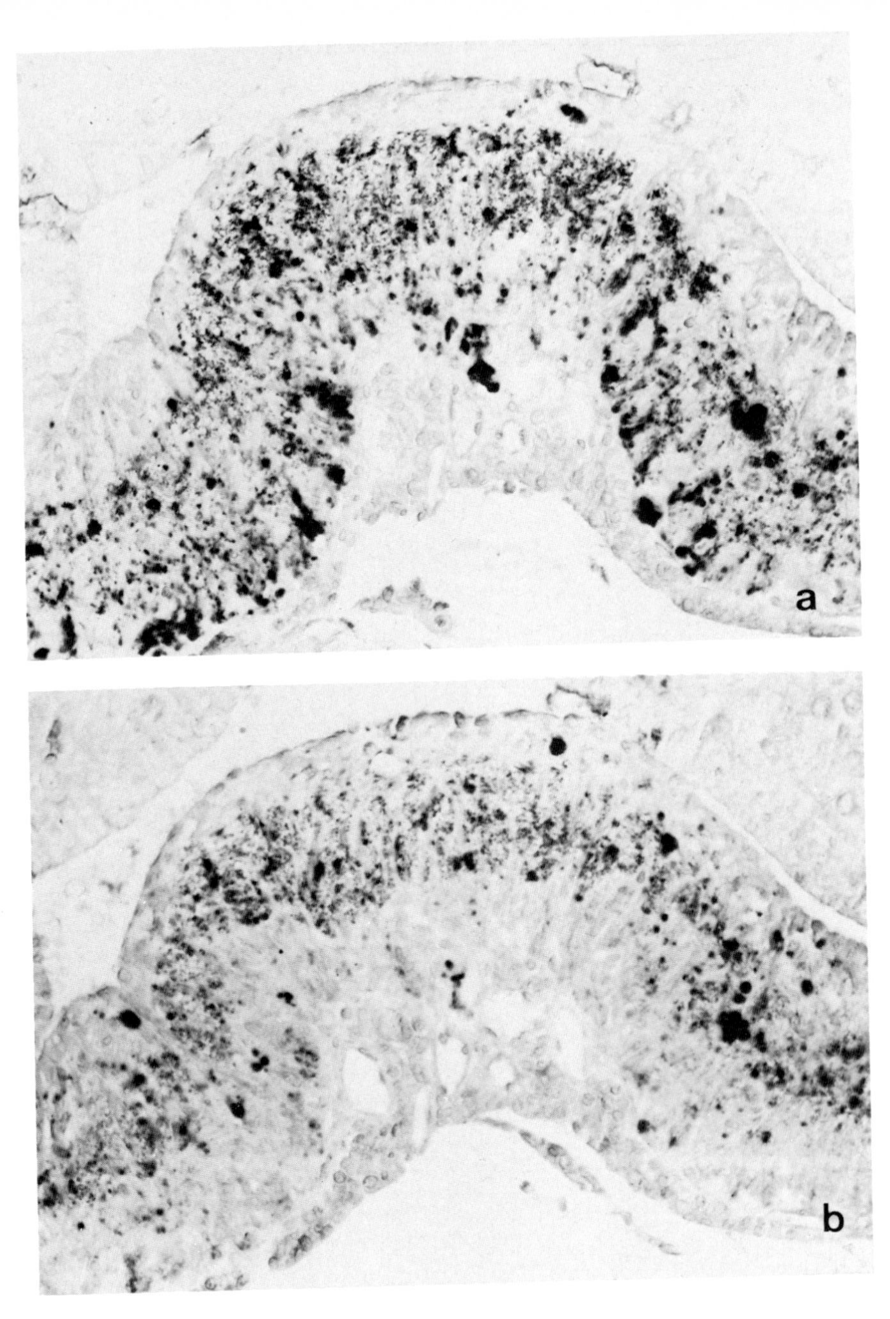

FIG. 10. Adjacent transverse sections of the median eminence of a bilaterally adrenalectomized rat: stained (a) with differentially adsorbed anti-vasopressin serum; (b) with differentially adsorbed anti-oxytocin serum. Compare the external region of the median eminence with that of Fig. 9. (From Dierickx *et al.*, 1976.)

TABLE III

NUMBER OF SELECTIVELY STAINED NERVE FIBERS IN THE EXTERNAL REGION OF THE RAT MEDIAN EMINENCE AFTER STAINING WITH DIFFERENTIALLY ADSORBED ANTI-VASOPRESSIN SERUM, DIFFERENTIALLY ADSORBED ANTI-OXYTOCIN SERUM, AND ANTISERUM AGAINST BOVINE NEUROPHYSIN I

Animal	Single staining			Double staining	
	Anti-vasopressin serum	Anti-oxytocin serum	Anti-neurophysin I serum	Anti-oxytocin + anti-vasopressin sera	Anti-oxytocin serum + anti-neurophysin I serum
Normal	±[a]	[±]	±		
Bilaterally adrenalectomized	++→+++	±→+	++→+++	++→+++	
Colchicine treatment subsequent to bilateral adrenalectomy	0→[±]	0→[±]	0→[±]		
Homozygous Brattleboro rat	0	±	±		± All reactive fibers are mixed stained

[a]Semiquantitative estimation of the number of immunoreactive fibers: 0, absent; [±], very few; ±, few; +, moderate number; ++, large number; +++, very large number.

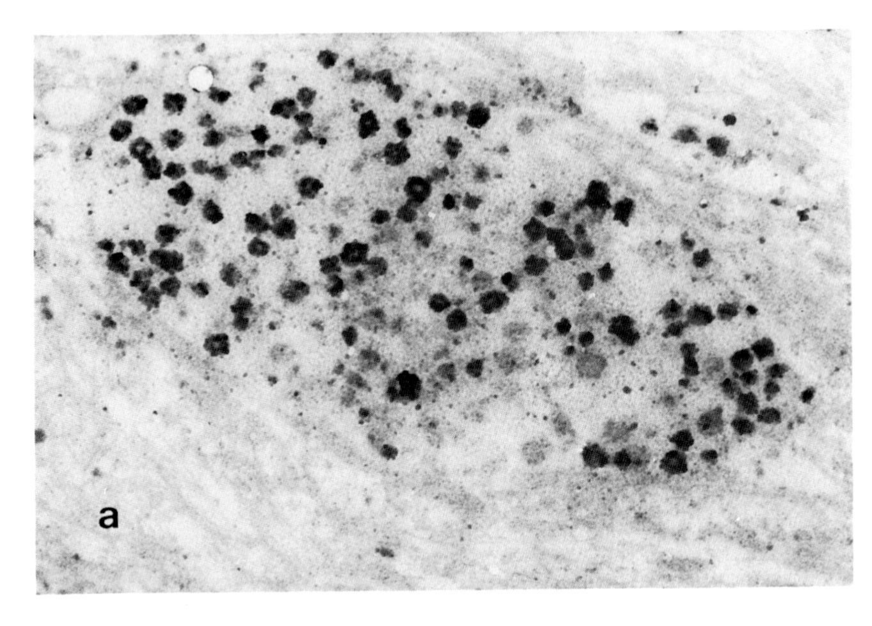

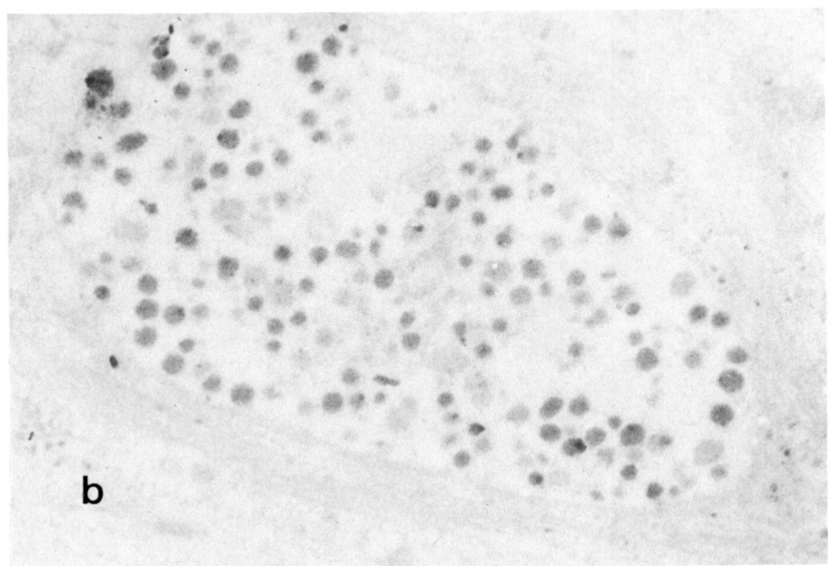

FIG. 11. Electron micrographs of two serial sections of the internal region of the median eminence of *Rana temporaria:* (a) axon stained with differentially adsorbed anti-vasotocin serum; (b) the *same* axon stained with differentially adsorbed anti-mesotocin serum. PAP method. ×15,000. (From Van Vossel *et al.*, to be published.)

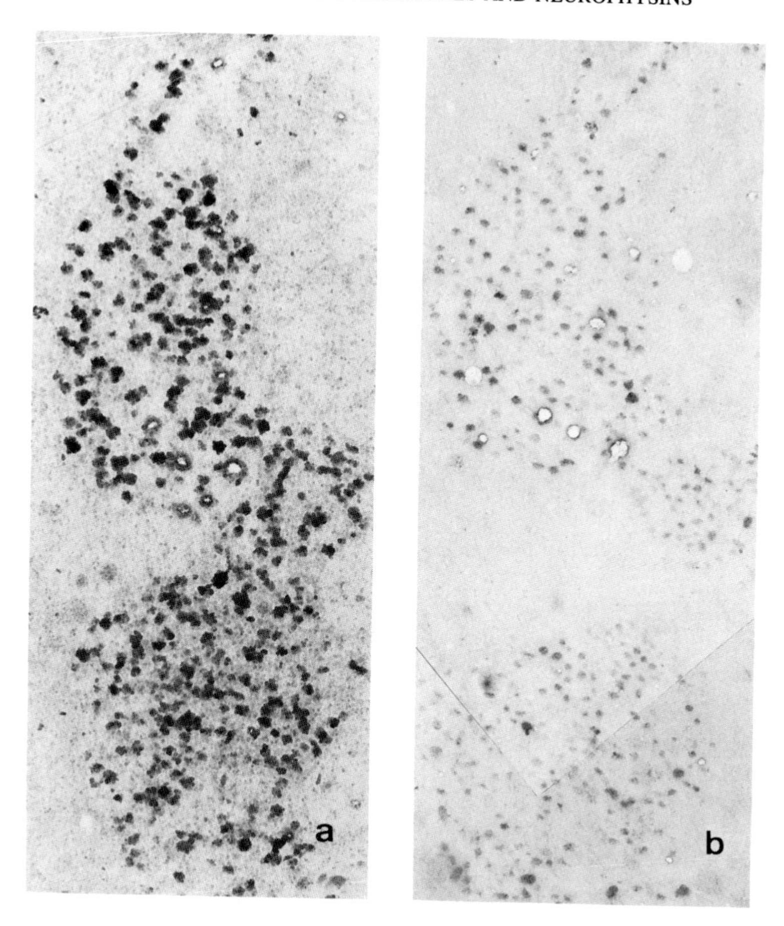

FIG. 12. Electron micrographs of two serial sections of the external region of the median eminence of *Rana temporaria:* (a) group of axons stained with differentially adsorbed anti-vasotocin serum; (b) the *same* axons stained with differentially adsorbed anti-mesotocin serum. PAP method. ×12,750. Compare mean size of vasotocin-containing granules with that of vasotocin-containing granules of Fig. 11. (From Van Vossel *et al.*, to be published.)

(Wittkowski *et al.*, 1970; Wittkowski and Bock, 1972) and in the guinea pig (Silverman and Zimmerman, 1975).

On the other hand, the external region of the avian anterior median eminence appears to contain exclusively vasotocinergic fibers (Goossens *et al.*, 1977a). Finally, the external region of the dipnoi median eminence contains separate vasotocinergic and mesotocinergic fibers (Goossens *et al.*, 1978). In all vertebrates examined, we found that the swollen endings of the neurohypophyseal

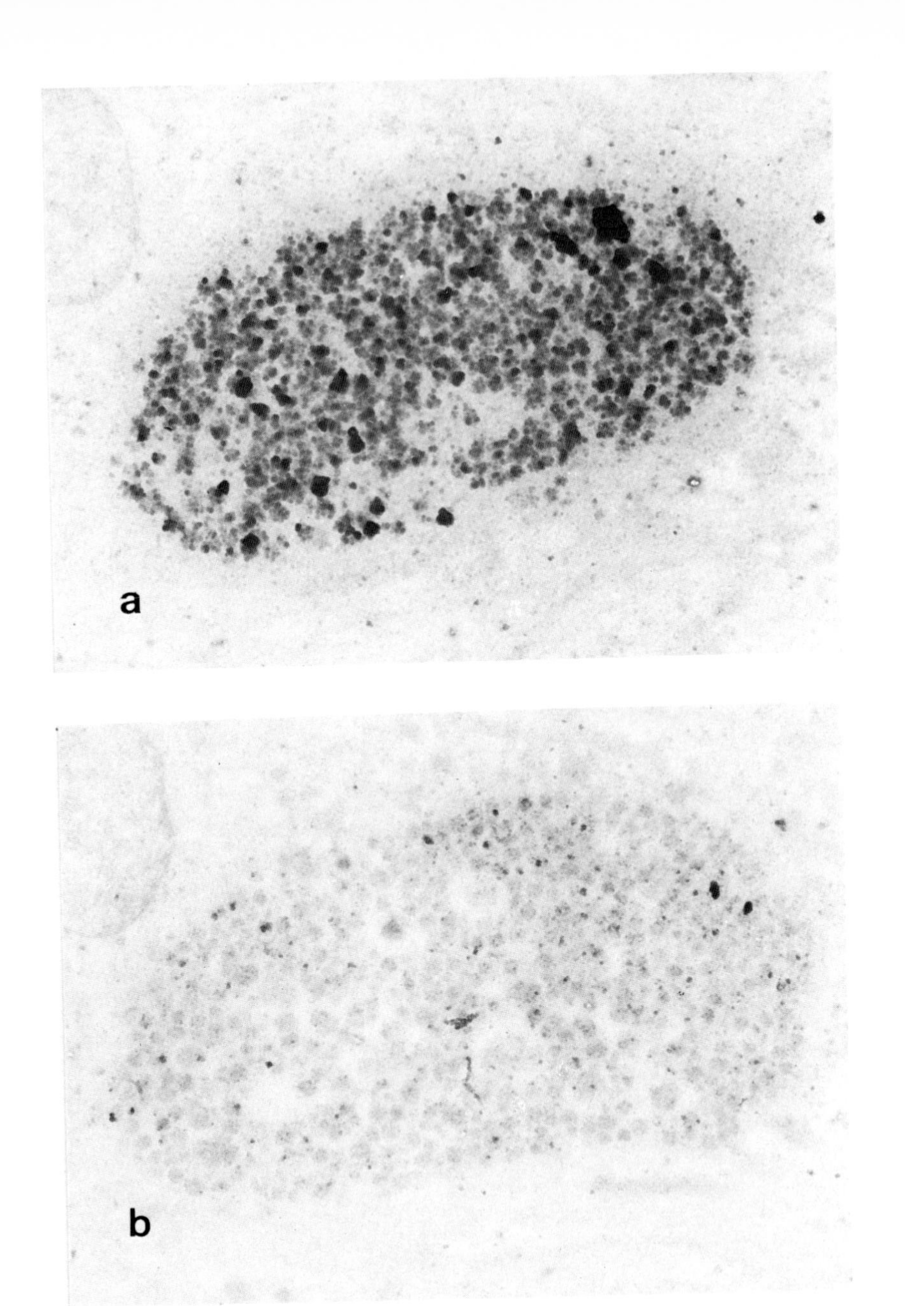

FIG. 13. Electron micrographs of two serial sections of the internal region of the median eminence of *Rana temporaria:* (a) axon stained with differentially adsorbed anti-mesotocin serum; (b) the *same* axon stained with differentially adsorbed anti-vasotocin serum. PAP method. ×15,000. (From Van Vossel *et al.*, to be published.)

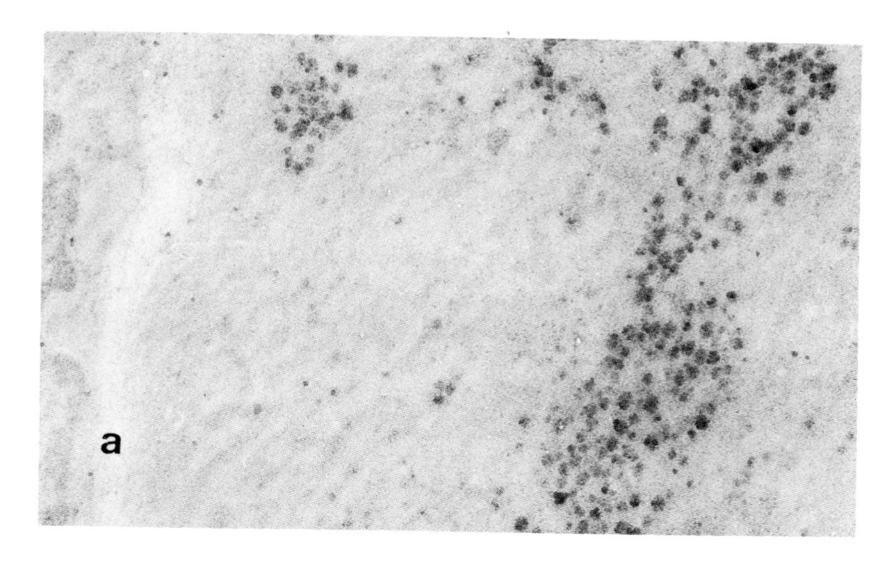

FIG. 14. Electron micrographs of two serial sections of the external region of the median eminence of *Rana temporaria:* (a) group of axons stained with differentially adsorbed anti-mesotocin serum; (b) the *same* group of axons stained with differentially adsorbed anti-vasotocin serum. PAP method. ×15,000. Compare mean size of mesotocin-containing granules with that of mesotocin-containing granules of Fig. 13. (From Van Vossel *et al.*, to be published.)

hormone-containing fibers accumulate around blood capillaries of the hypophyseal portal system.

C. The Origin of the Neurohypophyseal Hormone-Containing Fibers of the External Region of the Median Eminence

According to Zimmerman *et al*. (1973) and Robinson and Zimmerman (1973), some neurophysin (and neurohypophyseal hormone-)-containing fibers of the external region of the median eminence may be nerve fibers, while others could be processes of tanycytes. In our laboratory, this problem has been investigated in amphibians and mammals. The results of these studies, done on immunochemically single-stained and double-stained hypothalamic tissue sections, are the following.

In adult species of *Rana temporaria,* in which the hypothalamic magnocellular neurosecretory preoptic nuclei had been completely removed, the immunoreactive vasotocinergic and mesotocinergic fibers of the external region of the median eminence had disappeared. That the disappearance of these fibers was due to the complete removal of the magnocellular preoptic nuclei was shown by the fact that, in the external region of the median eminence of animals in which these nuclei had been incompletely removed, positively reacting fibers were still present. From these results it could be concluded that, at least the great majority of the vasotocinergic and mesotocinergic fibers of the external region of the amphibian median eminence are processes of neurosecretory perikarya located in the hypothalamic magnocellular preoptic nuclei. On the other hand, our results did not exclude the possibility that a minority of these fibers originate from small immunoreactive perikarya of neurons located in the tuber cinereum (Dierickx and Vandesande, 1977a).

From the results obtained in bilaterally adrenalectomized rats, in which total or subtotal destruction of both paraventricular hypothalamic nuclei had been performed, it appeared that neurons located in the paraventricular nuclei must be the origin of (nearly) all the vasopressinergic and oxytocinergic fibers of the external region of the rat median eminence (see Figs. 15 and 16; Table IV). Furthermore, the results of subtotal destruction of both paraventricular nuclei strongly suggested that both types of fibers originate from perikarya located in all parts of the paraventricular nuclei (Vandesande *et al.,* 1977). Studies done in monkeys with unilateral lesions of the paraventricular nucleus (Anthunes *et al.,* 1977),

Fig. 15. Sagittal sections of the median eminence stained with differentially adsorbed anti-vasopressin serum: (a) normal rat; (b) bilaterally adrenalectomized rat; (c) bilaterally adrenalectomized rat with total destruction of both paraventricular nuclei. PAP method. Compare the number of immunoreactive vasopressinergic fibers in the external region of the median eminence. (From Vandesande *et al.,* 1977.)

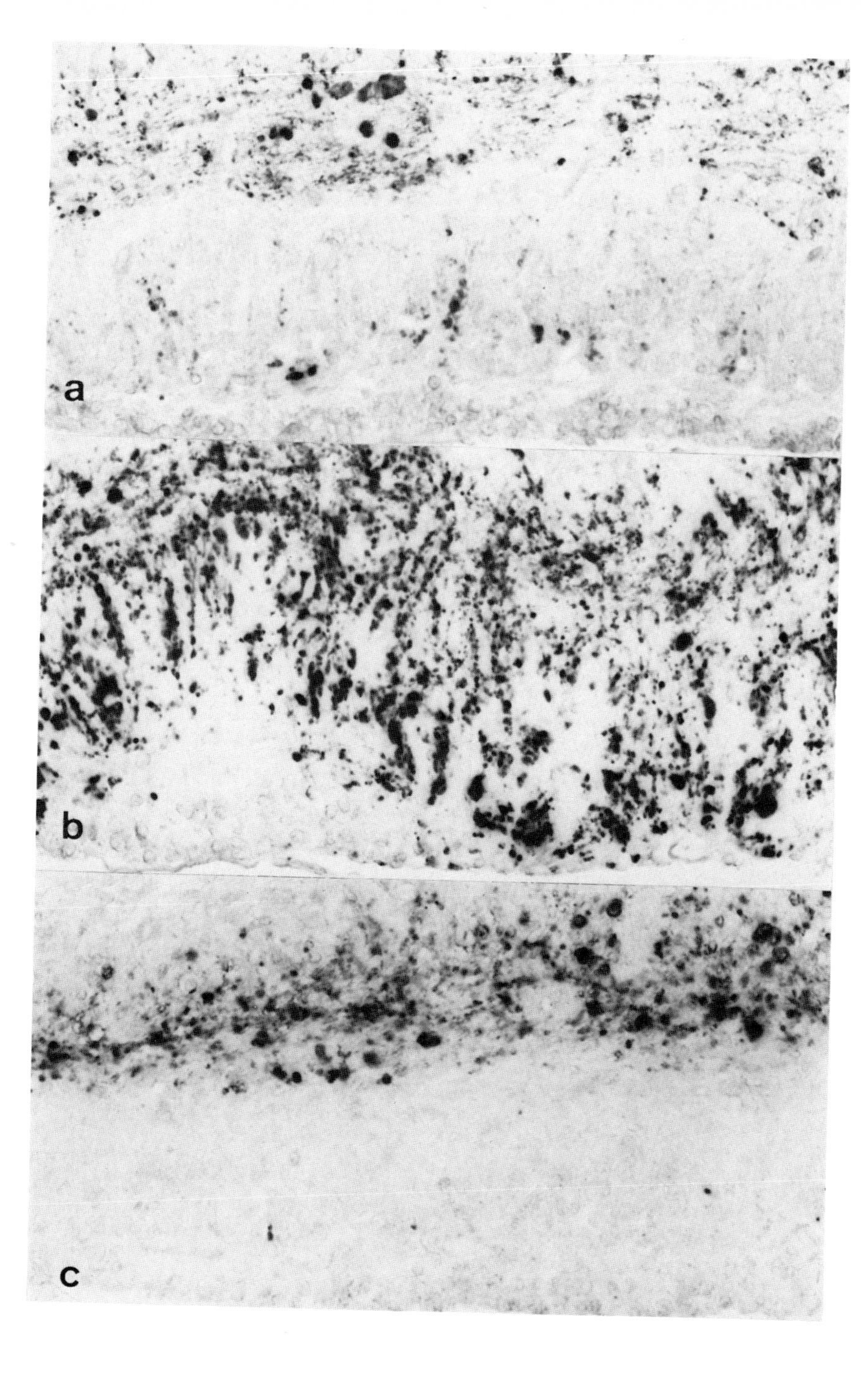
a
b
c

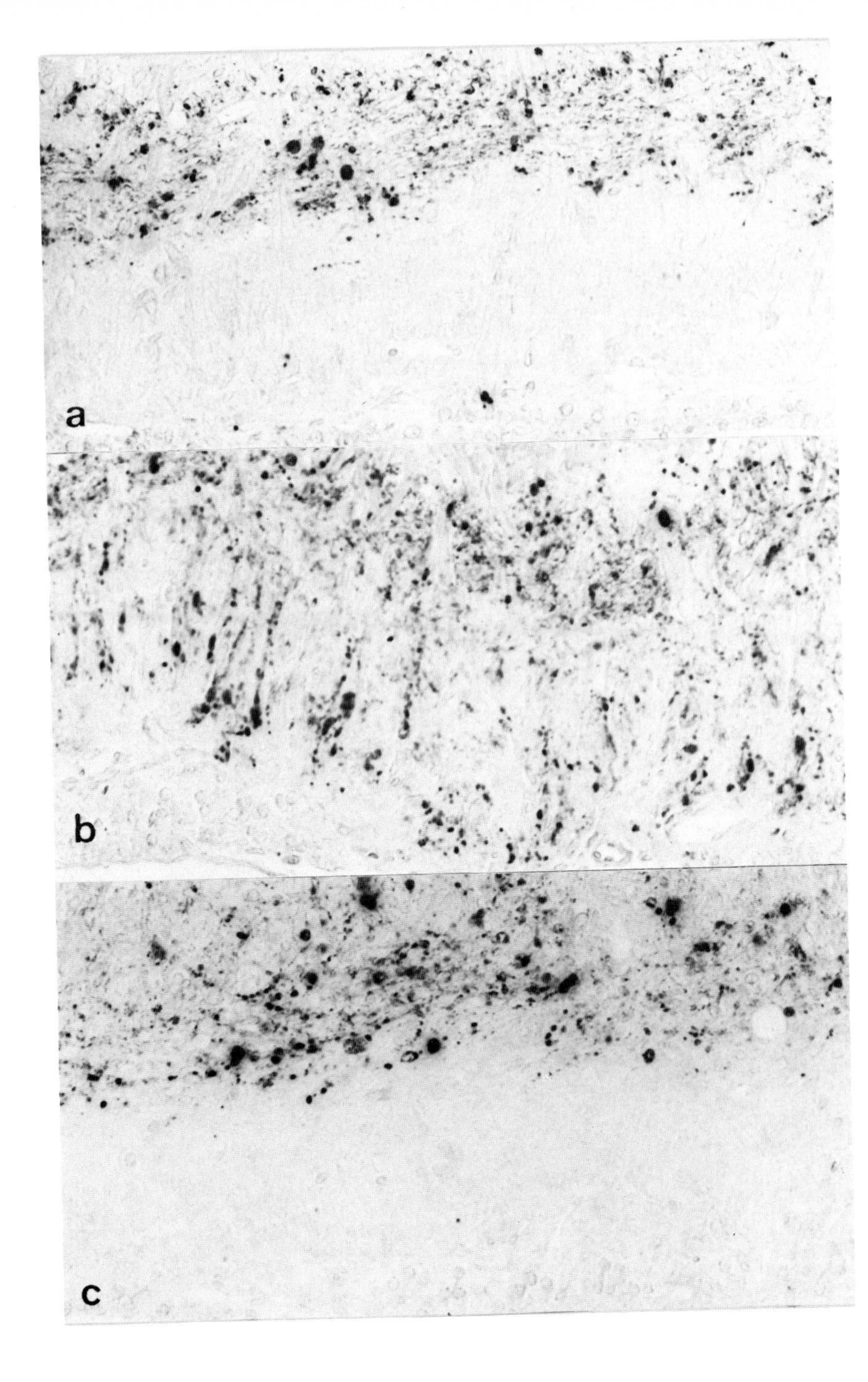
a
b
c

TABLE IV

NUMBER OF SELECTIVELY STAINED NERVE FIBERS IN THE EXTERNAL REGION OF THE MEDIAN EMINENCE OF NORMAL AND EXPERIMENTAL WISTAR RATS[a]

External region of median eminence			WAS		WAP		
	W	WA	WAST	WASP	WAPC	WAPS	WAPT
Oxytocinergic fibers	±[b]	+→++	+→++	+→++	+→++	±	0
Vasopressinergic fibers	+	++++	++++	++++	++++	±→+[+]	0

[a]Normal Wistar rats (W); bilaterally adrenalectomized rats (WA); bilaterally adrenalectomized rats with total (WAST) or partial (WASP) destruction of both suprachiasmatic nuclei (WAS); bilaterally adrenalectomized rats with total (WAPT) or subtotal (WAPS) destruction of both paraventricular nuclei (WAP); bilaterally adrenalectomized control rats with operative lesions near the paraventricular nuclei (WAPC).

[b]Semiquantitative estimation of the number of immunoreactive fibers: 0, absent; ±, very few fibers; +, small number; ++, large number; ++++, very large number.

and in rats with unilateral hypothalamic lesions (Bock and Jurna, 1977), have led to similar conclusions.

From all these observations, it may thus be concluded (1) that (practically) all the neurohypophyseal hormone-containing fibers of the external region of the median eminence are nerve fibers and (2) considering the results described in Section IV,B, that the one neuron–one hormone concept is also valid for the neurohypophyseal hormone-containing nerve fibers of the external region of the vertebrate median eminence.

It is still unknown whether there are separate neurons within the paraventricular (or the magnocellular preoptic) nucleus the axons of which extend only to the external region, or if the neurohypophyseal hormone-containing nerve fibers to the external region of the median eminence are collaterals of neurosecretory axons projecting to the neural lobe. However, a series of arguments plead in favor of the first possibility (see Bock and Jurna, 1977; Dierickx and Vandesande, 1977a; Anthunes *et al.*, 1977).

From heavy immunostaining of tanycytes located near the median eminence region, Zimmerman *et al.* (1973) and Robinson and Zimmerman (1973) have suggested that part of the immunoreactive neurophysin-containing fibers of the external region of the median eminence could be processes of tanycytes. How-

FIG. 16. Sagittal sections of the median eminence respectively adjacent to sections (a)–(c) of Fig. 15: (a) normal rat; (b) bilaterally adrenalectomized rat; (c) bilaterally adrenalectomized rat with total destruction of both paraventricular nuclei. PAP staining, using differentially adsorbed anti-oxytocin serum. Compare the number of immunoreactive oxytocinergic fibers of the external region of the median eminence. (From Vandesande *et al.*, 1977.)

ever, we have found that immunoreactive staining of tanycytes is due to aspecific, artificial staining and that all immunoreactive neurophysin- and neurohypophyseal hormone-containing fibers of the external region of the median eminence are to be considered as nerve fibers (Vandesande *et al.*, 1975c, 1977; see also Robinson, 1978). Thus, even if tanycytes play any (secondary) role in the transport of neurophysins and neurohypophyseal hormones, so far, there is no real immunocytochemical evidence in favor of this possibility.

D. The Nature of the Hormones of the "Neurohypophyseal Hormone"-Containing Fibers of the External Region of the Median Eminence

As is well known, immunocytochemical staining cannot exclude that, instead of neurohypophyseal hormones, the "neurohypophyseal hormone"-containing fibers of the external region of the median eminence could contain substances closely related to the neurohypophyseal hormones (see Watkins *et al.*, 1974; Bock *et al.*, 1976). Although this problem has not been definitely settled, an increasing body of evidence pleads in favor of the concept that it is the neurohypophyseal hormones and neurophysins themselves that are present in the fibers of the external region of the median eminence (see Dierickx *et al.*, 1976; Dierickx and Vandesande, 1977).

E. The Role of the Neurohypophyseal Hormone-Containing Fibers of the External Region of the Median Eminence

As described above, we found that the swollen endings of both types of neurohypophyseal hormone-containing nerve fibers of the external region of the vertebrate median eminence accumulate around blood capillaries of the hypophyseal portal system. This observation strongly suggests that both types of fibers release their content into the hypophyseal portal circulation. Thus, the vasopressin-containing fibers of the external region could be the source of the vasopressin which has been detected in the hypophyseal portal blood (Zimmerman *et al.*, 1973; Oliver *et al.*, 1977) and in the pars distalis of the hypophysis (Chateau *et al.*, 1974; Lutz *et al.*, 1974). On the other hand, evidence obtained in bilaterally adrenalectomized rats (see Bock, 1970; Brinkmann and Bock, 1973; Schwabedal *et al.*, 1977; Stillman *et al.*, 1977) and in amphibians (Notenboom *et al.*, 1976) indicates that the neurophysin–vasopressinergic (vasotocinergic) fibers of the external region of the median eminence could play a role in the control of the adrenocortical function. So far, however, conclusive evidence is lacking.

Nothing is known about the role of the neurophysin–oxytocinergic (mesotocinergic) nerve fibers of the external region of the vertebrate median eminence.

V. Immunocytochemical Localization of Neurohypophyseal Hormones and Neurophysins in Parvocellular Hypothalamic Nuclei

A. Suprachiasmatic Nucleus

In the medial (and dorsal) part of the rabbit suprachiasmatic nuclei, Gomori-positive parvocellular neurons have been described by Joussen (1970). As Gomori methods lack specificity, the significance of the Gomori-positive substance present in these suprachiasmatic nucleus neurons, remained unknown until we found that, in the rat, these neurons of the medial (and dorsal) part of the suprachiasmatic nucleus contain neurophysin. In addition, we showed that bilateral adrenalectomy causes a distinct increase in the number of immunoreactive neurophysin-containing suprachiasmatic nucleus neurons (Fig. 17; Vandesande *et al.*, 1974).

The demonstration of neurophysin-containing neurons in the suprachiasmatic nuclei prompted us, and other investigators, to examine whether these neurons also contain neurohypophyseal hormones. Therefore, we started studies on immunochemically single-stained and double-stained rat hypothalamic tissue sections. In the stainings, differentially adsorbed anti-neurohypophyseal hormone sera and anti-neurophysin sera were used. The following results have been obtained (Vandesande *et al.*, 1975b; Vandesande and Dierickx, 1976a; Dierickx *et al.*, 1976).

1. A moderate number of parvocellular neurons, located in the medial (and dorsal) part of the suprachiasmatic nuclei of the normal rat, contain vasopressin together with neurophysin (Figs. 17 and 18; Table V).
2. In the same region of the suprachiasmatic nuclei, the number of immunoreactive neurophysin–vasopressin-containing neurons distinctly increases after bilateral adrenalectomy (Table V).
3. The suprachiasmatic nuclei do not contain immunoreactive oxytocin-containing neurons (Fig. 18; Table V).
4. In the suprachiasmatic nuclei of the homozygous diabetes insipidus (Brattleboro) rat, vasopressin-, oxytocin-, and neurophysin-containing immunoreactive neurons are completely absent (see Table V).

In contradiction with the report of Defendini and Zimmerman (1978), we have shown that, also in the human suprachiasmatic nuclei, numerous parvocellular

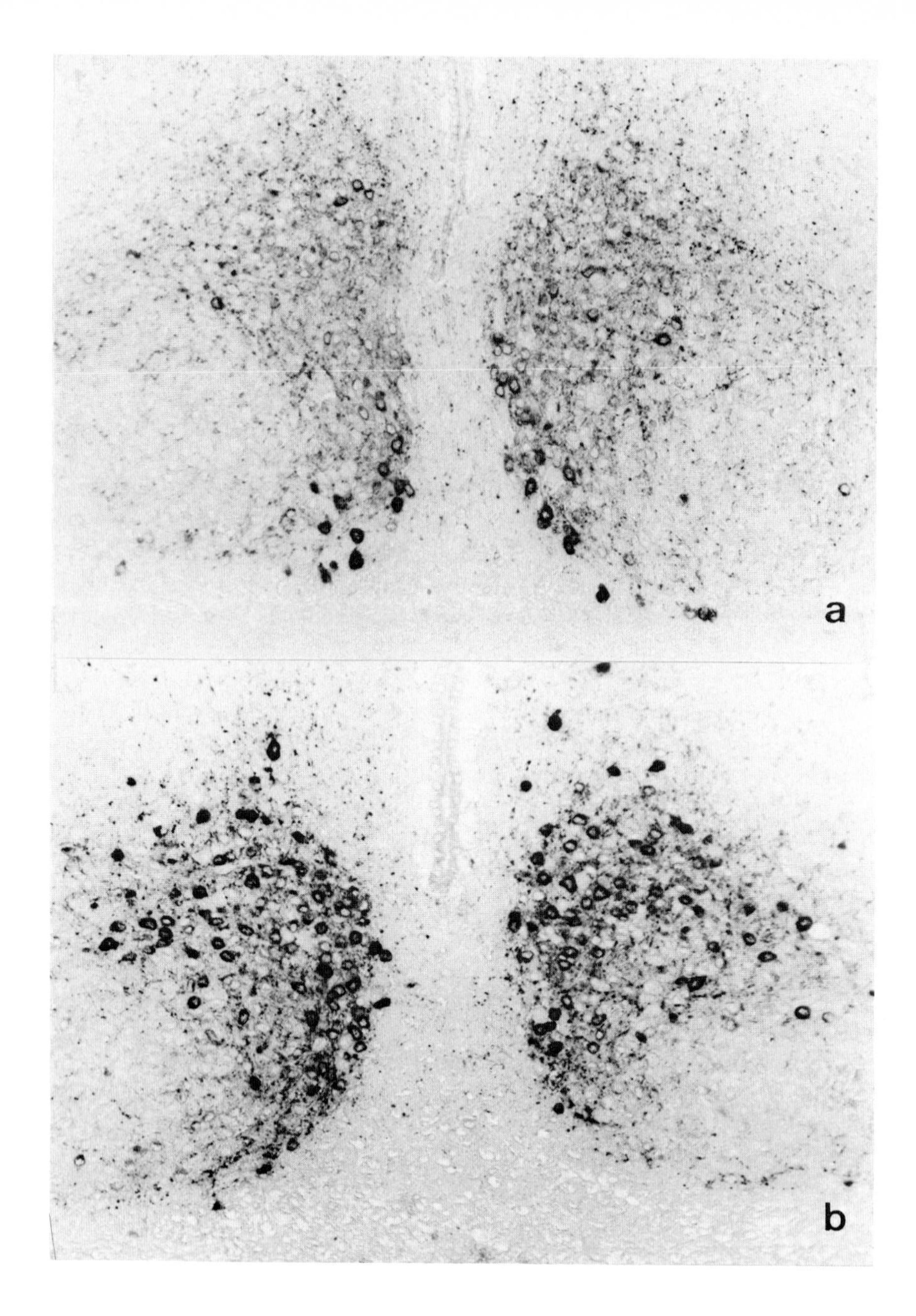

FIG. 17. Transverse section of the suprachiasmatic nuclei stained with anti-bovine neurophysin serum: (a) normal rat; (b) bilaterally adrenalectomized rat. Note the neurophysin-containing perikarya and processes. PAP method. (From Vandesande *et al.*, 1974.)

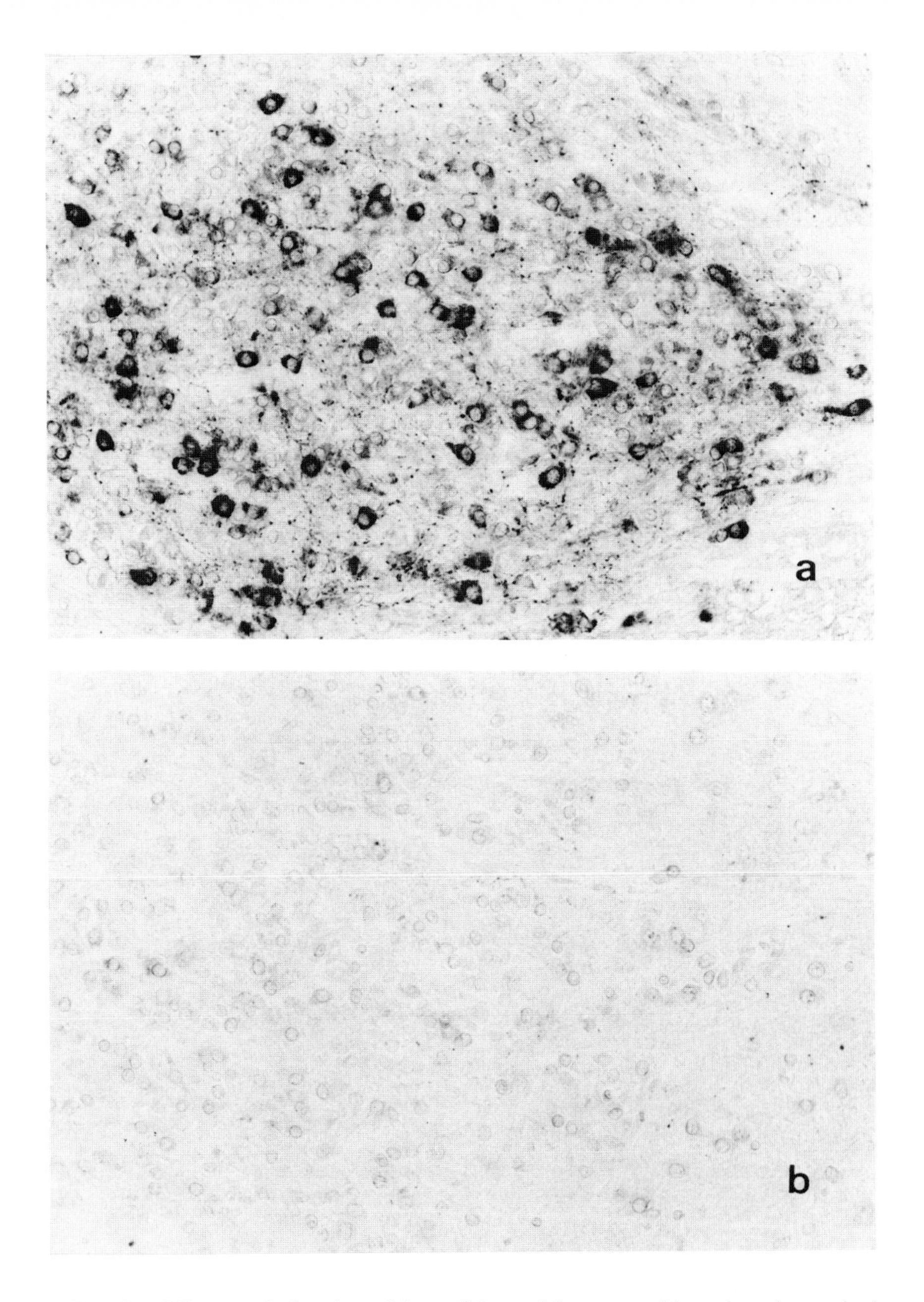

FIG. 18. Adjacent sagittal sections of the medial part of the rat suprachiasmatic nucleus: stained (a) with differentially adsorbed anti-vasopressin serum; (b) with differentially adsorbed anti-oxytocin serum. PAP staining. Note the vasopressin-containing perikarya in (a) and the absence of oxytocin-containing perikarya in (b). (From Vandesande *et al.*, 1975b.)

TABLE V
NUMBER OF SELECTIVELY STAINED NEURONS PRESENT IN THE SUPRACHIASMATIC NUCLEI AFTER SINGLE STAINING, USING DIFFERENTIALLY ADSORBED ANTI-VASOPRESSIN SERUM, DIFFERENTIALLY ADSORBED ANTI-OXYTOCIN SERUM, OR ANTISERUM AGAINST BOVINE NEUROPHYSIN I

Animal	Anti-vasopressin serum	Anti-neurophysin serum	Anti-oxytocin serum
Normal	±→+[a]	±→+	0
Bilaterally adrenalectomized	+→++	+→++	0
Colchicine treatment subsequent to bilateral adrenalectomy	+→++	+→++	0
Homozygous Brattleboro rat	0	0	0

[a]Semiquantitative estimation of the number of stained neurons. 0, Absent; ±, small number; +, moderate number; ++, large number.

neurons are present which contain vasopressin together with neurophysin. On the other hand, as in the rat, immunoreactive oxytocin-containing neurons are completely absent in the human suprachiasmatic nuclei (Dierickx and Vandesande, 1977b). The fact that the neurophysin–vasopressin-containing type of suprachiasmatic nucleus neuron exists not only in rodents, but also in man, strongly suggests that it is present in other mammals as well. Additional investigations are needed to settle this problem.

To be complete, it should be added that vasopressin-containing suprachiasmatic nucleus neurons have also been described in the rat by Burlet and Marchetti (1975), Swaab and Pool (1975), and Swaab *et al.* (1975b) and in the rat and mouse by Zimmerman *et al.* (1975). Moreover, it was found that, in the rat, the vasopressin-containing granules of the suprachiasmatic nucleus are significantly smaller than those of the magnocellular supraoptic nucleus (Van Leeuwen *et al.*, 1978).

As already mentioned, immunocytochemical staining cannot exclude the possibility that, instead of vasopressin and neurophysin, the neurophysin–vasopressin-containing suprachiasmatic nucleus neurons contain closely related substances. But the following arguments strongly suggest that these suprachiasmatic nucleus neurons really contain vasopressin and its associated neurophysin:

1. Like the vasopressin-containing neurons of the classical hypothalamic magnocellular neurosecretory system, the "vasopressin"-containing suprachiasmatic nucleus neurons are selectively stained with anti-vasopressin serum differentially adsorbed with Sepharose 4B–oxytocin. On the other hand, both

types of neurons do not stain either with differentially adsorbed anti-oxytocin serum or with differentially adsorbed anti-mesotocin serum.

2. In the homozygous Brattleboro rat, there is an absence of immunocytochemical staining for vasopressin in the suprachiasmatic nuclei as well as in the hypothalamic magnocellular neurosecretory system.

3. The human vasopressin-containing suprachiasmatic nucleus neurons are immunoreactive for vasopressin-associated neurophysin, while they do not stain for oxytocin-associated neurophysin (Dierickx and Vandesande, 1979).

4. The suprachiasmatic nuclei of the normal rat contain appreciable amounts of vasopressin, whereas oxytocin is absent. Moreover, the suprachiasmatic nuclei of the homozygous Brattleboro rat contain neither vasopressin nor oxytocin (George and Jacobowitz, 1975; George, 1976; George and Forrest, 1976).

So far, it has not been definitely settled whether the neurophysin–vasopressin-containing suprachiasmatic nucleus neurons either synthesize vasopressin and neurophysin or merely store these substances. The following arguments strongly plead in favor of the first possibility. First, the suprachiasmatic neurons show ultrastructural features of peptide hormone-producing cells (Suburo and Pellegrino de Iraldi, 1969). In addition, it is hardly conceivable that these neurons would show a concomitant uptake of vasopressin and its associated neurophysin, and would diffusely accumulate these substances in their perikarya, as well as in their processes.

Neither the efferent pathways nor the function of the vasopressinergic suprachiasmatic nucleus neurons are known. Recently, a vasopressin-containing pathway between the suprachiasmatic nucleus and the lateral habenular nucleus has been described in the rat (Buijs, 1978).

B. Infundibular (Arcuate) Nucleus

Immunocytochemical investigations have shown that the bovine infundibular (arcuate) nucleus contains parvocellular neurons which produce either neurophysin I or a closely related substance (De Mey *et al.*, 1975b). The significance of this finding is not known.

C. The Pars Ventralis of the Amphibian Tuber Cinereum

Immunocytochemical studies have demonstrated that the pars ventralis of the tuber cinereum of amphibians contains a few, separate vasotocin-containing and mesotocin-containing parvocellular neurons (Dierickx and Vandesande, 1977a). Additional investigations on the pathways of the processes and on the function of these neurohypophyseal hormone-containing cells are needed.

VI. Immunocytochemical Localization of Neurohypophyseal Hormones in the Pars Intermedia of the Hypophysis

Immunocytochemical studies, at the light and at the electron microscopic level, have shown that the pars intermedia of the amphibian hypophysis contains a diffuse intercellular network of separate fine mesotocinergic and vasotocinergic nerve fibers (Fig. 19). From results obtained in animals in which the hypothalamic magnocellular neurosecretory preoptic nuclei had been completely removed, it appeared that both types of fibers are axons of neurosecretory perikarya located in these hypothalamic nuclei. The role of the neurohypophyseal hormone-containing nerve fibers of the pars intermedia is still controversial (Dierickx and Vandesande, 1976; Van Vossel *et al.*, 1977).

Although there are indications that neurohypophyseal hormone-containing fibers are also present in the pars intermedia of other vertebrates (see Wingstrand, 1966; Bargmann *et al.*, 1967; Holmes and Ball, 1974), so far, immunocytochemical studies on such pars intermedia fibers in these animals are lacking.

VII. Immunocytochemical Localization of Neurohypophyseal Hormones and Neurophysins in Extrahypothalamic Pathways

Gomori-positive extrahypothalamic pathways, running to several parts of the brain (telencephalon, epithalamus, subfornical organ, mesencephalon), have been described in mammals as well as in lower vertebrates (Barry, 1956, 1961; Legait and Legait, 1957; Legait, 1959; Dierickx, 1962). Generally, in normal animals, these pathways are not, or scarcely, visible (Legait and Legait, 1957). They are more distinct when they accumulate "neurosecretory" material. Such accumulation occurs in hibernating animals and also in several experimental conditions. But, even then, the results are very variable (Legait and Legait, 1957; Legait, 1959; Dierickx, 1962). According to the authors mentioned above, all fibers of these pathways are processes of neurons located in the hypothalamic magnocellular neurosecretory nuclei. But, as Gomori methods lack specificity, the validity of this concept may be questioned. Therefore, immunocytochemistry has been used recently for the identification of the "neurosecretory" substances which are contained in the fibers of the Gomori-positive extrahypothalamic pathways. In the rat, neurophysin-containing fibers have been described running to the telencephalon (amygdala, nucleus of the diagonal band), to preganglionic parasympathetic cell groups located in the brain stem (Edinger–Westphal nucleus, nucleus ambiguus, dorsal motor nucleus of the vagus), and, at thoracic levels, to sympathetic (intermediolateral column) cell groups (Swanson, 1977;

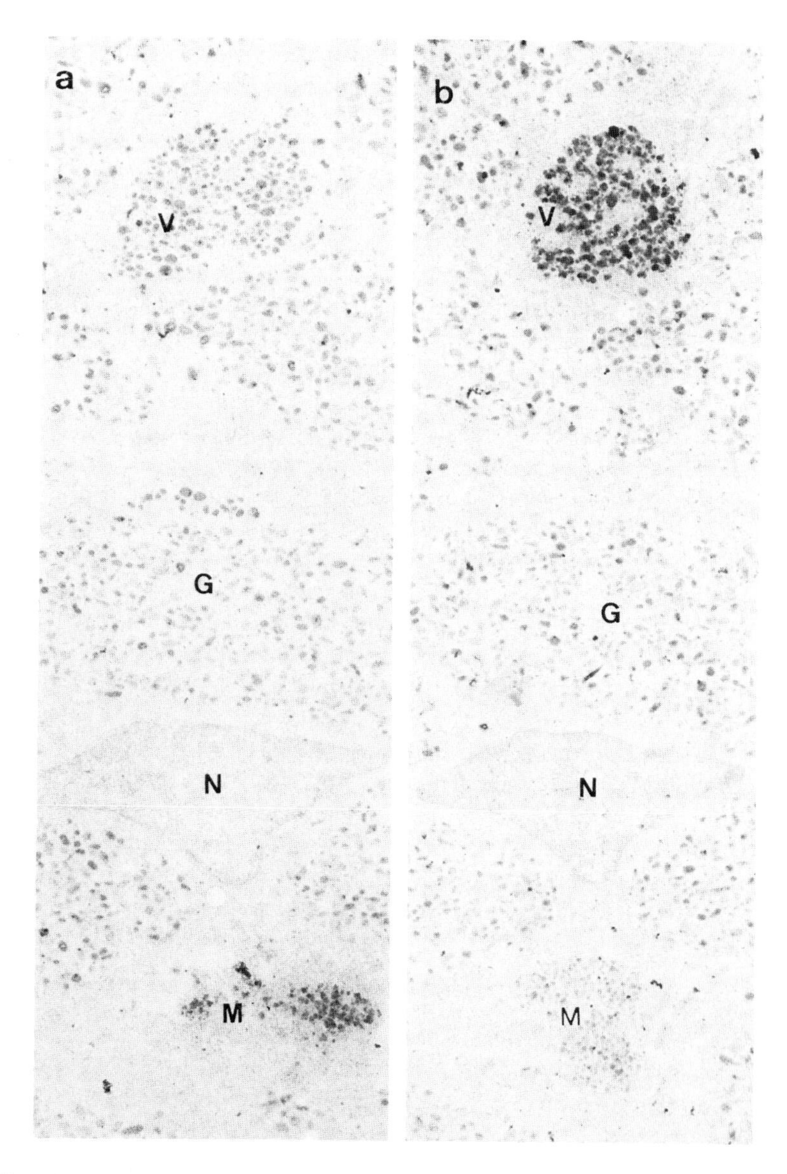

FIG. 19. Electron micrographs of two serial sections of the pars intermedia of the hypophysis of *Rana temporaria:* stained (a) with differentially adsorbed anti-mesotocin serum; (b) with differentially adsorbed anti-vasotocin serum. In (a), intense selective staining of the mesotocinergic axon and total absence of selective staining of the vasotocinergic axon. In (b), the staining of the *same* axons is exactly the opposite. M, Mesotocinergic axon; V, vasotocinergic axon; N, nucleus; G, secretory granules of pars intermedia cell. PAP staining. (From Van Vossel *et al.*, 1977.)

Sofroniew and Weindl, 1978). All these neurophysin-containing fibers would arise in the hypothalamic paraventricular nucleus. Via these pathways, the paraventricular nucleus could directly participate in the regulation of central autonomic cell groups (Swanson, 1977).

Fine neurophysin-containing fibers have been described, probably running from the suprachiasmatic nucleus to the lateral septum and dorsal thalamus (Sofroniew and Weindl, 1978). In addition, neurophysin-containing pathways have been traced from the rat hypothalamic supraoptic and paraventricular fields to the telencephalic choroid plexuses. This observation indicates a possible involvement of the neurosecretory system in the regulation of brain interstitial–ventricular cerebrospinal fluid dynamics (Brownfield and Kozlowski, 1977).

On the other hand, oxytocin- and vasopressin-containing extrahypothalamic pathways have been traced from the rat hypothalamic paraventricular nucleus to the dorsal and ventral hippocampus, the nuclei of the amygdala, substantia nigra and substantia grisea, nucleus tractus solitarius, nucleus ambiguus, and substantia gelatinosa of the spinal cord. (Buijs, 1978; Buijs *et al.*, 1978). In this connection, it may be mentioned that neurohypophyseal hormone-containing fibers have also been found in the organon vasculosum of the lamina terminalis of the rat (Buijs *et al.*, 1978) and of the monkey (Zimmerman and Anthunes, 1976).

From the above, it may be stated that immunocytochemistry of the extrahypothalamic neurohypophyseal hormone-containing pathways is only in its initial stage. So far, the available evidence concerning the origin and the function of these pathways is only inferential.

VIII. Immunocytochemistry and Morphology of the Hypothalamic Magnocellular Neurosecretory System

A. Anatomy of the Magnocellular Neurosecretory System

The immunohistochemical data concerning the anatomy of the vertebrate hypothalamic magnocellular neurosecretory nuclei and their pathways to the neural lobe (Fig. 20) are generally in good agreement with those obtained with Gomori methods (see Ortmann, 1951; Laqueur, 1954; Vandesande *et al.*, 1974; Goossens *et al.*, 1977a,c, 1978; and others). Even in man—and this in contradiction with the report of Defendini and Zimmerman (1978)—our immunohistochemical studies (Figs 21 and 22; Dierickx and Vandesande, 1977b) have confirmed the general shape and location of the paraventricular and supraoptic nuclei which have been determined using empirical staining methods (Greving, 1928; Le Gros Clark, 1936; Ingram, 1940; Christ, 1951; Diepen, 1962; Morton, 1969; Daniel and Prichard, 1975).

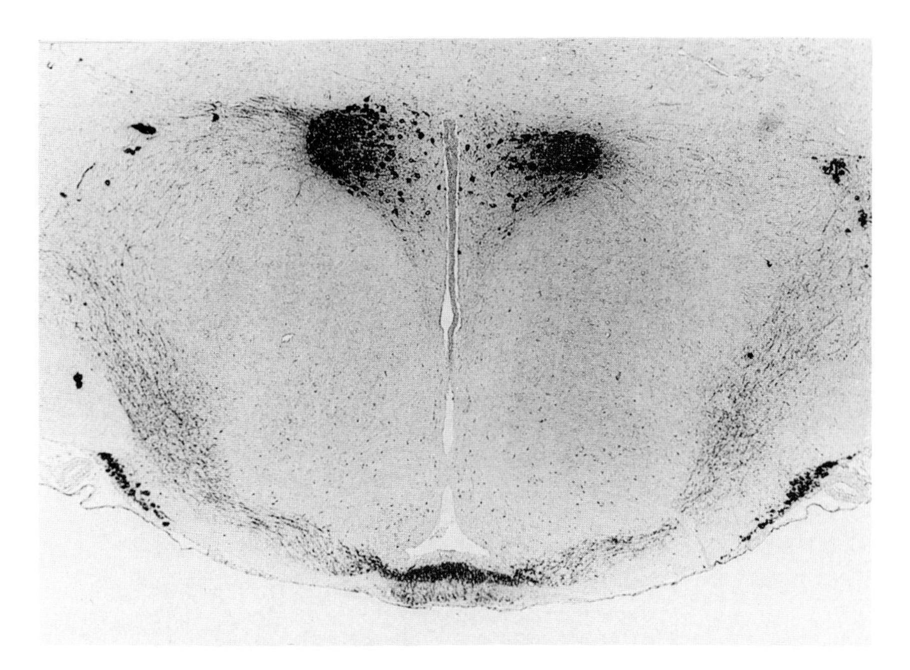

FIG. 20. Transverse section of the postchiasmatic region of the hypothalamus of a normal rat. PAP staining, using anti-bovine neurophysin serum. (From Vandesande *et al.*, 1974.)

On the other hand, immunohistochemistry has also added some new information to the knowledge derived from the Gomori stains. So, in the rat, the subependymally and medially located parvocellular part of the paraventricular nucleus consists of somatostatinergic neurons (see Section III,G; Fig. 7). In agreement with Bodian and Maren (1951), this parvocellular part may thus be considered as a separate cell group, morphologically and functionally distinct from the laterally located, neurohypophyseal hormone-producing magnocellular part of the rat paraventricular nucleus.

In man (Figs. 21 and 22), additional parvocellular neurohypophyseal hormone-containing cell groups have been found (1) in the roof of the optic recess and (2) between the lamina terminalis and the main mass of the paraventricular nucleus (Dierickx and Vandesande, 1977b).

In the neural lobe of some animals (e.g., rat, see Section III,C), studies on immunochemically double-stained sections have shown that the pattern of distribution of the two separate types of neurohypophyseal hormone-containing axons is partly different. Also in the neurosecretory pathways to the neural lobe, these two types of axons may be partly grouped in separate bundles, e.g., in the paraventricular-supraoptic tract of the rat (Dierickx and Vandesande, unpublished results). Finally, from immunohistochemical studies, a new concept is

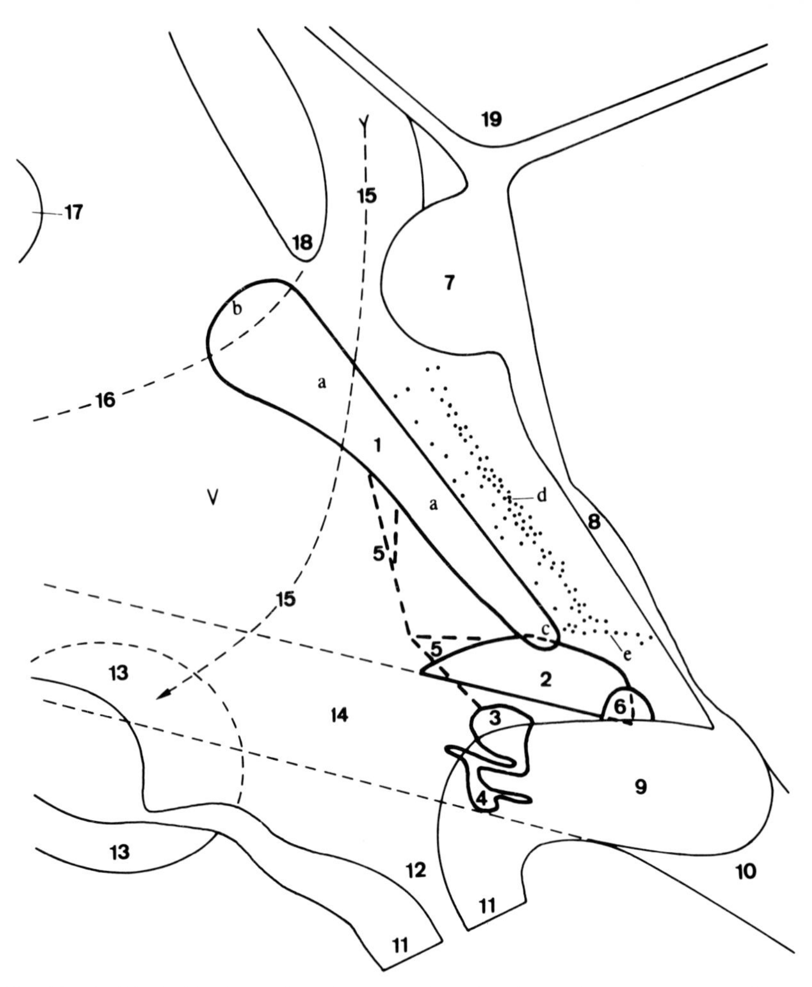

FIG. 21. Left half of the human hypothalamus projected on the mediosagittal plane. (1) Paraventricular nucleus with central part (a–a) and peripheral parts (b,c); (d,e) additional parvocellular neurohypophyseal hormone-containing cell groups; (2) dorsolateral part of the supraoptic nucleus; (3) dorsomedial part and (4) ventromedial part of the supraoptic nucleus; (5) interrupted lines indicating the location of the strands and small clusters of neurosecretory neurons extending between the supraoptic and paraventricular nuclei (= intersupraoptico − paraventricular islands); (6) suprachiasmatic nucleus as revealed by immunoreactive vasopressinergic neurons; (7) anterior commissure; (8) lamina terminalis; (9) optic chiasma; (10) optic nerve; (11) origin of pituitary stalk; (12) infundibular recess; (13) mamillary body; (14) optic tract (projected on the mediosagittal plane); (15) fornix + arrow indicating the central axis of the fornix; (16) hypothalamic sulcus; (17) interthalamic adhesion; (18) foramen of Monro; (19) septum pellucidum. V, Third ventricle. (From Dierickx and Vandesande, 1977b.)

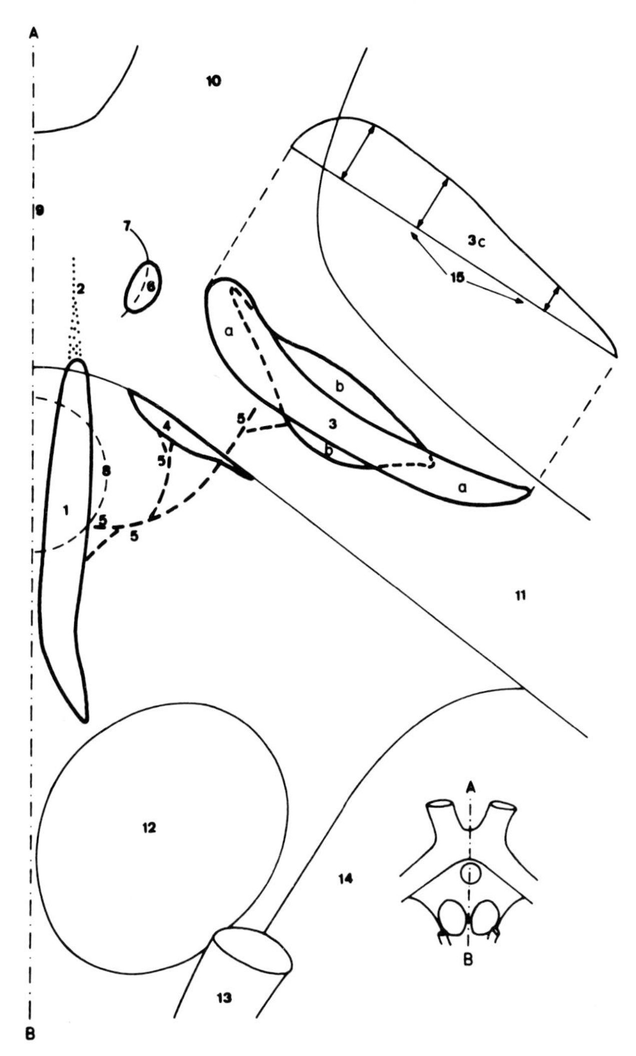

FIG. 22. Left half of the human hypothalamus projected on a horizontal plane (see inset). (1) Paraventricular nucleus; (2) strand of parvocellular neurosecretory neurons lying in the roof of the optic recess and extending from the ventrocaudal extremity of the paraventricular nucleus (PVN) toward the lamina terminalis; (3) dorsolateral part of the supraoptic nucleus (SON) represented by two horizontal sections, section a–a at the level of its maximal length and section b–b (located dorsal to section a–a) at the level of its maximal width; (3c) maximal dorsoventral dimensions of the dorsolateral part of the SON; (4) medial (= dorsomedial + ventromedial) part of SON, showing its maximal length and width; (5) interrupted lines indicating the location of the strands and small clusters of neurosecretory neurons, extending between the SON and PVN (= intersupraoptico-paraventricular islands); (6) suprachiasmatic nucleus as revealed by immunoreactive neurons; (7) limits of lateral extension of optic recess; (8) limits of the origin of pituitary stalk; (9) optic chiasma; (10) optic nerve; (11) optic tract; (12) mamillary body; (13) oculomotor nerve; (14) cerebral peduncle; (15) upper surface of optic tract. The interrupted line AB indicates the mediosagittal plane. (From Dierickx and Vandesande, 1977b.)

emerging about the anatomical and functional differentiation of the mammalian paraventricular nuclei. As has been mentioned (see Sections IV and VII), recent immunohistochemical findings indicate that only some neurons of the paraventricular nuclei send their axons to the neural lobe. Another, nonnegligible number of paraventricular nucleus neurons would be the origin of (1) the vasopressinergic and oxytocinergic fibers of the external region of the median eminence (Vandesande *et al.*, 1977; Anthunes *et al.*, 1977) and (2) the major part of the extrahypothalamic neurohypophyseal hormone (oxytocin?)-containing pathways (Swanson, 1977).

B. Cytomorphology, Distribution, and Ratio of the Two Types of Neurohypophyseal Hormone-Containing Neurons

1. *Introduction*

The neurons of the hypothalamic magnocellular neurosecretory system can be stained with suitable silver methods (Golgi silver stainings). Unfortunately, the results are precarious and incomplete (see Lefranc, 1966; Szentagothai *et al.*, 1968; Leontovich, 1970; Luqui and Fox, 1976). Other staining procedures are the Gomori and related methods. In contrast to the silver methods, the Gomori methods do not stain the neurons proper: The staining of perikarya and processes depends on the presence of neurosecretory material. There are striking differences between species with regard to the quantity of neurosecretory material which is found in the perikarya and along the fiber tracts. Moreover, in the same animal, the quantity of neurosecretory material depends on the rate of formation of neurosecretory granules by the neurosecretory cells and on the rate of discharge of the material from the neurohypophysis. Consequently, the Gomori methods also show important limitations.

As immunocytochemical procedures are also based on the presence of neurohypophyseal hormones and neurophysins, they show about the same limitations as those of the Gomori methods. On the other hand, they have the advantage of being more specific. Moreover, the PAP method is very sensitive.

2. *Immunocytochemical Results*

a. *The Cell Bodies.* Studies in our laboratory, done on immunochemically single-stained and double-stained hypothalamic serial tissue sections, indicate that the great majority of both types of vertebrate neurohypophyseal hormone-producing neurons are multipolar cells. Most of their perikarya are polygonal. Some perikarya (e.g., in the human supraoptic nucleus; in the preoptic nucleus of some fishes) are elongated, a number of them giving the false impression of bipolar neurons. This elongated shape appears to be an adaptation to the surrounding neuropil (see De Mey *et al.*, 1974; Vandesande and Dierickx, 1975;

Vandesande *et al.*, 1975a; Dierickx and Vandesande, 1977b; Goossens *et al.*, 1977a,b).

Studies on procion yellow-stained preoptic magnocellular neuroendocrine cells in the goldfish indicate that, in this species also, these neurons are multipolar (Hayward, 1974).

In some species (cow, man) a difference in mean size between the oxytocinergic and vasopressinergic perikarya has been described (De Mey *et al.*, 1974; Vandesande *et al.*, 1975a; Dierickx and Vandesande, 1977b). In the cow, the mean size of the oxytocinergic perikarya appeared to be larger than that of the vasopressinergic perikarya. In man, quite the reverse was found. Therefore, it may be questioned if this difference in mean size was not merely a transient physiological phenomenon, rather than a permanent feature. Studies in our laboratory, done on immunochemically single-stained and double-stained hypothalamic tissue sections have shown that both the mammalian supraoptic and paraventricular nuclei, and also the accessory neurosecretory islands, contain vasopressinergic and oxytocinergic perikarya. Moreover, the vasopressinergic and oxytocinergic perikarya may show a different preferential distribution pattern.

In the cow and rat, the dorsal region of the supraoptic part of the supraoptic nucleus consists mainly of oxytocinergic neurons, and its ventral region contains almost exclusively vasopressinergic neurons (Fig. 2). In the main mass of the pars dorsolateralis of the human supraoptic nucleus, of which more than 95% of the neuronal perikarya are vasopressinergic, the few oxytocinergic neurons are scattered among the vasopressinergic neurons. At its superiomedial margin, this main mass is covered with a strand of oxytocinergic perikarya. This strand is contiguous with the intersupraopticoparaventricular islands (Figs. 23 and 24).

In the rat paraventricular nucleus, the vasopressinergic neurons show a tendency to preferential grouping in the center of the nucleus, while the majority of the oxytocinergic neurons are located at the periphery of the nucleus. In the bovine and human paraventricular nucleus, the tendency to preferential grouping of oxytocinergic and vasopressinergic neurons is minimal (De Mey *et al.*, 1974; Vandesande *et al.*, 1975a; Vandesande and Dierickx, 1975, 1976a; Dierickx and Vandesande, 1977b).

So far, in lower vertebrates, no distinct regional distribution of the two types of neurohypophyseal hormone-producing neurons has been found (Vandesande and Dierickx, 1976b; Goossens *et al.*, 1977a,b, 1978).

Semiquantitative estimations have shown that there are striking differences between mammalian species with regard to the ratio of vasopressinergic to oxytocinergic neurons.

The human magnocellular neurosecretory nuclei are predominantly composed of vasopressinergic neurons: The great majority (more than 95%) of the neurons of the pars dorsolateralis of the supraoptic nucleus are vasopressinergic. In the

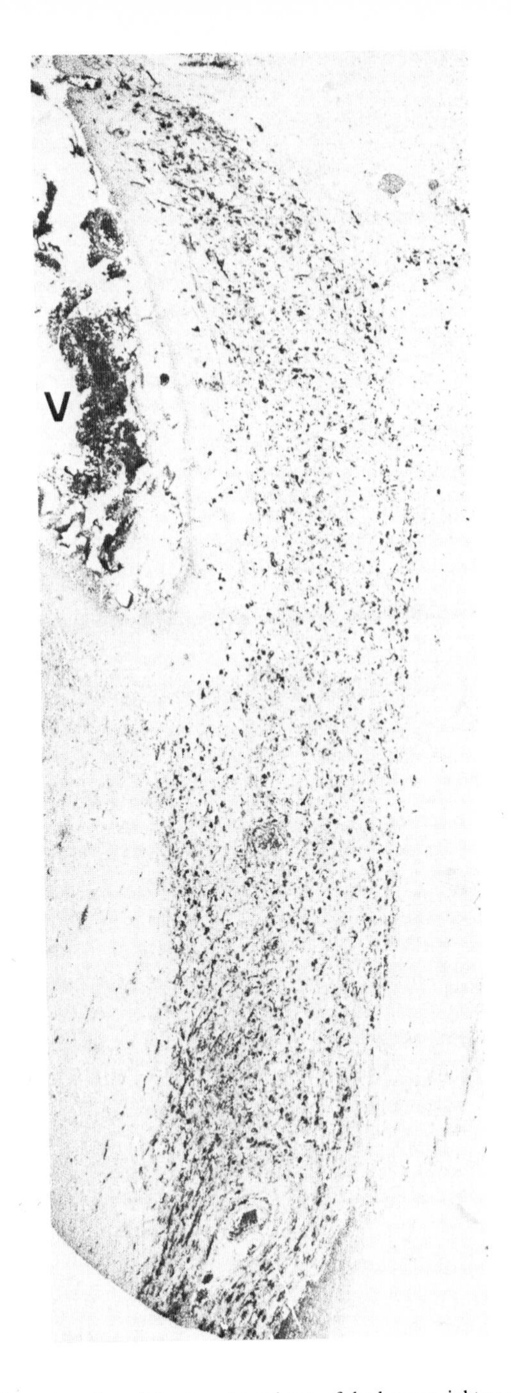

FIG. 23. Horizontal section of the dorsolateral part of the human right supraoptic nucleus corresponding to section a–a represented in Fig. 22. PAP staining, using differentially adsorbed anti-bovine neurophysin II serum. The adjacent section, stained with differentially adsorbed anti-vasopressin serum, showed a similar picture. V, Blood vessels of leptomeninx. (From Dierickx and Vandesande, 1979.)

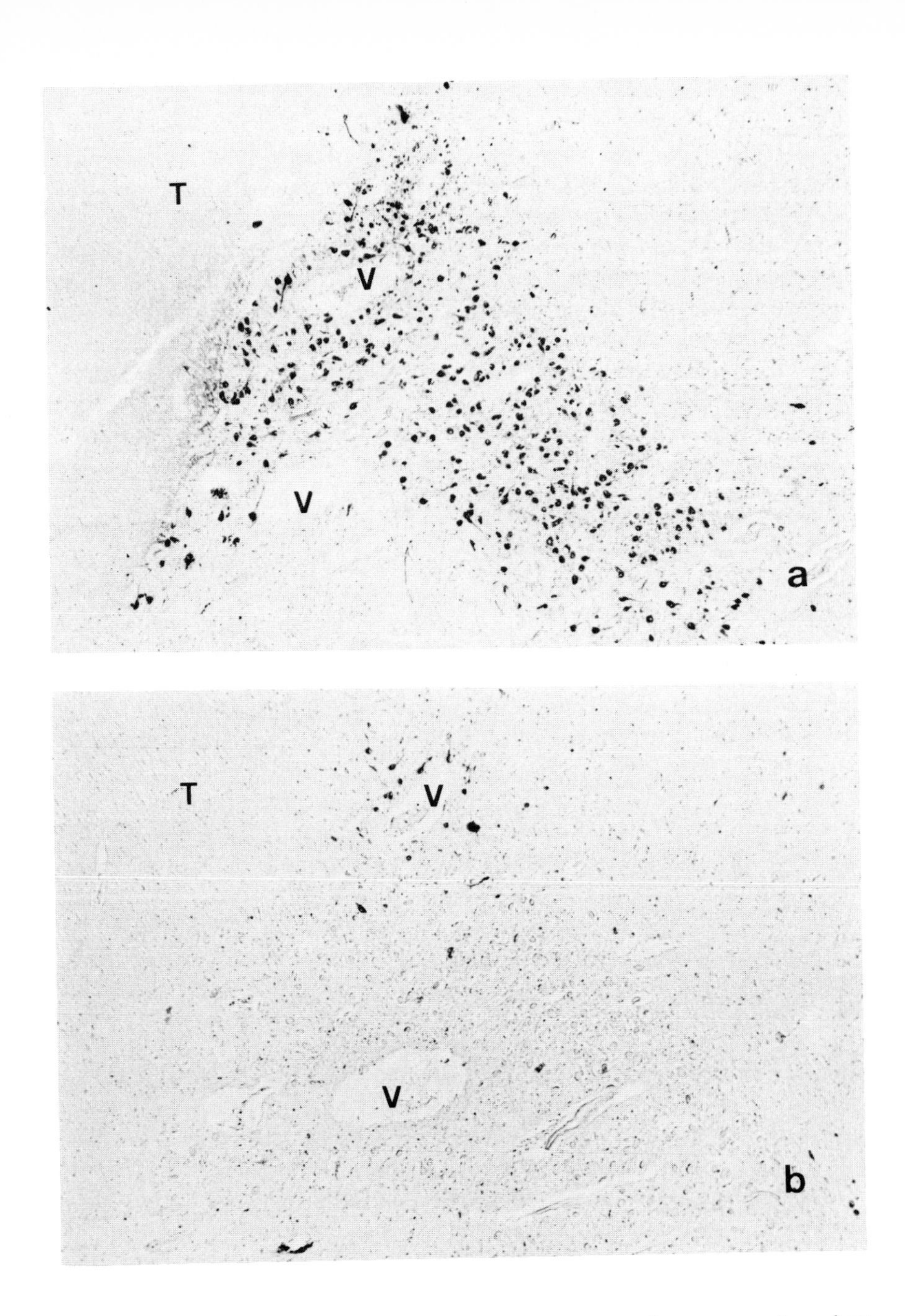

FIG. 24. Frontal (coronal) sections of the dorsolateral part of the human supraoptic nucleus: stained (a) with differentially adsorbed anti-bovine neurophysin II serum; (b) with differentially adsorbed anti-bovine neurophysin I serum. PAP staining. Similar pictures have been obtained when using differentially adsorbed anti-vasopressin serum and anti-oxytocin serum, respectively. Note that the great majority of the neurons are of the vasopressinergic type. V, Blood vessels; T, optic tract. (From Dierickx and Vandesande, 1979.)

pars medialis of the supraoptic nucleus, and in the greatest part of the main mass of the paraventricular nucleus, the vasopressinergic neurons are more numerous than the oxytocinergic neurons. Only in some limited small areas, the oxytocinergic neurons predominate in number: in a cell strand located at the superio(apico)-medial margin of the pars dorsolateralis of the supraoptic nucleus; in some internuclear islands; in a strand of small neurosecretory neurons, lying in the roof of the optic recess; and in a filiform cell aggregation located between the anterior commissure and the main mass of the paraventricular nucleus (Dierickx and Vandesande, 1977b).

In the bovine supraoptic nucleus, the vasopressinergic neurons are somewhat more numerous than the oxytocinergic neurons, whereas in the paraventricular nucleus, the oxytocinergic neurons largely predominate in number (De Mey *et al.*, 1974; Vandesande *et al.*, 1975a).

In the rat, we (Vandesande and Dierickx, 1975) found about equal proportions of vasopressinergic and oxytocinergic neurons in each of the two magnocellular nuclei. The report of Sokol *et al.* (1976) appears to confirm our estimations. On the other hand, Swaab *et al.* (1975b), using an immunofluorescence method, reported the following cell counts: "In the supraoptic nucleus, 52.8% of the cells appeared to contain vasopressin, 31.4% oxytocin, and 15.8% did not stain. In the paraventricular nucleus 50.6% of the cells contained vasopressin, 40.1% oxytocin, while 9.3% did not stain."

b. *Processes.* As has already been described, the great majority of the vertebrate hypothalamic magnocellular neurosecretory neurons are multipolar. On suitable sections of neurons, multiple processes leaving the perikaryon can be seen. Unfortunately, even in thick immunochemically stained tissue sections, the course of the individual processes can only be followed over a short distance. As with Gomori methods, this is principally due to the fact that only those fiber segments which contain adequate amounts of neurosecretory material are stained. The immunoreactive segments have the aspect of varicose fibers. It has to be noted that this varicose aspect is not merely typical for the classical neurosecretory fibers, since it is also found in monoaminergic fibers and in processes of releasing hormone-producing cells.

With the exception of a few myelinated axons to the neural lobe, so far, neither with immunocytochemistry nor with other methods (see Hayward, 1974) has it been possible to make a clear morphological distinction between "axons" and "dendrites" of magnocellular neurosecretory cells. It may even be questioned if it will ever be possible to make such a morphological distinction. Indeed, the magnocellular hypothalamic neurosecretory neurons (and also the parvocellular releasing hormone-producing neurons) appear to be a separate type of central autonomic neuron, quite different from the "ordinary" neurons of the central nervous system: Not only do they combine the characteristics of neurons with those of endocrine cells, but in addition—and this in contrast to "ordinary" central neurons—their, generally, nonmyelinated processes possess an enormous capacity for regeneration.

From the functional point of view, the neurosecretory processes to the neural lobe are to be considered as axons: They convey neurosecretory material and conduct action potentials to the neural lobe. Moreover, their endings in the neural lobe show the morphological phenomena of exocytosis (see review of Hayward, 1977).

The neurohypophyseal hormone-containing processes to the external region of the median eminence probably also function as axons: As has been described (see Section IV,E), their swollen neurohypophyseal hormone-containing pericapillary endings would release hormones into the blood capillaries of the hypophyseal portal system. Another category of processes are the extrahypothalamic fibers to the brain stem and spinal cord. Their considerable length pleads in favor of the axonal nature of these processes. According to Swanson (1977), they would be axons participating in the regulation of central autonomic cell groups.

In normal animals, only the pathways to the neural lobe can be rather easily

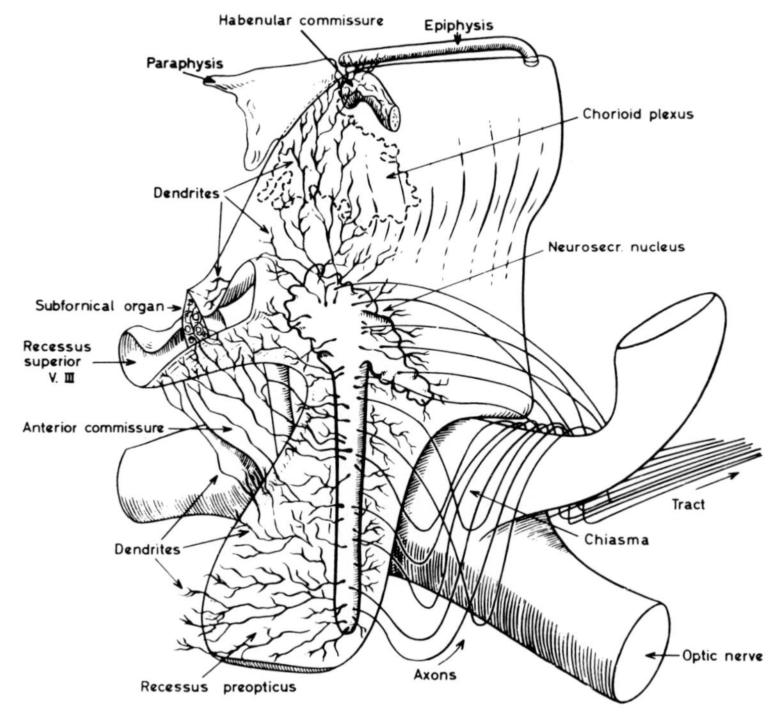

FIG. 25. Lateral view on the wall of the preoptic region of the third ventricle of the frog. General localization of the magnocellular neurosecretory preoptic nucleus, with its periventricular network of dendrites and its axons. (From Dierickx, 1962.)

traced. So far, the course of all other pathways is largely inferential. This shows that there must be a substantial difference in neurohypophyseal hormone–neurophysin content between the neurosecretory axons to the neural lobe, on the one hand, and all other neurosecretory fibers, on the other.

This lack of (adequate) neurohypophyseal hormone–neurophysin content is the reason that the morphology of the "dendrites" of the magnocellular neurosecretory neurons is largely unknown. In amphibians, we have shown that the neurons of the magnocellular preoptic nucleus possess an elaborate network of "dendrites," terminals of which penetrate into the ependymal lining and end, with an end bulb, in the cavity of the third ventricle (Figs. 25–27). In normal summer animals, this dendritic network does not contain neurosecretory material and therefore can not be visualized either with Gomori stains or with immunocytochemistry. On the contrary, in winter animals, and especially in experimental animals, in which the dendrites are filled with neurosecretory material, this dendritic network can be rather easily visualized (Dierickx, 1962; Vandesande and Dierickx, 1976b). Similar dendrites have been found in procion yellow-injected preoptic magnocellular endocrine neurons in the *goldfish* (Hayward, 1974).

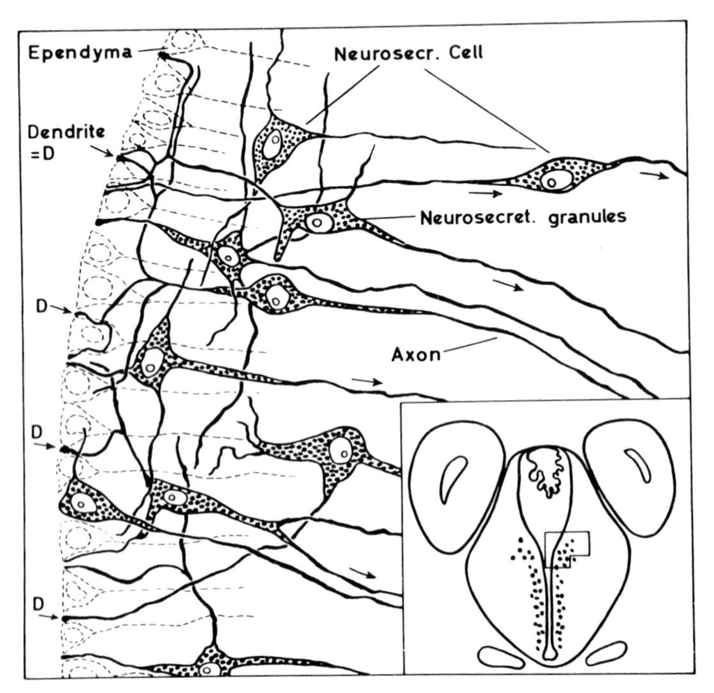

FIG. 26. Magnocellular preopticus nucleus of the frog: drawing of the neurosecretory neurons, their axons, and their periventricular dendrites. (From Dierickx, 1962.)

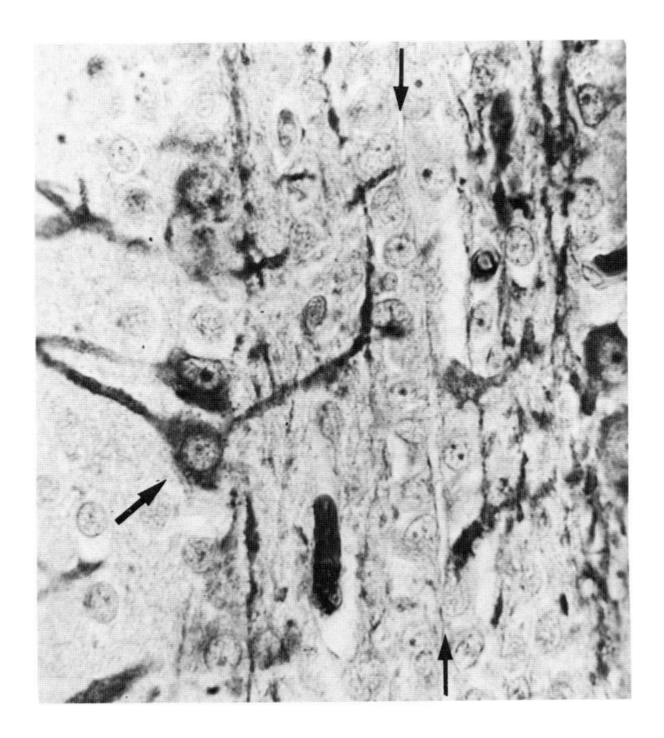

FIG. 27. Section of the magnocellular preoptic nucleus of the frog, showing neurosecretory perikarya and their periventricular dendrites. Note the neurosecretory cell (thick arrow), its axon, and its dendrite with ramifications to the third ventricle (fine arrows). (From Dierickx, 1962.)

Although their magnocellular neurosecretory neurons are also multipolar, so far it has been impossible to clearly visualize such a dendritic system in mammals. As in summer animals of amphibians, this could be due to the low neurohypophyseal hormone–neurophysin content of the mammalian neurosecretory dendrites. New marking techniques (e.g., intracellular injection of fluorescent dyes, retrograde transport techniques) should provide a means of tracing these mammalian processes.

From this brief overview, it may appear that, in the particular field of neurosecretory processes, there are many more questions than answers. A number of such questions are:

1. Are there separate neurons the axons of which extend only to the neural lobe, to the external region of the median eminence, and to extrahypothalamic areas, respectively, or are all these fibers collaterals of neurosecretory axons projecting to the neural lobe?
2. Do neurosecretory axons to the neural lobe have recurrent collaterals,

which could explain physiological evidence of recurrent inhibition and recurrent facilitation in magnocellular endocrine neurons (see review of Hayward, 1977)?

3. Do (some) magnocellular endocrine neurons have multiple axons, as suggested by observations of Hayward (1974)?

4. What is the nature of the morphological connections which have been described between neurosecretory "axons" and other cells (neurons, neuroglial cells, cells of the adenohypophysis)?

5. What are the connections between the neurosecretory dendrites and fibers of the surrounding neuropil?

6. What is the functional significance of the dendritic processes to the third ventricle?

New ideas and techniques are needed to resolve these, and other, important problems.

References

Acher, R. (1974). *In* "Handbook of Physiology" (E. Knobil and W. H. Sawyer, eds.), Vol. IV, pp. 119–130. Williams & Wilkins, Baltimore, Maryland.

Antunes, J. L., Carmel, P. W., and Zimmerman, E. A. (1977). *Brain Res.* **137,** 1.

Aspeslagh, M.-R., Vandesande, F., and Dierickx, K. (1976). *Cell Tissue Res.* **171,** 31.

Bargmann, W. (1968). *In* "Handbook of Experimental Pharmacology" (B. Berde, ed.), Vol. 23, pp. 1–39. Springer-Verlag, Berlin and New York.

Bargmann, W., Lindner, E., and Andres, K. H. (1967). *Z. Zellforsch. Mikrosk. Anat.* **77,** 282.

Barry, J. (1956). *C.R. Assoc. Anat.* **42,** 264.

Barry, J. (1961). *Ann. Sci. Univ. Besançon 2nd Ser.* **15,** 1.

Benoit, J., and Assenmacher, I. (1955). *J. Physiol.* **47,** 427.

Bigbee, J. W., Kosek, J. C., and Lawrence, F. (1977). *J. Histochem. Cytochem.* **25,** 443.

Bindler, E., Labella, F. S., and Sanwal, M. (1967). *J. Cell. Biol.* **34,** 185.

Bock, R. (1970). *In* "Aspects of Neuroendocrinology" (W. Bargmann and B. Scharrer, eds.), pp. 229–231. Springer-Verlag, Berlin and New York.

Bock, R., and Jurna, I. (1977). *Cell Tissue Res.* **185,** 215.

Bock, R., Salland, T., Schwabedal, P. E., and Watkins, W. B. (1976). *Histochemie* **46,** 81.

Bodian, D., and Maren, T. H. (1951). *J. Comp. Neurol.* **94,** 485.

Boudier, J.-A., Detieux, Y., Dutillet, B., and Luciani, J. (1972). *C.R. Acad. Sci. Paris Ser. D* **274,** 1051.

Brimble, M. J., and Dyball, R. E. J. (1977). *J. Physiol. (London)* **271,** 253.

Brinkmann, H., and Bock, R. (1973). *Naunyn-Schmiedeberg's Arch. Pharmacol.* **280,** 49.

Brownfield, M. S., and Kozlowski, G. P. (1977). *Cell Tissue Res.* **178,** 111.

Brownstein, M. J., Palkovits, M., Saavedra, J. M., and Kizer, J. S. (1976). *In* "Frontiers in Neuroendocrinology" (L. Martini and W. F. Ganong, eds.), Vol. 4, pp. 1–23. Raven, New York.

Buijs, R. M. (1978). *Cell Tissue Res.* **192,** 423.

Buijs, R. M., Swaab, D. F., Dogterom, J., and Van Leeuwen, F. W. (1978). *Cell Tissue Res.* **186,** 423.

Burford, G. D., Jones, C. W., and Pickering, B. T. (1971). *Biochem. J.* **124,** 809.

Burlet, A., and Marchetti, J. (1975). *C.R. Séances Soc. Biol.* **169,** 148.
Burlet, A., Marchetti, J., and Duheille, J. (1973). *C.R.Seances Soc. Biol.* **167,** 924.
Burlet, A., Marchetti, J., and Duheille, J. (1974). *In* "Neurosecretion—The Final Neuroendocrine Pathway" (F. Knowles and L. Vollrath, eds.), pp. 24–30. Springer-Verlag, Berlin and New York.
Calas, A. (1975). *In* "Brain-Endocrine Interaction. II. The Ventricular System in Neuroendocrine Processes" (K. M. Knigge and D. E. Scott, eds.), pp. 54–69. Karger, Basel.
Cameron, D. J., and Erlanger, B. F. (1977). *Nature (London)* **268,** 763.
Castel, M., and Hochman, J. (1976). *Cell Tissue Res.* **174,** 69.
Chateau, M., Burlet, A., and Marchetti, J. (1974). *J. Physiol.* **68,** 10.
Choy, V. J., and Watkins, W. B. (1977). *Cell Tissue Res.* **180,** 467.
Christ, J. (1951). *Dtsch. Z. Nervenheilk.* **165,** 340.
Cross, B. A. (1974). *In* "Neurosecretion—The Final Neuroendocrine Pathway" (F. Knowles and L. Vollrath, eds.), pp. 115–128. Springer-Verlag, Berlin and New York.
Cross, B. A., Dyball, R. E. J., Dyer, R. G., Jones, C. W., Lincoln, D. W., Morris, J. F., and Pickering, B. T. (1975). *In* "Recent Progress in Hormone Research" (R. O. Greep, ed.), Vol. 31, pp. 243–294. Academic Press, New York.
Daniel, P. M., and Prichard, M. M. L. (1975). *Acta Endocrinol. (Copenhagen)* **80,** Suppl. 201, 1.
Dean, C. R., Hope, D. B., and Kazic, T. (1968). *Br. J. Pharmacol.* **34,** 192P.
Defendini, R., and Zimmerman, E. A. (1978). *In* "The Hypothalamus" (S. Reichlin, R. J. Baldessarini, and J. B. Martin, eds.), pp. 137–154. Raven, New York.
De Mey, J., Vandesande, F., and Dierickx, K. (1974). *Cell Tissue Res.* **153,** 531.
De Mey, J., Dierickx, K., and Vandesande, F. (1975a). *Cell Tissue Res.* **157,** 517.
De Mey, J., Dierickx, K., and Vandesande, F. (1975b). *Cell Tissue Res.* **161,** 219.
Diepen, R. (1962). "Handbuch der Mikroskopischen Anatomie des Menschen," Vol. IV, pp. 1–525. Springer-Verlag, Berlin.
Dierickx, K. (1962). *Arch. Int. Pharmacodyn. Ther.* **140,** 708.
Dierickx, K., and Van den Abeele, A. (1959). *Z. Zellforsch. Mikrosk. Anat.* **51,** 78.
Dierickx, K., and Vandesande, F. (1976). *Cell Tissue Res.* **174,** 25.
Dierickx, K., and Vandesande, F. (1977a). *Cell Tissue Res.* **177,** 47.
Dierickx, K., and Vandesande, F. (1977b). *Cell Tissue Res.* **184,** 15.
Dierickx, K., and Vandesande, F. (1979). *Cell Tissue Res.* **196,** 203.
Dierickx, K., Vandesande, F., and De Mey, J. (1976). *Cell Tissue Res.* **168,** 141.
Dodd, J. M., Follett, B. K., and Sharp, P. J. (1971). *In* "Advances in Comparative Physiology and Biochemistry" (O. Lowenstein, ed.), Vol. 4, pp. 113–223. Academic Press, New York.
Dubois, M. P., and Kolodziejczyk, E. (1975). *C.R. Acad. Sci. Paris* **281,** 1737.
Elde, R., and Hökfelt, T. (1978). *In* "Frontiers in Neuroendocrinology" (W. F. Ganong and L. Martini, eds.), Vol. 5, pp. 1–33. Raven, New York.
Follett, B. K., and Heller, H. (1964). *J. Physiol. (London)* **172,** 74.
George, J. M. (1976). *Science* **193,** 146.
George, J. M. (1978). *Science* **200,** 342.
George, J. M., and Forrest, J. (1976). *Neuroendocrinology* **21,** 275.
George, J. M., and Jacobowitz, D. M. (1975). *Brain Res.* **93,** 363.
Ginsburg, M. (1968). *In* "Handbook of Experimental Pharmacology" (B. Berde, ed.), Vol. 23, pp. 286–371. Springer-Verlag, Berlin and New York.
Goldsmith, P. C., and Ganong, W. F. (1975). *Brain Res.* **97,** 181.
Goos, H. J. T. (1969). *Z. Zellforsch. Mikrosk. Anat.* **97,** 449.
Goossens, N., Blähser, S., Oksche, A., Vandesande, F., and Dierickx, K. (1977a). *Cell Tissue Res.* **184,** 1.
Goossens, N., Dierickx, K., and Vandesande, F. (1977b). *Gen. Comp. Endocrinol.* **32,** 371.
Goossens, N., Dierickx, K., and Vandesande, F. (1977c). *Cell Tissue Res.* **177,** 317.

Goossens, N., Dierickx, K., and Vandesande, F. (1978). *Cell Tissue Res.* **190,** 69.
Goossens, N., Dierickx, K., and Vandesande, F. (1979). *Cell Tissue Res.* **200,** 223.
Greving, R. (1928). *In* "Handbuch der Mikroskopischen Anatomie des Menschen" (W. Von Möllendorf, ed.), Vol. IV, pp. 918–1052. Springer-Verlag, Berlin.
Hayward, J. N. (1974). *J. Physiol. (London)* **239,** 103.
Hayward, J. N. (1977). *Physiol. Rev.* **57,** 574.
Hökfelt, T., Efendic, S., Hellerström, C., Johansson, O., Luft, R., and Arimura, A. (1975). *Acta Endocrinol.* **80,** Suppl. 200, 1.
Hökfelt, T., Elde, R., Fuxe, K., Johansson, O., Ljungdahl, A., Goldstein, M., Luft, R., Efendic, S., Nilsson, G., Terenius, L., Ganten, D., Jeffcoate, S. L., Rehfeld, J., Said, S., Perez de la Mora, M., Possani, L., Tapia, R., Teran, L., and Palacios, R. (1978). *In* "The Hypothalamus" (S. Reichlin, R. J. Baldessarini, and J. B. Martin, eds.), pp. 69–135. Raven, New York.
Holmes, R. L., and Ball, J. N. (1974). "The Pituitary Gland" (R. J. Harrison, R. M. H. McMinn, and J. E. Treherne, eds.), Cambridge Univ. Press, London and New York.
Husain, M. K., Frantz, A. G., Ciarochi, F., and Robinson, A. G. (1975). *J. Clin. Endocrinol. Metab.* **41,** 1113.
Ingram, W. R. (1940). *Res. Publ. Assoc. Res. Nerv. Ment. Dis.* **20,** 195.
Jones, C. W., and Pickering, B. T. (1969). *J. Physiol. (London)* **203,** 449.
Joussen, F. (1970). *Z. Zellforsch. Mikrosk. Anat.* **103,** 544.
Kalimo, H., and Rinne, U. K. (1972). *Z. Zellforsch. Mikrosk. Anat.* **134,** 205.
Laqueur, G. L. (1954). *J. Comp. Neurol.* **101,** 543.
Lederis, K. (1974). *In* "Handbook of Physiology" (E. Knobil and W. H. Sawyer, eds.), Vol. IV, pp. 81–102. Williams & Wilkins, Baltimore, Maryland.
Lefranc, G. (1966). *C.R. Acad. Sci. Paris* **263,** 976.
Legait, H. (1959). Ph.D. Thesis, Catholic University, Louvain.
Legait, H., and Legait, E. (1957). *Acta Anat.* **30,** 429.
Legros, J. J., Reynaert, R., and Peeters, G. (1974). *J. Endocrinol.* **60,** 327.
Legros, J. J., Reynaert, R., and Peeters, G. (1975). *J. Endocrinol.* **67,** 297.
Legros, J. J., Peeters, G., Marcus, S., De Groot, W., and Reynaert, R. (1976). *Arch. Int. Physiol. Biochem.* **84,** 887.
Le Gros Clark, W. E. (1936). *J. Anat.* **70,** 203.
Leontovich, T. A. (1970). *J. Hirnforsch.* **11,** 499.
Luqui, I. J., and Fox, C. A. (1976). *J. Comp. Neurol.* **168,** 7.
Lutz, B., Koch, B., and Mialhe, C. (1974). *C.R. Acad. Sci. Paris* **279,** 1903.
McNeill, T. H., Abel, J. H., Jr., and Kozlowski, G. P. (1975). *Cell Tissue Res.* **161,** 277.
Moore, G., Lutterodt, A., Burford, G., and Lederis, K. (1977). *Endocrinology* **101,** 1421.
Moriarty, G. C. (1973). *J. Histochem. Cytochem.* **21,** 855.
Moriarty, G. C., Moriarty, C. M., and Sternberger, L. A. (1973). *J. Histochem. Cytochem.* **21,** 825.
Morris, J. F. (1974). *J. Anat.* **118,** 376.
Morton, A. (1969). *J. Comp. Neurol.* **136,** 143.
Moses, A. M., and Miller, M. (1974). *In* "Handbook of Physiology" (E. Knobil and W. H. Sawyer, eds.), Vol. IV, pp. 225–242. Williams & Wilkins, Baltimore, Maryland.
Nairn, R. C. (1969). "Fluorescent Protein Tracing." Livingstone, Edinburgh and London.
Notenboom, C. D., Terlou, M., and Maten, M. L. (1976). *Cell Tissue Res.* **169,** 23.
Ochs, S. (1977). *In* "Hypothalamic Peptide Hormones and Pituitary Regulation" (J. C. Porter, ed.), pp. 13–40. Plenum, New York.
Oksche, A., and Farner, D. S. (1974). In "Advances in Anatomy, Embryology and Cell Biology" (A. Brodal, W. Hild, J. Van Limborgh, R. Ortmann, T. H. Schiebler, G. Töndury, and E. Wolff, eds.), Vol. 48/4, pp. 1–136. Springer-Verlag, Berlin and New York.
Oliver, C., Mical, R. S., and Porter, J. C. (1977). *Endocrinology* **101,** 598.

Orkand, P. M., and Palay, S. L. (1966). *Anat. Rec.* **154,** 396.
Orkand, P. M., and Palay, S. L. (1967). *Anat. Rec.* **157,** 295.
Ortmann, R. (1951). *Z. Zellforsch. Mikrosk. Anat.* **36,** 92.
Parry, H. B., and Livett, B. G. (1973). *Nature (London)* **242,** 63.
Paterson, J. A., and Leblond, C. P. (1977). *J. Comp. Neurol.* **175,** 373.
Pickering, B. T., and Jones, C. W. (1978). *In* "Hormonal Proteins and Peptides" (C. H. Li, ed.), Vol. V, pp. 103–158. Academic Press, New York.
Pickup, J. C., Johnston, C. I., Nakamura, S., Uttenthal, L. O., and Hope, D. B. (1973). *Biochem. J.* **132,** 361.
Poulain, D. A., and Wakerley, J. B. (1977). *J. Endocrinol.* **72,** 6P.
Poulain, D. A., Wakerley, J. B., and Dyball, R. E. J. (1977). *Proc. R. Soc. London B* **196,** 367.
Rinne, U. K. (1970). *In* "Aspects of Neuroendocrinology" (W. Bargmann and B. Scharrer, eds.), pp. 220–228. Springer-Verlag, Berlin and New York.
Robinson, A. G. (1975a). *J. Clin. Invest.* **55,** 360.
Robinson, A. G. (1975b). *Ann. N.Y. Acad. Sci.* **248,** 246.
Robinson, A. G. (1978). *In* "Frontiers in Neuroendocrinology" (W. F. Ganong and L. Martini, eds.), Vol. 5, pp. 35–59. Raven, New York.
Robinson, A. G., and Zimmerman, E. A. (1973). *J. Clin. Invest.* **52,** 1260.
Robinson, A. G., Zimmerman, E. A., and Frantz, A. G. (1971). *Metabolism* **20,** 1148.
Robinson, A. G., Haluszczak, C., Wilkins, J. A., Huellmantel, A. B., and Watson, C. G. (1977). *J. Clin. Endocrinol. Metab.* **44,** 330.
Rosenbloom, A. A., and Fisher, D. A. (1975). *Neuroendocrinology* **17,** 354.
Rurak, D. W., and Perks, A. M. (1974). *Gen. Comp. Endocrinol.* **22,** 480.
Rurak, D. W., and Perks, A. M. (1976). *Gen. Comp. Endocrinol.* **29,** 301.
Rurak, D. W., and Perks, A. M. (1977). *Gen. Comp. Endocrinol.* **31,** 91.
Sachs, H. (1970). *In* "Handbook of Neurochemistry" (A. Lajtha, ed.), Vol. IV, pp. 373–428. Plenum, New York.
Sawyer, W. H. (1965). *Gen. Comp. Endocrinol.* **5,** 427.
Sawyer, W. H. (1968). *In* "Handbook of Experimental Pharmacology" (B. Berde, ed.), Vol. 23, pp. 717–747. Springer-Verlag, Berlin and New York.
Sawyer, W. H., Munsick, R. A., and Van Dycke, H. B. (1961). *Endocrinology* **68,** 215.
Schwabedal, P. E., Bock, R., Watkins, W. B., and Möhring, J. (1977). *Anat. Embryol.* **151,** 81.
Silverman, A. J. (1976). *J. Histochem. Cytochem.* **24,** 816.
Silverman, A. J., and Zimmerman, E. A. (1975). *Cell Tissue Res.* **159,** 291.
Sloper, J. C. (1966). *In* "The Pituitary Gland" (G. W. Harris and B. T. Donovan, eds.), Vol. 3, pp. 131–239. Butterworths, London.
Sofroniew, M. V., and Weindl, A. (1978). *Endocrinology* **102,** 334.
Sokol, H. W., and Valtin, H. (1967). *Nature (London)* **214,** 314.
Sokol, H. W., Zimmerman, E. A., Sawyer, W. H., and Robinson, A. G. (1976). *Endocrinology* **98,** 1176.
Sternberger, L. A. (1974). "Immunocytochemistry" Prentice-Hall, New Jersey.
Sternberger, L. A., Hardy, P. H., Jr., Cuculis, J. J., Meyer, H. G. (1970). *J. Histochem.* **18,** 315.
Stillman, M. A., Recht, L. D., Rosario, S. L., Seif, S. M., Robinson, A. G., and Zimmerman, E. A. (1977). *Endocrinology* **101,** 42.
Stutinsky, F. S. (1970). *In* "The Hypothalamus" (L. Martini, M. Motta, and F. Fraschini, eds.), pp. 45–67. Academic Press, New York.
Suburo, A. M., and Pellegrino de Iraldi, A. (1969). *J. Anat.* **105,** 439.
Sunde, D. A., and Sokol, H. W. (1975). *Ann. N.Y. Acad. Sci.* **248,** 345.
Swaab, D. F., and Pool, C. W. (1975). *J. Endocrinol.* **66,** 263.
Swaab, D. F., Nijveldt, F., and Pool, C. W. (1975a). *J. Endocrinol.* **67,** 461.

Swaab, D. F., Pool, C. W., and Nijveldt, F. (1975b). *J. Neural Transm.* **36,** 195.
Swanson, L. W. (1977). *Brain Res.* **128,** 346.
Szentagothai, J., Flerko, B., Mess, B., and Halasz, B. (1968). ''Hypothalamic Control of the Anterior Pituitary. An Experimental-Morphological Study'' Akademiae Kiado, Budapest.
Tasso, F., Rua, S., and Picard, D. (1977). *Cell Tissue Res.* **180,** 11.
Valtin, H. (1977). *In* ''Disturbances in Body Fluid Osmolality,'' Vol. 9, pp. 197–215. Am. Physiol. Soc., Washington, D.C.
Valtin, H., Stewart, J., and Sokol, H. W. (1974). *In* ''Handbook of Physiology'' (E. Knobil and W. H. Sawyer, eds.), Vol. IV, pp. 131–171. Williams & Wilkins, Baltimore, Maryland.
Valtin, H., Sokol, H. W., and Sunde, D. (1975). *Rec. Progr. Horm. Res.* **31,** 447.
Vandesande, F., and Dierickx, K. (1975). *Cell Tissue Res.* **164,** 153.
Vandesande, F., and Dierickx, K. (1976a). *Cell Tissue Res.* **165,** 307.
Vandesande, F., and Dierickx, K. (1976b). *Cell Tissue Res.* **175,** 289.
Vandesande, F., De Mey, J., and Dierickx, K. (1974). *Cell Tissue Res.* **151,** 187.
Vandesande, F., Dierickx, K., and De Mey, J. (1975a). *Cell Tissue Res.* **156,** 189.
Vandesande, F., Dierickx, K., and De Mey, J. (1975b). *Cell Tissue Res.* **156,** 377.
Vandesande, F., Dierickx, K., and De Mey, J. (1975c). *Cell Tissue Res.* **158,** 509.
Vandesande, F., Dierickx, K., and De Mey, J. (1977). *Cell Tissue Res.* **180,** 443.
Van Oordt, P. G. W. J., Goos, H. J. TH., Peute, J., and Terlou, M. (1972). *Gen. Comp. Endocrinol.* Suppl. 3, 41.
Van Leeuwen, F. W., and Swaab, D. F. (1977). *Cell Tissue Res.* **177,** 493.
Van Leeuwen, F. W., Swaab, D. F., and Romijn, H. J. (1976). *In* ''First International Symposium on Immunoenzymatic Techniques INSERM Symposium No. 2'' (Feldmann *et al.,* eds.), pp. 345–353. North-Holland Publ., Amsterdam.
Van Leeuwen, F. W., Swaab, D. F., and de Raay, C. (1978). *Cell Tissue Res.* **193,** 1.
Van Vossel, A., Dierickx, K., Vandesande, F., and Van Vossel-Daeninck, J. (1976). *Cell Tissue Res.* **173,** 461.
Van Vossel, A., Van Vossel-Daeninck, J., Dierickx, K., and Vandesande, F. (1977). *Cell Tissue Res.* **178,** 175.
Van Wimersma Greidanus, T. B., Buys, R. M., Hollemans, H. J. G., and De Jong, W. (1974). *Experientia* **30,** 1217.
Watkins, W. B. (1973). *Z. Zellforsch. Mikrosk. Anat.* **145,** 471.
Watkins, W. B. (1976). *In* ''Progress in Neuropathology'' (H. M. Zimmerman, ed.), Vol. III, pp. 383–446. Grune & Stratton, New York.
Watkins, W. B., and Choy, V. J. (1976). *Neurosci. Lett.* **3,** 293.
Watkins, W. B., and Choy, V. J. (1977). *Cell Tissue Res.* **180,** 491.
Watkins, W. B., Schwabedal, P., and Bock, R. (1974). *Cell Tissue Res.* **152,** 411.
Wingstrand, K. G. (1966). *In* ''The Pituitary Gland'' (G. W. Harris and B. T. Donovan eds.), Vol. 3, pp. 1–27. Butterworth, London.
Wittkowski, W., and Bock, R. (1972). *In* ''Brain-Endocrine Interaction. Median Eminence: Structure and Function'' (K. M. Knigge, D. E. Scott, and A. Weindl, eds.), pp. 171–180. Karger, Basel.
Wittkowski, W., Bock, R., and Franken, C. (1970). *In* ''Aspects of Neuroendocrinology'' (W. Bargmann and B. Scharrer, eds.), pp. 324–328. Springer-Verlag, Berlin and New York.
Zimmerman, E. A. (1976a). *In* ''Subcellular Mechanisms in Reproductive Endocrinology'' (F. Naftolin, K. J. Ryan, and J. Davies, eds.), pp. 81–108. Elsevier, Amsterdam.
Zimmerman, E. A. (1976b). *In* ''Frontiers in Neuroendocrinology'' (L. Martini and W. F. Ganong, eds.), Vol. 4, pp. 25–62. Raven, New York.
Zimmerman, E. A., and Antunes, J. L. (1976). *J. Histochem. Cytochem.* **24,** 807.
Zimmerman, E. A., and Robinson, A. G. (1976). *Kidney Int.* **10,** 12.

Zimmerman, E. A., Carmel, P. W., Husain, M. K., Ferin, M., Tannenbaum, M., Frantz, A. G., and Robinson, A. G. (1973). *Science* **182,** 925.
Zimmerman, E. A., Robinson, A. G., Husain, M. K., Acosta, M., Frantz, A. G., and Sawyer, W. H. (1974). *Endocrinology* **95,** 931.
Zimmerman, E. A., Defendini, R., Sokol, H. W., and Robinson, A. G. (1975). *Ann. N.Y. Acad. Sci.* **248,** 92.
Zimmerman, E. A., Antunes, J., Carmel, P. W., Defendini, R., and Ferin, M. (1976). *Trans. Am. Neurol. Assoc.* **101,** 1.

INTERNATIONAL REVIEW OF CYTOLOGY, VOL. 62

Recent Progress in the Morphology, Histochemistry, Biochemistry, and Physiology of Developing and Maturing Mammalian Testis

SARDUL S. GURAYA

Department of Zoology, College of Basic Sciences and Humanities, Punjab Agricultural University, Ludhiana, Punjab, India

ISBN 0-12-364462-3

I. Introduction

The genetic or chromosomal sex of the individual is established after the zygote is formed. The sex-determining genes of the embryo begin to function in the undifferentiated or indifferent gonad. This results in the transformation of the gonad into a testis or ovary depending upon the sex. After their differentiation, the gonads begin to secrete some sex-specific hormonal substances which bring about the development, differentiation, and maturation of accessory sex organs and systems during the subsequent prenatal, postnatal, and prepubertal periods. Thus the developing mammalian testis and ovary undergo a long chain of successive morphological, histochemical, biochemical, and physiological changes during the embryonic, prenatal (or fetal), postnatal, and prepubertal periods before they become structurally and functionally mature. In the previous review, Guraya (1977a) has summarized, integrated, and discussed the results of recent ultrastructural, histochemical, biochemical, and physiological studies on the developing and maturing mammalian ovary. Since no detailed review dealing with similar aspects of the developing and maturing mammalian testis had been available previously, the present review with its multidisciplinary approach was, therefore, undertaken.

The details of changes could not be revealed more precisely in the earlier studies on the developing and maturing testis, which were mostly carried out using routine histological techniques (Van Wagenen and Simpson, 1965; Gier and Marion, 1970; Jost *et al.*, 1973). But the past few years have witnessed a great interest in the comparative study of origin, development, differentiation, structure, interrelationships, and physiology of different components of the mammalian testis during the embryonic, fetal, and postnatal periods as it has been extensively studied using the modern techniques of electron microscopy, histochemistry and biochemistry (including autoradiography and *in vitro* experiments), and physiology. The results of these comparative studies have not been integrated previously (Van Wagenen and Simpson, 1965; Gier and Marion, 1970; Jost *et al.*, 1973). However, Gondos (1977) has briefly discussed the morphological, biochemical, and physiological aspects of testicular development. The purpose of this review is, therefore, to summarize, integrate, and discuss in detail the cellular, subcellular, and molecular aspects of testicular development, differentiation, and maturation. Such integrated knowledge will be very useful for a better understanding and interpretation of the anomalies of testicular development as well as of the consequences of the influence of chemical and physical agents to which the mammalian testis is submitted at various stages of its development.

The classical experiments of fetal castration at different stages of development of the rabbit by Jost (1953) clearly demonstrated the crucial role of the fetal testis in accessory organ development. He showed that the presence of the fetal testis during Days 20 to 26 was essential for the establishment of the male type of development. The lack of influence of the fetal testis at this stage resulted in feminization. Subsequently, in several species of rodent the fetal testes have been shown to control the development of the male reproductive tract only a short time after gonadal sex differentiation (Jost, 1965, 1970; Price and Ortiz, 1965; Price *et al.,* 1969; Jost *et al.,* 1973). All these studies have indicated that some substance(s) is synthesized and secreted by the fetal testes at some stage of their development, which plays an important role in the development of male accessory organs at this crucial stage of development. The recent *in vivo* and *in vitro* biochemical studies have clearly demonstrated the synthesis and secretion of androgenic hormones by the fetal testes, which parallel the maximal occurrence and development of interstitial Leydig cells. In this review, the morphological (including ultrastructural), histochemical, biochemical, and physiological features of steroid-synthesizing cellular sites in the developing and maturing testes of mammals will be described and discussed in detail. An attempt will be made to compare and contrast them with those of developing mammalian ovary (Guraya, 1977a) and mature gonads of mammals and nonmammalian vertebrates (Christensen and Gillim, 1969; Guraya, 1971, 1973a–c, 1974a–d, 1976a,b, 1978; Neaves, 1975; Christensen, 1975). Such comparisons will be very useful in understanding the cellular sites of steroidogenesis and their regulation, characteristics, and functional significance in the developing testes of mammals during the embryonic, prenatal, postnatal, and prepubertal periods. The most striking common features of steroid-producing cells in the mature gonads of vertebrates, including mammals, are: (1) abundant diffusely distributed lipoproteins in the cytoplasm; (2) abundant membranes of smooth reticulum; (3) mitochondria with predominantly tubular cristae; (4) development of diffuse lipoproteins (or agranular endoplasmic reticulum), accompanied by various enzyme activities related to the biosynthesis of steroid hormones; (5) under certain physiological situations, stored cholesterol-positive lipid droplets; and (6) capacity to form a variety of steroids in biochemical experiments *in vitro* and *in vivo* (Guraya, 1971, 1976a,b, 1978).

Beside the steroidal secretions of the developing and maturing mammalian testis, the recent data in regard to the site of production of anti-müllerian hormone (AMH), androgen binding protein (ABP), and inhibin and the regulation of their synthesis and secretion will be described and discussed. Attempts will also be made to identify major gaps in current knowledge about various aspects of the development and maturation of mammalian testis.

II. Origin of Germ Cells

In the past there was a considerable controversy about the origin and continuity of the germ cell line in mammals (see reviews in Heys, 1931; Everett, 1945; Nieuwkoop, 1949). However, experimental (Ancel and Bouin, 1926; Everett,

1943; Mintz and Russel, 1957) and histochemical studies (McKay *et al.*, 1953; Chiquoine, 1954) have clearly demonstrated that the germ cell line in the mammal is a continuous one from the early embryo through the adult (Mintz, 1959; Witschi, 1962; Wartenberg *et al.*, 1971; Bustos-Obregon *et al.*, 1975). It is now well established that the mammalian primordial germ cells of the early embryo have an extragonadal origin and they are first detectable in the yolk sac splanchnopleure, especially in its endoderm long before formation of the genital ridges (Politzer, 1933; Everett, 1943; Witschi, 1948; Chiquoine, 1954; McKay *et al.*, 1953; van Wagenen and Simpson, 1965; Falin, 1969; Gier and Marion, 1970; Spiegelmann and Bennett, 1973; Clark and Eddy, 1975). From their place of origin, the germ cells migrate to the region of the hindgut where they can be seen in the wall of the primitive intestine (Figs. 1 and 2) or in the adjacent mesoderm (see also Merchant, 1975). Then they migrate to the presumptive gonads or genital ridges by amoeboid movements in the mesenteries of the gut or via the bloodstream (Everett, 1943; Witschi, 1948; Mintz, 1957; Blandau *et al.*, 1963). In all these previous studies, the migration of the primordial germ cells from the yolk sac to the genital or gonadal ridge was mostly investigated by light and phase-contrast microscopy and by histochemistry. But the recent studies carried out with the electron microscope have extended and confirmed the previous observations by showing the extragonadal origin of germ cells and their subsequent migration through the mesenteries (Spiegelman and Bennett, 1973; Merchant and Zamboni, 1973; Zamboni and Merchant, 1973; Pelliniemi, 1976a; Merchant, 1975; Clark and Eddy, 1975). The general concensus of opinion, however, is that the passage of germ cells via the bloodstream is a rare phenomenon in mammals (see Tarkowski, 1970).

Pelliniemi (1976a) using electron microscopy has investigated the primordial germ cells in 9-, 11-, and 12-mm human embryos (ages 36, 38, and 39 days). At 36 days, the germ cells are seen in the mesentery and in the adjacent posterior mesenchyme and by 38 days they are also seen in the surface epithelium and the underlying mesenchyme of the genital ridges. The primordial germ cells can be easily distinguished from the surrounding mesenchymal and epithelial cells by their large size and round shape. Merchant and Zamboni (1973) and Zamboni and Merchant (1973), using light and electron microscopy, have identified three phases in the life history of primordial germ cells, which include gut phase, migratory phase, and settlement phase in mouse embryos from Day 9 to Day 12 of intrauterine life. During the gut phase the germ cells are placed mainly in the ventral hemisphere of the intestine where they are dispersed among the epithelial cells of the wall without reaching the intestinal lumen. According to Eddy (1974), on Day 10 of fetal life primordial germ cells are also recognizable within the hindgut epithelium (endoderm) of the rat embryo (Fig. 2), although they are also found in the connective tissue surrounding the gut (see also Merchant, 1975; Clark and Eddy, 1975). The germ cells can be easily distinguished from the cells of intestinal epithelium due to their large size and different staining and morphological features (Figs. 1 and 2) (see also Pelliniemi, 1976a; Merchant, 1975). In the rat the primordial germ cells are also typically large, rounded cells with a somewhat darker-staining cytoplasm than surrounding somatic cells (Eddy, 1974; Merchant, 1975; Eddy and Clark, 1975). The germ cells show an overall round shape, a primitive-looking nucleus with an irregular or horseshoe shape, and very poor development of cytoplasmic organelles (Fig. 3). The electron microscopic observations of Merchant and Zamboni (1973) and Zamboni and Merchant (1973) on germs cells of the mouse have indicated that the spherical nucleus is occasionally indented by deep and regular invaginations as also observed by Clark and Eddy (1975). The round nucleus shows one or two or more prominent nucleoli with extensive nucleolonema (Fig. 3) (Eddy, 1974; Pelliniemi, 1976a). The cytoplasm of germ cells in the mouse, rat, and human exhibits simple organization, having large, spherical, primitive-looking mitochondria with a few cristae, poorly developed endoplasmic reticulum, and abundant free ribosomes (Fig. 3) (Zamboni and Merchant, 1973; Spiegelman and Bennett, 1973; Jeon and Kennedy, 1973; Merchant and Zamboni, 1973; Merchant, 1975; Eddy, 1974; Eddy and Clark, 1975; Clark and Eddy, 1975; Pelliniemi, 1976a). Other features typical of primordial germ cells are a prominent Golgi complex and sparse electron-dense small granules. Glycogen granules and lipid droplets are absent (Merchant and Zamboni, 1973; Eddy, 1974; Merchant, 1975). But according to Pelliniemi (1976a), lipid droplets are sometimes seen in

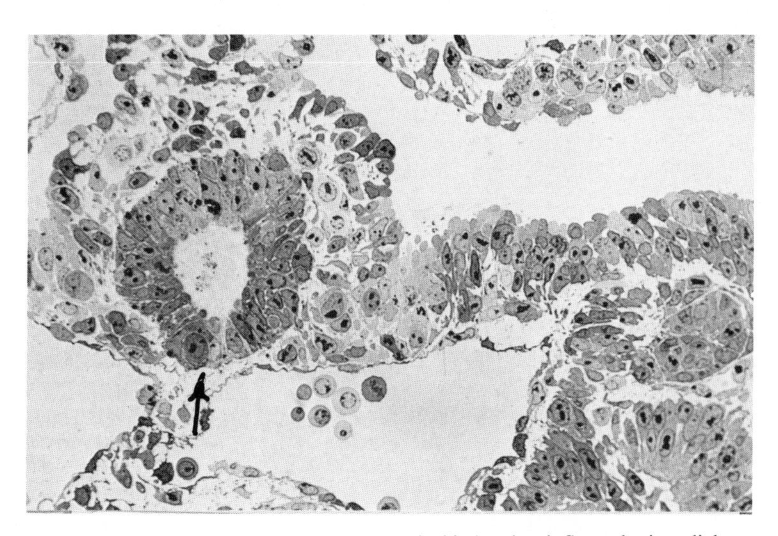

FIG. 1. General view of an 11-day rat embryo at the hindgut level. Several primordial germ cells can be observed in different locations (arrow). Light microscopy of a silver stained section. ×320. (From Merchant, 1975.)

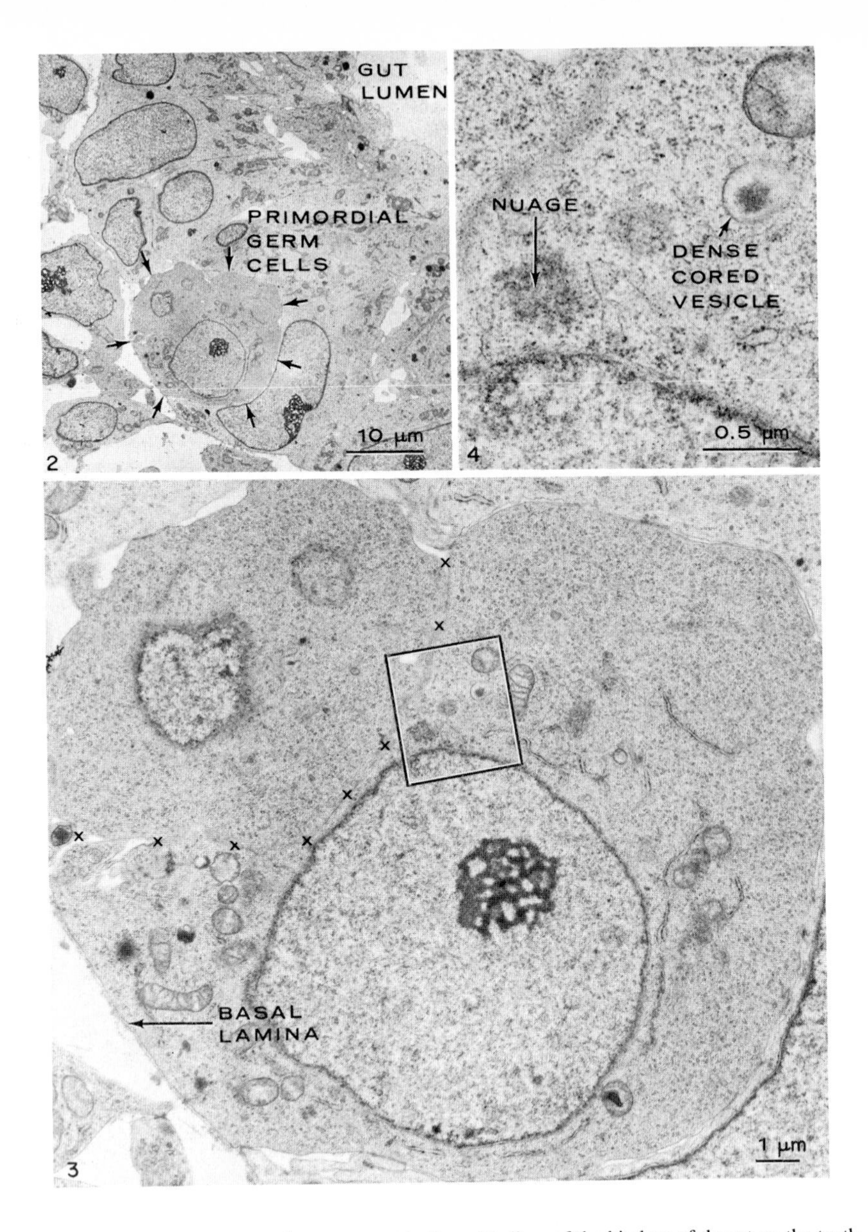

FIG. 2. Primordial germ cells are present in the epithelium of the hindgut of the rat on the tenth day of embryonic development. In this survey electron micrograph the primordial germ cells are recognizable because their cytoplasm is somewhat more intensely stained than that in surrounding somatic cells. ×1040. (From Eddy, 1974.)

FIG. 3. A higher magnification view of the primordial germ cells visible in Fig. 2 is depicted. The section passes through two primordial germ cells, cutting and apposing plasma membranes (**X**) obliquely. These cells are included by the basal lamina of the gut epithelium. ×6000. (From Eddy, 1974.)

human germ cells. The nuclei of germ cells are usually spherical with centrally placed nucleoli and lack the prominent rim of heterochromatin seen in adjacent cells (Eddy, 1974; Clark and Eddy, 1975). In addition to these general nuclear and cytoplasmic characteristics, Eddy (1974) has also demonstrated the presence of nuage which at this stage of development is present as a discrete accumulation of fibrous material usually situated in the perinuclear region (Figs. 3 and 4). Its presence has been confirmed in subsequent studies (Merchant, 1975; Clark and Eddy, 1975). The germ cells also contain dense-cored vesicles which are about 0.1 μm in diameter and number from one to four per micrograph of a germ cell in rat (Figs. 3 and 4). Their role is unknown. But Eddy (1974) believes that they serve as an additional useful marker for the identification of germ cells.

The structural organization of primordial germ cells during the migratory phase remains essentially unchanged. They may occasionally show amoeboid features and marked irregularities of the nuclear profile (Merchant and Zamboni, 1973; Zamboni and Merchant, 1973; Clark and Eddy, 1975; Merchant, 1975; Pelliniemi, 1976a). These morphological characteristics are consistent with the observations of Witschi (1948) who originally proposed that germ cells reach the genital ridges by means of amoeboid movement. Blandau *et al.* (1963) using a phase-contrast microscope equipped for time-lapse cinephotography also showed that the germ cell possessed pseudopodia and underwent amoeboid-like movements. Germ cells during the settlement phase have a structural organization which is consistent with their increased differentiation. Features of amoeboidism usually disappear (Merchant and Zamboni, 1973). The organelles are increased considerably and undergo marked development; mitochondria are increased in number and the Golgi complex is now organized in multiple tubular aggregates distributed in several cytoplasmic regions, showing more specialization of germ cells with time. But Pelliniemi (1976a) has not observed apparent ultrastructural differences in the primordial germ cells in different locations. A characteristics feature of germ cells in mouse at the settlement stage is their arrangement in pairs or clusters consisting of cells frequently connected by typical intercellular bridges (Merchant and Zamboni, 1973; Zamboni and Merchant, 1973). Various electron microscopic studies have also demonstrated that in all phases of their extragonadal life, germ cells remain closely associated with other cells (epithelial cells of the hindgut, mesothelial cells lining the mesentry and genital ridges, mesenchymal mesonephros, and endothelial cells) and the functional meaning of these morphological associations is still obscure

FIG. 4. A higher magnification view of the area in the box in FIG. 3 is shown. Features which are characteristic of germ cells of the rat help to identify the primordial germ cells at this stage. The nuage is present as a small discrete aggregation of dense fibrous material. Also, the dense-cored vesicle is characteristic of the differentiating germ cells, particularly during the premeiotic period. ×26,950. (From Eddy, 1974.)

(Zamboni and Merchant, 1973; Merchant and Zamboni, 1973; Clark and Eddy, 1975). Their consistent association with other cell types may be of some significance in providing exogenous substances for the metabolism of primordial germ cells which show a very simple cytological organization and appear to lack sufficient endogenous energy reserves such as glycogen and lipids (Merchant and Zamboni, 1973; Eddy, 1974; Merchant, 1975; Clark and Eddy, 1975). Such substances could be provided by active intercellular exchanges, a process which may account for the presence of intense alkaline phosphatase activity in the germ cells especially in association with their plasma membrane (McKay *et al.*, 1953, 1955; Essenberg *et al.*, 1955; Pinkerton *et al.*, 1961; Mayer, 1964; Clark and Eddy, 1975).

III. Genital Ridges

A. Development and Structure

It is now well established that the genital or gonadal ridges develop as paired mesenchymal primordia which project, on each side of the dorsal mesentery, from the medial aspect of the mesonephros. Each genital ridge has a mesenchymal core consisting of simple stellate or spindle-shaped cells and a covering of rapidly proliferating coelomic or surface epithelium (Fig. 5) (see also Merchant, 1975). Large primordial germ or sex cells can be seen migrating into the genital ridge from the gut endoderm or its associated mesoderm; these large germ cells can be characterized by their high alkaline phosphatase activity as already discussed (see also Baillie *et al.*, 1966a).

Genital ridges first become visible in bovine embryos at 28 days, in pig embryos at 26 days, and in dog embryos at 24 days, times at which all these embryos have differentiated approximately to 40 somites and their anterior limb buds are discernible (Gier and Marion, 1970). The genital ridge in bovine embryos is formed by a thickening of the coelomic epithelium and condensation of the underlying mesenchyme as also reported for rodents (Merchant, 1975).

According to Gier and Marion (1970), in both dog and bovine embryos multiple invaginations of the germinal (or surface) epithelium occur throughout its expanse within 12 hours after first visible detection of surface epithelium, and these invaginations give rise to primary epithelial cords. These cords, although beginning as tubular invaginations, quickly lose their lumina and become solid cords of 6 to 10 cells in cross section, surrounded by a periodic acid–Schiff-positive basement membrane. Gier and Marion (1970) could not identify the primordial germ cells within the mesenchyme adjoining the surface epithelium or between the epithelial cords as reported in the rat (Eddy, 1974; Merchant, 1975). During further development, the genital ridge grows in size by a rapid increase in

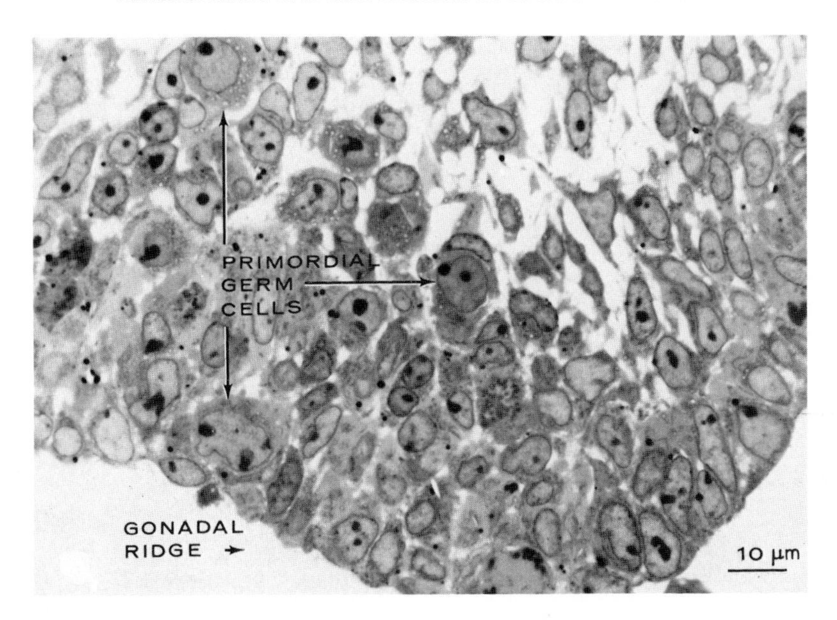

FIG. 5. This light micrograph illustrates the appearance of the gonadal ridge on Day 13 of development of the rat. Some primordial germ cells have migrated into this region. ×770. (From Eddy, 1974.)

its diameter but little increase in length, resulting in a globular gonad by 32 days in canine embryos and by 38 days in bovine embryos (Gier and Marion, 1970).

Grünwald (1942) and Pinkerton *et al.* (1961) put forth the hypothesis that the undifferentiated gonad develops as a blastema in which cells from the coelomic epithelium and mesenchymal cells are involved. Merchant (1975) has supported this hypothesis but he has also observed the participation of some mesonephric tubules in the formation of the blastema. There is an early phase in which cells from the surface epithelium and the mesenchyme and some mesonephric tubules placed on the urogenital fold start an active proliferation. This proliferation, which leads to the formation of the undifferentiated gonadal primordium, can take place in the absence of primordial germ cells or, at least, with only a few of them.

Pelliniemi (1975a) has produced details of ultrastructural changes which occur during the development of genital ridges in pig embryos. The gonads in the pig develop structurally similar in both sexes at the age of 21 to 24 days. The genital ridge consists of three different tissues such as the surface epithelium, the primitive cords, and the mesenchyme (Fig. 6). The surface epithelium consists varyingly of columnar cells in a single layer and smaller cuboidal cells arranged in two or three layers. Mitotic divisions result in its thickening and layered organization. In the human genital ridge, the surface epithelium also develops from a

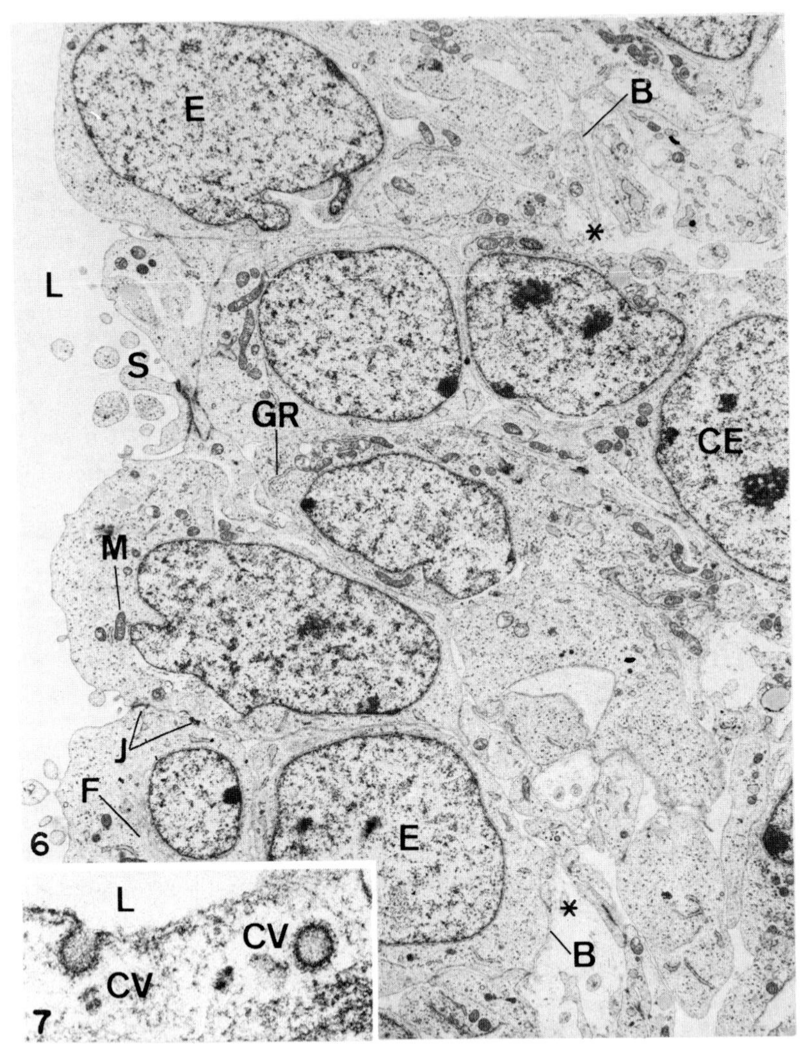

FIG. 6. Male pig embryo, age 21 days. A survey picture of the surface epithelium. A cord of epithelial cells directed toward the base of the ridge is seen in the middle of the right half of the figure between the asterisks. B, Basal lamina; CE, cord cell; E, surface epithelial cell; F, cytoplasmic filaments; GR, granular endoplasmic reticulum; J, junctional complex; L, coelomic cavity; M, mitochondrion; S, surface folds. ×4590. (From Pelliniemi, 1975a.)

FIG. 7. Female pig embryo, age 21 days. Coated vesicles (CV) in the luminal cytoplasm of a surface epithelial cell. L, Coelomic cavity. ×55,250. (From Pelliniemi, 1975a.)

simple to a stratified type, and the basal lamina at 38 to 39 days is discontinuous, especially in the medial half of the genital ridge (Pelliniemi, 1976a). The surface epithelial cells show elongate mitochondria with lamellar cristae, granular endoplasmic reticulum, Golgi complex, free polysomes, coated vesicles, and fine filaments (Figs. 6 and 7). The lateral cell membranes of the adjoining cells run smoothly parallel. But the intercellular space on the border toward the coelomic cavity is closed by a zonula occludens and a zonula adherens type of junction (Fig. 6); one or two desmosomes are often seen below these zones.

The surface epithelium remains continuous with the primitive cords (Fig. 6) which are composed of cells similar to those of the surface epithelium (Pelliniemi, 1975a; Merchant, 1975). The formation of primitive cords from the surface epithelium starts in the human genital ridge at 38 to 39 days of gestation (Pelliniemi, 1976a). The number and length of cords are increased to extend deeper into the mesenchyme. The adjoining cells in the cord are often connected with each other by zonulae adherens. The epithelial basal lamina consisting of amorphous material follows the cord surface, but covers it incompletely. Primordial germ cells are believed to enter the gonadal cords before these are surrounded by a continuous basal lamina (Pelliniemi, 1975a, 1976a). The temporal relationship between the arrival of primordial germ cells at the surface epithelium and the commencement of the proliferative differentiation of the latter has led Pelliniemi (1976a) to suggest that the primordial germ cells have a triggering effect on the epithelial proliferation.

The underlying mesenchymal tissue shows a meshwork-like structure with large intercellular spaces (Fig. 5) (Pelliniemi, 1975a; Merchant, 1975). Its cells are polymorphic having long and often narrow cytoplasmic processes, which are in contact with the neighboring cells by zonula adherens-like junctions or pure membrane apposition. Their nuclei also vary greatly in shape and size. The various organelles described for the surface epithelium are also seen in the cytoplasm of mesenchymal cells. The agranular endoplasmic reticulum is lacking. The blood vessels are situated in the basal layer capsule.

B. Histochemistry and Steroid Hormone Synthesis

Tbe presence of enzyme activities such as 3α-, Δ^5-3β-, 16β-, and 17β-hydroxysteroid dehydrogenases, in histochemically demonstrable amounts, has been reported in genital ridges (see Baillie *et al.,* 1966a). The presence of these enzyme activities, which are indicative of steroidogenesis, suggests that the genital ridge mesenchyme theoretically has the ability to produce a variety of androgens including testosterone, androsterone, and androstenedione from dehydroepiandrosterone (DHA). But the organelles specific to steroidogenic tissues (Christensen and Gillim, 1969) are lacking in its cells (Pelliniemi, 1975a; Merchant, 1975). Cholesterol biosynthesis is not established biochemically in the

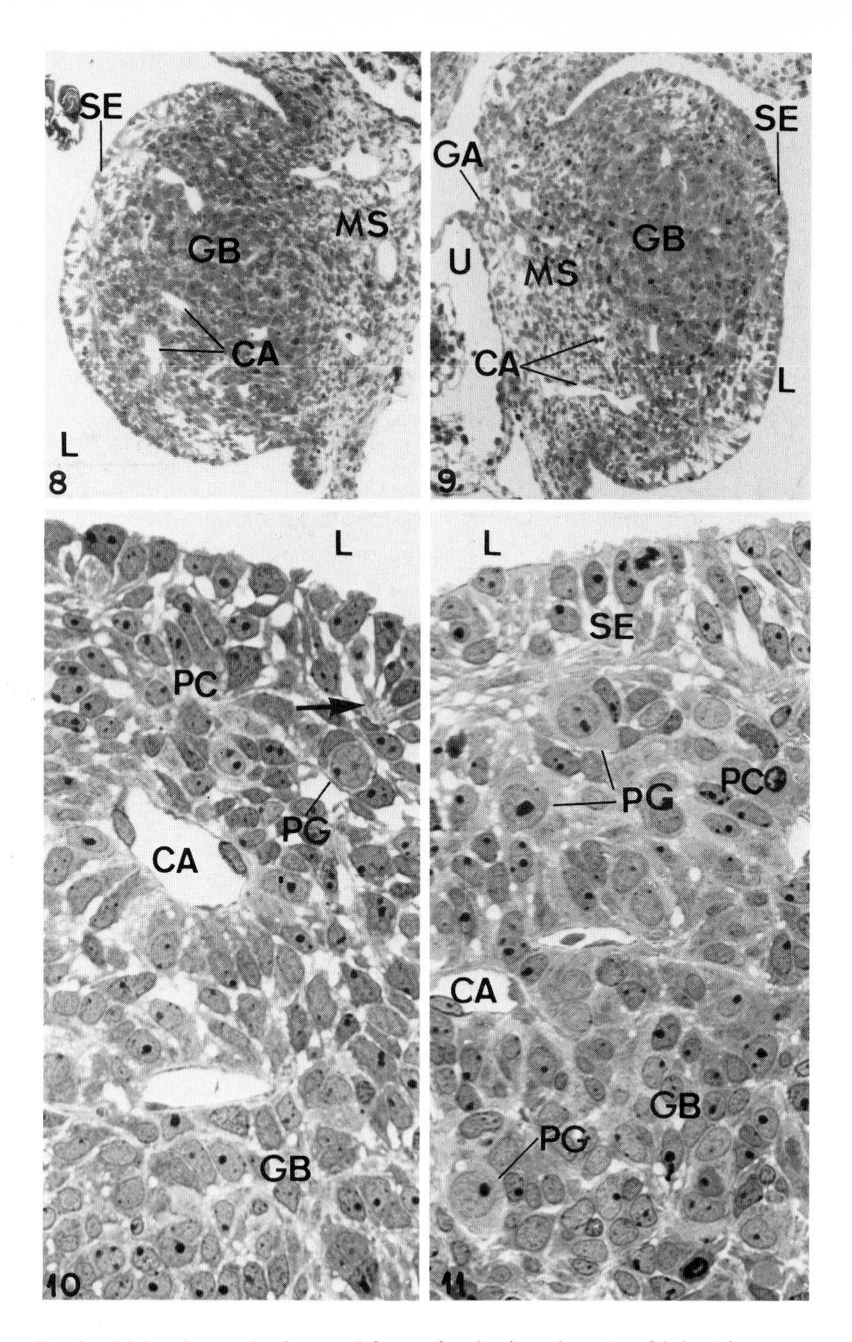

FIG. 8. Light micrograph of a gonad from a female pig embryo (age 24 days) in a transverse section from the middle region. The posterior direction is toward the top. The mesonephric glomerulus is on the right outside the picture. The surface epithelium (SE) consists of loosely arranged irregular columnar cells and the medulla of the tightly packed gonadal blastema (GB). CA, Capillary; L, coelomic cavity; MS, mesenchyme. ×128. (From Pelliniemi, 1976b.)

genital ridge. Baillie *et al.* (1966a) has suggested that the possible synthesis of steroid hormones, which may be androgens or estrogens, by the genital ridge mesenchyme is of great significance in early gonadal embryogenesis as the chemotactic agents believed to control germ cell migration (Witschi, 1948; Johnston, 1951) may be of steroidal nature. After reaching the genital ridge, the primordial germ cells undergo extensive and rapid mitosis (Mintz, 1957; Guraya, 1977a), and it may be that this mitotic activity, together with that known to occur in the coelomic epithelium at this time, may also be controlled by local concentrations of steroids in the genital ridge as suggested by Baillie *et al.* (1966a).

IV. Indifferent Gonad

A. Development and Structure

The indifferent gonads in both sexes are formed from the genital ridges. In earlier studies, their structural organization was mostly studied by light microscopy in pig (Allen, 1904; Black and Erickson, 1965; Moon and Raeside, 1972; Moon and Hardy, 1973), sheep (Mauléon, 1961), cattle (Gropp and Ohno, 1966), rabbit (Allen, 1904), rat (Jost *et al.,* 1973), and man (Gillman, 1948; van Wagenen and Simpson, 1965; Jirasek, 1971). The sex of embryos was reported in cattle and rat only. These studies indicated the absence of the tunica albuginea and of the sexually differentiated testicular cords in the indifferent gonads of pig embryos (Allen, 1904; Black and Erickson, 1965; Moon and Hardy, 1973). These findings on the indifferent embryonic gonads of the pig have recently been confirmed by electron microscopy (Pelliniemi, 1976b). Therefore, gonadal sex at the age of 24 days in the pig cannot be defined on a morphological basis (Figs. 8–11) (Jost *et al.,* 1973). The morphological studies have, however, supported the view that at the age of 24 days, the sex-determining factors (Hamerton, 1968; Boczkowski, 1971, 1973) have not as yet produced structural changes which could be determined by light and electron microscopy.

Recent studies have shown that indifferent gonads consist of surface

---

FIG. 9. Light micrograph of a gonad from a male pig embryo (age 24 days) with the mesonephric glomerulus on the left outside the micrograph, a view corresponding to that in Fig. 8. CA, Capillary; GA, glomerular capsule; GB, gonadal blastema; L, coelomic cavity; MS, mesenchyme; SE, surface epithelium; U, urinary space. ×128. (From Pelliniemi, 1976b.)

FIG. 10. Light micrograph of a gonad from a female pig embryo (age 24 days). A radiate pattern of surface epithelial cells is seen on the right border (arrow). Along the upper side of the capillary (CA), the columnar cells have a cordlike arrangement. GB, Gonadal blastema; L, coelomic cavity; PC, primitive cords; PG, primordial germ cells. ×510. (From Pelliniemi, 1976b.)

FIG. 11. Light micrograph of gonad from a male pig embryo (age 24 days). CA, Capillary; GB, gonadal blastema; L, coelomic cavity; PC, primitive cords; PG, primordial germ cell; SE, surface epithelium. ×510. (From Pelliniemi, 1976b.)

epithelium, primitive cords, mesenchyme, and germ cells (Gondos and Conner, 1973; Jost *et al.* 1974; Pelliniemi, 1975a,b; Merchant, 1975). According to Merchant (1975), the undifferentiated gonad in rat consists of packed cells which are gradually separated into two differently organized tissues separated by a basal lamina: epithelial layers and/or "cords" and the stromal tissue (or mesenchyme). The former, which holds the germ cells, separates early from the surface epithelium in testes and remains continuous with it in ovaries during the whole gestational life. The stromal tissue is formed by blood vessels and connective tissue where the first Leydig cells will appear in male gonads by the sixteenth day of gestation in the rat.

1. *Surface Epithelium and Primitive Cords*

The continuity of the surface epithelium with the primitive cords is believed to form a common feature in various mammals (Gier and Marion, 1970; Merchant, 1975; Pelliniemi, 1975a, 1976b). However, some workers have considered this morphological orientation to be indicative of the surface epithelium being the cellular source of the developing cords. The basal lamina of the surface epithelium usually continues on the primitive cords, but is discontinuous in the deeper layers, where the primitive cords are also continuous with the blastema proper (Pelliniemi, 1976b). The ultrastructure of the surface epithelial cells remains basically the same as that described in the genital ridges (Black and Christensen, 1969; Merchant, 1975; Pelliniemi, 1975a). They show some irregularly shaped cytoplasmic processes projecting into the peritoneal cavity.

2. *Blastema*

The blastema proper is the new structure in the indifferent gonads of the pig, which is considered to be responsible for the growth of the flat genital ridge into a longitudinal rounded protrusion (Pelliniemi, 1976b). The exact origin of its cells is still controversial. Two different theories have been proposed: the differentiation from the surface epithelium (Allen, 1904) and from the mesenchyme (Jirasek, 1971). Pelliniemi (1976b) has attributed the origin of blastema cells to both epithelial and mesenchymal cells because of their ultrastructural resemblance. Whatever the origin of its cells, the earlier workers described the gonadal blastema in the pig as cords (Allen, 1904; Moon and Hardy, 1973), in man as cords (Gillman, 1948; van Wagenen and Simpson, 1965) or as the gonadal blastema (Jirasek, 1971), and in cattle as the gonadal blastema (Gropp and Ohno, 1966; Ohno, 1967). The description of the gonadal blastema in man as given by Jirasek (1971) is essentially the same as recently reported by Pelliniemi (1976b) in the pig. The blastema in the cattle is similar but it contains no germ cells. According to Pelliniemi (1976b), the shape of the blastema cells varies greatly in the irregularly arranged areas and is usually columnar in the cordlike portions where the basal lamina is also formed. The nuclei of blastema cells show a prominent nucleolus with nucleolonema and pars amorpha. Their cytoplasm shows elongate mitochondria with lamellar cristae, solitary cisternae of granular endoplasmic reticulum, small Golgi complex, numerous polysomes, and some lipid droplets. The irregularly organized blastema cells in male and female embryos also show some circular profiles of smooth membranes (Pelliniemi, 1976b). The ultrastructural features of blastema cells are mostly indicative of protein synthesis in them. The interstitial gland cells or Leydig cells with organelles specific to steroidogenic tissue (Christensen and Gillim, 1975; Christensen, 1975; Neaves, 1975) are not yet present in the ovary or in the pig testis by the age of 27 days. Tubular endoplasmic reticulum with some patches of ribosomes has, however, been demonstrated in the interstitial cells and in the presumptive Sertoli cells in the indifferent gonad of the guinea pig fetus at the age of 22 to 23 days (Black and Christensen, 1969).

3. *Germ Cells*

The primordial germ cells in the indifferent gonads of pig embryos are located in primitive cords, among the blastema cells and among the mesenchymal cells in the mesogonadium and the subjacent mesenchyme (Figs. 10 and 11) (Pelliniemi, 1976b). This random distribution of the primordial germ cells corresponds to locations reported earlier in 2.0-cm embryos (Moon and Hardy, 1973). In contrast germ cells in cattle have been observed in the periphery of the gonad (Ohno, 1967). The ultrastructural morphology of primordial germ cells in the indifferent gonads basically remains the same as already discussed in the genital ridges (Gondos and Conner, 1973; Eddy, 1974; Merchant, 1975; Pelliniemi, 1976b). The nuclear chromatin is finely dispersed. The nucleolonema is well developed. The round mitochondria are often arranged in groups. The shape of the mitochondrial cristae varies from lamellar to tubulovesicular in the same cell. One to three separate Golgi complexes are often seen in the cytoplasmic region showing centrioles and also other organelles. The other cell components include the separate cisternae of the granular endoplasmic reticulum, free polysomes, coated vesicles and cisternae, some lipid droplets, and membrane-bound dense bodies of variable shape and size; dense-cored vesicles and nuage are also present in germ cells of the rat (Eddy, 1974; Merchant, 1975). In comparison to the early stages, the amount of organelles has apparently increased in the primordial germ cells of indifferent gonads, which may still show pseudopods (Pelliniemi, 1976b). Some germ cells begin to be eliminated through the degenerative process in the indifferent gonad from the age of 24 days onward in the pig (Pelliniemi, 1976b).

B. Histochemistry and Steroid Hormone Synthesis

Pregnenolone in the mesenchyme of the 11-day fetal mouse indifferent gonad gives a weak Δ^5-3β-hydroxysteroid dehydrogenase (3β-HSDH) reaction (Baillie and Griffths, 1964), although other 3β-hydroxysteroids including 17α-hydroxy pregnenolone and DHA are not histochemically utilized. These results have also been confirmed by Hart *et al.* (1966a) and the presence of a 16β-hydroxysteroid dehydrogenase has been established in the indifferent gonad mesenchyme. 3α-, 6β-, 11β-, 16α, and 17β-hydroxysteroids do not give histochemical reaction in the interstitium of indifferent mouse gonads (see Baillie *et al.*, 1966a). These histochemical results have indicated the possibility of steroidogenesis from the moment the gonad is first recognizable. The histochemical studies of the gonads in pig embryos (lengths 1.5–2.0 cm) have also shown a positive reaction for hydroxysteroid dehydrogenases in both sexes (Moon and Raeside, 1972). The ultrastructural changes in the endoplasmic reticulum of blastema cells in the pig (Pelliniemi, 1976b) may explain the development and presence of these enzyme activities related to steroid biosynthesis in its indifferent gonad. Sex differentiation and secretion of testosterone have been demonstrated in the indifferent gonads of 2.0-cm pig fetuses (age approximately 26 days), when cultured with or without rat ventral prostate tissue (Moon *et al.*, 1973; Moon and Hardy, 1973). The secretion of testosterone has also been reported in the nondifferentiated sheep gonad (Attal, 1969).

V. Embryonic Testis

A. Differentiation

The indifferent gonad in the male embryo is finally transformed into testis. The testicular differentiation has been investigated by electron microscopy in the rabbit (Gondos and Conner, 1973), rat (Jost *et al.*, 1974; Merchant, 1975), pig (Pelliniemi, 1975b), and man (Pelliniemi, 1970). According to Gondos and Conner (1973), development of a basal lamina around groups of Sertoli cells and germ cells in the rabbit is an early event that occurs between 15 and 16 days when the cell groups are still connected with the surface epithelium. But this contact is soon lost by the formation of a distinct tunica albuginea. Jost (1972a) found that between 12 days, when the gonadal primordium in the rat is still sexually indifferent, and 13 days, when testicular differentiation can be distinguished, one of the first processes identified is the differentiation of Sertoli cells which swell, make contact with each other, and surround individual germ cells. Merchant (1975) using electron microscopy has observed that testicular differentiation in the rat is indicated by the separation of the mesonephric tubules and the cell cord from the surface epithelium; these three tissues, however, show continuities in the indifferent gonad. At the age of 26 days, the blastema proper in the pig embryos begins to differentiate into testicular cordlike structures (Figs. 12 and 14) (Pelliniemi, 1975b) and this is the beginning of differentiation of the testis. In most of the earlier studies on the development of gonad in the pig, the tunica albuginea was believed to be the sex-specific feature (see references in Pelliniemi, 1975b). But according to Pelliniemi (1975b) the testicular cords differentiate earlier than the tunica albuginea. Thus the earliet morphologic effects of the male-determining factors in the pig become manifest at the age of 26 days in the differentiation of the testicular cordlike structures and the interstitium from the indifferent blastema proper. Other workers have also observed that the mammalian gonad remains morphologically at a sexually indifferent stage until the testicular cords and tunica albuginea develop in the male gonad (Black and Christensen, 1969; Mauléon, 1969; Gier and Marion, 1970; Merchant, 1975). At the time when these structural changes occur in the testes (Figs. 12 and 14), the morphology of the corresponding ovaries remains unchanged (Figs. 13 and 15) (see also Jost *et al.,* 1973).

B. Structure

The embryonic testis consists of various tissues and cell elements such as surface epithelium, primitive cords, testicular cordlike structures, interstitium, blood vessels, etc. (Figs. 12 and 14).

1. *Surface Epithelium*

The embryonic testis is covered by surface epithelial cells (Figs. 12 and 14) which derive from the coelomic epithelium as already discussed. Cell shape and size and number of cell layers in the surface epithelium of embryonic testis vary greatly in different regions (Vossmeyer, 1971; Pelliniemi, 1975b). The ultrastructural features of surface epithelium in the embryonic testis basically remain the same as those described in the genital ridges and indifferent gonads. In its location, the nucleus of surface epithelial cells varies from basal to apical. The mitochondria, Golgi complex, granular endoplasmic reticulum, and numerous polysomes, coated vesicles, and lipid droplets are present in the cytoplasm (Pelliniemi, 1975b; Merchant, 1975). The light and electron microscopic studies carried out so far have not revealed any appreciable sex differences in the structure of surface epithelium of early male and female gonads (Figs. 12–15) (Allen, 1904; Merchant, 1975; Pelliniemi, 1975b; Guraya, 1977a).

2. *Primitive Cords*

Very divergent views about the relationships between the surface epithelium and the primitive cords, as revealed by light microscopy, have been expressed in earlier studies (see reviews by Gier and Marion, 1970; Jost *et al.*, 1973). Recent electron microscopic studies have shown that the primitive cords are continuous with the surface epithelium in many places in both sexes in the pig (Figs. 12 and 13) (Pelliniemi, 1975a,b, 1976b). The basal lamina of the surface epithelium continues into the primitive cords but covers them incompletely in the deeper

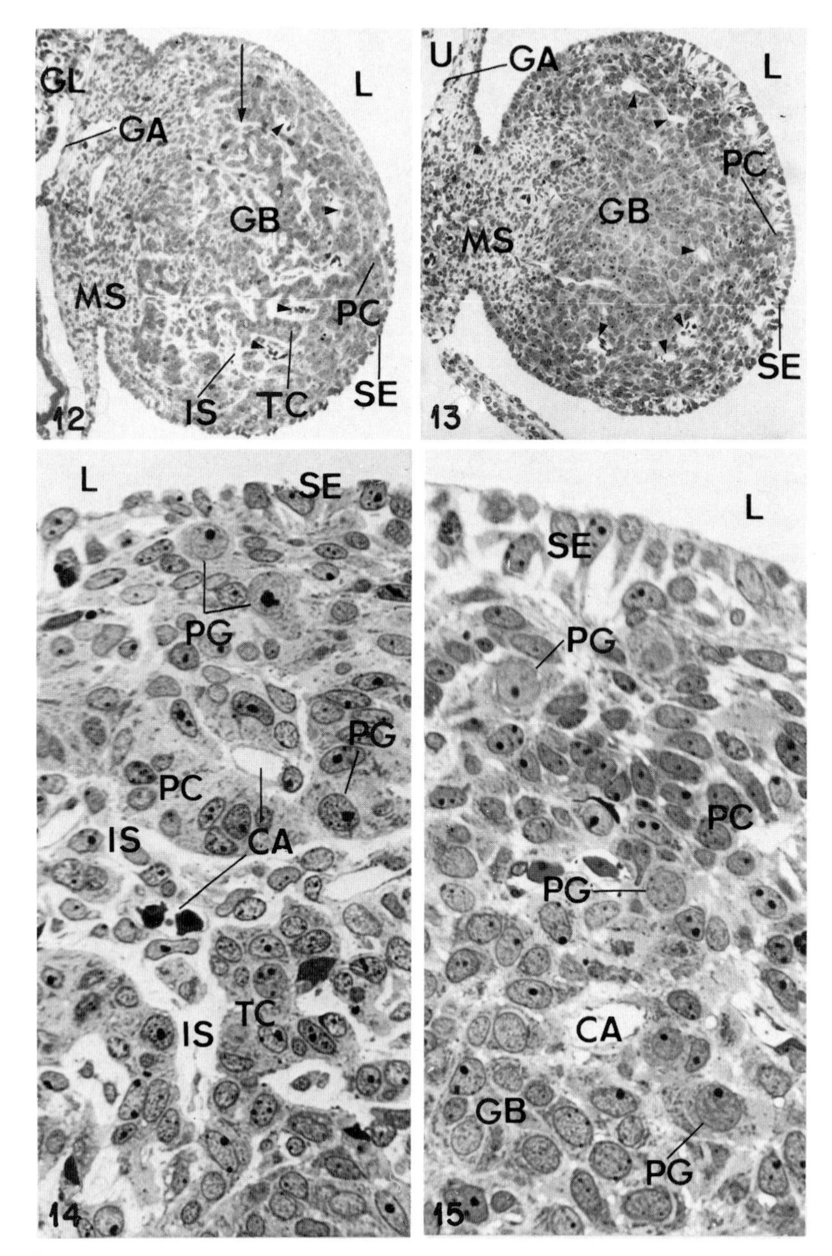

FIG. 12. Testis of pig embryo, age 26 days. A light micrograph of a transverse section. The characteristic testicular cords (TC) and interstitium (IS) are seen in the central region. Note the cross sections of large capillaries (arrowheads) at a relatively constant distance from the surface. The arrow indicates the direction of the sagittal section. The mark for the mesenchyme (MS) is in the mesotestis.

layers. The invaginations are not formed from the surface epithelium (Figs. 14 and 15). On the central side of the pig testis, the primitive cords are continuous with the testicular cords (Fig. 14) (Pelliniemi, 1975b). The shape and size of cells in the primitive cords as well as their nuclei vary greatly. In the testis, the cells lying directly under the surface epithelium are, sometimes, elongated in the plane of the gonadal surface, apparently representing the earliest signs of the differentiation of the tunica albuginea. The primitive cords at the ultrastructural level show no sex differences and resemble both the surface epithelial cells and the sustentacular cells (or Sertoli cell precursors), with the exception that they do not show free coelomic surface (Pelliniemi, 1975b).

The primordial germ cells in the pig enter the cords before these are bound by a continuous basal lamina (Pelliniemi, 1975b). The germ cells are present directly under the surface epithelium and also at all levels down to the testicular cords. The primordial germ cells of the both sexes, which are ultrastructurally similar, can be distinguished by their location only. They have a round nucleus with finely dispersed chromatin and one to three large nucleoli.

3. *Testicular Cords*

The formation and presence of testicular cords have been reported in earlier studies carried out with the light microscope (Allen, 1904; Whitehead, 1904; Grünwald, 1942; Vossmeyer, 1971; Moon and Hardy, 1973; Jost *et al.*, 1973). Jost *et al.* (1973) have explained the formation of sex cords by arrangement of precursors of Sertoli cells around germ cells. Pelliniemi (1975b) has observed that by 26 days the edge of the central region of the male gonadal blastema in pig embryos can be demarcated by the most peripheral capillaries and shows sex-specific differentiation into testicular cordlike structures (Figs. 12 and 14), which, peripherally, remain continuous with the primitive cords and, centrally, with the often, as yet undifferentiated, most central region of the gonadal blastema. The latter by 27 days is totally differentiated into testicular structures.

An area of undifferentiated gonadal blastema (GB) is seen in the center of the testis. GA, Glomerular capsule; GL, glomerulus; L, coelomic cavity; PC, primitive cords, SE, surface epithelium. ×100. (From Pelliniemi, 1975b.)

FIG. 13. Ovary of pig embryo, age 27 days. A light micrograph of a transverse section. The whole area of the gonadal blastema (GB) is still undifferentiated. Note the cross sections of large capillaries (arrowheads) at relatively constant distance from the surface. The mark for the mesenchyme (MS) is in the mesovarium. GA, Glomerular capsule; L, coelomic cavity; PC, primitive cords; SE, surface epithelium; U, urinary space. ×100. (From Pelliniemi, 1975b.)

FIG. 14. Testis of pig embryo, age 26 days. A light micrograph of a transverse section showing the surface epithelium (SE), primitive cords (PC), testicular cords (TC), and interstitium (IS), CA, Capillary; L, coelomic cavity; PG, primordial germ cell. ×510. (From Pelliniemi 1975b.)

FIG. 15. Ovary of pig embryo, age 26 days. A light micrograph of a transverse section showing the surface epithelium (SE), primitive cords (PC), capillary (CA), primordial germ cell (PG), and coelomic cavity (L). ×510. (Courtesy of Dr. L. J. Pelliniemi.)

The cells of gonadal blastema, which give rise to testicular cords, greatly vary in their shape (Pelliniemi, 1975b, 1976b). They show various organelles such as the elongated mitochondria often associated with cisternae of granular endoplasmic reticulum, numerous polysomes, coated vesicles, fine filaments, lipid droplets, etc. The cell membranes of adjacent blastema cells are smoothly parallel and sometimes attached by a patch of zonulae adherens. Some blastema cells possess long and narrow cytoplasmic processes. Pelliniemi (1975b) believes that the small area of the indifferent blastema seen in the center of the testis at the age of 26 days apparently represents the region of the future rete testis (Grünwald, 1934), which has also been identified in the human embryonic testis (Vossmeyer, 1971).

The early testicular cordlike structures (26–27 days) in the pig resemble more sheets or walls than cords (Figs. 12 and 14) (Pelliniemi, 1975b). Testicular or sex cords are also easily recognizable in the gonad of the male rat fetus by Day 15 of development (Eddy, 1974; Merchant, 1975). On Day 12, early testicular differentiation in the hamster embryos is indicated by the presence of branching cords containing gonocytes and Sertoli cell precursors (or sustentacular cells) (Gondos *et al.*, 1974). Mesenchymal cells in the interstitial regions are undifferentiated. Black and Christensen (1969) have observed that by about 23 days the sex cords of the developing testis in guinea pig embryos become closely packed and blend peripherally with cells along the inner surface of the surface epithelium. These cells, which are the precursors of the tunica albuginea, become elongated and oriented parallel with the surface of the gonad. By 29 days, the band of cells assumes the characteristics of loose connective tissue and this tunica albuginea separates the testicular cords from the surface epithelium. Both the tunica albuginea and the testicular cords in the guinea pig contain gonocytes before 26 days, but by 27 to 29 days the gonocytes are found chiefly in the cords. A few occur free in the interstitial tissue and probably correspond to those seen in the mesenchyme of undifferentiated gonad. In the human at the age of 8 weeks, which is the borderline between the embryonic and fetal periods, the testicular cords appear in sections as irregularly formed cell islets distinctly separated from the interstitial tissue by a basal lamina (Pelliniemi, 1970; Vossmeyer, 1971). The cords show two distinct cell types: supporting and primordial germ cells. Gropp and Ohno (1966) have observed that the so-called sex cords of the undifferentiated gonad in cattle originate in the condensation of the mesenchyme beneath the coelomic epithelium.

The number of cells per cross section of the testicular cordlike structures originating from the gonadal blastema in pig embryos varies (Pelliniemi, 1975b). The broadest cords are placed in the periphery and the narrowest in the central part. In the sagittal as well as transverse planes, by 27 days the testicular cords are irregularly shaped and anastomosing. Isolated roundish cords are seen very infrequently whereas rounded interstitial areas of that shape are frequent, indicating that the cords in a literal sense are not yet formed and that the sustentacular or supporting cells are actually organized as an irregular three-dimensional network of anastomosing sheets or walls of varying thickness around the interstitial space (Fig. 14).

The presence of testicular cords in man has been reported by different workers (Gillman, 1948; van Wagenen and Simpson, 1965; Pelliniemi and Niemi, 1969; Jirasek, 1971; Vossmeyer, 1971). Their structure is essentially similar to that of testicular cords of the pig. A stereological examination of the early human testis has shown that in reality the cords are initially plate-shaped (Elias, 1971, 1974) as also interpreted by Pelliniemi (1975b) in the pig. The exact mechanism by which the cords are formed from sheets has yet to be explained.

a. *Supporting or Sustentacular Cells.* The great majority of cells in the testicular cords of embryos are the somatic supporting cells or sustentacular cells, which are precursors of the Sertoli cells (Vossmeyer, 1971; Pelliniemi, 1975b; Merchant, 1975; Bustos-Obregon *et al.,* 1975). The supporting cells are believed to be derivatives of blastema cells from the previous stage (Pelliniemi, 1975a). Merchant (1975) has also observed that the sustentacular cells resemble the stromal or mesenchymal cells of the rat testis.

The size and shape of supporting cells and their nuclei vary greatly (Vossmeyer, 1971; Pelliniemi, 1975b; Merchant, 1975, 1976; Bustos-Obregon *et al.,* 1975). The nuclei are peripherally located and are dense in chromatin. One or sometimes two large nucleoli are present. The cytoplasm of sustentacular cells, which extends toward the central part of the sex cords, contains abundant rough endoplasmic reticulum, numerous free ribosomes, and small mitochondria (Fig. 16) (Pelliniemi, 1975b; Merchant, 1976). Cisternae of granular endoplasmic reticulum often lie in close association with the mitochondria. Numerous polysomes are scattered evenly in the cytoplasm of supporting cells. Some coated vesicles, a few lipid droplets, and a variable amount of fine filaments are oriented randomly in the cytoplasm. The varying location of the Golgi complex suggests that the supporting cells are not polarized. Agranular endoplasmic reticulum is not yet differentiated (Fig. 16). Merchant (1976) has, however, observed the specialized morphology of supporting cells at 16 days of gestation in rat. In these cells the rough endoplasmic reticulum is increased and becomes organized as numerous short cisternae loaded with homogeneous material (Fig. 16). He believes that these ultrastructural features are indicative of protein synthesis rather than steroid hormone synthesis. The smooth reticulum specific to steroidogenic tissue is not developed. The specific ultrastructural features of supporting cells in the testicular cords in guinea pig (Black and Christensen, 1969), man (Pelliniemi, 1970), rabbit (Gondos and Conner, 1973; Bjerregaard *et al.,* 1974), hamster (Gondos *et al.,* 1974), and rat (Merchant, 1976) are also an

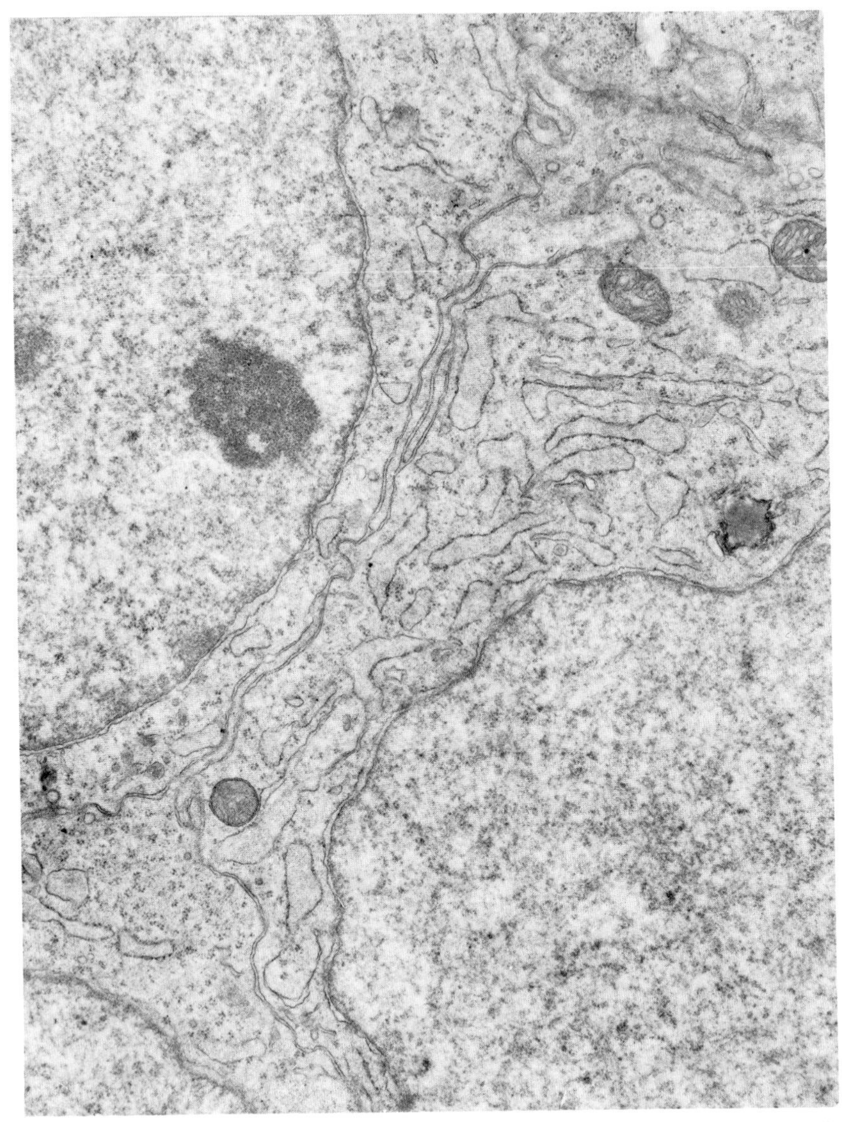

FIG. 16. Electron micrograph showing part of the cytoplasm of two precursors of Sertoli cells. The rough endoplasmic reticulum is well developed and organized as numerous short cisternae filled with a homogeneous material. Material is from a rat fetus at the sixteenth day of gestation. ×14,400. (From Merchant, 1976.)

irregular nuclear outline, numerous polysomes, rodlike mitochondria, and solitary cisternae of endoplasmic reticulum. The supporting cells retain this appearance until the establishment of spermatogenesis (Courot *et al.,* 1970).

The various cytoplasmic structures seen in the supporting cells were already present in columnar blastema cells in the cordlike portions of the indifferent pig gonad at the age of 24 days (Pelliniemi, 1975a), suggesting their origin from blastema cells which are believed to be mesenchymal cells (Merchant, 1975). Occasionally a large inclusion body is seen in the supporting cell which resembles a degenerating cell in its texture (Pelliniemi, 1975b).

The cell membranes of adjacent supporting cells run parallel and patches of zonulae adherens are occasionally seen between them (Pelliniemi, 1975b; Merchant, 1976). The cell membrane facing the interstitium is covered by a discontinuous basal lamina which is composed of loose fibrous material (Pelliniemi, 1975b). According to Merchant (1976) the cord cells remain continuous with the surrounding mesenchymal or interstitial cells at those places where the basal lamina is not yet developed.

b. *Gonocytes.* The germ cells within the cords of early testis are called gonocytes which constitute the second cell type and appear as large, roundish cells (Vossmeyer, 1971; Wartenberg *et al.,* 1971; Pelliniemi, 1975b; Merchant, 1976; Gondos, 1977). They are located singly among the supporting cells which completely surround them within the cords. The gonocytes in the human developing testis at the age of 8 weeks can also be distinguished from supporting cells by their larger size, roundish shape, and relatively pale cytoplasm (Pelliniemi, 1970; Vossmeyer, 1971; Wartenberg *et al.,* 1971). The round nucleus of gonocytes shows one to three nucleoli and chromatin arranged in an irregular network. The cytoplasm contains few roundish mitochondria, free ribosomes, and occasionally a very small amount of granular endoplasmic reticulum and Golgi complex (Fig. 17). These ultrastructural characteristics indicate the dormant state of gonocytes until the postnatal onset of spermatogenesis. Two gonocyte types have been distinguished in human material (Wartenberg *et al.,* 1971). The first type of gonocyte found up to the sixth week of pregnancy can be characterized by a round nucleus with a large nucleolus, mitochondria with tubular cristae, a peripheral cytoplasmic microfilamentous web, and microvillilike cell protrusions. The second type of gonocyte occurs up to the tenth week of pregnancy and shows abundant glycogen as a distinctive feature and more mitochondria. After 10 weeks, few glycogen-rich gonocytes are found.

Phagocytosis of degenerating gonocytes occurs in the rabbit (Gondos and Conner, 1973). Most of the gonocytes appear to take up their residency in the cords by Day 15 of rat development and are recognizable as larger, densely staining cells (Eddy, 1974; Merchant, 1976; Guraya and Uppal, 1977). They are also usually situated toward the center of the cords, which are without lumina at this stage of development. The gonocytes in the rat often lie in clusters contigu-

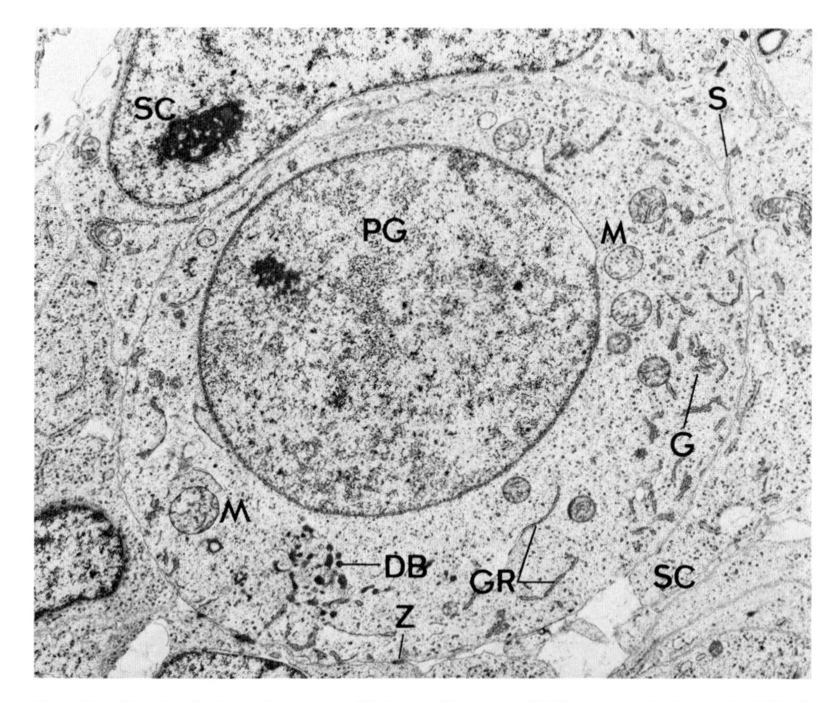

FIG. 17. Testis of pig embryo, age 27 days. Gonocyte (PG) in a testicular cord. DB, Dense bodies; G, Golgi complex; GR, granular endoplasmic reticulum; M, mitochondrion; S, surface folds; SC, sustentacular cell; Z, zonula adherens. ×6000. (From Pelliniemi, 1975b.)

ous with one another and shcw many ribosomes, few membranous organelles, small masses of nuage, and scattered dense-cored vesicles (Fig. 18) (Eddy, 1974).

4. *Interstitium*

The second derivative of the testicular blastema proper is the interstitial tissue which consists of undifferentiated fibroblast-like cells (Vossmeyer, 1971; Holstein *et al.*, 1971; Pelliniemi, 1975b; Merchant, 1975, 1976). At the age of 26 days in the pig embryos it is separated from the testicular cordlike structures by a large intercellular space and by the as yet incomplete, basal lamina of the testicular cords (Pelliniemi, 1975b). At some places no basal lamina is seen, and thus the interstitial cells are in direct contact with supporting cells by cell membrane apposition and patches of zonulae adherens (see also Merchant, 1976). On Day 12 of embryogenesis in the hamster, mesenchymal cells lying in the interstitial region are also undifferentiated (Gondos *et al.*, 1974). The interstitial cells are elongated, with oval nuclei occupying the major portion of the cell cytoplasm which contains scattered ribosomes, occasional strands of rough endoplasmic reticulum, and mitochondria with parallel cristae. The interstitial mesenchymal or fibroblast-like cells also show a similar ultrastructure in the developing testes of guinea pig (Black and Christensen, 1969), man (Pelliniemi and Niemi, 1969; Holstein *et al.*, 1971), rabbit (Gondos *et al.*, 1976), and pig (Pelliniemi, 1975b). The intercellular space is generally large and the interstitial cells are in contact with each other due to parallel apposition of the cell membranes and short contacts of the tips of cytoplasmic processes. The interstitial cells are similar to each other ultrastructurally and no agranular endoplasmic reticulum, a sign of the early differentiation of future Leydig cells (Christensen and Gillim, 1969; Christensen, 1975; Neaves, 1975), could be seen. Ultrastructurally, interstitial cells are not distinguishable from the mesenchymal cells located in the basal part of early testis in the pig (Pelliniemi, 1975b).

The interstitial cells originate from the mesenchymal cells (Vossmeyer, 1971; Holstein *et al.*, 1971; Black and Christensen, 1969; Merchant, 1975) as also described for the corresponding cells in the developing mammalian ovary (Guraya, 1977a). Their structural resemblance is a strong argument for this. Pelliniemi (1975b) has suggested that the direct contacts between interstitial cells and supporting cells in the early pig testis are consistent with a common origin of the supporting and interstitial cells as also observed by Merchant (1976) in the rat. It can be suggested here that both these cell types are of mesenchymal origin.

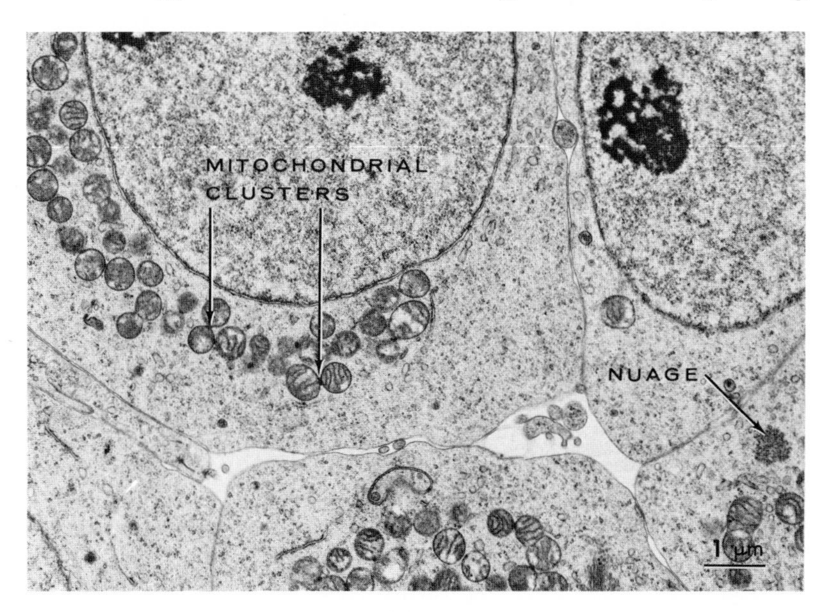

FIG. 18. On Day 17, most of the germ cells in the fetal testis of rat contain the nuage in small, solitary aggregates. However, some cells also contain clusters of mitochondria apparently cemented together by small amounts of dense material. ×7700. (From Eddy, 1974.)

This suggestion is supported by the fact that both the pregranulosa (or follicle) cells and the interstitial cells of developing mammalian ovary have also been shown to originate from the mesenchymal or fibroblast-like cells of gonadal blastema (Guraya, 1977a). A similar suggestion has also been made by Merchant (1975).

5. *Blood Vessels*

The vasculature of embryonic testis consists of capillaries of varying diameter which are distributed in the mesenchyme and interstitium (Fig. 12) (Vossmeyer, 1971; Pelliniemi, 1975b; Merchant, 1975). Surface epithelium, primitive cords, and testicular cords are completely devoid of vasculature. The most peripheral capillaries are located below the gonadal surface (Pelliniemi, 1975b; Merchant, 1975). In the testis they mark the border of the blastema proper that has differentiated into testicular cordlike structures and interstitium. Thus the vascular area in the embryonic testis overlaps with the area of the testicular cordlike structures and the interstitium, indicating a regulative relationship between the vascular pattern and the differentiation of gonadal blastema (Pelliniemi, 1975b,c). The interstitial cells are usually separated from the capillaries by an extracellular space of varying width. The mesenchymal capillaries are similar to those in the testicular interstitium confirming their close relationship.

C. Histochemistry and Steroid Hormone Synthesis

Moon and Raeside (1972) have demonstrated histochemically a low activity of 3β-HSDH in the gonads of 20-mm pig fetuses of undetermined sex and a high activity in the testes of 25-mm fetuses. In both cases, the enzyme activity is apparently present in the testicular cords. The activity of 17β-hydroxysteroid dehydrogenase (17β-HSDH) is very low at the former and not detectable at the latter stage. These enzyme activities are believed to be confined to the cords but not to the interstitium. Merchant (1976) has not observed the development and presence of organelles indicative of steroidogenesis in the cells of testicular cords in the rat. Histochemical results have indicated the possibility of androgen synthesis in the early developing gonad of pig embryos although the agranular endoplasmic reticulum is not yet developed in gonadal cells (Pelliniemi, 1975b). This suggestion is further supported by the results of observations on organ culture of fetal pig gonads at 26 and 30 days of age, which have demonstrated the secretion of androgens from the testis but not from the ovary (Moon and Hardy, 1973; Moon *et al.*, 1973). These androgens might be formed by those blastema cells which showed the beginning of early formation of agranular endoplasmic reticulum in the indifferent gonad (Pelliniemi, 1976b).

VI. Fetal, Postnatal, and Maturing Testis

The two important compartments of the embryonic testis are the interstitial cells and the seminiferous cords which undergo conspicuous morphological, histochemical, and biochemical changes during the fetal, postnatal, and prepubertal periods. Mills *et al.* (1977) have recently demonstrated correlative morphological and biochemical changes which occur during postnatal development and maturation of the rat testis. Many of the biochemical changes of proteins, RNA, and DNA are correlated to the differentiating cell population of the testis.

A. Interstitial Leydig Cells

1. *Origin, Differentiation, and Morphology*

It is now well established that the testicular interstitial Leydig cells of mature males develop through intermediate stages from fibroblast-like interstitial cells of mesenchymal origin (Hooker, 1944, 1948; Mancini *et al.*, 1963; Hatakeyama, 1965; Mancini, 1968; Christensen, 1970, 1975; Hayashi and Harrison, 1971;

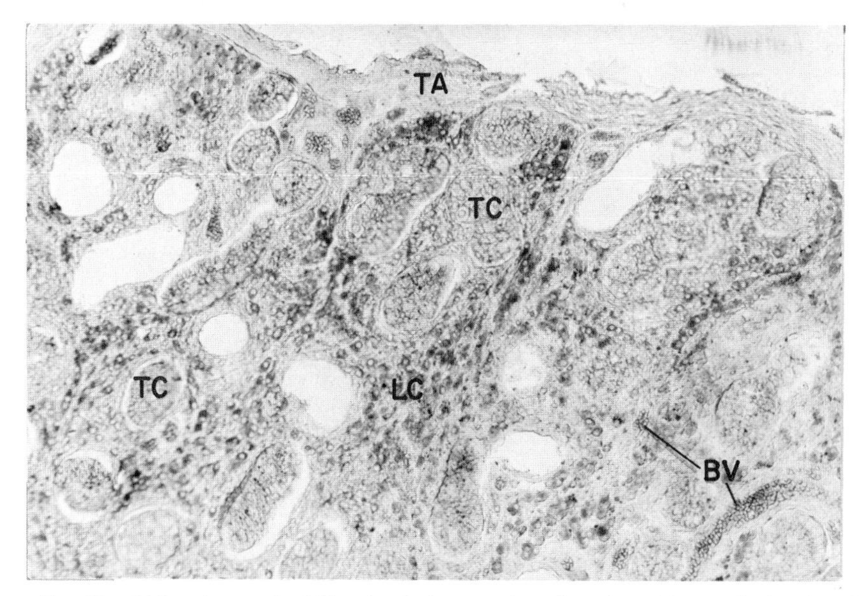

FIG. 19. Light micrograph of histochemical preparation of portion of human fetal testis, 18 weeks of age, showing distribution of interstitial Leydig cells (LC) containing sudanophilic lipids between testicular cords (TC). Note the origin of septula from tunica albuginea (TA). Blood vessels (BV) are also seen. ×29.

Gondos *et al.*, 1976, 1977). A similar sequence of steps in the formation of fetal Leydig cells has also been demonstrated in detailed studies on the rat (Narbaitz and Adler, 1967; Lording and de Krester, 1972; Roosen-Runge and Anderson, 1959), sheep (Baillie, 1960), man (Pelliniemi and Niemi, 1969; Vossmeyer, 1971; Holstein *et al.*, 1971; Guraya, 1974a; Gondos, 1975), guinea pig (Black and Christensen, 1969; Guraya, 1974b), pig (Moon and Hardy, 1973), mouse (Russo and de Rosas, 1971), rabbit (Gondos and Conner, 1972; Bjerregaard *et al.*, 1974; Catt *et al.*, 1975), hamster (Gondos *et al.*, 1974), horse (Gonzalez-Angulo *et al.*, 1971), monkey (Tseng *et al.*, 1975), and field rat *Millardia meltada* (Guraya and Uppal, 1977). The pattern of differentiation appears to be very similar in the different species studied. Initially, the fetal Leydig cells have been observed to originate and differentiate from the

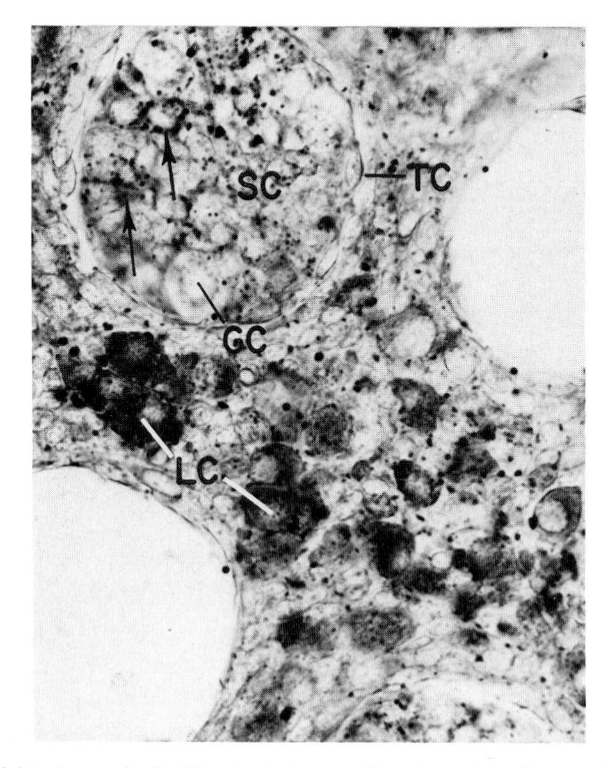

FIG. 20. Light micrograph of a histochemical preparation of a portion of human fetal testis, 17 weeks of age, showing abundant sudanophilic lipids in the cytoplasm of hypertrophied Leydig cells (LC). Such lipids are not much developed in the testicular cords (TC) which show Sertoli cells (SC) and germ cells (GC). Degenerating supporting cells accumulate deeply sudanophilic lipid droplets of coarse nature (arrows). ×380.

fibroblast-like interstitial cells of mesenchyme, which derive from the blastema proper and are distributed in the interstices of testicular cords (Figs. 19 and 20) as already discussed. The fibroblast-like cells are small cells with little cytoplasm. They are either roundish or slightly elongated cells. Electron microscope studies have revealed their simple morphological organization. The nuclear area in sections is amply more than one-half of the whole cell (Fig. 21). One to two nucleoli are usually present. They have a sparse rough endoplasmic reticulum consisting of long solitary straight or curved cisternae with patches of ribosomes on the surface. As is typical of embryonic cells, free ribosomes and polyribosomes are abundant (Fig. 21). The mitochondria are round or elongate. The cristae are of lamellar type. The mitochondria have either a relatively dense matrix (Black and Christensen, 1969), or a matrix staining like the surrounding cytoplasm (Pelliniemi and Niemi, 1969). Generally no inclusions are seen between the cristae of mitochondria in the undifferentiated interstitial cells. Black and Christensen (1969) have also reported the presence of membrane-bound dense bodies, abundant small filaments, and microtubules in addition to the other well-known organelles. The cell surface may form several ingrowths and irregular processes, which constitute the contact points between the adjacent cells over the wide extracellular space without any attachment specialization or stable differentiations. Sometimes, solitary flagella may be seen in the fibroblast-like interstitial cells. The extracellular space does not show any electron-positive

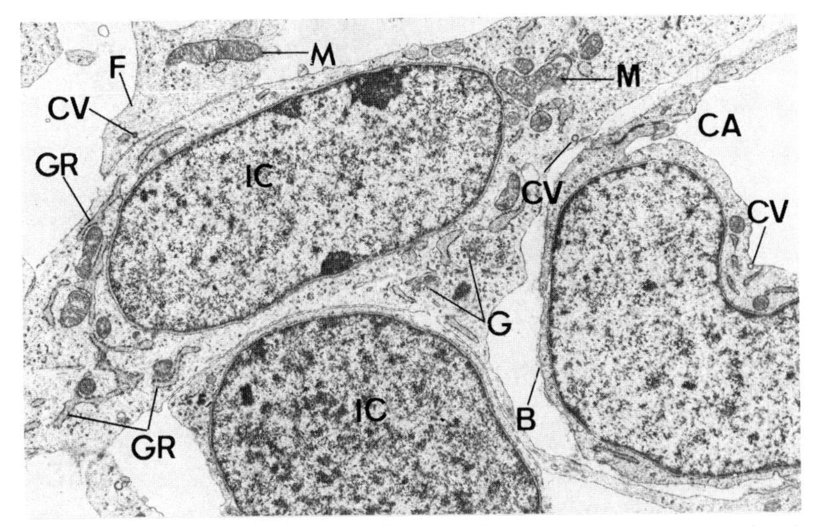

FIG. 21. Testis of pig embryo, age 27 days. Two interstitial cells (IC) on the left and one capillary endothelial cell on the right. B, Basal lamina; CA, capillary lumen; CV, coated vesicles; F, cytoplasmic filaments; G, Golgi complex; GR, granular endoplasmic reticulum; M, mitochondrion. ×6160. (From Pelliniemi, 1975b.)

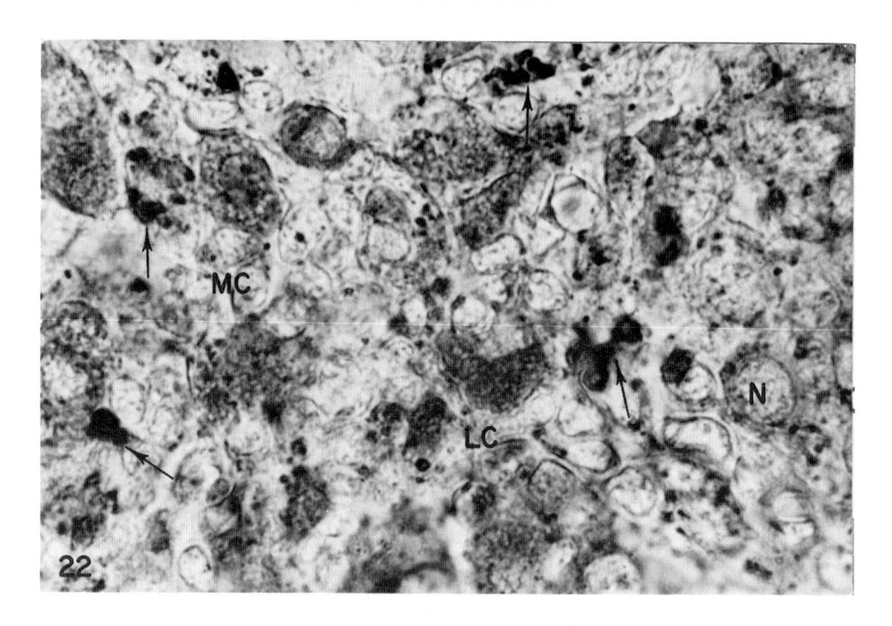
MC
N
LC
22

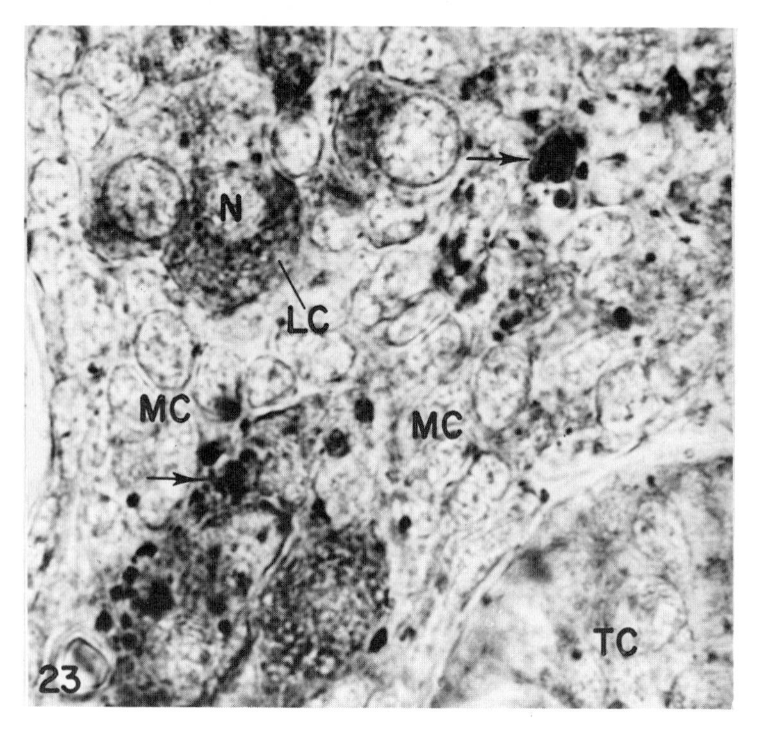
N
LC
MC
MC
TC
23

material except in the vicinity of the testicular cords where a basal lamina and some fibers have already developed.

In the developing and maturing testis are seen some intermediate or immature or partially differentiated interstitial Leydig cells which in their histochemical and ultrastructural features resemble partly the undifferentiated fibroblast-like cells and partly the fully differentiated and mature interstitial Leydig cells (Baillie, 1960; Pelliniemi and Niemi, 1969; Black and Christensen, 1969; Holstein *et al.*, 1971; Gonzalez-Angulo *et al.*, 1971; Russo and de Rosas, 1971; Gondos and Conner, 1972; Moon and Hardy, 1973; Merchant, 1976; Gondos *et al.*, 1974, 1976, 1977; Guraya, 1974a,b; Guraya and Uppal, 1977). In the developing and maturing testis, the study of differentiation of the Leydig cells is complicated by the fact that not all the cells differentiate at the same time (Gondos *et al.*, 1976, 1977). This results in a heterogeneous population of interstitial cells which are at different stages of their differentiation (Fig. 22). Gondos *et al.* (1977) have observed that the differentiation process of Leydig cells in the rabbit occurs over several weeks postnatally as opposed to only a few days in the fetus. But with differentiation of fibroblast-like cells into interstitial Leydig cells, an increase is seen in cell size, and the shape begins to become smoothly roundish (Fig. 23). A general trend of increasing amount and complexity of smooth endoplasmic reticulum becomes evident (Fig. 24) (see details in Holstein *et al.*, 1971). Changes are also seen in other organelles and inclusions (Fig. 24). The nucleolus becomes round and is gradually located eccentrically in the nucleus. The chromatin is condensed in the periphery of the nucleus. Centrioles are rare. Mitochondria increase in size and number. Their cristae become tortuous lamellae and tubules and the matrix contains no inclusions. Fenestrated smooth cisternae and whorls of fenestrated cisternae occur more frequently in the differentiating Leydig cells, but the tubular form of the endoplasmic reticulum continues to predominate (Fig. 24). At the same time, there occurs a relative decrease in the number of polysomes seen on the tubules of the smooth endoplasmic reticulum. Meanwhile, the cisternae of the rough endoplasmic reticulum become less numerous and tend to be segregated into parallel stacks (Fig. 24). Nuclei become spherical and assume an eccentric location. Corresponding to these ultrastructural changes in the endoplasmic reticulum and mitochondria, there is a progressive increase in the amount of diffusely distributed sudanophilic lipids (lipoproteins) in the cytoplasm of differentiating interstitial Leydig cells as demonstrated with the histo-

---

FIGS. 22 and 23. Light micrographs of a histochemical preparation of portions of human fetal testis, 18 weeks of age, showing abundant diffusely distributed sudanophilic lipids and deeply sudanophilic lipid granules in cytoplasm of hypertrophied Leydig cells (LC). Such lipids are not seen in the nucleus (N) or undifferentiated stromal or mesenchymal cells (MC), or in cells of testicular cord (TC). Partially differentiated Leydig cells show relatively less sudanophilic lipids. Degenerating Leydig cells (arrows) accumulate deeply sudanophilic lipid droplets of coarse nature. Fig. 22, ×740; Fig. 23, ×960.

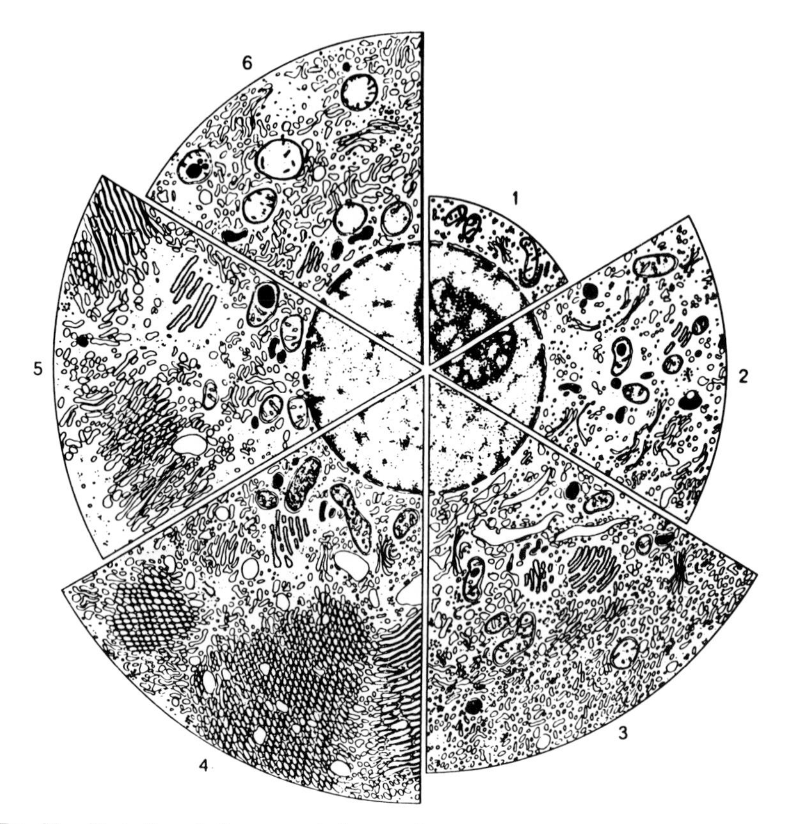

FIG. 24. Illustration of ultrastructural changes of cytoplasmic organelles during the differentiation of fibroblast-like stromal cells into interstitial Leydig cells (1 to 4) during the development of human testis. Note the accumulation of abundant agranular endoplasmic reticulum, alterations in the cristae of mitochondria, hypertrophy of Golgi complex, sparsely scattered ribosomes, and some elements of granular endoplasmic reticulum. With the involution of Leydig cells (5 and 6) the various organelles undergo conspicuous changes in their morphology and amount. (From Holstein *et al.*, 1971.)

chemical techniques (Figs. 22 and 23) (Guraya, 1974a,b; Guraya and Uppal, 1977), suggesting a close relationship between the development of increasing membranes of smooth reticulum and diffusely distributed lipoproteins. The number and size of the dense bodies and highly sudanophilic lipid droplets are increased (Figs. 20, 22, and 23). Particulate glycogen is decreased. Microtubules and filaments become less apparent as the cytoplasm becomes increasingly packed with smooth membranes of endoplasmic reticulum. The Golgi complex increases in size in the fully differentiated Leydig cells (Fig. 24).

According to Pelliniemi and Niemi (1969), the cell membrane of a differentiating Leydig cell does not develop any external coats, attachment specializations, or other differentiations except some surface folds. But Black and Christensen (1969) have observed that Leydig cells of the developing guinea pig testis are generally separated from one another by the usual space of 200 Å or more. Occasional indistinct desmosomes, and a few punctate tight junctions, can be seen between interstitial cells of 25 to 26 days. In older testes, the number of desmosomes and tight junctions increases (Figs. 25 and 26) (see also Gondos *et al.*, 1974, 1976). Some of the tight junctions become very extensive and surround the projections from adjacent cells. These projections, which are quite common, vary in size and depth of penetration, some nearly reaching the Golgi complex. The cytoplasm within the projection usually has higher density.

In summary, it can be stated that the fully hypertrophied Leydig cells of fetal and postnatal testes in all the mammalian species investigated so far are round to oval, with variable amounts of smooth endoplasmic reticulum, large mitochondria with predominantly tubular cristae and some granules, prominent lipid droplets in variable amount, and well-developed Golgi complexes (Figs. 25 and 26) (Black and Christensen, 1969; Pelliniemi and Niemi, 1969; Gonzalez-Angulo *et al.*, 1971; Holstein *et al.*, 1971; Gondos and Conner, 1972; Gondos *et al.*, 1974, 1976; Tseng *et al.*, 1975). Prominent areas of the loosely organized tubular endoplasmic reticulum are seen near the nucleus (Black and Christensen, 1969). This reticulum is predominantly smooth surfaced, but it has scattered patches of polyribosomes on some areas of its surface. The rest of the Leydig cell shows rough-surfaced reticulum which is sparsely distributed and is in both the cisternal and the saccular form. It is freely interconnected with the tubular reticulum. Some free ribosomes and polyribosomes are present. The large Golgi complex, which appears to have been missed by Pelliniemi and Niemi (1969) in human fetal Leydig cells, is usually associated with coated vesicles (Holstein *et al.*, 1971). In addition to the lipid droplets, lysosome-like bodies develop in the mature Leydig cells which may contain some myelin figures also. For their ultrastructural features, the fully differentiated Leydig cells of fetal and postnatal testes closely resemble those in the adult (Neaves, 1975; Christensen, 1970, 1975).

2. *Histochemistry*

Histochemical techniques for lipids applied to frozen sections of fetal testes in the field rat (Guraya and Uppal, 1977), guinea pig (Guraya, 1974b), and man (Guraya, 1974a) have revealed that the actively steroid-secreting interstitial Leydig cells show abundant diffusely distributed sudanophilic lipids (lipoproteins) and a few highly sudanophilic lipid droplets (Figs. 20, 22, 23, and 27). The lipid droplets consist of either phospholipids, or phospholipids and triglycerides; no cholesterol and/or its esters could be demonstrated. The diffusely distributed lipoproteins of hypertrophied fetal Leydig cells have been presumed to derive from ultrastructural membranes of smooth reticulum. With the regres-

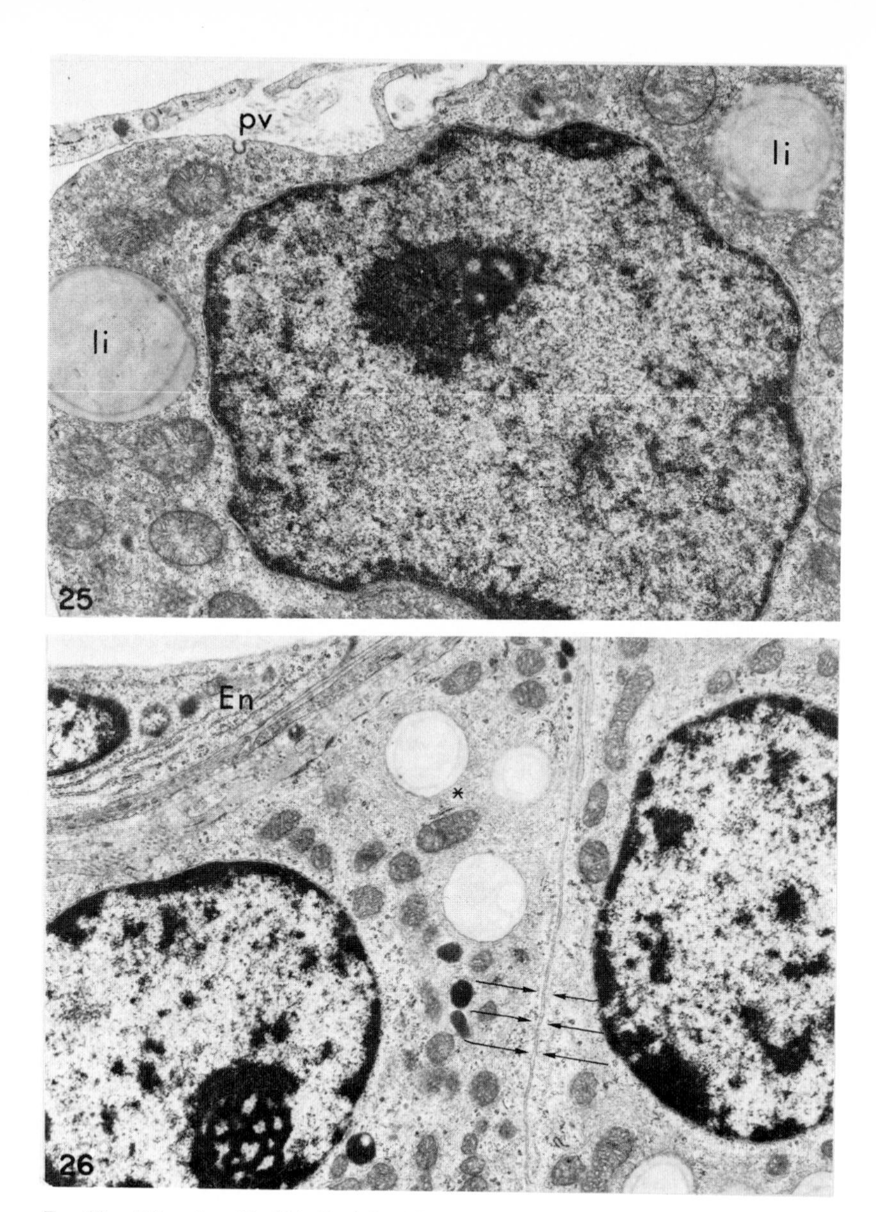

FIG. 25. Thirty-day-old rabbit. Partially differentiated interstitial Leydig cell, ovoid type, showing close association of lipid (li) and mitochondria, and association of ribosomes with endoplasmic reticulum. Pinocytotic vesicles (pv) are present. Perfusion fixation. ×15,940. (From Gondos *et al.*, 1976.)

FIG. 26. Sixty-day-old rabbit. Portions of two fully differentiated Leydig cells, with plasma membranes in close apposition (arrows). Abundant smooth endoplasmic reticulum is present, with occasional continuities between rough- and smooth-surfaced membranes evident (asterisk). Free surface of Leydig cell at left lies near endothelial cell (En) lining interstitial blood vessel. Perfusion fixation. ×9240. (From Gondos *et al.*, 1976.)

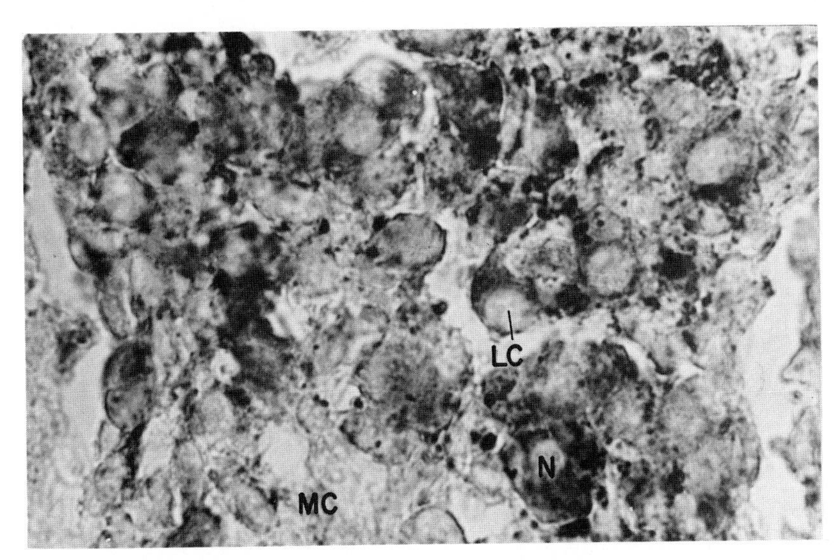

FIG. 27. Light micrograph of a histochemical preparation of a portion of guinea pig fetal testis at 40 days of gestation, showing diffusely distributed sudanophilic lipids and deeply sudanophilic lipid granules in the cytoplasm of hypertrophied Leydig cells (LC). Such lipids are not seen in the nucleus (N) or in undifferentiated mesenchymal cells (MC). Some cells intermediate between fully differentiated cells and mesenchymal cells show some sudanophilic lipids. ×740.

sion of Leydig cells, the highly sudanophilic lipid droplets accumulate (Figs. 22 and 23). They consist of cholesterol and/or its esters, triglycerides, and some phospholipids.

Histochemical techniques for localizing enzymes, which are involved in steroid hormone synthesis, have clearly demonstrated their presence in the Leydig cells of fetal and postnatal testes in different mammalian species (Bloch *et al.*, 1962; Niemi and Ikonen, 1961, 1963; Hitzeman, 1962; Chieffi *et al.*, 1964; Turolla *et al.*, 1965; Baillie, 1965; Ikonen, 1965; Price *et al.*, 1964; Baillie *et al.*, 1965, 1966a,b; Cavallero *et al.*, 1965; Hart *et al.*, 1966a,b; Goldman *et al.*, 1966; Ortiz *et al.*, 1966; Picon and Ottowicz, 1967; Schlegel *et al.*, 1967; Niemi *et al.*, 1967; Knorr *et al.*, 1970; Tramontana *et al.*, 1967; Lording and de Kretser, 1972; Gondos, 1977). The hydroxysteroid dehydrogenases are enzymes that function in a specific manner at different steps in the biosynthesis of the steroid hormones (Christensen and Gillim 1969; Neaves, 1975; Christensen, 1975). Of all the dehydrogenases, 3β-HSDH has been more extensively studied by histochemical methods in the steroid gland cells (see Baillie *et al.*, 1966a). It is involved in the transformation of pregnenolone to progesterone and of DHA to androsterone, precursors of androgens, estrogens, and corticosteroids. 3β-HSDH has been extensively reported in the testes of different mammalian species at the stage immediately following sex differentiation. But the fetal guinea pig testes at 22 to 30 days of gestation (this period covers the formation of the undifferentiated gonads) show an androgenic effect on rat prostate cultured *in vitro* whereas 3β-HSDH can only be found at 20 days (Price *et al.*, 1964; Ortiz *et al.*, 1967). This indicates that the enzymes required for steroidogenesis are synthesized earlier. However, in the fetal rabbit testis, the development of 3β-HSDH activity correlates well with the differentiation and maturation of Leydig cells as well as with the onset of androgen formation (Goldman *et al.*, 1972). The appearance of various enzyme activities involved in steroid biosynthesis usually parallels the development of agranular endoplasmic reticulum (or diffuse lipoproteins) and mitochondria with complex cristae in the Leydig cells. Studies in the human testis have also shown glucose-6-phosphate dehydrogenase activity in the Leydig cells during the postnatal prepubertal period (Wolfe and Cohen, 1964).

3. *Biochemistry*

a. *Fetal Testis.* Biochemical studies of developing testes in different mammalian species have produced a conclusive evidence that the fetal testis secretes steroid hormones; the main steroid hormone is testosterone. Synthesis of these steroid hormones begins shortly after testicular differentiation at about the time of Leydig cell formation; there is then a sharp rise to maximal levels and a decrease in steroid hormone secretion correlating with the regression of Leydig cells (see Blackshaw, 1970; Hall, 1970). Testosterone has been quantified in the fetal testis of guinea pig (Sholl, 1974), sheep (Attal, 1969), monkey (Resko, 1970a), man (Huhtaniemi *et al.*, 1970; Reyes *et al.*, 1975), rat (Warren *et al.*, 1973; Bloch *et al.*, 1974), and more recently rabbit (Catt *et al.*, 1975; Veyssiere *et al.*, 1975, 1976). Testosterone has also been measured in the fetal blood of some species, such as monkey (Resko *et al.*, 1973), cattle (Challis *et al.*, 1974), pig (Reyes *et al.*, 1973, 1974), and rabbit (Veyssiere *et al.*, 1975, 1976). Biochemical studies have shown that fetal testes of different mammalian species, utilizing biosynthetic pathways basically similar to those of the adult testes (Tamaoki *et al.*, 1969; Neaves, 1975; Christensen, 1975), are capable of synthesizing testosterone and androstenedione from labeled precursors such as progesterone, pregnenolone, cholesterol, or sodium acetate (Lipsett and Tullner, 1965; Solomon, 1966; Noumura *et al.*, 1966; Ikonen and Niemi, 1966; Bloch, 1967; Rice *et al.*, 1966; Haffen, 1970; Resko, 1970a; Weniger *et al.*, 1967; Serra *et al.*, 1970; Mathur *et al.*, 1972; Tsujimura and Matsumoto, 1974; Ahluwalia *et al.*, 1974; Villee, 1974; Tayler *et al.*, 1974; Gondos, 1977; Sanyal and Villee, 1977). The main compounds found in the midgestational testis are testosterone and pregnenolone. Progesterone is not present, but several compounds with 3β-hydroxy-Δ^5 structure can be found, showing that androgens are synthesized via the Δ^5 pathway (Huhtaniemi *et al.*, 1970; Mathur *et al.*, 1972). Sulfate-congugated as well as neutral steroids are also present. Lamont *et al.*

(1970) have demonstrated sulfatase activity in the fetal testis *in vitro* with formation of testosterone from pregnenolone sulfate. The ability of the human fetal testis to cleave pregnenolone sulfate, which occurs in high amounts in the fetal circulation (Huhtaniemi and Vikho, 1970), is important (Payne and Jaffe, 1975). But there is no sulfurylation of testosterone in the testis (as there is in the adrenal gland), permitting testosterone to be kept in its active form (Jaffe and Payne, 1971). According to Sanyal and Villee (1977), the synthesis of testosterone by the rat fetal testes in the latter half of gestation (Days 17–21) is increased by the addition of precursors: 15-fold by the addition of pregnenolone, 20-fold by progesterone, and 47-fold by dehydroepiandrosterone. But the increase is marginal in the presence of sulfates of pregnenolone or dehydroepiandrosterone. These data have suggested that Δ^5-isomerase-3β-HSDH and 17β-HSDH are active in the fetal testes; in contrast, the activity of steroid sulfatase is relatively low.

Usually, for the study of biosynthetic pathways, the fetal testes especially of rodents were obtained during the second half of pregnancy. Two studies have been carried out on the younger stages of the fetal development in the rat (Noumura *et al.*, 1966) and the rabbit (Lipsett and Tullner, 1965). Progesterone is transformed by the fetal testis into 17-hydroxyprogesterone, androstenedione, and testosterone, respectively, under the action of a 17α-hydroxylase, a 20-desmolase and a 17β-HSDH (Haffen, 1970). These enzymes split the side chain of 21-steroids giving rise to C19-steroids. The investigations of Noumura *et al.* (1966) on rat embryos do not reveal androgen biosynthesis by the fetal testis at the time morphological differentiation takes place (13.5 to 14.5 days of gestation). But this capacity of androgen synthesis increases especially from 15.5 days, the stage when the male genital system and the Wolffian ducts and their derivatives develop. It reaches a plateau between 19.5 and 20.5 days of gestation when the fetal Leydig cells also attain their maximum development (see Guraya and Uppal, 1977). Pregnenolone is also converted into androgens by the fetal testes of different mammalian species (see Haffen, 1970). The formation of testosterone and androstenedione in the fetal testis shows the occurrence of the following enzymatic reactions: side-chain splitting; isomerization with shifting of double bond from C-5 to C-4 and dehydrogenation in the C-3 position; and 17β-hydroxydehydrogenation (see Haffen, 1970).

There is a strong evidence that in the rabbit the differentiation of fetal Leydig cells is closely associated with the development of a capacity for androgenic activity (Lipsett and Tullner, 1965; Wilson and Siiteri, 1973). Lipsett and Tullner (1965) have also shown that the rabbit fetal testis acquires the capacity to synthesize androgens from pregnenolone at 18 days of gestation when Wolffian ducts and their derivatives are developing. This conversion can be demonstrated prior to the appearance of Leydig cells which become visible on Day 19 of gestation when Müllerian ducts begin to undergo involution. It will be significant to mention here that as early as Day 16 of gestation, the rabbit fetal testis transforms androstenedione into testosterone under the influence of a 17β-HSDH; this conversion increases with age of gestation which closely corresponds to the development and differentiation of Leydig cells (Gondos and Conner, 1972). Wilson and Siiteri (1973) have studied the steroidogenic ability of fetal rabbit testis as a function of the age of embryos, using different precursors, and have demonstrated a parallelism between the events during masculinization of the male genital tract and the pattern of testosterone synthesis by the fetal testis. They have concluded that the enzymes 3β-HSDH and isomerase crucial in the pathway of testosterone biosynthesis are acquired between Days 17 and 19, whereas all the enzymes involved in the synthesis of testosterone appear on Days 21 to 23, indicating that enzymes appear at different stages of testicular differentiation. In the sheep, androgen content of the fetal testis has been demonstrated on the thirtieth day of gestation, 5 days before the onset of sexual differentiation (Mauleon, 1961; Attal, 1969).

Acevedo *et al.* (1963) and Ikonen and Niemi (1966) have investigated the relative importance of two biosynthetic pathways in the human fetal testis, starting from progesterone or from 17α-hydroxypregnenolone and pregnenolone. According to Ikonen and Niemi (1966), testosterone and androstenedione are produced in greater quantities from 17α-hydroxypregnenolone than from progesterone. These results have indicated that the synthesis of testosterone proceeds through DHA and the conversion from DHA into testosterone goes through androstenediol. According to Acevedo *et al.* (1963), 17α-hydroxylase is the most active enzyme in the biosynthesis from progesterone and pregnenolone. DHA is also an intermediate step in the biosynthesis of androstenedione and testosterone from pregnenolone. Rice *et al.* (1966), while investigating the synthesis of androgens from sodium acetate in organ culture, have demonstrated the formation of more pregnenolone and 17α-hydroxyprogesterone than of androstenedione and testosterone; DHA could not be demonstrated with certainty. They have inferred a deficiency in 20-desmolase rather than in 3β-HSDH activity under experimental conditions. Weniger *et al.* (1967) have detected the formation of androstenedione and testosterone from sodium [1-^{14}C]acetate in culture media 24 hours after explantation of embryonic testes from 3.5-month-old calves and in culture media of mouse testes (at 12 to 19 days of gestation), respectively. MacArthur *et al.* (1967) have recovered radioactive DHA in their study on the formation of steroids from preparations of an equine fetal testis. The yield of DHA from sodium [1-^{14}C]acetate greatly exceeded that of testosterone and androstenedione. Its isolation from fetal testes and ovaries of horse points to a possible role for DHA as a precursor in the placental formation of estrogens in the mare as recently suggested by Raeside (1976).

Abramovich and Rowe (1973) have found high levels of plasma testosterone in early to midterm human male fetuses (12–18 weeks) which parallel the maximal development and occurrence of Leydig cells (Figs. 19, 20, 22, and 23) (Pelliniemi and Niemi, 1969; Holstein *et al.*, 1971; Guraya, 1974a) and of their 3β-HSDH activity in the fetal testis (Baillie *et al.*, 1965; Niemi *et al.*, 1967). At this point of development, substrates such as NADH, NADPH, and glucose 6-phosphate can be used by the testes (Niemi *et al.*, 1967). Similar results have also been obtained with older fetuses (Mancini *et al.*, 1963). Maximum quantities of NADH-tetrazolium reductase and 3β-HSDH are found when the Leydig cells are at a maximum; in addition, evidence for substrate specificity of the 3β-HSDH in fetal human testes has been shown (Baillie *et al.*, 1965). Similar observations have also been made by Reyes *et al.* (1973). These studies have indicated that for a period of about 2 weeks after the formation of external genitalia, there occur high levels of plasma testosterone associated with Leydig cell hyperplasia (Abramovich and Rowe, 1973). The various organ culture experiments have shown that guinea pig fetal testes are capable of secreting enough androgen to maintain the normal secretory histology of rat ventral prostate (Ortiz *et al.*, 1966).

From these various studies, it can be concluded that the mammalian fetal testis is capable of synthesizing various steroid hormones as is the adult testis (Tamaoki *et al.*, 1969; Neaves, 1975; Ewing and Brown, 1977), when a radioactive precursor is supplied. This biosynthetic ability appears at the time when the Wolffian ducts and their derivatives are developing. During normal embryogenesis their development and differentiation occur under the influence of the male hormone secreted by the fetal testis, as has been demonstrated by Jost (1947, 1965), Reynaud and Frilley (1947), Reynaud (1950), and Wells (1950). The Leydig cells are the site for the synthesis of male hormones in the fetal testis as revealed by correlation of various ultrastructural, histochemical, and biochemical findings. The androgens formed in and released from embryonic and fetal Leydig cells are also believed to influence the multiplication and differentiation of germ cells (Holstein *et al.*, 1971) as Leydig cell development in the human testis precedes germ cell development by 14 days and reaches its maximum between the twelfth and fourteenth week of gestation.

b. *Maturing Testis.* The use of sensitive methods of estimation of steroids has revealed that in the rat (Resko *et al.*, 1968; Warren *et al.*, 1973; Miyachi *et al.*, 1973), mouse (McKinney and Desjardins, 1973; Berger *et al.*, 1975), guinea pig (Rigaudiere, 1973), ram (Attal *et al.*, 1972; Crim and Geschwind, 1972), monkey (Resko, 1970a, 1977), man (Frasier *et al.*, 1969; Winter and Faiman, 1972; Resko, 1977), and bull (Rawlings *et al.*, 1972; Secchiari *et al.*, 1976; Lacroix *et al.*, 1977), but not in sheep (Cotta *et al.*, 1975), the initiation of puberty in the male is characterized by an increase in plasma and testicular testosterone. Biochemical studies on steroid biosynthesis in the postnatal rat testis have revealed that the principal androgen pathway prior to puberty is from pregnenolone to androstenedione which is reduced to 5α-androstanedione (see Inano *et al.*, 1967; Tamaoki *et al.*, 1969). With maturation of the testis, 17β-HSDH increases in activity and the reductases decrease in activity so that androstenedione is effectively converted to testosterone. The formation of C19-steroids from progesterone is believed to take place primarily in the interstitial tissue, with additional activity in germ cells and Sertoli cells (Yamada *et al.*, 1973; Dorrington and Fritz, 1973, 1975). According to Yamada and Matsumoto (1974) the α-reduced pathway is the major pathway for androgen synthesis in the testis of immature rat. 5α-Reductase activity is maximal at 20 to 30 days when testosterone production is lowest (Nayfeh *et al.*, 1966; Ficher and Steinberger, 1971). It is still not known more precisely whether the higher reductase activity is present in the interstitial tissue (Yamada *et al.*, 1972; Folman *et al.*, 1973; Matsumoto and Yamada, 1973), or in the seminiferous tubules (Rivarola *et al.*, 1972), or in both (van der Molen *et al.*, 1975). Nayfeh *et al.* (1975a) have demonstrated that 5α-reductase of the immature rat testis is present at least in two compartments and is stimulated by either FSH and/or LH. However, Yamada *et al.* (1973) and Dorrington and Fritz (1973) have shown that the reductase activity present in the tubules is much higher in the immature than in the mature rat. Dorrington and Fritz (1975) have further determined that germ cells, especially spermatocytes, are capable of metabolizing testosterone to dihydrotestosterone, whereas Sertoli cells produce androstanediol as the major product.

There occurs a gradually increasing testosterone:androstenedione ratio during the pubertal development and maturation of testis in different mammalian species such as the rat (Ficher and Steinberger, 1971; Tsang *et al.*, 1973), bull (Linder, 1961; Rawlings *et al.*, 1972), pig (Elsaesser *et al.*, 1972), and guinea pig (Resko, 1970b). A similar change has also been noticed in the plasma of prepubertal and pubertal boys (Vermeulen, 1973). However, testosterone and androstenedione rise in parallel during pubertal maturation of the mink (Nieschlag and Bienieh, 1975) and ram (Attal, 1970).

Goldman and Klingele (1974) have observed that steroidogenic enzymes appear in three stages in the rat testis. The first stage is characterized by the onset of 5α-reductase activity which starts at 15 days of age and increases to a maximum level between 30 and 40 days. The activities of 17α-hydroxylase, C_{17}-20-lyase, and 17β-HSDH appear between 20 and 30 days during the second stage. During the third stage, from 40 to 60 days, 17-ketoreductase develops and increases to adult levels while inversely 5α-reductase and 17β-HSDH activities decrease to adult levels. HCG stimulates the activities of various enzymes involved in the biosynthesis of testosterone, with the exception of 17β-HSDH which in the mouse is increased by prolactin treatment (Musto *et al.*, 1972).

Berger *et al.* (1976a) have reported the development patterns of plasma and testicular testosterone in rabbits from birth to 90 days of age. Plasma testosterone varies little from birth to 40 days and increases thereafter until 60 days. This level, characteristic of puberty, remains unchanged until 90 days and increases thereafter. Testicular testosterone also follows a similar pattern. It remains steady from 1 to 20 days. From 20 days to 50 days it rises slowly. Between 50 and 60 days it increases rapidly and reaches its maximum at 60 days. It remains at the same level until 90 days. These patterns of plasma and testicular testosterone in rabbit correlate with the morphological data of Gondos *et al.* (1976, 1977) who have revealed corresponding fluctuations in the development and differentiation of interstitial Leydig cells during the maturation of testis. Testicular androgens in the rabbit do not decrease from birth to puberty as they do in the rat (Resko *et al.*, 1968), monkey (Resko, 1970a), and man (Forest *et al.*, 1974).

Cholesterol and cholesterol esters constitute the precursor material for the biosynthesis of steroid hormones in the gonads of vertebrates (Neaves, 1975; Christensen, 1975; Guraya, 1971, 1976a,b). Their amounts change with the prepubertal development of mammalian testis. The concentrations of cholesterol esters increase before puberty in the mouse, attaining maximum levels at 31 to 44 days, after which a gradual decrease occurs (Bartke, 1971); free cholesterol contents increase until 12 days of age, then decline and remain low until 44 days. Renston *et al.* (1975) have observed relatively constant amounts of free cholesterol throughout the prepubertal development of rabbit testis; the levels of cholesterol esters remain highest early in development, then decline to their lowest values after 60 days of age. The values of esterified cholesterol in mouse testes at 50 and 60 days of age remain significantly higher in early maturing than in late maturing strains (Bartke *et al.*, 1974). These observations have suggested that the selection for early and late sexual maturation can be related to varying rates of synthesis of androgenic steroid from cholesterol ester precursors used in steroid hormone synthesis. Coniglio *et al.* (1975), while studying the lipids of human testes in specimens obtained at autopsy, have observed changed fatty acid composition in infants in comparison to adults and higher free and esterified cholesterol levels in infants. Ahluwalia *et al.* (1975) have found that the fatty acid composition of the fetal testis at 12 to 16 weeks of gestation is similar to that of the adult testis and nervous system.

4. *Correlation of Morphological, Histochemical, and Biochemical Data in Different Species*

Some mammalian species have been investigated relatively in more detail for the morphological (including the ultrastructural), histochemical, and biochemical changes of interstitial Leydig cells during development and maturation of their testes. The results of these studies have been reported in isolation in different papers dealing either with ultrastructure, or with histochemistry, or with *in vivo* and *in vitro* biochemical analysis of steroid synthesis and secretion. Therefore, it will be very useful to make correlations between the results of these various studies in order to provide a deeper insight into the functional significance of ultrastructural, histochemical, and biochemical alterations of interstitial Leydig cells during the development and maturation of the testis in different mammalian species. In making such correlations, the repetition of some results discussed earlier is unavoidable.

a. *Hamster.* Testicular differentiation in the hamster occurs at 11.5 days (Price and Ortiz, 1965). Leydig cells make their appearance between Days 12 and 13 and are present as groups of large cells concentrated around interstitial blood vessels (Gondos *et al.*, 1974). The cells have been identified as Leydig cells by the presence of characteristic features of steroid-producing cells; abundant smooth endoplasmic reticulum, large mitochondria with tubular cristae, lipid droplets, and prominent Golgi complexes as already discussed. Cells of this type are present throughout the remainder of gestation and for the first few days after birth; regression begins on about Day 4. Very little or no histochemical and biochemical work has been carried out on the developing and maturing testes of hamster.

b. *Rat.* In the rat fetal testis, typical Leydig cells are developed around Day 17 of gestation (Merchant, 1976). Their maximum development occurs on fetal Day 19 (Roosen-Runge and Anderson, 1959). The Leydig cells show poor development on postnatal Day 4, showing their regression after birth (Roosen-Runge and Anderson, 1959; Niemi and Ikonen, 1963). *In vitro* studies of steroid bioconversion have shown that rat testis is capable of rapid conversion of acetate or pregnenolone or progesterone to testosterone during the later stages of fetal development (Ficher and Steinberger, 1968; Bloch, 1967; Bloch *et al.*, 1971a, Warren *et al.*, 1972, 1973, 1975; Sanyal and Villee, 1977), when there is seen maximum development of Leydig cells possessing the features typical of steroidogenic tissue (see Guraya and Uppal, 1977). The capacity for testosterone biosynthesis then declines as the testes develop so that at 20 days only trace amounts are synthesized from progesterone (Nayfeh *et al.*, 1966; Inano *et al.*, 1967; Ficher and Steinberger, 1971; Coffey *et al.*, 1971). A sharp decline in testosterone synthesis occurs immediately after birth and has been attributed to the regression of Leydig cells (Guraya and Uppal, 1977). As the rats continue to mature, the synthesis of testosterone from progesterone returns gradually and increases to reach adult levels at around 90 days (Coffey *et al.*, 1971; Ficher and Steinberger, 1971; Podesta *et al.*, 1974; Sowell *et al.*, 1974). Biochemical analysis of testicular and spermatic vein testosterone has also revealed low levels between 20 and 40 days with continually increasing concentrations between 40 and 90 days (Knorr *et al.*, 1970). The testosterone levels in the plasma show a sharp increase starting at 30 to 40 days, reaching a maximum at 60 to 70 days, followed by a slight decrease (Knorr *et al.*, 1970; Grota, 1971; Miyachi *et al.*, 1973; Lee *et al.*, 1975; Brown-Grant *et al.*, 1975; Pahnke *et al.*, 1975). The alterations in testosterone levels may be due to changes in endogenous LH concentrations which are highly variable and often exceed adult values between Days 10 and 32 (Payne *et al.*, 1977). After Day 22, a steady rise is observed, which reaches adult values between Days 37 and 42. Enhancement by HCG of 5α-androstane-3β-hydroxysteroid dehydrogenase and alcoholic dehydrogenase activity between 20 and 30 days correlates well with the timing and effect of HCG on the proliferation of Leydig cells (Frowein, 1973). The fluctuations in testosterone production studied in *in vitro* experiments appear to correspond to evolution and involution of Leydig cells at different stages of development and maturation of rat testis as demonstrated by histological, histochemical, and electron microscope studies. In the rat, spermatogenesis begins shortly after birth (Clermont and Perey, 1957) and evidence for the existence of two distinct generations of Leydig cells is not conclusive due to overlap of the fetal and adult generations of Leydig cells during the second and third postnatal weeks (Lording and de Kretser, 1972). In the field rat (*Millardia meltada*) the full development of Leydig cells is seen between Days 19 and 21 of fetal life (Guraya and Uppal, 1977). Immediately after birth, Leydig cells show an abrupt decrease in their number. But after Day 4 of postnatal life, there is a slow decline in their number. Hayashi and Sandoval (1976) have observed that the testes of rat (*Rattus norvegicus albinus*) during intrauterine and extrauterine life show Leydig cells with lipid droplets. The percentage varies from one testis to another; it increases at 20 days of intrauterine life and between 60 and 90 days of extrauterine life. The percentage of lipid droplets in the cytoplasm of Leydig cells increases at 20 days of intrauterine life, at 15 days of extrauterine life, and near puberty. The percentage of mitochondria is in inverse ratio to that of lipid droplets until the fortieth day of life. These alterations in lipid droplets and mitochondria may be related to fluctuations in testosterone production, which have been revealed in various biochemical studies on the laboratory rat.

c. *Mouse*. The fetal testis of the mouse has the capacity to convert pregnenolone to testosterone as early as 12.5 days (Weniger and Zeis, 1972) and also at 14.5, 17.5, and 19.5 days and on the day of birth (Bloch *et al.*, 1971b). In the male fetal mice testosterone levels also increase slowly from Day 13 to Day 17, and drop dramatically afterwards (Pointis *et al.*, 1978). In one strain of mice used by Hooker (1948), Leydig cells virtually disappear at the time of birth. Spermatogenesis in the mouse begins in the early neonatal period (Nebel *et al.*, 1961) and Leydig cells are believed to remain "quiescent" for 2 to 3 weeks after birth (Blackburn *et al.*, 1973). Blackburn *et al.* (1973) have observed that in the mouse, regressive changes occur in the neonatal period, although the interstitial cells at this time show relatively large quantities of lipid, glycogen, and large round mitochondria as well as smooth and rough varieties of endoplasmic reticulum (Ichihara, 1970; Russo, 1970). In the rat and mouse, the pattern of androgen production reflects a corresponding persistence of functional activity of Leydig cells during the postnatal and prepubertal periods (Miyachi *et al.*, 1973; Lee *et al.*, 1975; Tsujimura and Matsumoto, 1974).

d. *Rabbit*. In the rabbit, the testicular differentiation begins in the fetus on

Day 16 (Gondos and Conner, 1973), Leydig cells appear on Day 18, and regression of Leydig cells occurs during the first week after birth (Gondos and Conner, 1972; Bjerregaard *et al.,* 1974). Gondos *et al.* (1976, 1977) have recently observed that after the gradual disappearance of rabbit fetal Leydig cells in the first week after birth, the interstitial tissue of the postnatal testis consists mainly of undifferentiated mesenchymal cells between 1 and 5 weeks of age. Also present during this time are scattered partially differentiated interstitial cells with oval-shaped nuclei, prominent nucleoli, and abundant cytoplasm. These partially differentiated cells show some of the cytoplasmic characteristics of steroid-secreting cells, but extensive development of smooth endoplasmic reticulum and the grouped perivascular arrangement characteristic of fully differentiated Leydig cells do not occur. The latter features develop at 5 weeks of age (Fig. 25), indicating the formation of mature Leydig cells at that time. By 7 weeks, the bulk of the interstitial tissue consists of Leydig cell aggregates, typical of the appearance in adult testis (Fig. 26). These ultrastructural studies have indicated that in the rabbit the fetal and adult generations of interstitial cells derive from the same cell pool which is established during fetal development. Since spermatogonial mitoses first appear at 7 to 8 weeks of age (Gondos *et al.,* 1973), the findings of Gondos *et al.* (1976, 1977) and Berger *et al.* (1976a) have indicated that Leydig cell differentiation and testosterone secretion precede the onset of spermatogenesis. The biphasic pattern of Leydig cell differentiation in the rabbit correlates well with the alterations in testosterone production, which is high during the fetal stage of Leydig cell development (Lipsett and Tullner, 1965; Wilson and Siiteri, 1973; Veyssiere *et al.,* 1975, 1976), falls to undectectable or negligible levels in the first week after birth, and first rises significantly at 6 weeks of age (Chubb *et al.,* 1975; Berger *et al.,* 1976a). The development of the capacity for testosterone synthesis by the rabbit fetal testis appears to be due to the acquisition of 3β-HSDH and isomerase activity (Wilson and Siiteri, 1972). Throughout fetal life, the rates of synthesis of testosterone from progesterone are much lower than those from pregnenolone (Wilson and Siiteri, 1973). But the rates of synthesis of testosterone from the two precursors are approximately equal in the newborn mouse (Goldstein and Wilson, 1972) and human fetus (Siiteri and Wilson, 1974).

e. *Guinea Pig.* Development of the fetal guinea pig testis after Day 26 is characterized by changes in the area occupied by the interstitial tissue and tubules (Black and Christensen, 1969). An increase is seen in the relative volume of interstitial tissue until Day 29. This increase in the interstitial tissue from Days 26 to 30 is closely accompanied by the appearance of distinct Leydig cells as polygonal cells having the ultrastructural and histochemical features of steroid-secreting cells (Black and Christensen, 1969; Guraya, 1974b). After Day 30, the Leydig cells continue to show the cytological and histochemical characteristics of actively steroid-secreting cells, which consist of the presence of abundant diffuse lipoproteins and a few lipid droplets consisting mainly of phospholipids (Fig. 27) (Guraya, 1974b). From these cytological and histochemical features, Guraya (1974b) has suggested that the Leydig cells of the guinea pig testis are active steroid secretors during fetal life. This suggestion is strongly supported by biochemical observations which have indicated that the guinea pig testis acquires the capacity to convert cholesterol to pregnenolone between Days 30 and 35 in the fetus, with a gradual increase in activity up to the day of birth (Sholl, 1974). After birth the Leydig cells of guinea pig postnatal testis appear to become quiescent in androgen production as judged by the storage of highly sudanophilic lipid droplets in their cytoplasm (Fig. 28) which consist of cholesterol and/or its esters, triglycerides, and some phospholipids (Guraya, 1974b); the storage of cholesterol-positive lipid droplets in steroid gland cells is believed to be indicative of low steroidogenic activity (Guraya, 1978). Therefore, the accumulation of cholesterol-positive lipid droplets has suggested that immediately after birth the fetal Leydig cells of guinea pig testis become relatively inactive in steroidogenesis after performing their active function of steroid production during fetal life. This is in contrast to the observations of Black and Christensen (1969) who did not observe degenerative changes in the Leydig cells of postnatal testes up to Day 13.

f. *Pig*. Bouin and Ancel (1903) revealed the presence of well-developed interstitial cells in the testes of pig fetuses of 3.0-mm crown–rump length (CRL)

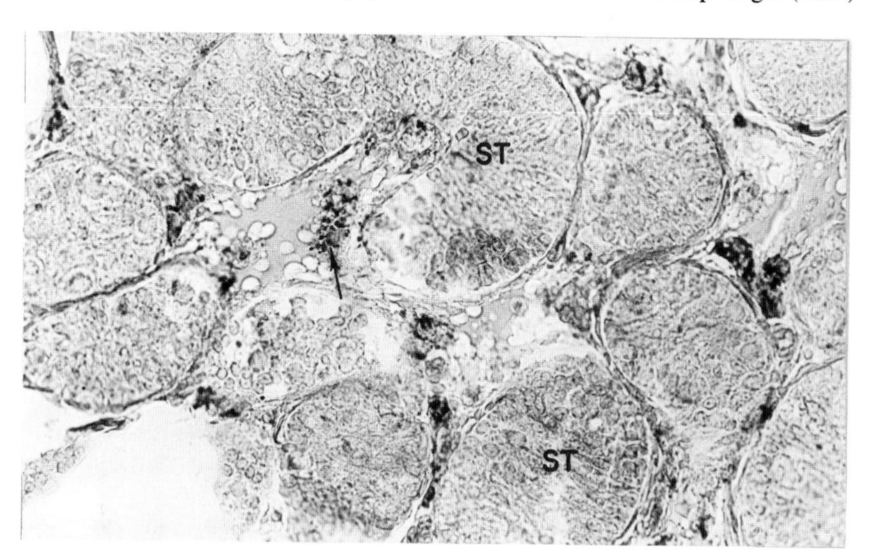

FIG. 28. Light micrograph of a histochemical preparation of a portion of guinea pig testis at Day 3 of postnatal life, showing the accumulation of deeply sudanophilic lipid droplets of coarse nature in the involuting or quiescent Leydig cells (arrows). Seminiferous tubules (ST) do not show any appreciable development of sudanophilic lipids. ×19.

at which stage the development and differentiation of the genital tract occur. They suggested that the gonads of the male fetus might be influencing sexual differentiation. Whitehead (1904), studying a limited number of fetuses of unspecified breed, observed that the earliest stage at which his "mature" Leydig cells developed in the fetal pig was at 2.4-cm CRL. These Leydig cells attained their maximal size and number by 3.5-mm CRL, and then at 5.5-cm CRL there began a conspicuous reduction both in number of cells and in amount of cytoplasm per cell which persisted at 14.0 CRL. In the 20.0-cm-CRL fetus, the Leydig cells had entered into a second phase of growth, which reached a second peak, just prior to term, at 28.0-cm CRL. The recent observations of Moon and Hardy (1973) on a more extensive series of fetuses have confirmed Whitehead's findings inasmuch as one phase of mature Leydig cell differentiation occurs after gonadal sex differentiation and another at a later stage of fetal life, although Moon and Hardy (1973) have demonstrated the first peak of mature Leydig cell size and number occurs at 4.5-cm CRL instead of at 3.5 cm. These observations have indicated that, unlike all other mammals studied, which show first a prenatal phase and second a postnatal phase of Leydig cell differentiation, the pig has two prenatal phases and one postnatal phase (see also Dierichs *et al.*, 1973). A histochemical study of some hydroxysteroid dehydrogenase enzymes in the fetal pig testis (Moon and Raeside, 1972) supports the idea of a separate role of the first mature Leydig cells. However, it is still not known whether the Leydig cells of different periods are synthesizing the same steroid hormones. Dufour (1967) has shown that the second period of high frequency of differentiated Leydig cells in the pig extends for a short time after birth. According to Moon and Hardy (1973), despite a superficial resemblance, the mature Leydig cells of the pig in early fetal and late fetal life differ in many morphological details even at the light microscopic level. Van Straaten and Wensing (1978) have recently investigated the development of the Leydig cells from the fetal period to sexual maturity. During this period, the Leydig cells show two developmental and activity phases, one perinatally and other from 13 weeks postpartum onward. Before the perinatal phase Leydig cells are scarce and poorly developed. The highly developed intertubular Leydig cells and the relatively less-developed peritubular Leydig cells have been distinguished during the perinatal period and undergo regression after 2 to 3 weeks postpartum. From 13 weeks postpartum onward, the peritubular Leydig cells constitute the predominant type. In rats the fetal and adult Leydig cells also show some cytological differences (Roosen-Runge and Anderson, 1959; Lording and de Kretser, 1970).

g. *Rhesus Monkey*. In rhesus monkeys, interstitial Leydig cells can be recognized within intertubular connective tissue at 50 days and they attain maximal density at about 85 to 90 days of gestation; thereafter they fall off sharply (van Wagenen and Simpson, 1965). This correlates with the development of steroidogenic activity as Resko (1970a) found that the fetal rhesus monkey testis is able to convert pregnenolone to testosterone and androstenedione as early as 45 days of age. This is the earliest demonstration of androgen biosynthesis in this species. In the same study, preformed testosterone and androstenedione were found in the testes of 14-week-old fetuses. These studies have indicated early development in the fetal testes of the capacity both to biosynthesize androgen *in vitro* and to preform these hormones *in vivo*. The quantities of these androgens studied by gas–liquid chromatography or by radioimmunoassay (Resko *et al.*, 1973) decline between 96 and 150 days and drop to low or unmeasurable levels shortly after birth. The basal level of testosterone recently observed by Huhtaniemi *et al.* (1977) also agrees with the observations of Resko *et al.* (1973) in late gestation. In primates, there is enough evidence to suggest that fetal testes secrete testosterone early in gestation corresponding to the maximum development of Leydig cells (see Resko, 1977). Testosterone secretion is not maintained at high levels in males throughout gestation, especially in the human fetus. The reason for the decrease in testosterone secretion in late gestation is not known. Resko (1977) has suggested that this decrease may be related to mechanisms that control testicular secretion.

h. *Human.* In the human fetus, testicular differentiation begins at 6 weeks (van Wagenen and Simpson, 1965; Resko, 1977). Leydig cells begin to differentiate from mesenchymal cells at the beginning of 8 weeks (Jirasek, 1967; Pelliniemi and Niemi, 1969; Holstein *et al.*, 1971). Their considerable increase in size from 9 to 12 weeks of gestation is closely accompanied by the development of smooth vesicular and tubular endoplasmic reticulum, by an increase in the number of mitochondria with tubular cristae and of Golgi complexes, and by the formation of rough-surfaced cisternae, numerous cytosomes, and dense bodies (Fig. 24) (Holstein *et al.*, 1971). Thus, from 8 weeks of gestation onward, Leydig cells show features typical of steroid-producing cells. This time corresponds to histochemical studies in which 3β-HSDH activity appears in testes (Baillie *et al.*, 1965; Niemi *et al.*, 1967). A decrease in the relative and absolute volume of Leydig cells occurs after 17 weeks of gestation. According to Pelliniemi and Niemi (1969), the regression of Leydig cells begins at 18 weeks of gestation. According to Guraya (1974a), who subjected human fetal testis 17 and 18 weeks of age to histochemical techniques, the major portion of steroidogenic tissue is constituted by many patches of actively steroid-secreting Leydig cells (Figs. 19, 20, 22, and 23) as shown by abundant diffuse lipoproteins and some phospholipid granules. Besides these actively secreting Leydig cells, there are also some degenerating ones (Figs. 22 and 23) which have accumulated coarse, highly sudanophilic lipid droplets consisting mainly of triglycerides, cholesterol and/or its esters, and very few phospholipids. During the later stages of their regression, the lipid droplets begin to lie free in the interstitium (Figs. 22 and 23), suggesting the disappearance of the cell membrane and other cytoplasmic machinery. The various results obtained with electron microscope and histochemical techniques also correlate well with biochemical observations on steroid hormone synthesis as testosterone formation from pregnenolone is first detectable at 8 weeks, peaks at 12 to 13 weeks, and falls to barely detectable levels after 21 weeks (Bloch, 1964; Huhtaniemi *et al.,* 1970; Siiteri and Wilson, 1974). Examinations of testicular testosterone by isotope displacement assay have revealed similar peak activity at 11 to 14 weeks and a gradual decline to negligible levels by 24 weeks (Reyes *et al.,* 1973). The development and regression of Leydig cells also correlate well with circulating testosterone levels which show a peak during the second trimester and then fall sharply after 18 weeks (Abramovich and Rowe, 1973; Diez D'Aux and Murphy, 1974; Reyes *et al.,* 1974).

It has been demonstrated that the human testis shows a distinct biphasic pattern of differentiation, the fetal generation of Leydig cells undergoes an extensive regression at the end of the second trimester (Fig. 24) (Pelliniemi and Niemi, 1969; Holstein *et al.,* 1971; Guraya, 1974a), and the adult generation makes its appearance only around the time of puberty (Charny *et al.,* 1952; Mancini *et al.,* 1963; Vilar, 1970) at which time there is also a sharp rise in testosterone production, especially between 10 and 12 years of age (August *et al.,* 1972; Winter and Faiman, 1972). Actually, our information about steroid hormone synthesis during the maturation of human testis is still meager. However, Payne and Jaffe (1975) observed a small amount of testosterone synthesis from pregnenolone in testicular material obtained from a 3-week-old infant; no such activity was found at 8 weeks, 60 weeks, and 9 years. Forest (1978) has observed that human testes remain very active at birth. Testosterone Δ^4-androstenedione and 17-hydroxyprogesterone after showing a rapid decrease during the first week of life increase to peak levels at 1–3 months of age, decrease thereafter, and reach prepubertal values by 6 months of age. Plasma testosterone levels in children have been measured (Boon *et al.,* 1972; August *et al.,* 1972; Winter and Faiman, 1972; Forest *et al.,* 1973; Vermeulen, 1973). Testosterone levels show a sharp rise in a sleep-related pattern correlated to corresponding changes in LH levels at the time of puberty (Judd *et al.,* 1974).

From the results of various studies as summarized and integrated above, it can be concluded that there occur conspicuous species variations in the time of appearance, differentiation, and involution of Leydig cells during the fetal, postnatal, and prepubertal periods, which reflect alterations in testosterone synthesis and secretion related to some important differences in the time of establishment of sexual differentiation and maturation in the reproductive ducts and central nervous system. The fetal Leydig cells remain in a fully differentiated state for a specific period of time, then undergo regressive changes, which start shortly before birth in some species such as the mouse, rat, hamster, and rabbit. In species (e.g., man) having a longer gestation period, the regressive changes generally occur well before birth. But there is some sort of a continuum from the fetal to adult Leydig cells with an intervening quiescent stage found in the postnatal and prepubertal periods. This conclusion is supported by the fact that the number of Leydig cells per testis in the rat increases steadily from 20 to 40 days and then decreases from 50 to 90 days (Knorr *et al.*, 1970). McKinney and Desjardins (1973) have observed that the Leydig cells in the mouse increase from 10 to 60 days when the maximal level is achieved and then decrease about 35% by 100 days. Dierichs *et al.* (1973) found a multiplication of Leydig cells in the neonatal period up to Day 14 in the pig, followed by marked regression, with a second pubertal stage of proliferation starting at 3 to 4 months of age.

B. Seminiferous Tubules

The development of seminiferous tubules has been studied in detail in different mammalian species (see references in Steinberger and Steinberger, 1975). The seminiferous epithelium in the fetal testis shows two distinct cell types which include the gonocytes (primordial germ cells) and the supporting cells (precursors of the Sertoli cells). They show active mitotic activity during fetal life. With development and differentiation of the testis, fetal testicular cords undergo conspicuous morphological, histochemical, and biochemical changes to form the seminiferous tubules of adult testis. Two distinct cell lines, germ cells and Sertoli cells, continue to be present in the seminiferous tubules of developing and maturing mammalian testis (Clermont and Perey, 1957; Mancini *et al.*, 1960; Sapsford, 1962; Courot, 1962; Attal and Courot, 1963; Huckins and Clermont, 1968; Pelliniemi, 1970; Vossmeyer, 1971; Gondos and Conner, 1973; Wartenberg *et al.*, 1971; Merchant, 1976; Guraya and Uppal, 1977). Pelliniemi (1970) has observed that during the first week of the human fetal period the testicular cords become regularly cylindrical forming smoothly arched loops retaining their solid structure without a tubular lumen (Figs. 19 and 20). In the cords or tubules, the number of germ cells and Sertoli cells increases per cross section as reported for the rat (Merchant, 1976) and field rat (Figs. 29A–D) (Guraya and Uppal, 1977). According to Guraya and Uppal (1977) from Day 17 to Day 19 of fetal life in the field rat, the tubules attain a diameter of 46 to 50 μm. They are composed of precursors of Sertoli cells and large gonocytes (Fig. 29A). The Sertoli cells show a small irregular nucleus with coarse chromatin.

1. *Basal Lamina*

The seminiferous tubule is surrounded by the basement membrane or basal lamina which appears as soon as the sex cords are formed (see Jost *et al.*, 1973; Courot, personal communication). It consists of a framework of fibers and cells which give support to the germinal epithelium (see Burgos *et al.*, 1970). Various electron microscope studies have revealed the development of the peritubular (boundary) tissue in the rat (Leeson and Leeson, 1963; Hovatta, 1972; Kormano

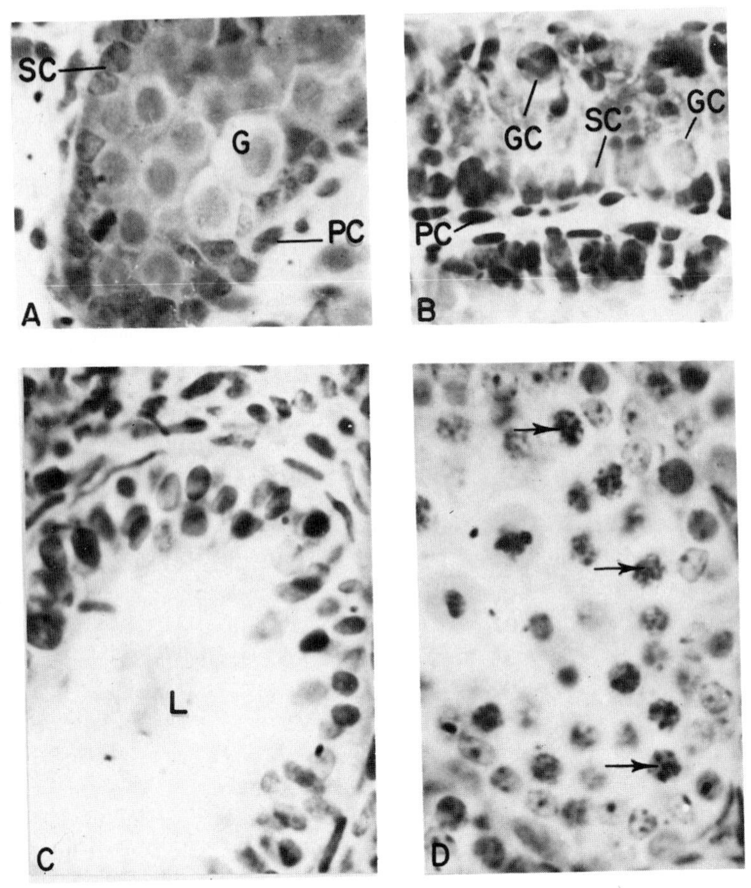

FIG. 29. (A) Light micrograph of testicular cord from the fetal field rat testis at 17 days of gestation showing gonocytes (G), fetal Sertoli cell (SC), and peritubular cells (PC). (B) Light micrograph of a histological preparation of seminiferous tubules from the field rat testis on Day 3 of postnatal life, showing germ cells (GC), Sertoli cells (SC), and peritubular cells (PC). Tubular lumen is not yet developed. (C) Light micrograph of histological preparation of a portion of seminiferous tubule from the field rat testis on Day 10 of postnatal life, showing the placement and distribution of nuclei of spermatogonia and Sertoli cells adjacent to the basement lamina. Tubule lumen (L) is well developed. (D) Light micrograph of a histological preparation of a portion of seminiferous tubule from the field rat testis on Day 23 of postnatal life, showing spermatogonia and primary spermatocytes at one stage of meiosis (arrows). (A)–(D), ×850.

and Hovatta, 1972), mouse (Ross, 1967; Bressler and Ross, 1972), pig (Dierichs and Wrabel, 1973), and lamb (Bustos-Obregon and Courot, 1974). The peritubular cells resemble fibroblasts in the early stages (Figs. 29A and B) and then thin filaments appear in their cytoplasm transforming them into myoid cells (Leeson and Leeson, 1963; Hovatta, 1972). The bundles of cytoplasmic filaments are anchored to dense zones under the plasma membrane (Bustos-Obregon and Holstein, 1973). There are five to six concentric layers of cells in the lamina propria of human seminiferous tubules. In tangential sections they are stellate-shaped and their extensions come close to one another. They may also display a zonula adherens-like contact, although infrequently. Outwardly, typical fibroblasts and connective tissue elements are found. The maturation and functioning of the peritubular tissue are under androgenic control (Hovatta, 1972; Bressler, 1976). The differentiation of the peritubular tissue of the lamb occurs at a considerably earlier period in testicular maturation than in rodents, apparently under the effect of fetal testosterone (Bustos-Obregon and Courot, 1974).

2. *Germ Cells*

a. *Infantile Period.* In earlier studies, except that of Ancel and Bouin (1926), giving an experimental demonstration of the origin of definite germ cells from gonocytes, it could not be determined more precisely whether gonocytes are precursors of definitive germ cells in the adult testis. Many workers were of the opinion that gonocytes degenerate in the neonatal testis and the definitive germ cell line is formed from undifferentiated epithelial-like, supporting cells (see Steinberger and Steinberger, 1975 for references). But recent light and electron microscopic observations of germ cells in the fetal and postnatal testes of rat (Clermont and Perey, 1957; Franchi and Mandl, 1964; Hilscher, 1970; Hilscher *et al.*, 1972; Mauger and Clermont, 1974), mouse (Sapsford, 1962), bull (Nicander *et al.*, 1961), lamb (Courot, 1971), man (Vilar, 1970; Pelliniemi, 1970; Gondos and Hobel, 1971; Vossmeyer, 1971; Wartenberg *et al.*, 1971; Gondos, 1975; Bustos-Obregon *et al.*, 1975), rabbit (Gondos and Conner, 1973; Gondos *et al.*, 1973), and field rat (Guraya and Uppal, 1977) have clearly demonstrated the continuity of development and differentiation from fetal gonocyte to adult spermatogonia (Fig. 30), although a portion of the germ cells is eliminated through a degenerative process. The maturation and differentiation of gonocytes into spermatogonia occur over a given time which greatly varies with mammalian species (Courot *et al.*, 1970; Bustos-Obregon *et al.*, 1975). The early development and differentiation of germ cells before the initiation of spermatogenesis are generally referred to as prespermatogenesis (Hilscher, 1970). Prespermatogenic differentiation consists of an initial fetal stage of multiplication, a quiescent period during which germ cell mitosis probably stops, and a second mitotic stage postnatally just before the start of spermatogenesis. The quiescent period is not evident in some species such as the sheep (Courot, 1962, 1971). Courot (personal communication) is of the view that prespermatogenic differentiation is a progressive event from testicular differentiation to the beginning of spermatogenesis with the more conspicuous changes occurring at the time of initiation of spermatogenesis. In the laboratory rat and field rat gonocyte multiplication stops at 17 to 18 days of gestation (Beaumont and Mandl, 1963; Huckins and

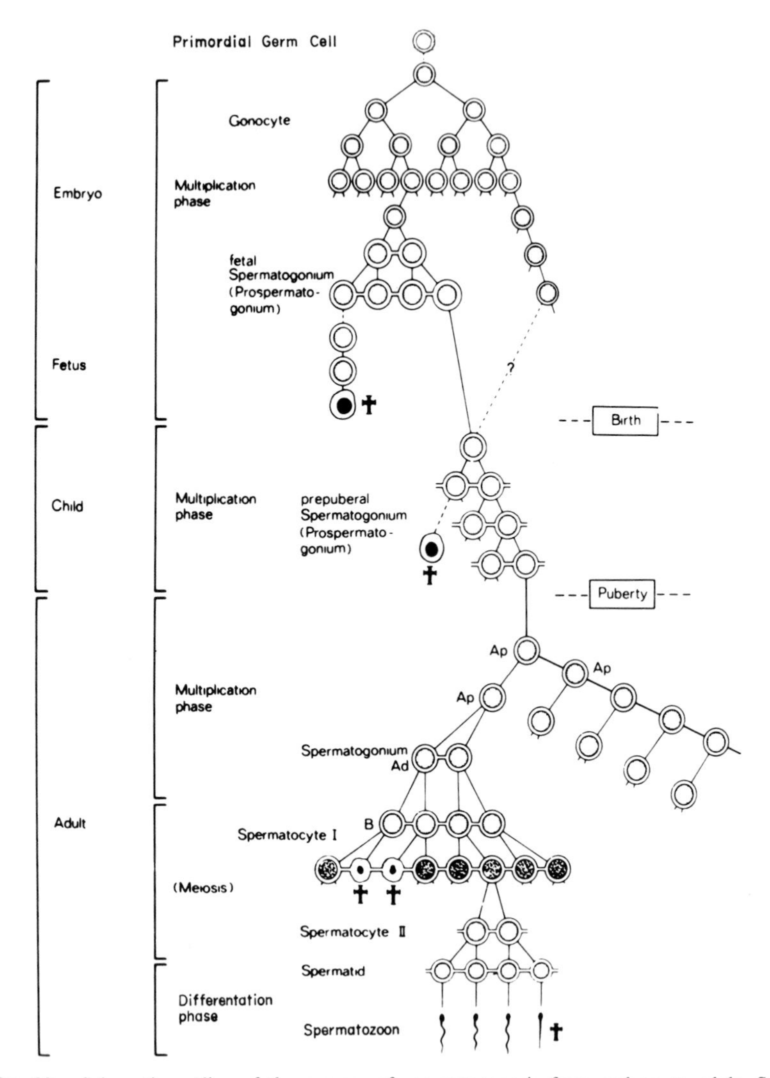

FIG. 30. Schematic outline of the process of spermatogenesis from embryo to adult. Some propositions (size of cell clones, filiation of spermatogonia in childhood) are only tentative. It is uncertain whether the prepubertal spermatogonia derive from gonocytes directly or from fetal spermatogonia. But the last possibility is more probable. Degenerating germ cells are shown as dark nuclei, marked by a cross. The nomenclature presented here arises from the concept of the continuity of the spermatogenic processes shown in this figure. (After Bostos-Obregon *et al.*, 1975.)

Clermont, 1968; Guraya and Uppal, 1977) and is followed by a quiescent period lasting 1 week until the fourth postnatal day when type A spermatogonia first appear (Clermont and Perey, 1957; Guraya and Uppal, 1977). Hilscher *et al.* (1972) have also observed that the cessation of mitosis occurs between Days 18 and 19 in the rat fetus, and division does not resume while 1 week later, 4 days after birth. In the rabbit, mitosis stops shortly after birth and does not begin until 7 weeks of age (Gondos *et al.*, 1973). In the lamb, gonocyte multiplication is low at the time of birth and the prespermatogonial stage lasts 2 months (Courot, 1962, 1971).

In the fetal testicular cords, the germ cells, which are generally called gonocytes, are randomly distributed as single large cells among many Sertoli cell precursors (Figs. 20 and 31A and B). Various electron microscope studies have shown that the gonocytes in the rat (Eddy, 1974; Mauger and Clermont, 1974; Merchant, 1975, 1976), sheep (Courot, 1962, 1971), calf (Attal and Courot, 1963), man (Gondos and Hobel, 1971; Wartenberg *et al.*, 1971; Vossmeyer, 1971; Bustos-Obregon *et al.*, 1975), and pig (Pelliniemi, 1975b) have their regular round to oval shape and a large centrally placed spherical nucleus with evenly distributed chromatin and multiple reticular nucleoli. Their cytoplasm appears clearer than that of supporting cells (Fig. 32). It shows loosely distributed ribosomes, conspicuous mitochondria with a dense matrix and regular cristae, fairly well-developed endoplasmic reticulum, and most frequently smooth, an occasionally well-developed Golgi complex, and granular inclusions, presumably lysosomes (Figs. 31D and F and 32). These cytoplasmic organelles are relatively sparsely distributed as compared to the adjacent supporting cells (Figs. 31C–F and 32), indicating a limited cytoplasmic machinery for metabolic activities on the part of gonocytes. The gonocytes also lose alkaline phosphatase activity temporarily after birth in man (Mancini *et al.*, 1960) and in certain rodents (Dalcq, 1967). Neither dehydorgenases, sulfatases, cytochrome oxidases (Baillie, 1964), nor acid phosphatases (Dalcq, 1967) have been observed. On the other hand, these cells become rich in ribonucleoproteins, glycogen, and lipids (Mancini *et al.*, 1960).

During the development of seminiferous tubules, movement of gonocytes toward the basal lamina has been suggested based on various electron microscopic studies (Franchi and Mandl, 1964; Gondos and Hobel, 1971; Gondos and Conner, 1973). Some workers have suggested that the germ cells might also move in the opposite direction (Gier and Marion, 1970). In view of the progressive increase in peripherally located germ cells and the eventual disappearance of centrally located germ cells (Figs. 29A–C), the primary movement seems to be toward the periphery as suggested by Gondos and Conner (1973). The mechanism by which germ cells are shifted adjacent to the basal lamina is still not clear. Primitive germ cells are considered to be capable of amoeboid movement (Witschi, 1948), even after incorporation into the developing gonad (Blan-

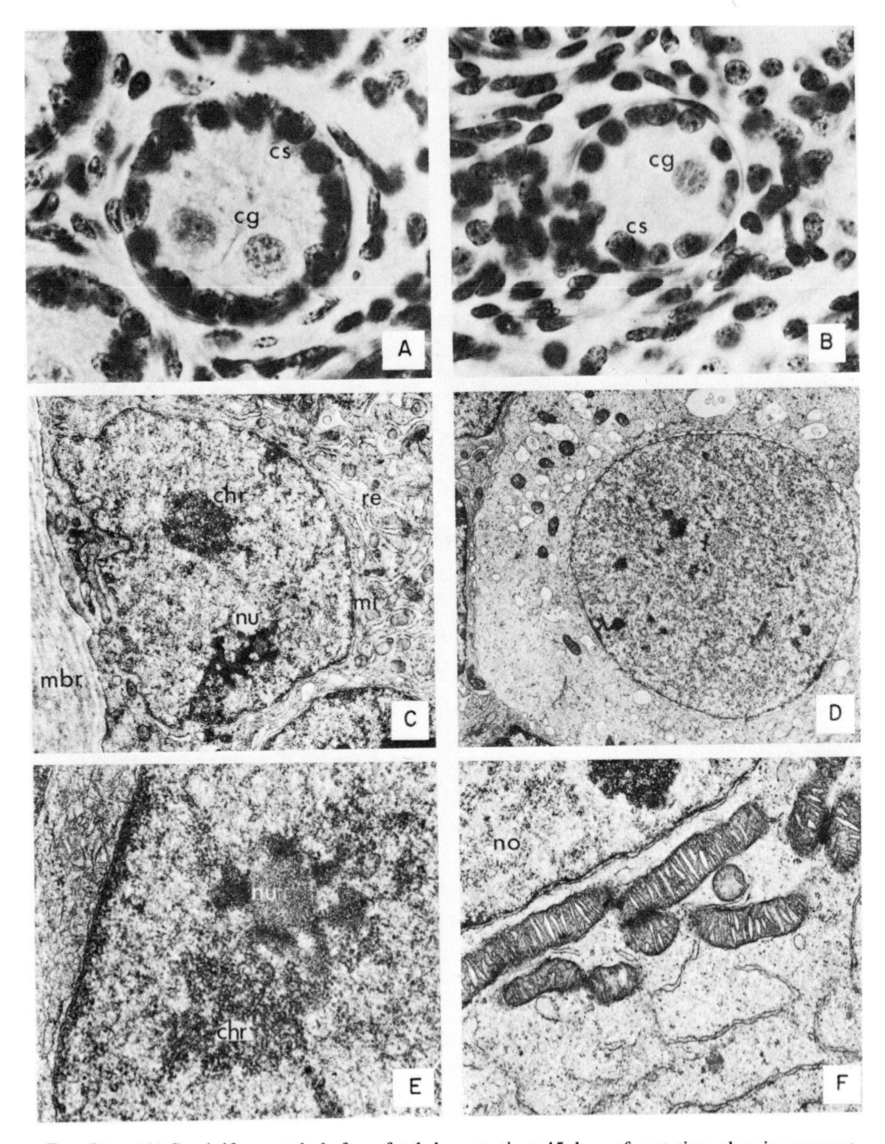

FIG. 31. (A) Seminiferous tubule from fetal sheep testis at 45 days of gestation, showing supporting cells (cs) and gonocytes (cg). ×960. (B) Seminiferous tubule from sheep testis on Day 3 of postnatal life, showing supporting cells (cs) and gonocyte (cg). ×560. (C,E) Ultrastructure of supporting cells from sheep testis (50 days postnatal) showing nucleolus (nu) and heterochromatin (chr) in the nucleus. The cytoplasm shows abundant granular endoplasmic reticulum (re) in relation to mitochondria (mi). Basement membrane (mbr) of the tubule is also seen. (C), ×6120; (E), ×18,000. (D,F), Ultrastructure of gonocytes from sheep testis (150 days postnatal), showing the distribution and relationships of various cytoplasmic organelles such as mitochondoria, endoplasmic reticulum, ribosomes, etc. Nucleus (no) shows nuclear envelope. (D), ×5100; (F), ×14,400. (Courtesy of Dr. M. Courot.)

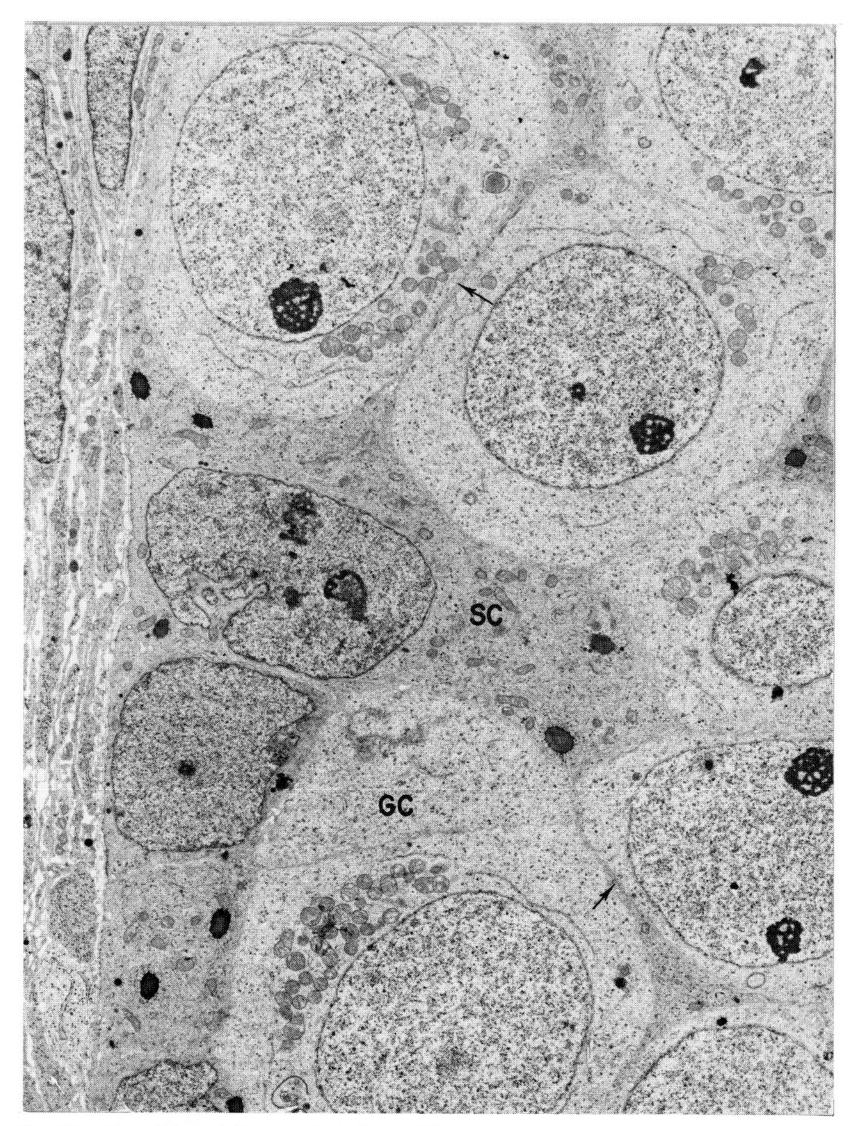

FIG. 32. Part of a seminiferous tubule from a 20-day-old rat fetus. At this stage of development the Sertoli cells (SC) appear compressed at the periphery of the tubule, leaving thin cytoplasmic projections which separate clusters of germ cells (GC, arrows). ×3240. (From Merchant, 1976.)

dau, 1969; Pelliniemi, 1976b). Electron microscope studies have revealed the presence of microfilaments in the cytoplasm of germ cells of the human fetal testis (Wartenberg *et al.*, 1971), which may be involved in germ cell motility. The movement of germ cells toward the periphery is also believed to have a protective function (Franchi and Mandl, 1964), since many of the centrally situated gonocytes are destined to degenerate. However, some peripheral cells also undergo degeneration. Steinberger and Steinberger (1969) have suggested that maturation of gonocytes to spermatogonia might be under the control of testosterone. Movement of germ cells to the periphery brings them nearer to the testosterone-producing interstitial Leydig cells, possibly an important prerequisite for further maturation (Neaves, 1975).

After their gradual movement to the periphery of the cell cords where they are arranged in pairs adjacent to the basal lamina, the gonocytes begin to resemble spermatogonia. During this period of prespermatogenic differentiation, the germ cells have been designated by different names such as type II gonocytes, prospermatogonia, primitive type A spermatogonia, A_0 spermatogonia, prespermatogonia (Fig. 33), and fetal spermatogonia. Hilscher *et al.* (1972), on the

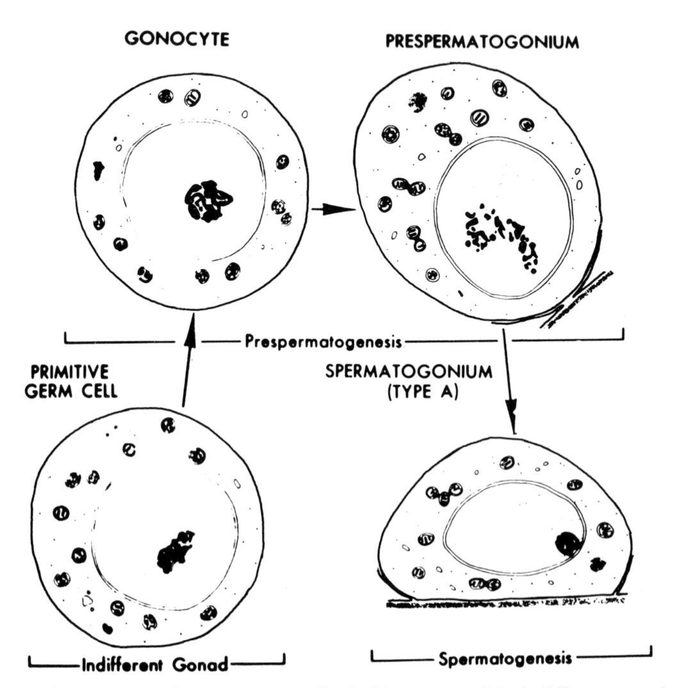

FIG. 33. Diagram comparing appearance of primitive germ cell in indifferent gonad, gonocyte, and prespermatogonium in prespermatogenic testis, and type A spermatogonium associated with initiation of spermatogenesis. Drawings are based on electron microscopic observations in the rabbit. (From Gondos, 1977.)

basis of their kinetic studies called those germ cells undergoing mitosis in the fetal testis of rat, type 1 gonocytes, whereas the cells that resume division on postnatal Days 4 and 5 just before the onset of spermatogenesis were named type II gonocytes. Hilscher *et al.* (1974) referred to these two stages as M-prospermatogonia and T-prospermatogonia. Mancini *et al.* (1960), using cytometric, histological, and histochemical techniques, observed that the gonocytes form type A spermatogonia shortly after birth in man. Most of these cells degenerate; a small proportion forms a series of morphologically dissimilar spermatogonia of which most, in turn, also degenerate and only a few enter the process of spermatogenesis. According to Steinberger *et al.* (1970), who used organ culture technique, the transition from gonocytes to spermatogonia occurs via intermediate cells called as primitive type A spermatogonia (see also Steinberger and Steinberger, 1975). Prespermatogonia are also called A_0 spermatogonia (Courot *et al.*, 1970). In the mature animal, A_0 spermatogonia have been described for the rat (Clermont and Bustos-Obregon, 1968; Dym and Clermont, 1969), mouse (de Rooij, 1970), monkey (Clermont, 1969), and bull (Hochereau-de Reviers, 1970). They are believed to form reserve stem cells for other kinds of spermatogonia. Mauger and Clermont (1974), using the electron microscope, could not identify a second stage of prespermatogenic germ cell development intermediate between gonocytes and spermatogonia. Gondos and co-workers, also using the electron microscope, have revealed the presence of two separate stages in the rabbit, a scattered individual arrangement in the fetus (Gondos and Conner, 1973), and paired prespermatogonia near the tubular periphery in the postnatal period (Gondos *et al.*, 1973). The term, prespermatogonia (see Courot *et al.*, 1970) indicates the similarity of the paired prespermatogenic cells to spermatogonia in placement and pattern of association. Prespermatogonia differ from spermatogonia in other aspects such as their larger size, more uniform spherical contour, and looser chromatin distribution (Fig. 33). Prespermatogonia also remain separated from the tubular periphery by narrow strips of intervening Sertoli cell cytoplasm and thus are not directly flattened against the basal lamina; in contrast, the spermatogonia are placed very close to the basal lamina (Fig. 33). Wartenberg *et al.* (1971) have used the term, fetal spermatogonia, to describe human germ cells which essentially show the same features as those indicated for prespermatogonia. At about the third month of gestation in man, in addition to the gonocytes, fetal spermatogonia are found in the testis (Wartenberg *et al.*, 1971; Wartenberg, 1974; Bustos-Obregon *et al.*, 1975). The fetal spermatogonia or prespermatogonia in the human testis contain no glycogen but can be recognized by their light nucleus and clusters of mitochondria with transverse, septa-like cristae (Wartenberg *et al.*, 1971; Bustos-Obregon *et al.*, 1975). Some mitochondria are connected by a dense cementing substance. Intercellular bridges are occasionally present. In this they resemble the adult spermatogonia.

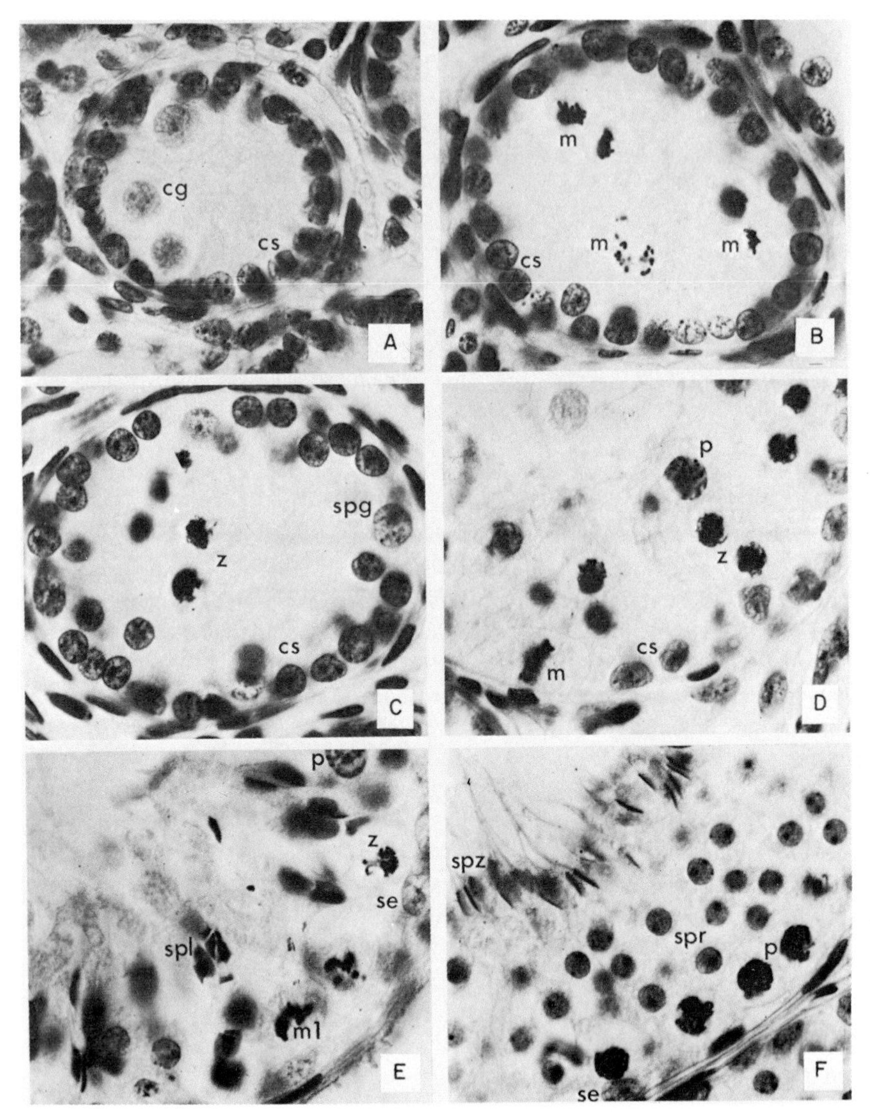

FIG. 34. (A–F) Successive stages in the initiation of spermatogenesis in the lamb. (Courtesy of Dr. M. Courot.) (A) Increase in the number of germ cells (cg) in the tubules obtained from a 90-day-old lamb. Supporting cells (cs) maintain the same appearance as in the preceding days. (B) Spermatogonial multiplication (m, mitosis). Supporting cells (cs) still maintain the same appearance as in the preceding days. (C) First meiotic prophases. Notice a few zygotene spermatocytes (z), and spermatogonia (spg) along the basement membrane. No clear change in the supporting cells (cs). (D) Initiation of spermatogenic activity. The most advanced generation of germ cells (elongated spermatids) is missing in this stage 3 of the seminiferous epithelium. Pachytene (p) and zygotene (z) primary spermatocytes are present. m, Mitosis; cs, supporting cells. (E) Initiation of spermatogenic

Prespermatogenic germ cells are especially sensitive to radiation damage. Type II gonocytes in the bull have been observed to be highly susceptible to the nectrotizing effects of ionizing radiation whereas type I gonocytes are refractory to radiation effect (Erickson *et al.*, 1972). The irradiation of rat testis at Day 18 of fetal life or on the day of birth causes almost complete disappearance of germ cells from the seminiferous tubules but irradiation on Day 10 of postnatal life has little effect (Beaumont, 1960; Hughes, 1962; Matsumoto, 1971). It be interesting to point out here that the damaging effects of radiation during the quiescent period (e.g., Day 19) are greater than at the time of mitotic activity between 15 and 17 days of gestation (Erickson and Martin, 1972). Supporting cells are adversely affected by low doses of radiation but not the Sertoli cells (Courot, 1963).

Atresia normally affects the germ cells during prespermatogenic differentiation (Fig. 30) (Wartenberg *et al.*, 1971; Bustos-Obregon *et al.*, 1975); the degenerating germ cells are phagocytized by adjacent Sertoli cells (Black, 1971; Gondos and Hobel, 1971; Gondos and Conner, 1973). It is not clear whether prepubertal (and adult) spermatogonia originate from the gonocytes or from the fetal spermatogonia in man (Bustos-Obregon *et al.*, 1975).

b. *Period of Initiation of Spermatogenesis.* The initiation of spermatogenesis occurs with the transformation of gonocytes into spermatogonia (Figs. 30 and 34A and B). This transformation appears in a very progressive way, apparently at random, throughout the testis (Courot *et al.*, 1970). Stem spermatogonia can be distinguished from the gonocytes by their cytological characteristics. The type A spermatogonia are placed near the basal lamina (Fig. 33). Their chromatin is more dustlike than that of the gonocytes, and only one large nucleolus often surrounded by an accumulation of chromatin is present. The gonocytes have smaller nucleoli which vary in number and are without the associated chromatin. Their nuclei are larger than those of spermatogonia (Sapsford, 1964). In the rabbit, where it has been possible to study in detail the changes occurring just before the onset of spermatogenesis, prespermatogonia undergo a diminution in cell size, nuclear condensation, and a marked increase in nucleolar complexity just before the formation of spermatogonia (Fig. 33) (Gondos *et al.*, 1973). Similar changes also occur in the mouse and rat (Sapsford, 1962) and in the prepubertal human testis (Vilar, 1970). Nucleolar development may be related to the high rate of RNA synthesis present in type A spermatogonia (Monesi, 1965). Between 7 and 8 weeks, type A spermatogonia appear in the

---

activity (stage 4 of the cycle). Elongated spermatids (spl) are present. Maturation divisions (ml) are also seen. Zygotene I spermatocytes (z) are few in number. Sertoli cells (se) are now differentiated. p, Pachytene spermatocytes. (F) Last stage of initiation of spermatogenesis (stage 8 of the cycle). Spermatozoa (spz) are about to be released into the lumen of the tubule. Pachytene (p) primary spermatocytes and spermatids (spr) and Sertoli cells (se) are also seen. (A–F) ×600

rabbit. Appearance of type A spermatogonia is associated with resumption of mitosis and initiation of spermatogenesis. The period of accelerated nucleolar activity possibly involving RNA synthesis is related to the maturation of prespermatogonia to spermatogonia from 26 to 31 days; germ cells become aligned in rows. With transformation of gonocytes into type A spermatogonia the cytoplasm becomes more eosinophilic (Baillie, 1964). Degenerative changes also become visible in some gonocytes. Different types of spermatogonia characteristic of the progress of the spermatogenic line are subsequently formed. Their morphological characteristics have been discussed in previous excellent reviews (Ortavant *et al.,* 1969; Courot *et al.,* 1970).

In the rabbit, prespermatogonia are already present in the last week of fetal development (Gondos and Conner, 1973). The appearance of type A spermatogonia at the end of the second month initiates a spermatogenic cycle similar in duration to that of the adult rabbit (Swierstra and Foote, 1965). On postnatal Day 1, prespermatogonia showing mitotic activity are present among the gonocytes in the field rat (Guraya and Uppal, 1977). As spermatogonia increase in number by mitosis the gonocytes correspondingly disappear (Figs. 29A–C). Primary spermatocytes are seen on postnatal Day 8. During this period, the Sertoli cells form a palisade layer along the basement membrane. Up to Day 23, spermatogonia and primary spermatocytes are seen in the field rat testis (Fig. 29D). Corresponding to the multiplication of spermatogenic cells, the tubule diameter is also increased.

Pelliniemi (1970) has not observed any apparent changes in the structure and number of mitochondria of germ cells in the seminiferous tubules of human fetal testis until the end of 27 weeks. But the germ cells begin to show regularly granular endoplasmic reticulum in clusters of one to ten membranous pairs. In early stages of testicular development, the mitochondria show a random distribution in the germ cells. They arrange in pairs and groups as they increase in number later on (Gondos and Hobel, 1971; Wartenberg *et al.,* 1971; Gondos and Conner, 1973). Wartenberg *et al.* (1971) have also observed the formation of intermitochondrial granules in later stages of human fetal development which are not seen in rabbit germ cells during the period described (Gondos and Conner, 1973) but which have been reported in adult spermatogonia (Nicander and Ploen, 1969). The ultrastructure of spermatogonia and spermatocytes in the initial spermatogenic cycle is similar to that in postnatal and adult testis of the rabbit (Nicander and Ploen, 1969; Gondos *et al.,* 1973). By Day 17 of development in the rat, most germ cells show patches of nuage but in some, it is associated with clusters of mitochondria, apparently "cementing" them together (Fig. 18) (Eddy, 1974). This relationship is maintained throughout early spermatogenesis (leptotene to zygotene) but from the late pachytene phase of meiosis on, the nuage is again found as discrete aggregations.

Prominence of nucleoli and their lack of attachment to the inner nuclear membrane have been observed in developing germ cells of mouse (Baillie, 1964), rat (Novi and Saba, 1968), man (Gondos and Hobel, 1971), and rabbit (Gondos and Conner, 1973) testes. All these findings support the view that attachment of the nucleolus to the inner nuclear membrane, a characteristic of type A spermatogonia in the adult (Burgos *et al.,* 1970), is a late development. In the rabbit (Gondos and Conner, 1973), there is also evidence of increased nucleolar size and complexity during fetal differentiation, but the nucleolus continues to maintain a central or slightly eccentric position. There is seen a progressive increase in chromatin density during testicular differentiation. In the rabbit (Gondos and Conner, 1973), the arrangement of chromatin at the end of the prenatal period is still not as dense as in adult spermatogonia (Nicander and Ploen, 1969).

Charny *et al.* (1952) suggested that spermatogonia in the human testis are very rare at birth and remain so until about 10 years of age, when they begin to divide actively and shortly afterward enter the spermatogenic process. The morphology and behavior of the germ cells in childhood are poorly understood (Bustos-Obregon *et al.,* 1975). Vilar (1970) described two types of germ cells in the infantile testis, which include large and small cells. The large germ cells show degenerative changes in the central part of the sex cords. The smaller germ cells having intercellular bridges are located adjacent to the basement membrane and are similar to the more evolved spermatogonia. Seguchi and Hadziselimovic (1974) have studied the fine structure of testicular biopsies from children 3.5 months to 13 years of age. They have observed spermatogonia which resemble types A_p (Type A pale) and A_d (type A dark) in the adult. Type A_p occurs more frequently up to the age of 13 years; type A_d is the second most frequent. Fetal spermatogonia as well as a transitional type toward the type A occur until 4 years of age. The gonocytes were not present in the specimens used. Type B spermatogonia and primary spermatocytes appear first in 4-year-old boys, probably in pathological conditions.

With the appearance of spermatogonia, maturation to spermatocytes follows and the initial spermatogenic cycle is finally established. To start with the pictures of meiotic prophase are few and occur in a few tubules only. Then primary spermatocytes accumulate progressively in all the tubules. They evolve into different stages of the meiotic prophase (Figs. 34C–F) which have been described in detail for the adult animal (Courot *et al.,* 1970; Guraya and Bilaspuri, 1976a–c). When primary spermatocytes reach the end of the pachytene stage, a new generation of young primary spermatocytes is formed in their vicinity. Finally, spermatids develop in the tubules. Their formation progressively extends to all the tubules as they develop by undergoing the normal process of spermiogenesis (Figs. 34E and F) which ends in the release of the first spermatozoa. Simultaneously, other generations of spermatids are formed from new generations of primary spermatocytes. This leads to the formation of the cellular associations in the seminiferous tubules, which are similar to those of the adult (see Ortavant *et al.*, 1969; Courot *et al.*, 1970; Clermont, 1972; Guraya and Bilaspuri, 1976a,b,c). These reviews and papers contain all the details of the kinetics of spermatogenesis, which will, therefore, not be described here. From the very beginning, the rate of spermatogenesis in the lamb (Courot, 1971), calf (Attal and Courot, 1963), and rabbit (Gondos *et al.*, 1973) is the same as in the adult, although many germ cells degenerate in the first spermatogenetic cycle; the quantitative and qualitative characteristics of this cell loss are species specific (Roosen-Runge, 1973). In man the remaining germ cells multiply (Mancini *et al.*, 1960). The initial spermatogenic cycles in the rat (Knorr *et al.*, 1970) and shrew (Hasler *et al.*, 1975) appear to be comparable to those in the adult. In the mouse, the frequency of cell associations is equivalent in 21- and 100-day-old animals (McKinney and Desjardins, 1973). The initiation of spermatogenesis through the cellular multiplications leads to changes in the testis structure: increase in the tubule diameter, length, and relative volume. All this involves an increase in testis weight.

3. *Intercellular Bridges*

The development of intercellular bridges connecting developing germ cells in both male and female gonads has been reported in different mammalian species (see Burgos and Fawcett, 1955; Gondos, 1973; Mauger and Clermont, 1975; Guraya, 1977a). Intercellular bridges connecting pairs of prespermatogenic germ cells have been observed as early as 18 days of gestation in the rabbit (Gondos and Conner, 1973), 12 weeks of age in the human fetus (Wartenberg *et al.*, 1971), and 2 days postnatally in the rat (Mauger and Clermont, 1974). In the rabbit, a few scattered bridges are found on Day 22. Their number is progressively increased on succeeding days. The presence of multiple closely spaced bridges on Days 28 through 31 results in an appearance resembling the pattern of multiple interconnections involving large numbers of spermatogonia in the adult (Dym and Fawcett, 1971; Moens and Go, 1972). These cellular interconnections are believed to be formed by incomplete cytokinesis during germ cell mitosis. In the rabbit fetus, the initial formation of intercellular bridges occurs during a period of accelerated mitotic activity, indicating a temporal relationship between mitosis and bridge formation (Gondos and Conner, 1973; Gondos, 1973). Gondos and Conner (1973) have stated that since bridges appear only after the differentiation of germ cells has begun, it appears that some degree of maturation must be achieved before intercellular bridge formation can take place. Dym and Fawcett (1971) are of the opinion that once incomplete cytokinesis is initiated in the germ cell line, this mode of division continues. The progressive increase in the number of bridges observed during fetal development supports this view. Gondos (1973) has discussed the possible functions of intercellular bridges in germ cell differentiation. Bridges connecting large numbers of germ cells in the adult testis are believed to have an important role in the synchronization of spermatogenic maturation. But their exact significance during prespermatogenic differentiation remains to be revealed. However, the syncytial nature of the germ cells is believed to have important implications for the dynamics of cell movement within the spithelium and for the mechanism of sperm release. The ascent of preleptotene spermatocytes from the basal to the adluminal compartment obviously does not involve independent movement of single cells, but of long chains linked by intercellular bridges. Since there is little evidence for the active motility of spermatocytes, it has been suggested that their upward displacement is brought about by synchronous formation, dissolution, and reformation of the junctional complexes between Sertoli cells (Ross, 1976; Russel, 1977a,b, 1978; Dym and Cavicchia, 1977). The formation and significance of these Sertoli junctional complexes will also be discussed later.

4. *Sertoli Cells*

Recent studies have shown that the Sertoli cells partially control the development of seminiferous epithelium and spermatogenetic processes in mammals (Hochereau-de Reviers and Courot, 1976, 1978). Hochereau-de Reviers and Courot (1978) have observed that hypophysectomy of impubertal lambs results in a decrease in the total number of Sertoli cells per testis while the number of stem cells per testis is maintained. Type A_0 spermatogonia divide but few type A_1 spermatogonia are seen. There is no development of seminiferous epithelium. In the prepubertal rat or lamb where mitotic activity of supporting cells has almost stopped, hypophysectomy results in a decrease in Sertoli cell number per testis. Stem spermatogonia number per testis and the efficiency of spermatogenesis are highly decreased. Positive correlations between the total number of Sertoli cells per testis, the total length of seminiferous tubules per testis, and the total area of seminiferous epithelium have been observed in adult males of different species (rat, ram, bull). The total number of Sertoli cells per testis and of stem cell stocks per testis have also been correlated. After hypophysectomy of adult animals (rat, or ram) Sertoli cell stocks per testis do not vary while spermatogonia stock and the efficiency of the spermatogenetic process are greatly affected. These quantitative relationships demonstrated between number of Sertoli cells per testis and number of renewing spermatogonia per testis in the adult (bull, ram, rat) have provided evidence of quantitative regulation of adult spermatogenesis by the population of Sertoli cells which is established during the impubertal period of testicular development. Thus sperm production in the adult is programmed (or controlled) a long time in advance during the development of testes (Hochereau-de Reviers and Courot, 1976). The development and maturation of Sertoli cells will be reviewed during the prenatal, postnatal, and pubertal periods.

a. *Prenatal Period.* The supporting cells of the embryonic testis present the earliest stage of Sertoli cell differentiation and give rise to fetal Sertoli cells which have been investigated by electron microscopy in the guinea pig (Black and Christensen, 1969), man (Pelliniemi, 1970), rabbit (Bjerregaard *et al.*, 1974), sheep (Courot, 1971), rat (Jost *et al.*, 1974; Merchant, 1976), mouse (Dyche, 1975), and pig (Pelliniemi, 1975b). Their cytoplasm is generally characterized by the presence of many elongate and pleomorphic mitochondria with lamellar cristae and a dense matrix, granular endoplasmic reticulum, abundant free ribosomes, polyribosomes, glycogen, and variable amounts of lipid and scattered dense bodies (Figs. 31C and 32); their Golgi complexes are composed of flattened cisternae and numerous smooth vesicles. The nucleus shows one large nucleolus and electron-dense chromatin masses (Figs. 31C and E and 32). The number of Sertoli cells is increased throughout the fetal period due to continual mitotic activity (Courot, 1971). In general, they are not affected by degeneration. But according to Pelliniemi (1970), in the fourth month the human fetal testis develops darkly staining Sertoli cells, in addition to the light ones, showing shrinkage of cytoplasm and unequal folding of nuclear envelope. Pelliniemi has interpreted these cells as the degenerative forms of Sertoli cells. They are seen until the end of 27 weeks. With the arrest of germ cell mitosis, the seminiferous cords show a marked preponderance of Sertoli cells in the neonatal period. As the supporting cells differentiate into Sertoli cells mitoses cease (Clermont and Perey, 1957; Courot, 1962; Attal and Courot, 1963; Steinberger and Steinberger, 1971; Nagy, 1972). Simultaneously adult spermatogenesis starts and the seminiferous tubules are crowded with primary spermatocytes (Figs. 29D and 34D–F).

A close relationship is gradually established between Sertoli cells and germ cells during fetal development (Fig. 32) (Gondos and Conner, 1973), suggesting a possible role for Sertoli cells in support of early germ cell development. Their close anatomical association with the seminiferous cords and the relatively small amount of organelles in early germ cells (Fig. 32) suggest the possible dependence of the latter on the organelle-rich Sertoli cells. The phagocytosis of degenerated germ cells by the Sertoli cells is well established (Black, 1971). The exact physiological significance of this phagocytotic activity in relation to promotion of germ cell development in the fetal testis is not known, but the presence of a similar phenomenon in the developing ovary involving germ cells and granulosa cells (Guraya, 1977a) indicates this as a characteristic feature of early gonadal development.

Besides the supporting and phagocytic functions during fetal life, Black and Christensen (1969) and Bjerregaard *et al.* (1974) have also implicated the Sertoli cells of the fetal guinea pig and rabbit, respectively, in steroid hormone synthesis as evidenced by the presence of some elements of agranular endoplasmic reticulum. Bjerregaard *et al.* (1974) have speculated about the possibility that this steroid hormone of Sertoli cell origin is responsible for the regression of the Müllerian ducts in the male embyro. But the suggestion in regard to steroid hormone synthesis by the fetal Sertoli cells is not supported by the electron microscope studies of Pelliniemi (1970) and Merchant (1976) who have not observed agranular endoplasmic reticulum in them at any phase of the fetal period, which forms the main organelle of steroid gland cells (Christensen and Gillim, 1969; Neaves, 1975; Christensen, 1975). Similarly, Guraya (1974a,b) has not observed any appreciable development of diffuse lipoproteins in the Sertoli cells of human and guinea pig fetal testes, which are abundantly developed in the Leydig cells of the same testes; diffuse lipoproteins demonstrated with histochemical techniques in different steroid gland cell species are believed to derive from the membranes of agranulanr endoplasmic reticulum (Guraya, 1971, 1976a,b). However, human fetal testis shows high activity of oxidative enzymes in Sertoli cells (Jirasek, 1971). In agreement with Pelliniemi (1970), Guraya (1974a,b) has stated that the Sertoli cells of tubules in the fetal testis may have no or very limited steroidogenic activity. Similarly, Merchant (1976) has denied the presence of steroidogenic activity in the Sertoli cells of early rat testis.

b. *Postnatal and Prepubertal Periods.* The Sertoli cells undergo conspicuous morphological and chemical changes with the maturation of testis during the postnatal and prepubertal periods. Flickinger (1967) produced the details of ultrastructural changes of Sertoli cells during the postnatal life of the mouse. Since then, several other studies have been made in the lamb (Courot, 1971), rat (Kormano, 1967; Vitale *et al.*, 1973), rabbit (Gondos *et al.*, 1973), and mouse (Bressler, 1975). All these studies have revealed the presence of several ultrastructural changes in Sertoli cells with the initiation of spermatogenesis, which include increase in cell size, extension of cytoplasmic processes, change of nucleolus to a tripartite form, appearance of smooth endoplasmic reticulum, formation of filaments and microtubules, and development of specialized intercellular junctions (or occluding Sertoli junctions) near the base of the seminiferous epithelium. Fawcett (1975) has discussed in detail the general cytological characteristics of Sertoli cells in the adult testis. Junctions also form between adjacent Sertoli cells *in vitro* (Eddy and Kahri, 1976) as they do in the testis. Kormano (1967) and Vitale *et al.* (1973) have observed in the rat between 15 and 18 days of postnatal life, these specialized tight junctions which have been associated with the regulation of the blood–testis barrier. According to Means (1974a) and Vitale (1975), development of the occluding Sertoli cell junctions is not dependent on the presence of germ cells despite the fact that the blood–testis barrier is established shortly after the initial wave of germ cell multiplication begins. The Sertoli junctions appear to constitute one of the tightest and most resistant epithelial permeability barriers in the body. The compartmentation of the seminiferous epithelium that is achieved by them serves to isolate germ cells from the general extracellular space of the testis, permitting the Sertoli cells to maintain in the adluminal compartment a microenvironment favorable for the continuing differentiation of the germ cells in the adult animal (see Dym and

Fawcett, 1970; Fawcett, 1975; Ross, 1976; Russell, 1977a,b, 1978; Dym and Cavicchia, 1977). Despite the evidence for the unusual stability of the Sertoli cell junctions, it appears that the occluding Sertoli junctions cannot endure unchanged for longer than one cycle of the seminiferous epithelium, for then it becomes essential for the next generation of spermatocytes to move from the basal to the adluminal compartment. Recent electron-opaque tracer studies by Dym and Cavicchia (1977) have demonstrated that type A and B spermatogonia and preleptotene and leptotene spermatocytes reside in the basal compartment of the seminiferous epithelium and that the more differentiated germ cells are located in the adluminal compartment. Late leptotene or early zygotene spermatocytes ascend in the epithelium from one compartment to the next and this occurs in stages IX to X of the cycle of the seminiferous epithelium. New Sertoli cell junctional complexes reform beneath these migrating germinal elements prior to the dissolution of the existing junctions above the cells. This temporal difference serves to seal the barrier during the germ cell migration. The observations of Russel (1977a) have indicated that desmosome-like junctions formed between germ cells and Sertoli cells are strong adhesive sites. Russel (1977b) has stated that at no time in the process of germ cell movement toward the lumen do these cells show evidence of amoeboid movement or lose desmosome-like contacts with the surrounding Sertoli cells. From these observations, he has concluded that the Sertoli cells play an active role in the transfer of spermatocytes to the adluminal compartment. A transient-intermediate compartment of the seminiferous tubule allows for the continual maintenance of the blood–testis barrier during transit of spermatocytes from the basal to the abluminal compartment without disrupting the continuity of the blood–testis barrier (see Russel, 1978). The tracer experiments of Dym and Cavicchia (1977) with lanthanum and peroxidase do not support this concept of a separate "intermediate" compartment. Their results reinforce the concept of two separate compartments within the seminiferous epithelium: (1) basal, and (2) adluminal. These results have also indicated that the initiation of meiosis occurs while the early spermatocytes (preleptotene and leptotene) are still in the basal compartment of the epithelium and may not require the special microenvironment provided by the Sertoli cells in the adluminal compartment. These results agree with observations on the development of the blood–testis barrier in the immature rat (Vitale *et al.*, 1973). Early spermatocytes, including leptotene cells, appear in the seminiferous tubules on Days 11 to 14 prior to the onset of the development of Sertoli cell junctional complexes at Days 16 to 18. Therefore, during development, the initiation of meiosis also does not depend upon the prior formation of the blood–testis barrier.

The development of Sertoli junctions does not appear to take place under the direct control of gonadotrophins (Vitale *et al.*, 1973). The local control of the orderly process of modification of the Sertoli junctions during the cycle of the seminiferous epithelium presents a challenging unsolved problem. It is possible that the formation and dissolution or closing and opening of the junctions must somehow depend upon signals originating from the germ cells when they reach a certain stage of their differentiation (see Ross, 1976; Russell, 1977b). However, Connell (1977) has observed the effect of HCG on pinocytosis within the inter-Sertoli cell tight junction in prepubertal and adult dogs. Stimulation with HCG results in an increased number of pinocytotic vesicles on the membrane faces enclosed by tight junctions within the inter-Sertoli cell junctions.

It is generally believed that Sertoli cells have an important role in the early maturation of the seminiferous epithelium (Flickinger, 1967; Courot, 1970). But the nature of their role in this regard is also still obscure. Lacy (1962) has suggested that phagocytosis of residual lipid body may stimulate the Sertoli cell to synthesize a substance capable of influencing germ cell maturation. But according to others, the Sertoli cell lipid appears to be formed by the breakdown of ingested material (Brokelmann, 1963; Flickinger, 1967; Guraya, 1968). From the presence of phagocytized material and lipid droplets during the prespermatogonial stage in the Sertoli cells, Gondos *et al.* (1973) have suggested that phagocytosis of degenerating germ cells may represent a source of metabolites to be utilized by the Sertoli cells. How such material might contribute to support of germ cell maturation requires further detailed investigation.

Postnatally interstitial Leydig cells undergo regression (Gondos and Conner, 1972). Gondos *et al.* (1973) believe that Sertoli cells, therefore, represent possible alternate source of steroid hormones in the prepubertal period. Lacy (1974) has also suggested that the Sertoli cells are the main site of conversion of progesterone to testosterone by the immature testis *in vitro*. Cameron and Markwald (1975), by applying electron microscope and histochemical techniques to maturing rat testes, have suggested that Sertoli cells develop the potential to participate in some event(s) of steroid modifications as evidenced by the development of smooth endoplasmic reticulum and some steroidogenic enzymes. Satyaswaroop and Gurpide (1975) have observed that the seminiferous tubules from immature rat testis are capable of androgen synthesis in the absence of precursors provided by Leydig cells and that testosterone is formed mainly by the intermediacy of androstenedione. Armstrong *et al.* (1975) have provided the first direct evidence of estradiol-17β and estrone synthesis from exogenous testosterone by Sertoli cells of 18- to 20-day-old rats. They have also offered evidence that synthesis of this steroid is regulated at the level of the aromatizing enzyme system by FSH and cAMP; the latter is mediator of this action of FSH (see references in Armstrong and Dorrington, 1977). Armstrong and Dorrington (1977) have proposed that testosterone formed by the interstitial cells under the influence of LH is transported to the Sertoli cells which in the presence of FSH convert this substrate to estrogen. Recent studies have shown that it occurs only in immature rat, as aromatizing activity decreases sharply after 20 days of age in the rat.

Testicular androgen-binding protein (ABP) has been shown to be a product of the Sertoli cell (Hagenas *et al.*, 1975; Tindall *et al.*, 1975, 1978; Fritz *et al.*, 1976; Means, 1977; Steinberger and Steinberger, 1977). ABP disappears following hypophysectomy but reappears after the chronic administration of FSH (Hansson *et al.*, 1973; Tindall *et al.*, 1974; Vernon *et al.*, 1974; Elkington *et al.*, 1975; Sanborn *et al.*, 1975a; Fritz *et al.*, 1976; Griswold *et al.*, 1977). Means (1977), after reviewing the previous data, has suggested that testosterone itself might be the primary hormone responsible for the regulation of ABP. Recent studies have also suggested that Sertoli cells form the site for the production of inhibin (Setchell *et al.*, 1977). Its production normally occurs in response to a signal from the more advanced germ cells; and somehow under certain circumstances the Sertoli cells are misled into thinking that these cells are still present. The presence of differentiated germ cells in the seminiferous epithelium enhances Sertoli cell differentiation and secretion of ABP and inhibin, while their absence in sterile or cryptorchid testis provokes a sharp increase in plasma FSH (De Krester *et al.*, 1973).

In summary, the Sertoli cells, with the maturation of testis, form permanent nondividing elements of the seminiferous tubules and begin to perform following different functions during the reproductive life of the male: (1) They are involved in the architecture of the seminiferous epithelium as their cytoplasmic processes extend throughout its thickness around the germ cells. (2) They play an important role in the metabolic exchange with the germ cells and the secretion of the tubular fluid in the luminal compartment in relation to the blood–testis barrier. (3) They coordinate spermatogenesis in relation to the release of spermatozoa and phagocytosis of the residual bodies. (4) They are involved in the production of steroid hormone, ABP, and inhibin. Fawcett (1975) has discussed in detail the ultrastructural aspects of these functions in relation to Sertoli cells of adult testis.

5. *Enzyme Histochemistry*

Some enzymes show alterations in activity with the postnatal maturation of mammalian testis (Bishop, 1969; Blackshaw, 1973; Fritz, 1973; van der Molen *et al.*, 1975; Hodgen, 1977). Males and Turkington (1971) have observed a decrease in β-glucuronidase activity in the rat testis during the postnatal period as also described in other studies (van der Molen *et al.*, 1975). It is present only in Sertoli cells and spermatogonia. β-Glucuronidase shows a high specific activity in the immature rat, which decreases during further development (Males and Turkington, 1971; van der Molen *et al.*, 1975). Males and Turkington (1971) have suggested that the change in activity of this lysosomal enzyme, having detectable activity only in Sertoli cells and spermatogonia, reflects the increase of testis tissue through the development of the spermatocytes. According to van der Molen *et al.* (1975), the specific activities of β-glucuronidase and γ-glutamyltranspeptidase reach a maximum on Days 28 to 30. During this time the first pachytene spermatocytes complete meiosis I and II and give rise to the first spermatids. The increase in testis weight is almost linear during this period. Therefore, the change in enzyme activities at Day 28 has been related to a change in the enzyme profile of some cell types or as a result of a change in the number of different cell types in the germinal epithelium. Uridine diphosphatase activity becomes high during the differentiation of spermatogonia and spermatocytes and then shows a sharp decrease during the development of spermatids, in which the enzyme activity could not be demonstrated with histochemical techniques (Xama and Turkington, 1972). Various enzymes show a progressive increase during the initial spermatogenic cycles. They include acid phosphatases (Males and Turkington, 1971; Vanha-Perttula and Nikkannen, 1973), aminopeptidase III (Vahna-Perttula, 1973), sorbitol dehydrogenase (Mills and Means, 1972; Hodgen and Sherins, 1973), adenyl cyclase (Hollinger, 1970), carnitine acetyl transferase (Vernon *et al.*, 1971), and adenosine 3′,5′-monophosphate diesterase (Monn *et al.*, 1972). In general, the level of their activity is correlated to the stage of the germinal epithelium. But the localization of enzymes at the cellular level is still to be determined more precisely. At least, some of these enzymes may be present in the Sertoli cells and the others seem to show correlations with specific germ cell populations. According to Hodgen and Sherins (1973), lactate dehydrogenase (LDH) activity is a specific marker for spermatocytes. It is high in the neonatal rat testis, decreases to a nadir at 25 days, and then increases 2-fold to adult levels. The increase in total LDH activity after Day 25 is also closely related to the appearance of LDH-X which is a unique isoenzyme specific to testis. LDH-X activity appears at a time when the pachytene primary spermatocytes are formed in the rat and mouse (Blackshaw and Elkington, 1970a,b). Montamat and Blanco (1976) have found that LDH-X in the adult rat testis is associated with a special type of mitochondria found in cells of the spermatogenic line and in the mitochondrial sheath of spermatozoa, and it is also distributed in the cytoplasmic soluble phase. The isozyme X is located in the matrix of mitochondria. Hyaluronidase activity appears with the differentiation of spermatids in the rat (Males and Turkington, 1970). It could not be demonstrated until 33 days of age, when it makes its first appearance and shows a rapid 400-fold increase. The development of hyaluronidase activity is closely correlated to the formation of the acrosome in spermatids. Similarly, the appearance of specific isoenzymes of β-galactosidase and *N*-acetyl-β-glucosaminidase is closely related to development of the acrosome (Majumder *et al.*, 1975); the gonadotrophins or testosterone appear to be involved in the induction of these enzymes.

The enzyme systems of Sertoli cells appear to undergo changes during the maturation of testis (Steinberger and Steinberger, 1977). Hodgen and Sherins (1973) have observed an increase in γ-glutamyl transpeptidase (γ-GTP) activity between Days 15 and 20 in the postnatal rat testis which is apparently related to the Sertoli cell development. Van der Molen *et al.* (1975) have also made similar observations but in their experiments GTP activity increased up to Day 29. This difference might be due to the different strains of rats used by them. Lu and Steinberger (1977) have observed that in the 30-day-old rat testis the γ-GTP activity is associated mainly with the Sertoli cells. Its specific activity is increased with the culture age of Sertoli cells, showing a 5-fold increase after 3 days of incubation and a 20-fold increase after 7 days of culture. The physiological significance of γ-GTP activity in Sertoli cells and its increase under culture conditions remains obscure, but this enzyme may be playing a role in the synthesis of specific proteins known to be secreted by the Sertoli cells. Catalase activity is high in rabbit testis during the early postnatal period, when Sertoli cells are predominantly present (Ihrig *et al.,* 1974); the activity decreases rapidly to adult levels with the start of spermatogenesis. MacIndoe and Turkington (1973) have also observed similar alterations in ornithine decarboxylase activity during the maturation of rat testis. Kuehl *et al.* (1970) reported the presence in the testis of a phosphodiesterase which specifically hydrolyzed cAMP. Monn *et al.* (1972) studied the level of phosphodiesterase in testis during postnatal maturation. Testes of immature animals contained low levels of phosphodiesterase. However, as the age of the animal increased, a specific isozyme of this enzyme developed and reached maximal levels at approximately 35 days of age. Christiansen and Desautel (1973) have reported that the specific phosphodiesterase isozyme disappears from the testis following hypophysectomy. Means and Huckins (1974) have suggested that phosphodiesterase may play a role in the regulation of the action of FSH in the mature rat.

VII. Hormonal Control

The role of gonadotrophins in the development, differentiation, and function of the fetal testis has been discussed previously in some excellent reviews (Jost, 1953, 1966, 1971, 1972b; Jost *et al.,* 1973; Prasad, 1974) and thus will be described very briefly here. The factors which regulate the biosynthesis of testosterone in the fetal testis and initiation of puberty are still to be determined more precisely. However, various experimental studies have produced evidence for the existence of a pituitary-testicular axis during the fetal, postnatal, and prepubertal life of different species of mammals. The gonadotrophins secreted by the pituitary (or placenta in primates) influence the development, differentiation, and function of Leydig cells, germ cells, and Sertoli cells.

A. Leydig Cells

Tbe fetal pituitary has been shown to produce a gonadotrophic substance(s) which stimulates the fetal testis optimally during a limited but critical period of development and differentiation of interstitial Leydig cells; the latter produce androgenic hormones which control the development, differentiation, and maturation of accessory organs and also cause some changes in the nervous system which may be concerned with the development of sexual behavior and feedback mechanisms.

The stimulation of Leydig cells in the human and primate fetal testis has been attributed to the action of gonadotrophins, secreted by the fetal pituitary and/or HCG (see references in Guraya, 1974a; Prasad, 1974; Resko, 1977), which appear to ensure doubly the normal sexual development and differentiation by stimulating steroid production in the fetal testis. Holstein *et al.* (1971) have discussed that the development of Leydig cells depends on the secretion of HCG. Actually, their differentiation occurs at a time when HCG is produced in very large quantities. This is especially likely in view of the normal sexual differentiation in human fetuses without pituitaries (Blizzard and Alberts, 1956) and in anenocephalic fetuses (Bearn, 1968). These observations are further supported by the fact that the human fetal testis responds to HCG stimulation by converting acetate and cholesterol into testosterone, apparently using cAMP as a secondary messenger (Ahluwalia *et al.*, 1974), testosterone synthesis being enhanced by increasing dibutyryl cAMP levels. Human fetal testes maintained in organ culture have also been shown to synthesize increasing amounts of testosterone after treatment with HCG (Abramovich *et al.*, 1974; Villee, 1974). HCG levels are maximal during the time when Leydig cells differentiate and consequently testosterone formation starts at 8 weeks. These results have indicated that HCG receptors must be present in the fetal testes for it to be effective, but the earliest demonstration of an HCG receptor in the testis was at 26 weeks of gestation (Frowein and Engel, 1974). However, Rajaniemi and Niemi (1974) are of the opinion that during pregnancy maternal gonadotrophins possibly are unable to penetrate into the fetuses and thus do not influence the development of the fetal gonads. Human fetal pituitaries, on the other hand, contain gonadotrophins (Levina, 1968; Grumbach and Kaplan, 1973) and biosynthesize FSH and LH in culture by 13 weeks of gestation (Groom *et al.*, 1971). Reyes *et al.* (1974) and Clements *et al.* (1976) have demonstrated significant amounts of pituitary gonadotrophins in the fetal circulation at a time when testosterone biosynthesis shows peak levels. Testosterone formation shows a decline in the second half of gestation which seems to be accompanied by a corresponding regression of pituitary gonadotrophs (Pasteels *et al.*, 1974). The data of Huhtaniemi *et al.* (1977) have indicated that the testes of fetal rhesus monkeys during late gestation are capable of androgen biosynthesis *in vivo* and *in vitro* and can bind and respond to gonadotrophin (HCG) stimulation. Furthermore, the pituitary gonadal axis in the fetal male monkey is capable of responding to gonadotrophin-releasing hormone stimulation at this stage of gestation.

At present it is accepted that the development and maintenance of testicular Leydig cells in the testes of different mammalian species depend upon luteinizing hormone (LH) (see references in Neaves, 1975; Chemes *et al.*, 1976a). Studies on decapitated fetuses have revealed that the hypothalamus begins to govern gonadotrophic control of Leydig cell differentiation and hence of steroidogenesis before birth (Eguchi *et al.*, 1975, 1978). In fetal rats, brain control of pituitary Leydig cell system begins to operate from Day 18 of gestation, when the day following overnight mating is designated as Day 1 of gestation (Eguchi *et al.*, 1978). Sanyal and Villee (1977) have observed that as little as 10 mIU of HCG added *in vitro* significantly increases androgen synthesis in the rat fetal testis on Day 19, at which time Leydig cells also show their maximum development as reported by Guraya and Uppal (1977). Bovine, ovine, or human LH, but not bovine or ovine FSH, also stimulates the synthesis of androgens by the rat fetal testis. Feldman and Bloch (1978) have observed that rat testicular development from 14.5 to 20.5 days is characterized by an increased capacity to synthesize testosterone and respond to LH. Both parameters appear on Day 15.5 of gestation. These results have indicated that testosterone could be under gonadotrophic regulation. Warren *et al.* (1975) have measured testosterone formation in fetal and neonatal (2 hours–20 days) rat testes incubated with and without 100 IU HCG. At all stages investigated testosterone production was significantly elevated in the presence of HCG. Neonatal testes showed an absolute drop in testosterone production with increasing age (Warren *et al.*, 1975) which correlates very well with the decrease in the number of Leydig cells demonstrated by histological and histochemical techniques (Guraya and Uppal, 1977). The percentage increase in testosterone production above baseline produced by HCG stimulation as observed by Warren *et al.* (1975) supports the suggestion that the decrease in testosterone production in the neonatal testes is due to a relative reduction in the number of Leydig cells with increasing age (Guraya and Uppal, 1977) rather than to a lowered sensitivity of HCG stimulation. This suggestion is also supported by the results of previous studies on the immature rat in which the disappearance of the first generation of Leydig cells (Niemi and Ikonen, 1961, 1962) is followed by a regression in testicular enzyme activities (Niemi and Ikonen, 1963) and a reduction in the biosynthesis of steroids (Steinberger and Ficher, 1968; Hall, 1970). Weniger and Zeis (1975) have recently reported the stimulatory effect of gonadotrophins on androgen (or testosterone) synthesis by the testes of mouse and rat embryos. Other studies have demonstrated that LH stimulates androgen synthesis *in vitro* in the fetal rat testis (Feldman and Bloch, 1974, 1978); various studies have demonstrated the specific binding of LH and HCG to interstitial Leydig cells (see Dufau and Means, 1974; Catt *et al.*, 1975). LH elicits a rapid increase in cAMP in Leydig cells, leading to a rise in the synthesis of testosterone (Moyle and Ramachandran, 1973).

Chemes *et al.* (1976a) believe that the elevation of plasma testosterone in immature rats treated with HCG is probably determined first by the stimulation of preexisting Leydig cells and then by an increase in the total number of hormone-secreting cells; an increase in the number of mitoses in Leydig cells in developing rats was reported earlier by Roosen-Runge and Anderson (1959). These data, showing not only a histological response but also a hormonal stimulation, strongly indicate that the interstitial Leydig cells of the immature rat are able to respond to gonadotrophic stimulation, a potentiality denied by some other workers (Hashimoto and Suzuky, 1966; Steinberger and Steinberger, 1972). The results of Chemes *et al.* (1976a) have also indicated the close interdependence between the two effects of LH, one to promote secretion of testosterone and the other to stimulate Leydig cell development and growth, even though Steinberger and Steinberger (1972) described the independence of these two actions. Lee *et al.* (1975) have observed a parallel rise in LH and testosterone levels occurring from Day 30 to sexual maturity in the rat, which closely corresponds to the development of the adult generation of interstitial Leydig cells. Similar parallelism in LH and testosterone levels has been observed in the impubertal and prepubertal lamb (Cotta *et al.*, 1975). Lacroix *et al.* (1977) have demonstrated temporal fluctuations of plasma LH and testosterone in charolais male calves from birth to 1 year of age. Plasma LH increases from 1 to 5 weeks of age, then fluctuates widely from 5 to 20 weeks (3–5 ng/ml) and finally decreases to about 2 ng/ml and remains steady up to 1 year of age. Plasma testosterone increases very slowly from 1 to 20 weeks of age and then increases more rapidly with large fluctuations up to 1 year. These findings have suggested that testicular sensitivity to LH is not established between 1 and 5 months of life. The high pituitary activity during the first month of life appears to be a signal for puberty.

For the past several years, the control of steroid hormone synthesis in the prepubertal testis has been studied to reveal the mechanisms which regulate sexual maturation (Odell *et al.*, 1973; Grumbach *et al.*, 1974). But the nature of the mechanisms is still not known more precisely. It remains to be established whether the rising levels of testosterone secreted by the pubertal testis are due to the enhanced secretion of gonadotrophins by the pituitary as a result of decreasing hypothalamic sensitivity to negative feedback (Grumbach *et al.*, 1974), or to a cumulative effect whereby continued exposure to FSH causes a gradual increase in sensitivity to the effect of LH (Odell *et al.*, 1973). The presence of some other factors within the maturing testis cannot be ruled out. Brown-Grant *et al.* (1975) have recently found that inhibition of FSH and LH concentrations after treatment of neonatal rats causes a decrease in testosterone secretion and delays sexual maturation; the significant factor appears to be the decrease in FSH secretion. The HCG receptors are already present in the testes of newborn rats but hormone sensitivity appears at the end of the second postnatal week (Frowein and Engel, 1974; Engel and Frowein, 1974), thus correlating with the onset of the adult phase of Leydig cell differentiation. The neonatal rat testis possess the ability to recognize and bind HCG specifically; although Leydig cells are insensitive (Frowein and Engel, 1974), HCG or LH causes marked hyperplasia very early during the neonatal period (Arai and Serisawa, 1973). Odell *et al.* (1974) have observed that in the rat, the testicular response to LH regularly increases from 21 to 40 days, whereas from 20 to 60 days minimal changes are observed in the receptivity of the rabbit testis to HCG (Berger *et al.*, 1976b). This discrepancy may be related either to the species or to a difference in the chronology of FSH secretion. The latter suggestion is supported by the fact that continual exposure of the Leydig cells to FSH enhances sensitivity to LH (Odell *et al.*, 1974). Van Beurden *et al.* (1976) have observed that hypophysectomy of immature rats results after 5 days in an almost complete loss of LH sensitivity of isolated Leydig cells. But daily administration of FSH for 5 days starting immediately after hypophysectomy maintains the LH responsiveness of isolated Leydig cells of immature rats. FSH administration starting on Day 5 after hypophysectomy restores LH responsiveness. Estradiol benzoate injected simultaneously with FSH abolished the FSH-induced responsiveness. The HCG binding capacity of rat testis also progressively increases with age up to 60 days accompanied by a rise in serum and testicular testosterone concentrations (Pahnke *et al.*, 1975). This increase can be correlated to a corresponding increase in Leydig cell number; the specific binding capacity per Leydig cell does not change from Day 30 to Day 110. These observations have suggested that the critical factor for increased testosterone production is the increase in Leydig cell number during prepubertal development of testis. Besides the gonadotrophins, testosterone itself may also influence the maturation of Leydig cells (Neaves, 1975). Parlow *et al.* (1973) found a 5-fold increase in serum testosterone concentration within 15 minutes after a single intravenous injection of rat LH and an elevation to adult testosterone levels 1 hour after injection. LH stimulates steroidogenesis, indicating the marked responsiveness of the prepubertal rat testis. Berger *et al.* (1976b) have investigated the response of the testis to HCG stimulation as a function of age in the immature rabbit. Their results have indicated that although HCG significantly stimulates the synthesis and secretion of testosterone by the immature testis, the response obtained, expressed as a percentage of basal level, was always much lower than in adults (see Berger *et al.*, 1974). After HCG stimulation of the immature testis, plasma testosterone concentration reached values similar to the levels seen in adult controls as also described for the rat (Parlow *et al.*, 1973), monkey (Bennett *et al.*, 1973), and man (Frasier *et al.*, 1969). The increase in the synthesis of testosterone after HCG stimulation in the immature rabbit (Berger *et al.*, 1974, 1976b) indicates that receptors for HCG are present in the testis and functional in the early stages of development, whereas in the rat, they are present but not functional (Frowein *et al.*, 1973). There is no explanation for the fact that the response of immature testis to HCG is always less than that in the adult as observed by Berger *et al.* (1974, 1976b). They have suggested that it may be due to the number and maturation of specific HCG receptors, the development of testicular enzymes involved in steroidogenesis, or the endogenous secretion of gonadotrophins. But the lesser reactivity of the immature testis in the rabbit and rat has been related to the low ability to concentrate HCG (Frowein and Engel, 1975; Berger *et al.*, 1976b).

B. Germ Cells

The two important characteristics of normal development of male germ cells are (1) that the gonocytes do not enter meiosis early as the female germ cells do (see Guraya, 1977a), and (2) that their maturation to spermatogonia and initiation of spermatogenesis occur at a time corresponding to the period of pubertal development. The regulation of these aspects of germ cell differentiation is still not understood precisely. However, some evidence about possible regulatory factors has been obtained. Jost (1970) suggested the presence of some factor in the male gonad which inhibits early meiosis. Such a factor, which is lacking in the female gonad, is responsible for the difference in the timing of oogenesis and spermatogenesis. According to Jost *et al.* (1974), the fetal Sertoli cells, which enclose the germ cells in the developing testicular cords, may be inhibiting the meiosis in the testis. Experimental evidence for their premature entry into meiosis has been obtained in some studies (Ozdzenski, 1972) but the nature of the mechanisms responsible is not known. It is possible that some kind of interference with Sertoli cell development causes the premature entry of germ cells into meiosis in these experiments. Male germ cells are also triggered to enter meiosis when an undifferentiated male gonad is cultivated together with older ovaries containing germ cells progressing through meiotic prophase (Bysko and Saxen, 1976; O and Baker, 1976).

Very divergent views have actually been expressed about the mechanism of action of gonadotrophins and steroids in the initiation of first spermatogenic cycle, as well as about the ratios between the two gonadotrophins required to assure its normal development (Bergada and Mancini, 1973). According to Courot (1970, 1971), the initiation of spermatogenesis in the lamb depends on the presence of anterior pituitary hormones as its ablation shortly after birth prevents differentiation of germ cells. This effect can be reversed by administration of either FSH, or LH, or synergistically by both gonadotrophins, but not by testosterone. The latter is unable to support the transition from reverse (A_0) to renewing (A_1) stem spermatogonia in the ram (Monet-Kuntz *et al.*, 1977; Courot *et al.*, 1978). The effect of the gonadotrophins seems to be mediated by the Sertoli cells (Courot, 1970, 1971). Lostroh (1969) believes that FSH and LH apparently act synergistically in the rat. On the basis of cultures of seminiferous tubules and experimental evidence from immature rats, Steinberger and Steinberger (1967) and Steinberger and Duckett (1965), following Cutuly and Cutuly (1940) and Tonutti (1954), have made a suggestion that no hormones are needed for spermatogenesis to proceed from the early spermatogonial steps up to those of late pachytene, and that FSH and testosterone are needed only for the final stages of the spermatogenic cycle. Thus, initiation of spermatogenesis in the rat is believed to be mediated by testosterone (Steinberger, 1971), or other androgens (Chowdhury and Steinberger, 1975), and does not require gonadotrophins. With the initiation of gonadotrophins by administration of estradiol benzoate to newborn rats, spermatogenesis is initiated and on administration of testosterone meiosis is completed with the formation of spermatids (Steinberger, 1971). Other workers have suggested that androgens, LH, and other pituitary hormones (growth hormone (GH) and thyroid-stimulating hormone (TSH)) are needed for the spermatogenesis (Simpson and Evans, 1946; Woods and Simpson, 1961; Mancini, 1968; Mills and Means, 1972). But there are still other workers who believe that the spermatogonial renewal and the early steps of meiosis are under the control of FSH (Greep *et al.*, 1936; Courot, 1970, 1976; Courot *et al.*, 1971; Davies, 1971; Mills and Means, 1972; Ortavant *et al.*, 1972; Huckins *et al.*, 1973; Means, 1974b). However, various hormones including testosterone show no stimulatory influence on the initiation of spermatogenesis *in vitro* (see Steinberger *et al.*, 1970). The finding of precocious spermatogenic maturation in a 6-year-old boy having a functioning Leydig cell tumor has indicated that high levels of local androgens are capable of initiating spermatogenesis in some cases (Steinberger *et al.*, 1973; Neaves, 1975). The actions of HCG and pregnant mare serum gonadotrophin (PMSG) for different periods and the action of testosterone on the immature rat testis have recently been investigated by Chemes *et al.* (1976b). Short-term administration of HCG (1–3 days) induces an early meiotic and postmeiotic stimulatory effect but a decrease in spermatogonial numbers. Administration of HCG for longer periods (10 days) causes a reduction in numbers of all cell types. Treatment with HCG+PMSG reduces the amount of inhibition, while PMSG alone results in histological and hormonal signs of stimulation of the interstitial tissue and the meiotic and postmeiotic stages; the numbers of spermatogonia are not affected. Testosterone causes stimulation of the meiotic and postmeiotic stages and a reduced number of spermatogonia. From these results, Chemes *et al.* (1976b) have concluded that while PMSG directly stimulates spermatogonia, HCG acts through testosterone secretion at the meiotic and postmeiotic stages. The early inhibitory effects of HCG and testosterone on spermatogonial numbers could be ascribed to the inhibition of endogenous FSH by androgens. Courot *et al.* (1978) have suggested that in the adult ram, the differentiation of renewing stem spermatogonia is under LH control and that the last stages of spermatogonial multiplication, from intermediate to B spermatogonia and to primary spermatocytes, are under the control of the FSH-like activity of PMSG.

C. SERTOLI CELLS

Various studies have indicated the effect of gonadotrophins on testis weight, and Sertoli cell number and development (Fig. 35). Hypophysectomy causes a 40 to 50% decrease in the number of supporting cells or precursors of Sertoli cells in the newborn lamb (Courot, 1971); this decrease can be prevented if ovine LH is injected. FSH also evokes a response in hypophysectomized lambs, but not as strong as that of LH. When LH and FSH are administered together, there is a great synergistic effect on the increase in the number of Sertoli cells (Hochereau-de Reviers and Courot, 1978). FSH given to immature mice starting on Day 6 causes an increase in the numbers of Sertoli cells as well as in their cytoplasm (Davies, 1971). These effects seem to indicate enhanced protein synthesis (Davies *et al.*, 1975). Bressler (1975), while carrying out experiments on implantation of testes into normal and hypophysectomized mice, has observed that the pituitary gland affects the postnatal maturation of Sertoli cells, with specific actions on their nucleolar development and the development of intercellular junctions. However, replication of Sertoli cells *in vitro* is not affected by FSH or LH, either in normal or in hypophysectomized rat (Steinberger, 1971). Hochereau-de Reviers and Courot (1978) have recently observed that hypophysectomy of immature rat induced a decrease in the number of Sertoli cells and this can be avoided by gonadotrophins (FSH). Griswold *et al.* (1976) have investigated the incorporation of [^{3}H]thymidine into nuclear DNA in cultured Sertoli cells prepared from testes of 20-day-old rat. Addition of FSH or cAMP to the culture medium greatly increased incorporation. It is possible that

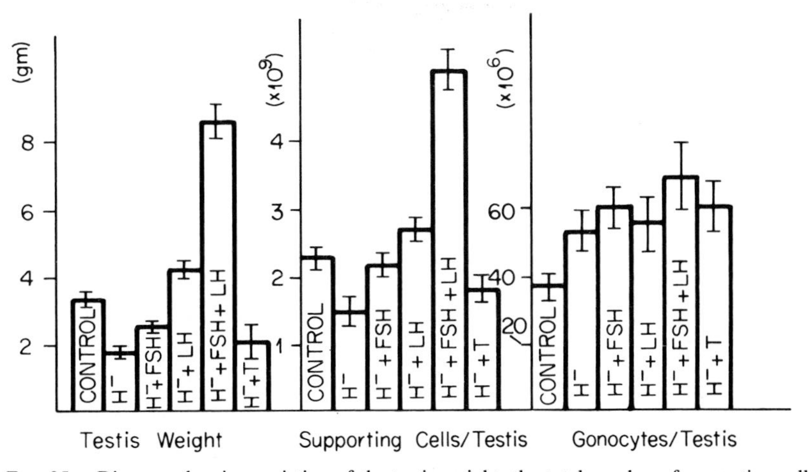

FIG. 35. Diagram showing variation of the testis weight, the total number of supporting cells, and the total number of gonocytes per testis after hormonal supplementation in the hypophysectomized lamb. (From Ortavant *et al.*, 1977.)

this DNA synthesis could be a replication of total DNA associated with cell mitosis, but with an extended cell cycle time.

The cells isolated from testes of 20-day-old rats, maintained in primary culture in a defined medium, respond to FSH or cyclic 3:5-AMP with characteristic morphological changes (Tung and Fritz, 1975) or production of ABP (Steinberger *et al.*, 1975a,b). No response is seen in cells treated with LH or 5′-AMP. The cells form a monolayer, and have been identified structurally by electron microscopy as presumptive Sertoli cells which do not undergo mitosis. The presumptive Sertoli cells have been shown to be morphologically and functionally different from peritubular fibroblasts which have high rate of mitosis and do not respond to FSH. Dorrington and Fritz (1974) observed that Sertoli cells are FSH responsive. Cameron and Markwald (1975) have observed that FSH stimulates an increase in the amount of smooth endoplasmic reticulum and some steroidogenic enzymes in the Sertoli cells of maturing rat testis. These observations have suggested that this gonadotrophin enhances (and possibly initiates) the steroidogenic potential of Sertoli cells. But this has been disputed by other workers (Christensen and Mason, 1965; Hall *et al.*, 1969; Cooke *et al.*, 1972). Vilar (1973) demonstrated a strong effect of FSH on Sertoli cells in the human testis; this effect is operative in the immature state. LH also appears to influence the developing Sertoli cells in the human testis and can by itself bring about initiation of spermatogenesis (Vilar, 1973). Bressler (1976) has demonstrated that maturation of the Sertoli cells is dependent on normal pituitary function. Armstrong and Dorrington (1977) have recently proposed that rat Sertoli cells in the presence of FSH convert testosterone of Leydig cell origin to estrogen only before 20 days of age.

FSH causes activation of adenylate cyclase in the seminiferous cords of newborn rats (Kuehl *et al.*, 1970; Braun and Sepsenwal, 1974; see also Dufau and Means, 1974, for other papers) and influences protein and RNA synthesis in a testis slice (Means and Hall, 1971) and ABP production in culture of Sertoli cells (Fritz *et al.*, 1974; Steinberger *et al.*, 1975a,b; Sanborn *et al.*, 1975b, 1976). Hypophysectomy causes a decline in protein kinase levels which can be restored by administration of FSH (see Dufau and Means, 1974). Stimulation of protein kinase by FSH has been shown to be dependent upon the age of the animal (see Means and Huckins, 1974). FSH causes a marked stimulation of protein kinase in seminiferous tubules isolated from rats 16 days of age. But the magnitude of this response decreases with age until finally in rats 30 days old or older no stimulation of protein kinase can be shown. Actually, all biochemical effects which have been shown to be stimulated by FSH in the testis are age dependent (Means and Huckins, 1974). Binding of 3β-FSH to testicular material from mature rats is about 60% of that observed in immature rats (Means and Vaitukaitis, 1971). FSH binding is mainly associated with the plasma membranes in the tubules of 16-day-old rats (Means, 1973); the interaction of FSH with the specific binding site appears to correlate temporally with activation of membrane-bound adenylate cyclase (see Dufau and Means, 1974). The data of Means and Huckins (1974) have revealed the following three of the earliest events after the administration of a single dose of FSH to immature rats: (1) its binding to receptors present on cells of seminiferous epithelium; (2) the resulting stimulation of membrane-bound adenylate cyclase and an associated increase in the intracellular accumulation of cAMP; and (3) activation of cAMP-dependent protein kinase (Means, 1977). The kinetics of stimulation of these three events indicate a coupled system. The experiments of Dorrington *et al.* (1974) have suggested that FSH does not elicit an increase in adenylate cyclase activity of interstitial cells or germinal cells, but demonstrate that Sertoli cells are FSH responsive (Steinberger *et al.*, 1975a,b).

Hansson *et al.* (1973, 1975) have observed a profound effect of FSH on testicular levels of androgen-binding protein (ABP) in immature animals. ABP synthesis *in vitro* by immature testis tissue is also stimulated by addition of human postmenopausal gonadotrophin, FSH, or testosterone to the medium (Ritzer *et al.*, 1975). Its production and secretion are also increased after cAMP treatment (Fritz *et al.*, 1976) ABP is produced by Sertoli cells (Dufau and Means, 1974; Hansson *et al.*, 1975; Fritz *et al.*, 1976; Steinberger and Steinberger, 1977). By stimulating ABP production in Sertoli cells FSH causes local accumulations of androgen and thus may be of great significance in the initiation of spermatogenesis in the rat. Hypophysectomy causes a fall in ABP levels, which can be restored by administration of FSH not LH (see references in Dufau and Means, 1974; Fritz *et al.*, 1976). Destruction of the germinal epithelium with X rays causes no loss of ABP, confirming a nongerminal cell (Sertoli cell) source for this protein. Hypophysectomy causes a more rapid change in the level of ABP in the immature rat than in the mature animal (Sanborn *et al.*, 1975a,b). Morphologically the Sertoli cell represents a logical candidate for the primary target cell for FSH.

From the various evidences reviewed above, it seems reasonable to conclude that LH/HCG exert their effects on the Leydig cells and that FSH and testosterone act on the seminiferous epithelium of developing and maturing testis. FSH is well known to stimulate and maintain synthetic activity of Sertoli cells (Fakundig *et al.*, 1976; see review by Means, 1977). Bressler and Lustbader (1978) have demonstrated that high local levels of testosterone accelerate seminiferous tubule development in the rat and indicate that tubule diameter may not be a valid basis for estimating development of the testis. They have suggested that testosterone exerts this effect through its actions on the Sertoli cells.

VIII. General Discussion and Conclusions

The results of various morphological (including ultrastructural), histochemical, biochemical, and physiological studies, as integrated here, have revealed three broad phases in the development, differentiation, and maturation of mammalian testis, which, however, overlap to a variable degree. The first phase involves the formation, growth, and differentiation of genital ridges by the proliferation and thickening of the surface epithelium, mesenchyme, and possibly some mesonephric tissue; meanwhile germ cells begin to be lodged in the genital ridges. During the second phase there occur the shifting and rearrangement of these cell types to form the various components of early testis, which are apparently brought about by morphogenetic movements; there occur important changes in the interrelationships of different cell types during this stage. During the third phase, the processes of differentiation and maturation are initiated in different structures of the testis. There occur several species variations in the timing of appearance of these three broad phases of testicular development, differentiation, and maturation, which seem to be controlled by genetic, mechanical, hormonal, biochemical, thermal, vascular, and nutritional factors to a large extent.

A. Origin and Movement of Germ Cells

The recent histochemical and ultrastructural data as summarized and correlated here have further confirmed and extended the extragonadal origin of germ cells in the yolk sac endoderm from where they subsequently migrate to the genital ridges by their amoeboid movements in the mesenteries. It is still not known more precisely whether the germ cells originate from the endodermal cells or whether they originate elsewhere and subsequently enter the endoderm (Clark and Eddy, 1975). Germ cell migration is an active process involving pseudopodia and some lytic enzymes with which the cells apparently manage to pass through the basement membrane of different tissues, or between cells within a tissue. The final movements of germ cells to the correct sites seem to be directed by some chemotactic inductor substances secreted by genital ridges (Witschi, 1948, 1951; Dubois, 1964, 1965, 1966). The nature of these substances is still to be determined. However, Baillie *et al.* (1966a) believe that these substances may be of steroidal nature as the mesenchyme of genital ridge shows the enzyme systems related to biosynthesis of steroid hormones (see also Moon and Raeside, 1972). On this basis, Baillie *et al.* (1966a) and Moon and Raeside (1972) have suggested that this tissue may be synthesizing androgens and other steroids. Recent correlative morphological and biochemical studies have shown that the hydroxysteroid dehydrogenases in most steroid gland cells are located in the microsomal fraction derived from the abundant agranular endoplasmic reticulum (Neaves, 1975; Christensen, 1975); cholesterol side-chain cleavage occurs in the mitochondria. The recent electron microscope studies of the genital ridges have revealed that all its cell types lack agranular endoplasmic reticulum and mitochondria with complex internal structure, which form specific organelles of steroid-synthesizing cellular sites (Christensen and Gillim, 1969; Christensen, 1970, 1975; Neaves, 1975; Guraya, 1976a,b). On this basis, it can be concluded that there is no ultrastructural evidence for the synthesis and secretion of steroid hormones in the early genital ridge. But from the histochemical demonstration of hydroxysteroid dehydrogenases, it can be suggested that the enzymes related to steroid biosynthesis are possibly synthesized before the membranes of the agranular endoplasmic reticulum or mitochondria with tubular cristae are prepared.

The nature of the factors which direct the movement of germ cells toward the genital ridges still needs to be determined more precisely in future studies. But it is significant that the germ cells, while migrating toward the genital ridges, show a close morphological association with various cell types. These close morphological relationships must be of great physiological significance especially in providing nutrient substances to the actively migrating germ cells which do not show appreciable endogenous energy reserves in the form of glycogen and lipids. The intense alkaline phosphatase activity demonstrated histochemically in the germ cells may be involved in the transport of nutrient substances from the adjoining cells. This suggestion is strongly supported by the fact that such a function has already been assigned to the presence of alkaline phosphatase activity in different tissues and cells (Pearse, 1968). The alkaline phosphatase reaction of primordial germ cells appears to be the most specific (Clark and Eddy, 1975) as it has been demonstrated that mutant mice characterized by the absence of germ cells do not show the reaction found in their normal counterparts (Mintz, 1957; Mintz and Russel, 1957). Primordial germ cells during the migratory phase show a very simple structure as they have some mitochondria with few cristae, abundant ribosomes, and prominent Golgi complex; the endoplasmic reticulum shows poor development. But they continue to become more specialized with time (see Clark and Eddy, 1975).

B. Formation, Composition, and Differentiation of Genital Ridges

The genital ridges are formed by the thickening of the coelomic epithelium and by the progressive condensation of the underlying mesenchyme; Merchant (1975) has also observed the contribution of some cells from the mesonephric tubules lying close to the developing gonad. The specific histogenetic roles of these three cell types in the formation of gonads still continue to be speculative as recent electron microscope studies on the development and differentiation of early gonad have not provided convincing evidence for the precise cellular source of different organs in the testis (Pelliniemi, 1975a,b, 1976b; Merchant, 1975) and ovary (Guraya, 1977a). This is due to the fact that the various cell types of the genital ridges and indifferent gonads show, more or less, similar ultrastructure, possibly due to their embryonic nature. Therefore, it is very difficult to follow their transformation into specific tissues or organs of the testis with the electron microscope. The continuity between the surface epithelium and the primitive cords in the testis as demonstrated in some studies does not form sufficient basis for attributing the origin of the latter to the former as the primitive cords on the other side are also continuous with the cells of the blastema proper which is of mesenchymal origin. It is, therefore, possible that the mesenchymal cells themselves are organized in the form of cords outer to the blastema proper. According to Merchant (1975), in smaller male embryos where several primordial germ cells have still not been incorporated into the cords, interruptions of the basal lamina are seen. At these places direct cell to cell contacts between the surface epithelium and mesenchyme are present, suggesting an active incorporation of the mesenchymal cells in the surface epithelial tissue. His studies have, therefore, indicated the movement of mesenchymal cells toward the surface epithelium.

C. Differentiation of Indifferent Gonad into Testis

In some earlier studies, the cellular components of indifferent gonad were believed to have a triple origin (Gillman, 1948); (1) Proliferation of the coelomic epitheoium forms the sex cords and granulosa cells; (2) proliferation of the underlying mesenchyme forms the medulla, stroma ovarii, interstitial Leydig cells, and thecal cells (see also Merchant, 1975); and (3) the germ cells have an extragonadal origin. Recent morphological and histochemical data have demonstrated that subjacent mesenchyme is relatively of greater importance in the formation of various cell types in the developing ovary (Guraya, 1977a) as was suggested earlier by Gropp and Ohno (1966). The surface epithelium does not make much contribution to the histogenetic and organogenetic processes of the developing ovary (Guraya, 1977a). Due to some induction and morphogenetic processes, the mesenchymal cells of the developing male gonad are also apparently rearranged in such a fashion as to give rise to tunica albuginea, primitive cords, testicular cords, interstitial tissue, etc. This suggestion is strongly supported by the fact that the mesenchymal or blastema cells and their derivatives (primitive and testicular cord cells, interstitium, etc.) show similar ultrastructural features. Pelliniemi (1975b) has suggested that the organization of cell groups, but not the internal structure of individual cells, is changed during sexual differentiation of the testis. The similarity in the ultrastructure of the organelles in various cell types of indifferent gonads has indicated that the differences are mainly at the molecular level and are manifested in the regulation of cell shape and position. It is suggested here that some specific staining and biochemical techniques including autoradiographic techniques at the subcellular and molecular levels continue to be used to follow more precisely the movements and histogenetic roles of the surface epithelium, mesenchyme, and possibly mesonephric tissue in the developing testis and ovary. The germ cells assume a cortical position to give rise to an ovary and a medullary position to form the testis.

The developmental changes characteristic of the gonad are believed to be determined mainly by genetic factors (Beatty, 1960, 1964; Ohno, 1967; Hammerton, 1968; Boczkowski, 1971, 1973; Short, 1972), although intragonadal or external factors such as hormones, etc. may also play an important role (Brambell, 1930; Witschi, 1951; Wolff, 1962; Wells, 1962). It is not known how the genetic determinants are translated into specific morphogenetic and endocrine changes. The primordial germ cells do not play any active role in the differentiation of gonad (Merchant, 1975) as normal testicular structures can develop even in their absence (Ohno, 1967). The distribution of the primordial germ cells in all compartments of the early gonad in the pig (Pelliniemi, 1975b) further supports their passive role in the differentiation of gonad. The factors controlling gonadal sex differentiation, especially the early formation of cell cords and subsequent establishment of functioning Leydig cells, are still to be determined more precisely.

In the male the sex-determining factors are apparently first activated in the somatic blastemal cells of the indifferent gonad, as the morphological changes, including establishment of specific cell contacts and simultaneous formation of a basal lamina, are occurring in these cells. The differentiation of an indifferent gonad into an identifiable testis is initiated without any cells which may show the ultrastructural features of mature, differentiated Leydig cells (Christensen and Gillim, 1969; Christensen, 1975; Neaves, 1975). Elger *et al.* (1971) have suggested that the Leydig cells and androgens do not play any specific role in the early development of testicular cords. But Attal (1969) has shown the presence of androgens in the undifferentiated gonad of sheep. Jost *et al.* (1973) believe that the supporting cells organizing themselves in the form of cords around primordial germ cells act as the initiator of gonadal sex differentiation. The beginning of synthesis of the basal lamina also does not seem to be an initiator of cord formation as the formation of the basal lamina is not yet completed when the testicular cords have already separated from the interstitium (Pelliniemi, 1975b). The overlapping of the peripheral borders of the vascular bed and the area of the testicular cords seems to be significant as the differentiation begins in the vascular periphery and the central cords are formed last (Pelliniemi, 1975b). The blastemal cells adjacent to the capillaries differentiate into interstitial cells and anastomosing channels of loose interstitial tissue form along the capillaries. The rest of the blastema remains as anastomosing sheets of compact tissue, known thereafter as the testicular cords. But it is still difficult to decide which of the two tissues, the interstitium or the testicular cords, is the primary organizer in testicular morphogenesis. However, the role of the vasculature in the regulation of gonadal sex differentiation deserves special attention.

D. Interstitial Leydig Cells and Steroid Hormone Synthesis

Sex differences in the timing of primary interstitial cell differentiation remain to be explained. The development of testicular interstitial Leydig cells takes place considerably earlier than in the ovary (Jirasek, 1967; Black and Christensen, 1969; Gondos and Hobel, 1973; Guraya *et al.*, 1974; Guraya, 1974b, 1977a,b; Resko, 1977). In the guinea pig testis, the structural differentiation of interstitial cells or Leydig cells starts as early as in 22- to 24-day-old fetuses, i.e., 3 weeks earlier than in the female guinea pig (Guraya *et al.*, 1974; Guraya, 1977b). Similarly, the time of initial appearance of interstitial cells in the human fetal ovary at 12 weeks is later than the appearance of Leydig cells in the testis at 8 weeks (Jirasek, 1967; Holstein *et al.*, 1971; Gondos and Hobel, 1973; Resko, 1977). Similarly, in the field rat, the testicular interstitial Leydig cells differentiate much earlier than the ovarian interstitial cells (Guraya and Uppal, 1977, 1978). But it is now well established that in both sexes the primary interstitial cell system develops from the fibroblast-like stromal cells entirely independently of the germ cells (see also Guraya, 1977a,b). Guraya (1977a) has suggested that the differences in timing of interstitial cell development may, somehow, be related to the very early differentiation of the testis. The time difference in regard to the development of interstitial cells probably represents responses to quite different stimuli of hormonal nature. Or they probably indicate a difference in the response of competent cells to hormonal substances. Milewich *et al.* (1977) have recently observed that the time of appearance of the enzymatic capacity to convert testosterone to estradiol in the fetal rabbit ovary is similar to the onset of the enzymatic activity to form testosterone by the fetal testis. From these observations they have suggested that the acquisition of the enzymatic activities that allow specific endocrine function by these two tissues may be regulated by the same or similar factors during development. They have hypothesized that enzymatic programming is determined at a genetic level and that other factors regulate the timing of the expression of the capacity to serve as an endocrine organ. This supposition does not imply that specific steroid-synthesizing enzymes are unique to one organ or the other but only that the relative rates of formation of androgens and estrogens are characteristic.

The correlation of results of various morphological (including ultrastructural), histochemical, biochemical, and physiological studies on developing and maturing testes has clearly indicated that Leydig cell differentiation in most mammals usually occurs in two phases: a fetal stage and an adult stage. Electron microscopic observations on different mammalian species have demonstrated that fetal Leydig cells differentiate at a specific time of development shortly after testicular differentiation. Despite many species variations in the timing of the appearance of Leydig cells, it is well established that they develop through intermediate or partially differentiated stages from fibroblast-like cells of mesenchymal origin.

The process of Leydig cell differentiation is closely associated with cellular enlargement, extensive proliferation of smooth endoplasmic reticulum or diffuse lipoproteins, appearance of large round mitochondria with tubular cristae, development of a prominent Golgi apparatus, and appearance of enzyme activities related to steroidogenesis. Similar ultrastructural, histochemical, and biochemical changes also occur during the differentiation of Leydig cells in the adult testis (Neaves, 1975; Christensen, 1970, 1975). The general pattern that has emerged from various electron microscope studies on Leydig cell differentiation is that mesenchymal cells first give rise to partially differentiated cells which are ultimately transformed into Leydig cells by accumulating more organelles related to steroidogenesis (see also de la Balze *et al.*, 1960). Leydig cell differentiation is generally a rapid process in fetal testis, whereas in the prepubertal testis, it is prolonged (Gondos *et al.*, 1977), indicating some differences either in the response of cells or in the nature of factors which cause interstitial Leydig cell development.

The interstitial Leydig cells of fetal testis do not constitute a stable population of cells as they involute after performing their function at a specific stage of development (Guraya and Uppal, 1977). Similarly, histochemical studies by Guraya (1977a,b) have clearly shown that the primary interstitial cells of the developing mammalian ovary involute with the initiation of postnatal life in the guinea pig. Meanwhile, the secondary interstitial gland cells of thecal origin begin to differentiate and accumulate with the initiation of follicular atresia. The nature of the factors which determine the regression of fetal interstitial gland cells of both testis and ovary after they have reached such an advanced stage of differentiation is not known.

The interstitial Leydig cells of developing and maturing testes possess the cytological, histochemical, and biochemical features indicative of steroid hormone biosynthesis in them. Their most conspicuous common characteristics related to steroid hormone biosynthesis are: (1) abundant diffuse lipids (lipoproteins) in the cytoplasm; (2) abundant membranes of smooth reticulum; (3) mitochondria with a complex system of internal cristae which become predominantly tubular; (4) development of abundant agranular endoplasmic reticulum accompanied by various enzyme activities indicative of the biosynthesis of steroid hormones; (5) under certain physiological conditions stored cholesterol-positive lipid droplets in the cytoplasm; (6) capacity to form steroid hormones under *in vivo* and *in vitro* conditions, which has been extensively studied in the fetal testes of different mammalian species. The interstitial gland cells of developing ovaries have also been shown to possess similar cytological, histochemical, and biochemical characteristics related to steroid hormone synthesis (Guraya, 1977a). In their cytological, histochemical, and biochemical features, the interstitial gland cells of developing and maturing mammalian testis and ovary closely resemble the well-established steroid gland cells of sexually mature mammalian and nonmammalian gonads (Christensen and Gillim, 1969; Guraya, 1971, 1973a–c; 1974a–d, 1976a,b; Christensen, 1970, 1975; Neaves, 1975). In general, the results of cytological, histochemical, and biochemical studies on interstitial Leydig cells of developing and maturing testes agree well with each other and are compatible with the known endocrine function of Leydig cells. From this discussion, it can also be concluded that the cytoplasmic machinery required for carrying out steroid hormone biosynthesis is basically very similar in the steroid-producing cells of mature and immature gonads of mammals and submammalian vertebrates (see also Guraya, 1974a,b). Biochemical studies have revealed that the pathways of steroid hormone synthesis are also basically similar in the immature and mature gonads of mammals and nonmammalian vertebrates (see also Tamaoki *et al.*, 1969; Neaves, 1975; Guraya, 1976a,b, 1977a; Ewing and Brown, 1977). But differences in detail do exist, depending upon the sex and stage of development and maturation of the gonad.

Cytological and histochemical features common to the interstitial gland cells of developing and maturing testis in mammals, which are related to their endocrine function as concluded above, do not show any appreciable development in the undifferentiated mesenchymal cells, which show mostly elements of granular endoplasmic reticulum, free ribosomes, and mitochondria with simple internal structure. With the transformation of fibroblast-like cells into interstitial gland cells in the developing mammalian testis and ovary, organelles, lipid droplets, and enzyme systems specific to steroidogenic tissue (Christensen and Gillim, 1969; Guraya, 1971, 1973a–c, 1974a–d, 1976a,b; Christensen, 1970, 1975; Neaves, 1975) are developed in abundance. The exact nature of the factors that bring about the cytoplasmic and biochemical changes typical of steroid hormone biosynthesis is still poorly understood for the developing mammalian testis and ovary. The gonadotrophic hormones (especially LH or ICSH) seem to be the key to these changes (see also Neaves, 1975; Christensen, 1975; Guraya, 1976a,b, 1977a). The nature and patterns of secretion of gonadotrophins in the developing and maturing mammalian species should be worked out in future studies, as relatively very little work has been carried out previously along these lines (Prasad, 1974; Guraya, 1977a). However, the use of exogenous gonadotrophins, especially those containing the luteinizing factor, has been found to stimulate the development of cytoplasmic organelles and enzyme systems related to steroidogenesis in the interstitial gland cells of developing mammalian testis. It has also been shown in the chick that LH administration in doses known to stimulate testosterone synthesis produces an increase in rough endoplasmic reticulum (Connell, 1972), which is subsequently transformed into smooth reticulum. HCG given to immature mice (Aoki, 1970) and guinea pig (Merkow *et al.*, 1968) has been found to stimulate the formation of large areas of smooth endoplasmic reticulum in interstitial cells. Androgens may also be needed for Leydig cell maturation (Blackburn *et al.*, 1973; Neaves, 1975).

It has been shown recently in various studies that the development of agranular endoplasmic reticulum (or abundant diffuse lipoproteins) concomitant with alterations of mitochondrial structure and storage or depletion of lipid droplets in the steroid gland cells of mature gonads is influenced by gonadotrophins (Armstrong, 1968; Armstrong *et al.*, 1969; Carithers and Green, 1972a,b; Christensen and Gillim, 1969; Guraya, 1971; Hilliard *et al.*, 1968; Zarrow and Clark, 1969; Guraya, 1975a–c; Neaves, 1975; Christensen, 1975). The development in the developing mammalian testis of interstitial gland cells, which have been shown and discussed to possess the cytological and histochemical features of well-established steroid gland cells, also suggests the strong possibility of secretion of some gonadotrophic substances (especially LH) during the fetal life of mammals (see also Prasad, 1974). These results have indicated the establishment of hypothalamic-pituitary-gonad interrelationships at a very early stage (see also Foster *et al.*, 1972; Foster, 1974; Eguchi *et al.*, 1975, 1978). However, the development, differentiation, and activation of Leydig cells in both man and rhesus monkey correlate well with the secretion of HCG (see Resko, 1977). Experimental data further support this possibility that the developing testis and ovary gradually become sensitive to exogenous gonadotrophic stimulation, both *in vivo* and *in vitro,* and produce steroid hormones (Moger and Armstrong, 1974; Engel and Frowein, 1974; see papers in Dufau and Means, 1974; French *et al.*, 1975; Picon, 1976; Feldman and Bloch, 1978; Sanyal and Villee, 1977). Various recent studies have indicated that gonadotrophins stimulate processes involved in spermatogenesis, follicle development, and secretory activity of developing and maturing testis and ovary (Guraya, 1977a). If differentiation of the interstitial gland cells in developing gonads actually depends on gonadotrophic stimulation, sex differences in the timing of interstitial cell differentiation remain to be explained as already discussed.

The specific physiological roles of diffuse lipoproteins or abundant membranes of agranular reticulum, mitochondria having tubular cristae, and cholesterol-positive lipid droplets in relation to steroid biosynthesis in the steroid gland cells of mature gonads have been discussed in detail in previous reviews (Christensen and Gillim, 1969; Guraya, 1971, 1976a,b; Christensen, 1970, 1975; Neaves, 1975). Therefore, their detailed discussion will be omitted here. The much more extensive diffuse lipoproteins or membranes of agranular endoplasmic reticulum form sites for the enzymes involved in the synthesis of steroid hormones. In addition to acting as a source of the synthesizing enzymes involved in the biosynthesis of steroid hormones, diffuse lipoproteins (or membranes of smooth reticulum) synthesize and store cholesterol in their lipids, which in turn may be acting as a precursor in the biosynthesis of steroid hormones. The enzyme activity necessary for splitting the cholesterol side chain usually resides in the mitochondria and particularly in the inner mitochondrial membrane while most of the other steroid-converting enzymes are localized in the membranes of smooth reticulum (Christensen, 1975). The mitochondria of steroid gland cells, including Leydig cells, are unusual in their internal structure in that the inner membrane forms tubular or vesicular invaginations into the matrix instead of the foliate cristae commonly found in other cell types (Christensen and Gillim, 1969; Christensen, 1970, 1975; Neaves, 1975). Whether this unusual inner membrane configuration is specifically related to the capacity of these mitochondria for cholesterol side-chain cleavage is still not clear; but LH, which influences this important-rate limiting step in the biosynthesis of testosterone, also causes subtle changes in the internal structure of the mitochondria (Christensen and Gillim, 1975; Christensen, 1975; Neaves, 1975). Alterations in the internal structure of mitochondria (i.e., the development of tubular cristae) corresponding to the appearance of ultrastructural smooth membrane suggest close morphological and functional similarities in steroid hormone biosynthesis in the steroid gland cells of immature and mature gonads in mammals and nonmammalian vertebrates (Guraya, 1971, 1976a,b). The limited areas of rough endoplasmic reticulum and their associated ribosomes are concerned with the protein synthesis required for the maintenance and renewal of cytoplasmic constituents including steroidogenic enzymes (Fawcett *et al.,* 1969). The roles of Golgi complex, microbodies, lysosomes, etc. in steroid biosynthesis and secretion are still to be determined more precisely (see Christensen and Gillim, 1969; Neaves, 1975; Christensen, 1970, 1975).

The results of recent studies as integrated here have provided enough evidence that Leydig cell ultrastructure, especially the formation of abundant smooth endoplasmic reticulum and alteration in the internal structure of mitochondria, directly reflects the capacity for androgen production. The gonadotrophins with luteinizing hormone activity increase the quantities of smooth endoplasmic reticulum and alter the internal structure of mitochondria as already discussed. There also exists a close correlation between fetal and postnatal ultrastructural and histochemical differentiation of Leydig cells and testosterone production (see also Lee *et al.,* 1975).

Cholesterol-positive lipid droplets in the steroid gland cells store potential precursor material for the biosynthesis of steroid hormones (Guraya, 1971, 1973a–c, 1974a–d, 1976a,b; Christensen, 1970, 1975; Neaves, 1975). A similar function can also be assigned to the cholesterol-containing lipid droplets of interstitial gland cells of developing mammalian testis Cholesterol destined for conversion to androgenic steroids may be synthesized in advance and stored as cholesterol esters in the lipid droplets. The physiological and biochemical mechanisms involved in the storage and depletion of cholesterol-containing lipid droplets in steroid gland cells have been discussed in previous reviews and papers (Armstrong, 1968; Garren *et al.,* 1971; Flint and Armstrong, 1972; Neaves, 1975; Christensen, 1975; Guraya, 1975a–c). Our knowledge is still very meager in this regard. Experiments with prolactin have suggested that the hormone may favor intracellular accumulation of cholesterol (Armstrong, 1968; Neaves, 1975). Synthesis of steroid hormone in the testis first requires retrieval of cholesterol by hydrolysis of the ester. This important step is stimulated by LH both *in vivo* and *in vitro* (Neaves, 1975). Administration of LH results in a dramatic reduction of lipid droplets and a significant depletion of cholesterol ester as determined by biochemical analysis (see Neaves, 1975). Lipids, cholesterol, and cholesterol esters show alterations with the maturation of testis (Bartke, 1971; Bartke *et al.*, 1974; Renston *et al.*, 1975; Coniglio *et al.*, 1975; Ahluwalia *et al.*, 1975), indicating their involvement in steroid hormone synthesis and possibly in the growth processes of maturing testis.

It is well established now that gonadotrophins with luteinizing hormone activity (LH, HCG, PMSG) increase the synthesis and secretion of testicular testosterone *in vivo* and *in vitro* (Ewing and Brown, 1977; Feldman and Bloch, 1978). The mechanism by which gonadotrophin stimulates testosterone secretion has been clarified by autoradiographic and biochemical studies (see Dufau and Means, 1974; French *et al.*, 1975; Neaves, 1975; Ewing and Brown, 1977). The radioiodine-labeled LH binds to the membrane of Leydig cells where it activates adenyl cyclase, which in turn causes intracellular accumulation of cAMP and release of active kinase. These intracellular conditions trigger hydrolysis of cholesterol esters, increasing available precursor and stimulating steroidogenesis by accelerating the rate-limiting first step in the biosynthetic processing of the cholesterol. The mechanism whereby LH stimulates cholesterol side-chain cleavage in Leydig cells needs to be clarified. The functional significance of structural variation in the organelles and inclusions involved in steroidogenesis is still not known. The detailed localization of steroid biosynthetic steps within the organelles should be determined in future cytochemical studies. The exact nature and amount of steroid hormones synthesized by the Leydig cells at different stages of developing and maturing testes in different mammalian species are still to be worked out more precisely under both *in vivo* and *in vitro* conditions, as relatively very little work has been carried out previously in this regard, especially for the developing ovary (Haffen, 1970; Guraya, 1977a).

E. Functions of Testicular Androgens during Sexual Development

It is well established that fetal differentiation of Leydig cells correlates closely with acquisition of the capacity to synthesize testosterone (see also Dufau and Means, 1974; French *et al.*, 1975) and with important androgen-dependent changes occurring during early development (Jost, 1970; de Moor *et al.*, 1973). The testicular steroidal androgens are responsible for the differentiation of the Wolffian ducts and urogenital sinus (Bloch *et al.*, 1971a,b; Jost *et al.*, 1973; Price *et al.*, 1975) as well as for the sexual differentiation of the central nervous system (Nakai *et al.*, 1971; Arai and Serisawa, 1973; Neill, 1973; Davidson,

1974; Arai *et al.*, 1975; Resko, 1977). Veyssiere *et al.* (1976) have claimed that their data provide the first evidence for testicular testosterone secretion at the time of genital differentiation in the rabbit (between 20 and 25 days according to Jost, 1947). It is well established that testosterone not only virilizes the Wolffian duct but also serves as a prohormone for dihydrotestosterone, which masculinizes the urogenital sinus and external genitalia. The essential role of the endocrine function of the testes for male phenotypic development has been established by endocrine ablation and hormone replacement studies (Jost, 1953, 1966, 1970; Jost *et al.*, 1973; Price *et al.*, 1975), by characterization of several mutations in hormone synthesis or hormone action that cause abnormal or incomplete male sexual development (Wilson *et al.*, 1975), and by inducing a high proportion of bilateral cryptorchidism in the male offspring of a mother treated with estradiol during pregnancy (Jean *et al.*, 1975). Neaves (1975) has emphasized that the primary function of Leydig cells may be provision of the androgen stimulus required for maintenance of spermatogenesis in the germinal epithelium. Armstrong and Dorrington (1977) have proposed that the testosterone formed by Leydig cells also serves as a substrate for the synthesis of estrogen in the Sertoli cells under the influence of FSH. Estrogen synthesis is at a maximum during the first wave of spermatogenesis in the rat, which occurs before 10 days of age (Clermont and Perey, 1957). This suggest the possibility that estrogens may be involved in regulation of this process. Although the timing of the onset of hormone production by the fetal testis has been charted in detail (Feldman and Bloch, 1978; Pointis *et al.*, 1978), the factors that regulate testicular function during embryonic life are much less clear (see Catt *et al.*, 1975). Future progress will depend on finding ways to study circulating gonadotrophin levels at early stages of differentiation and to follow intracellular pathways of steroid synthesis secretion. George *et al.* (1978) have suggested that the initiation of androgen synthesis at the time of male phenotypic differentiation in the rabbit embryo is independent of extragonadal hormone stimulation. They have attributed the structural and functional development of fetal Leydig cells to inherent genetic programming rather than to hormonal or substrate induction at the time of gonadal differentiation. Pointis *et al.* (1978) have suggested a negative feedback of testosterone on fetal pituitary LH secretion.

F. Alterations in Steroidogenic Enzymes during Testicular Maturation

Wiebe (1976) has suggested that during sexual maturation the testicular biosynthesis of active 5-ane androgens may proceed via 5-ane precursors with the help of age-dependent 5-ane-3β-hydroxysteroid dehydrogenases. The present state of knowledge might suggest that Leydig cells contain enzymes for several steroid biosynthetic pathways: (a) those for the production of 5-ane-C_{19}-steroids.

via the reduction of pregnenolone or 5-ane-C_{21} derivatives, and (b) those for the production of testosterone and DHT via progesterone or 17α-hydroxyprogesterone (see Wiebe, 1976). Wiebe (1978) has suggested that sexual maturation of the testicular biosynthesis of active 5-ane androgen(s) proceeds via 5-ane precursors with the help of age- and gonadotropin-dependent 5-ane-3β-oxidoreductases. Yoshizaki *et al.* (1978) have indicated that in 29- to 35-day-old rat testes, 5α-reduction of C_{19}-Δ^4-3-ketosteroids and the formation of 5α-reduced C_{21}- and C_{19}-steroids from pregnenolone takes place largely in the microsomes of interstitial tissue. Nayfeh *et al.* (1975b) have demonstrated the hormonal regulation of 5α-reductase. There may be several enzymes (or enzyme classes) some present in greater amounts than others. Further understanding of the nature and number of testicular steroid oxidoreductases during sexual maturation is expected from enzyme purification and kinetic studies currently in progress. Booth (1975) has recently observed a high concentration of testosterone in the testes of 84-day-old fetuses of boar which is considered to be significant in the differentiation of male behavior. The amount of testosterone exceeds that of androstenedione during postnatal development, and the observation of high concentrations of DHT and 5-androstenediol, free and in sulfate form, particularly in postpubertal boars suggests that the 5-ane pathway for the synthesis of testosterone may be important, and it may differ with the species. The changes in various enzymes of the testis can be used as marker proteins for testicular differentiation and maturation (Bishop, 1969; Blackshaw, 1973; Fritz, 1973; Payne *et al.*, 1977).

The pattern and significance of Leydig cell differentiation in the postnatal period are also not as clearly defined for different mammalian species as they are in the fetus. More work is needed in this regard. But Gondos *et al.* (1976), after studying the postnatal differentiation of Leydig cells in the rabbit, believe that the timing of Leydig cell differentiation is related to the duration of the postnatal prespermatogenic period.

G. Differentiation and Maturation of Germ Cells and Sertoli Cells

Testicular cords are transformed into seminiferous tubules which show two distinct cell lines, germ cells and Sertoli cells. Recent light and electron microscope studies have clearly demonstrated the continuity of development and differentiation of germ cells from fetal gonocytes to adult spermatogonia through prespermatogonia (Fig. 30). Prespermatogonia undergo a diminution in cell size and a marked increase in nucleolar complexity before the appearance of spermatogonia. This nucleolar development is a prerequisite for the high rate of RNA or ribosomes synthesis present in type A spermatogonia whose appearance is associated with resumption of mitosis and initiation of spermatogenesis. There occur some structural changes in the germ cells during fetal and postnatal development of the testis, which include accumulation of cytoplasmic organelles and an increase in nucleolar complexity and nuclear condensation. The organelles, which show their further accumulation, include mitochondria, granular endoplasmic reticulum, and patches of dense material or nuage; the latter may develop morphological association with the mitochondria. On the basis of ultrastructure, it is possible to distinguish the gonocytes, prespermatogonia, and various types of spermatogonia (type A, intermediate, type B) in rats 2 to 37 days of age (see also Mauger and Clermont, 1974) and in man (Bustos-Obregon *et al.*, 1975). Some enzymes show conspicuous changes in activity during the initial spermatogenic cycles. But their localization at the cellular level is still to be determined more precisely.

The regulation of germ cell development is poorly understood. The nature of the mechanisms which affect and maintain an arrest of germ cell mitosis following a period of active multiplication is not known. These changes occur at a time when the hypothalamic-pituitary-gonadal axis is already established and their proper explanation will require an understanding of very complex hormonal actions. The mechanisms by which germs cells are placed adjacent to the basal lamina is also not known more precisely. The development of intercellular bridges connecting developing germ cells in both male and female gonads has been demonstrated in different mammalian species (Burgos and Fawcett, 1955; Gondos, 1973; Guraya, 1977a). They are believed to promote the developmental synchrony observed between cells within a single cluster. The maturation and differentiation of gonocytes into spermatogonia occur over a given time which greatly varies with the mammalian species. Similarly, the timing of follicular development in the mammalian ovary during the prenatal and postnatal periods also varies greatly with the mammalian species (Guraya, 1977a). Guraya (1977a) has suggested that these differences in follicular development might, somehow, be related to the variable length of gestation in different mammals. A similar suggestion can also be made to explain the species differences in the timing of appearance of different types of spermatogenic cells in the seminiferous tubules of different mammals. The results of recent studies have indicated that gonadotrophins, especially FSH, stimulate spermatogonia and that testosterone acts at the meiotic and postmeiotic stages (see also Choudhury and Steinberg, 1975). Actually, the receptors for androgens have been demonstrated in interstitial cells and Sertoli cells (see papers in French *et al.*, 1975). Androgen exchange activity has been suggested to exist between Sertoli and germ cells. The results of various studies as summarized in French *et al.* (1975) have indicated that the mechanism of action of androgens in the testis becomes quite complex during its maturation. The functional significance of androgen action on interstitial cells remains to be clarified (Neaves, 1975). Further studies should also be carried out on the receptors for hormones in spermatogenic cells to determine quantitatively and qualitatively the stages of spermatogenesis controlled by testosterone and FSH (see

Galena *et al.*, 1974). Attention should also be paid to differences between species. Such studies will be possible because different types of spermatogenic cells can now be separated using techniques which are now being developed and standardized (Meistrich and Trostle, 1975). Studies along these lines may help us to determine more precisely the various factors responsible for the establishment of the initial spermatogenic cycle about which our knowledge is still very meager.

Sertoli cells originate from the supporting cells of early testis. They develop complex morphological relationships with each other and with germ cells during the maturation of testis, and thus undergo maturation processes as evidenced by their morphological and chemical changes. With the differentiation of testis, Sertoli cells develop elongated cytoplasmic processes extending around adjacent spermatogenic cells. They show large numbers of mitochondria, abundant endoplasmic reticulum, and well-developed Golgi structures. The correlation of ultrastructural and histochemical features of Sertoli cells in the fetal testis has suggested that they have a very limited steroidogenic activity. In the early stages of testicular differentiation the Sertoli cells show a well-developed rough endoplasmic reticulum suggesting active synthesis of proteins that can be exported (Merchant, 1976). If the tubules make any contribution to steroid hormone synthesis, it is quantitatively insignificant compared to that of the Leydig cells which form the principal site of steroidogenesis in developing and maturing testes as already discussed. However, recent biochemical studies have indicated that testosterone formed by the Leydig cells under the influence of LH is transported to the Sertoli cells, where in the presence of FSH, it is converted to estrogen during the first 3 weeks after birth in the rat. Clarification of the relative roles of Leydig cells and Sertoli cells in the support and regulation of spermatogenesis is needed. Recent studies have produced evidence of quantitative regulation of adult spermatogenesis by the population of Sertoli cells established during testicular development. Thus sperm production in the adult is programmed (or controlled) a long time in advance during the development of testis (Hochereau-de Reviers and Courot, 1976, 1978). The role of Sertoli cells in the early maturation of the seminiferous epithelium is also still to be determined more precisely. But it is well established that they play an important role in the blood–testis barrier and in disposing of the degenerating germ cells from the seminiferous epithelium of developing and maturing testes (see Russell, 1978). The possible function of such phagocytic activity in promoting germ cell development is unknown but the presence of similar activity in the fetal ovary involving germ cells and granulosa cells (Guraya, 1977a) shows that this is a common feature of early development of the testis.

Evidence for the presence of Müllerian-inhibiting substance or anti-Müllerian hormone (AMH) has been obtained in the testes of different mammalian species during early development (Picon, 1970, 1971; Josso, 1971, 1974; Jost *et al.*,

1972). It is interspecific and nondialyzable (Josso, 1971, 1972). Experimental work in mammals (Josso, 1970, 1973, 1974) and birds (Stoll *et al.,* 1973) points to the sex cords or testicular cords as the source of AMH which is protein synthesized in the Sertoli cells (see Josso, 1974, 1972; Merchant, 1976). Müllerian regression factor is still an incompletely characterized substance.

Various morphological and biochemical evidences have indicated the influence of gonadotrophins on Sertoli cell development and function (Dufau and Means, 1974; French *et al.,* 1975). FSH has a conspicuous effect on testicular levels of androgen-binding protein (ABP) which is produced by Sertoli cells especially in immature animals (see also Hansson *et al.,* 1973) The effect of FSH on ABP production by Sertoli cells causing local accumulations of androgen may be particularly important in the initiation of spermatogenesis. Similarly, the effect of FSH on the conversion of testosterone to estrogen in Sertoli cells may also be of significance in the regulation of spermatogenesis. Krueger *et al.* (1974) have provided new evidence for the role of the Sertoli cells and spermatogonia in feedback control of FSH secretion in male rats. The production of inhibin has also been attributed to the Sertoli cells.

It can be concluded that Sertoli cells show specific patterns of differentiation and specialization during fetal, postnatal, and prepubertal periods; in the adult testis, they show a complicated structure. The various changes of Sertoli cells show that their functions may be multiple, relating both to gametogenesis and to endocrine activity. Future studies using various techniques including *in vitro* experiments will provide much more information about their specific functions during development and maturation of the testis.

H. Hormonal Changes during Sexual Maturation

Resumption of germ cell division and Leydig cell differentiation (both morphological and biochemical) at the time of sexual maturation constitutes the important changes occurring under a variety of possible influences of intratesticular and extratesticular origin. In some species, especially the rat, it appears that the hypothalamic-hypophyseal-testicular axis may be functional from birth onward and during prepubertal period (Goldman and Gorski, 1971; Debeljuk *et al.,* 1972; Payne *et al.,* 1977; Goomer *et al.,* 1978). The blood level of steroids is determined by at least two factors: the plasma concentration of pituitary gonadotrophins and the testicular responsiveness to these hormones. Ramirez and McCann (1965) and McCann *et al.* (1974) believe that a change in the hypothalamic sensitivity to steroids occurs at puberty in the rat that would induce a rise in the blood level of gonadotrophins and thus testosterone. But it is still not known more precisely whether the rise in blood testosterone at puberty, as demonstrated in different mammalian species, is due to an increase in plasma LH, because in some species such as sheep (Cotta *et al.,* 1975), guinea pig

(Donovan *et al.*, 1975), and bull (Lacroix *et al.*, 1977), the circulating LH level remains relatively stable during sexual maturation, and in the rat, it increases very little at puberty (Swerdloff *et al.*, 1971, 1972). Odell *et al.* (1973) have suggested that in sexual maturation, a change in the FSH-induced testicular sensitivity to LH may occur as was recently confirmed in the rat (Odell *et al.*, 1974). Berger *et al.* (1974) have demonstrated that the testis of 30-day-old immature rabbits is able to respond significantly to HCG stimulation and that the response was identical from 30 to 120 minutes after the treatment. Berger *et al.* (1976b) have recently shown that the response of the immature rabbit testis to HCG is much less than that of the adult. From this discussion, it can be concluded that a dynamic hypothalamic-pituitary-gonadal relationship is established early in development, with important changes occurring during prepubertal and pubertal periods (see also Gupta, 1978). However, the problems with regard to the nature of the gonadal-pituitary interactions that change from infancy to puberty (Payne *et al.*, 1977) and the influence of secretions of developing testes on these interactions are still to be solved (Donovan, 1974; Grumbach *et al.*, 1974; Goomer *et al.*, 1978). The data of Reiter *et al.* (1976) have shown the greater Leydig cell responsitivity to transient increases of endogenous gonadotrophin in pubertal boys. Their data also suggest that there may be a relationship between adrenal androgen production and maturation of the hypothalamic-pituitary-gonadal system. Payne *et al.* (1977) have investigated hypothalamic-pituitary and testicular function during sexual maturation of the male rat. Gonadotrophin-releasing hormone (GnRH), LDH, and FSH show fluctuations in their concentrations, on different days of sexual maturation, of 5-ane-3β-HSDH isomerase activity of the testis, with a rapid increase between Days 12 and 19 followed by an even greater rate of increase between Days 19 and 32, when adult levels are attained. 17β-HSDH is very low between birth and Day 22. Enzyme activity begins to increase at Day 22 with a rapid increase in activity observed between days 37 and 58. The increase in the capacity to synthesize testosterone closely follows the increased 17β-HSDH activity. These observations have indicated that during sexual maturation in the male rat changes in serum LH and FSH do not reflect changes in hypothalamic GnRH. The appearance of Leydig cells as monitored by 5-ane-3β-HSDH isomerase activity precedes by approximately 20 days the increase in the testicular capacity to synthesize testosterone *in vitro*. The latter coincides with an increase in 17β-HSDH activity which is probably a limiting factor in the activity of the testis to respond to LH stimulation.

IX. Addendum

Since this review was completed and submitted for publication, a number of additional papers and some related reviews have come to my attention. They are outlined and discussed below. This has also been done to provide the reader with more recent references and a brief account of significant developments as the research in this area is developing very fast in both breadth and depth.

A. Development and Differentiation of Testis

Recent studies have indicated that the mesonephros play an important role in the development of indifferent gonads in mouse and sheep as their somatic cells derive from the cellular components of the giant mesonephric glomerulus and its tubules (Upadhyay *et al.*, 1978; Zamboni *et al.*, 1978; Mauleon, 1978). Merchant-Larios (1978), using an autoradiographic approach, has not been able to solve the problem of the origin of the first inner epithelial cells which soon become independent of the coelomic epithelium and form the seminiferous cords in the male gonads. The seminiferous cords will be continuous with some mesonephric cords which in turn become the rete testis (personal communication). Byskov (1978a) has briefly described the development of testicular cords in the mouse. Rohloff and Mulling (1977) have made a comparative study of the testicular development in domestic and wild pigs. Pelliniemi *et al.* (1979) have described the ultrastructural aspects of sexual differentiation and prenatal development of the gonads in human and pig embryos. Testis is identifiable in man in the seventh week and in the pig at the age of 26 days. The differentiation of the Leydig cells occurs in the male gonad by the end of the eighth week in man and by the age of 30 days in the pig. No ultrastructural sexual differences are seen in the gonads prior to the formation of the testicular cords. Ohno (1976, 1978) has discussed the major sex-determining genes.

Recent studies have indicated that the gonocytes in the fetal testis of mouse and man are prevented from reaching the leptotene stage of meiosis presumably as a result of a meiosis-preventing substance (MPS) secreted by the morphologically sex-differentiated testis (Luciani *et al.*, 1977). Studies using organ cultures of gonads with and without the rete system and/or mesonophros (which almost certainly have a common origin) have also suggested the presence of a meiosis-inducing substance (MIS) (Byskov, 1974, 1975; Rivelis *et al.*, 1976). Byskov (1978a,b) has observed that around the time when the mouse gonads are being morphologically sex differentiated the rete system of both sexes appears to secrete MIS which triggers the germ cells to enter preleptotene chromosome condensation. Only those gonocytes of the developing testis which are left outside the testicular cords and placed close to or within the rete system will proceed to leptotene and zygotene stages. Gonocytes lying within the testicular cords are prevented from entering meiosis by MPS which may be secreted by the Sertoli cells. The enclosure of gonocytes within cords together with MPS prevents the action of MIS. O and Baker (1978) have recently discussed the influence of somatic cells affecting the differentiation and development of germ cells in gonadal differentiation. Both male and female germ cells can be induced to enter meiosis under the influence of the rete system. The results on artificial chimeric gonad have shown that XY germ cells can survive in the XX somatic environment and are induced to enter meiosis precociously, whereas germ cells fail to survive in an XY somatic environment.

Ayala *et al.* (1977) have studied the effects of polyunsaturated fatty acids of the a-linolenic series on the development of rat testicles. Arshan *et al.* (1978) have investigated the effect of gonadotrophins and LH-RH on functional differentiation of immature monkey testis. Hall-Martin *et al.* (1978) have observed that gonadal hypertrophy occurs in the late fetal life of giraffe, *Giraffe camelopardalis*. Spermatogenesis begins at 3 to 4 years of age and coincides with a rapid increase of testicular weight and seminiferous tubule diameter. Pelliniemi *et al.* (1977) have studied the differentiation of embryonic pig testis. Differentiation into testis occurs by the age of 26 days. Leydig cells first appear at the age of 30 days and attain ultrastructural maturity at the age of 34 days. Their differentiation from the undifferentiated interstitial cells is gradually accompanied by the accumulation of agranular endoplasmic reticulum. The amount of agranular endoplasmic reticulum is increased considerably in the Leydig cells of human embryos during the ninth week of development (Kellokumpu-Lehtinen *et al.*, 1979). Sholl and Goy (1978) have observed that fetal guinea pig testes contain the necessary enzymes for *de novo* androgen synthesis at a time during development when anatomical and behavioral phenotypes are differentiated. Gulyas *et al.* (1977) have demonstrated the dependence of fetal testis on endogenous secretions of the fetal pituitary for its normal development and growth in the rhesus monkey.

B. Leydig Cells

The demonstration of receptors for human chorionic gonadotrophin (HCG) and the positive effect of HCG on testosterone synthesis in human fetal testis (Huhtaniemi *et al.*, 1977), as well as histochemical localization of HCG in rat fetal Leydig cells (Childs *et al.*, 1978), confirm the role of HCG in the regulation of Leydig cell function. However, a thorough study of the mechanism of HCG action is required to determine more precisely whether HCG induces the synthesis of the androgen pathway enzymes or activates preexisting enzymes. It is widely accepted now that LH acts on the Leydig cell which is the only testicular cell that binds labeled LH specifically (Catt and Dufau, 1976). LH elicits a rapid increase in cAMP formation in Leydig cells, causing a rise in the synthesis of testosterone (Catt and Dufau, 1976).

From the results of various recent studies, it is becoming increasingly clear that Leydig cell function of developing and maturing testis may be influenced by several hormones (Van Beurden *et al.*, 1978). Besides the presence of LH receptors on Leydig cell membranes, it has also been shown that Leydig cell contains receptors for androgens (Sar *et al.*, 1975; Wilson and Smith, 1975), estrogens (Mulder *et al.*, 1974), glucocorticoids (Evain *et al.*, 1976), and prolactin (Charreau *et al.*, 1977). This suggests that these hormones may also influence the function of Leydig cells at least during certain phases of testicular development. Purvis and co-workers have studied the biological and biochemical characteristics of rat Leydig cells with special reference to the behavior of their receptors under different experimental conditions (Purvis *et al.*, 1976a,b, 1977b,c, 1978a–c). Their studies have shown that hormone interaction with the Leydig cell is extremely complex, not only because of the number of hormones which are potentially involved, but also because of the multiplicity of sites of action where these influences can be exerted. Purvis and co-workers have suggested that while interpreting the effects of hormones on Leydig cells, one must keep in mind the following: Leydig cell population is not functionally synchronized. Developmental alterations or hormone treatment may induce some qualitative changes in Leydig cell secretion (by changing metabolism) without affecting the total quantity of steroid hormone produced. While determining the integrity of the LH receptor–adenyl cyclase apparatus by a steroid response, a more nonspecific parameter of steroidogenic function than one specific component of secretion must be used. Hormones or drugs may exert an influence on the LH receptor population without modifying the maximum response of the Leydig cell to HCG/LH. The biology of Leydig cells with specific emphasis on their hormone receptors also forms the subject of several other recent studies (Hsueh *et al.*, 1976; Sharpe, 1976; Sinha *et al.*, 1978). Sinha *et al.* (1978) have suggested that the spare receptors play an important role in modulating Leydig cell sensitivity to gonadotrophins (LH and HCG) during sexual maturation. The influences of prolactin on Leydig cell function have also been indicated (Bartke, 1976; Purvis *et al.*, 1978c). Cyproterone and dihydrotestosterone propionate have no effect on the LH receptor population but decrease Leydig cell function their effect on the steroidogenic enzymes (Purvis *et al.*, 1978c).

Chakravarti *et al.* (1978) have investigated serum testosterone, Leydig cell population, and activities of marker enzymes during sexual maturation in the rat. Testosterone decreases to a minimum by Day 25 and thereafter increases progressively to the maximum level by Day 90. Serum LH does not show any significant differences at any age. But the relative proportion of Leydig cell per testis increases progressively throughout maturation. Concurrent to decrease in serum testosterone in early postnatal life, both 3β-HSDH and alcohol dehydrogenase (ADH) activities decrease. By Day 30, 3β-HSDH activity is comparable to that in adults. The increasing testosterone levels during pubertal development have been related to an increase in Leydig cell sensitivity.

C. Seminiferous Tubules

In agreement with the various observations discussed in Section VI,B of this review, Gilula *et al.* (1976) have observed that conspicuous structural changes occur in the Sertoli cell during the early periods of postnatal development. Bousquet *et al.* (1978) have studied the ultrastructure of the Sertoli cell in normal and dysgenesic rat testis during fetal and postnatal life. Meyer *et al.* (1977) have observed that Sertoli cells of the seminiferous tubules in immature rats (1–20 days) show gap and tight junctions in different stages of development as seen in freeze-fracture replicas. Tight junctions assume a variety of configurations including linear, macular, and extensive occluding complexes. A decrease in the frequency of gap junctions and a corresponding increase in the number of tight junctions have been demonstrated quantitatively. According to Ramaley (1979), the key features just prior to the initiation of mature spermatogenesis are the closure of the blood–testis barrier and the entrance of leptotene spermatocytes into the adluminal compartment of the seminiferous tubule at around Day 20, the initiation of mature androgen secretion, and the concomitant conditioning of the epididymis for support of "ex foliate" maturation of the sperm once they are released into the duct system around Day 40.

Alterations in ultrastructure similar to those occurring *in vivo* have also been demonstrated in Sertoli cells obtained from testes of 10-day-old rats and cultured for varying periods of time up to 2 weeks (Solari and Fritz, 1978). These ultrastructural changes include the formation of a large nucleolus within an irregularly shaped nucleus, the development of an extensive smooth endoplasmic reticulum, and the appearance of specialized tight junctions between cells. Freeze-fracture electron microscopic study has also revealed the presence of extensive junctional complexes in the cultured Sertoli cells (Fritz *et al.*, 1978). The maturation-like morphological changes can be easily studied in Sertoli cells cultured in the presence of FSH and or cAMP (Tung *et al.*, 1975; Fritz *et al.*, 1978). Fritz *et al.* (1978) have also suggested the appearance of unique antigenic determinant on the plasma membrane of Sertoli cells at a time corresponding approximately to the time of development of the blood–testis barrier, which may serve as a maturation marker for Sertoli cells to modulate interactions with germ cells.

Bressler (1978) using different experimental approaches has demonstrated that the pituitary gland is essential for normal development of the peritubular myoid cells and Sertoli cells. The ability of testosterone to partially replace the effects of the pituitary has shown that LH is involved by way of Leydig cell stimulation. A direct effect of testosterone on tubule development has been observed. Preliminary electron microscope studies have indicated that Sertoli cell junctions are more abundant in testes containing pellets of crystalline testosterone than in contralateral or sham-operated control testes. Daily injections of progestin

(Depo-Provera), alone or with FSH and LH, severely retard canalization of the seminiferous cords. Madhwa Raj and Dym (1976) have observed that seminiferous cords of rats treated with LH antiserum fail to develop a lumen. But similar treatment with FSH antiserum does not inhibit canalization of cords. These findings complement the observations of Bressler and co-workers (see references in Bressler, 1978) in that lumen formation does not occur under conditions of reduced testosterone levels but is unaffected by the absence of FSH. All these observations further support the suggestion that testosterone is the major factor in postnatal development of the seminiferous tubules in the mouse and rat. Sivashankar *et al.* (1977) using highly purified antiserum to FSH have shown that FSH is necessary for the maintenance of the cellular integrity of the seminiferous epithelium during completion of the first wave of spermatogenesis. Meiotic divisions are completed even in the absence of FSH. Ramaley (1979) has suggested that the closure of the Sertoli cell barrier and termination of Sertoli cell divisions and the cessation of Sertoli cell FSH responsiveness and estradiol synthesis are key events associated with the development of the mature rate of sperm production.

In agreement with various observations already discussed in Sections VI,B and VIII,G of this review, FSH binds specifically to Sertoli cells but not to Leydig cells (Means and Vaitukaitis, 1972; Steinberger *et al.*, 1974; Steinberger *et al.*, 1977). FSH increases cAMP production in Sertoli cells, but not in Leydig cell preparations (Dorrington and Fritz, 1974; Dorrington *et al.*, 1975; Steinberger *et al.*, 1977); LH does not elicit this effect. Neither LH nor FSH has been shown to act directly on germinal cells or on peritubular myoid cells. Fritz *et al.* (1978) have briefly reviewed the previous work on the biochemical responses of Sertoli cells to different hormones. They have also made some new observations on the properties of Sertoli cells during different stages of maturation in rats. A hypothesis presented by them is that somatic cells of the testis, appropriately stimulated by hormones, produce the microenvironment within the seminiferous tubule needed for the development of germ cells which do not respond directly to androgens or gonadotrophins.

FSH greatly increases several biochemical processes in cultures of Sertoli cells prepared from testes of immature rats ranging in age from 5 to 30 days (Fritz *et al.*, 1978). Thus observations on the stimulation by FSH of cAMP production (Dorrington and Fritz, 1974; Steinberger and Steinberger, 1976; Means *et al.*, 1976; Steinberger *et al.*, 1977), ABP synthesis (Fritz *et al.*, 1975, 1976a,b; Hansson *et al.*, 1975, 1976a,b; Rommerts *et al.*, 1978; Kotite *et al.*, 1978), 17β-estradiol synthesis from testosterone (Dorrington and Armstrong, 1975; Dorrington *et al.*, 1976a,b; Rommerts *et al.*, 1978), etc. by Sertoli cells during maturation have demonstrated greatest responsiveness in cells from immature rats. The addition of cAMP or other cAMP derivatives duplicates all of the late responsiveness in Sertoli cells elicited by FSH, and choleratoxin elicits similar effects (Fritz *et al.,* 1975, 1976a). These findings further support the suggestion that FSH actions on Sertoli cells are mediated by cAMP (see Section VII,C of this article).

The various biochemical responses either decrease or become nondetectable in Sertoli cells obtained from testes of rats 30 days old or older (Dorrington *et al.,* 1975, 1976a,b; Steinberger and Steinberger, 1976; Fritz *et al.,* 1978). Fritz *et al.* (1978) have suggested that during Sertoli cell maturation, complex and coordinated changes take place which reduce the capacity of the cell to increase cAMP production in response to FSH. FSH or androgens do not stimulate initially the production of ABP by cultured Sertoli cells prepared from 5-day-old rats (Fritz *et al.,* 1978). But both hormones increase its production during the third to seventh day of culture.

The various findings of Hansson and co-workers (Hansson *et al.,* 1975, 1976b) have provided further convincing evidence that FSH is important in stimulating Sertoli cell production and secretion of ABP in hypophysectomized rats, confirming that the Sertoli cell is the target for FSH in the testis (see also Kotite *et al.,* 1978). However, it is still to be determined more precisely whether additional target sites exist in other cell types (immature Leydig cells) of the testis or if all the testicular effects of FSH can be accounted for by the responses induced in the Sertoli cell.

In addition to FSH, androgens have also been shown to affect the metabolism of the Sertoli cell (Hansson *et al.,* 1974, 1976b) as its sensitivity to FSH, as measured by ABP response, decreases dramatically after hypophysectomy but can be maintained by treatment with high doses of testosterone propionate (Hansson *et al.,* 1974). These findings have provided an evidence that in prepubertal animals androgens greatly influence the sensitivity of the Sertoli cell to FSH. The secretory function of the Sertoli cell during this time period appears to be dependent on FSH and androgens. Kotite *et al.* (1978) have demonstrated, qualitatively, different effects of FSH and androgens on Sertoli cell function in the immature rat. Initiation of Sertoli cell function of ABP has an absolute dependence on FSH, just as does the initiation of spermatogenesis. It has been shown that testosterone is not essential for normal Sertoli cell function and spermatogenesis, as long as a sufficient amount of dihydrotestosterone is present in the testis (Purvis *et al.,* 1977a). Testosterone acts directly on the testis to stimulate Sertoli cell production of ABP (Elkington *et al.,* 1975; Means *et al.,* 1976; Louis and Fritz, 1977). Hansson *et al.* (1978) have suggested that when androgen treatment maintains spermatogenesis it also maintains Sertoli cell function. The fact that the hormonal requirement for normal Sertoli cell function is similar to that required for normal spermatogenesis emphasizes the significance of the Sertoli cell in the regulation of the spermatogenic process (Hansson *et al.,* 1976b).

Fritz (1977) in a recent review has discussed the sites of action of androgens and FSH on cells of the seminiferous tubules. He has included extensive data on the biology, biochemistry, and hormone responsiveness of Sertoli cells, which have permitted the conclusion that somatic cells of tubules are probably the exclusive targets in the testis for androgens. Possible means by which Sertoli cells could influence germinal cell development are actively being investigated in various laboratories. Steinberger and Steinberger (1976) have described the action of gonadotrophins in testis organ and cell culture. According to Hansson *et al.* (1978) the Sertoli cell is the main regulator of spermatogenesis and its function is regulated by both FSH and androgen. This conclusion is based on the correlation and discussion of recent data on the biology and biochemistry of Sertoli cells from various laboratories.

Louis and Fritz (1977) have observed that Sertoli cells are certainly under the influence of FSH and androgens. The presence of a receptor specific for androgens (including testosterone) has been demonstrated in Sertoli cells (Mulder *et al.*, 1975; Grootegoed *et al.*, 1977). But in the isolated germinal cells (spermatocytes and round spermatids) only a negligible amount of testosterone is bound to nuclear material. Rommerts *et al.* (1978) have demonstrated androgen receptors in Sertoli cells but not in germinal cells obtained from the same testicular tissue of rats 30 to 35 days of age. Isolated Sertoli cells from 21- to 24-day-old rats can be maintained in culture for more than 25 days and remain responsive to FSH and testosterone for more than 6 days. These cells when incubated with [^{3}H]leucine secrete, in addition to ABP, many different labeled proteins. The secretion of these proteins is stimulated in the presence of testosterone, FSH, or cAMP. Sertoli cells also appear to secrete an inhibin-like substance. The fetal Sertoli or sustentacular cells are also probably the source of anti-müllerian hormone (Josso *et al.*, 1977) as already discussed in Section VIII,G of this review.

Sanborn *et al.* (1978) have observed that Sertoli cells possess a cytoplasmic form of androgen receptor which is distinguishable from ABP. After their incubation with labeled hormones a nuclear form of receptor can be isolated. The chromatin from cultured Sertoli cells shows a limited number of high-affinity acceptor sites for the cytoplasmic androgen–receptor complex. These data have provided further evidence that Sertoli cells are primarily target cells for androgen action in the testis. From the discussion of all recent data, it can be concluded that most, if not all, the hormonal effects on spermatogenesis are mediated through the Sertoli cell. Whether or not androgens also act directly on the germ cells remains to be established. Ramaley (1979) has recently suggested that the conditioning of Leydig cell responsiveness to LH by a gradual process of prior stimulation by several pituitary hormones including FSH and GH may be responsible for the rising titer of testosterone secretion and the initiation of ABP synthesis as a marker for mature Sertoli cell function.

Steinberger (1978) and Tcholakian and Steinberger (1978) have recently reported that Sertoli cells prepared from testes of 80- and 36-day-old rats fail to aromatize appropriate precursors including testosterone to estrogens even in the presence of FSH. In fact, total metabolism of the testosterone is suppressed in the presence of FSH. A steroid metabolite chromatographed like estradiol is actually not estradiol. Sertoli cells are capable of converting progesterone to androgens and this conversion is small in comparison to that of interstitial cells. Sertoli cells also show considerable 5α-reductase activity. The peritubular cells show only marginal steroid metabolic activity. Wiebe and Tilbe (1978) have studied the effects of age and FSH on the capacity of Sertoli cells from immature rats to convert progesterone to 20α-hydroxypregn-4-en-3-one, 3α-hydroxy-5α-pregnan-20-one, and 5α-pregnan-3,20-dione. The peak steroidogenic activity and FSH sensitivity of Sertoli cells appear to be related to the onset of gametogenesis. Musto *et al.* (1978) have observed that rats with the H^{re} gene have a metabolic defect which results in a progressive Sertoli cell dysfunction. This produces an environment which is not suitable for germ cell development.

Clausen *et al.* (1978) have made micro-flow fluorometric (MFF) DNA measurements of isolated testicular cells from 17- to 150-day-old rats. The DNA distribution patterns consist of two to three peaks of fluorescence, the relative size of which changes with increasing age. Up to Day 25, two peaks of fluorescence were observed. Treatment with estradiol benzoate prevents the completion of meiosis in immature rats. From the results obtained, it has been concluded that MFF offers a practical and sensitive technique for investigating selected aspects of spermatogenesis, which involve changes in the DNA frequency distribution pattern of testicular cells. MFF may also be used to assess the purity of Leydig cell-enriched suspensions. Mills *et al.* (1977) have studied the histones of rat testis chromatin during early postnatal development and their interactions with DNA. This study has shown a tighter binding of the new histones to the DNA of meiotic cells. Kula (1977) has followed the completion of spermatogenic cells in the course of spermatogenesis in immature rats.

According to Mills and Means (1977) the metabolically active cells (supporting cells or immature Sertoli cells, spermatogonia, and primary spermatocytes) contain many high molecular weight nonhistone proteins, whereas less active spermatids show predominantly smaller species. The results of this study have suggested that the high molecular weight protein may represent enzymes or structural proteins needed for mitosis, meiosis, and cell growth. Vanha-Perttula (1978) has reviewed the results of recent studies on specific hydrolytic enzymes of the testis. He has placed special emphasis on their role in and association with spermatogenic maturation. Many of these enzymes are characterized as biochemically distinct from similar enzymes in other tissues as already discussed in Section VI,B of this review. They can be employed to demonstrate many special metabolic and physiologic features of the spermatogenic process. Sarkar *et al.* (1978), using gel electrophoresis, have studied the patterns of lactate dehydrogenase isoenzymes during gonadogenesis in the rat.

D. Hormonal Changes during Sexual Development and Maturation

Gross and Baker (1979) have made a study of developmental correlation between hypothalamic gonadotropin-releasing hormone (GnRH) and hypophysial LH in fetal mouse. The close temporal relationship between the developmental appearance of GnRH and its target cell, the gonadotrope, has provided further evidence that the potential for neuroendocrine control of gonadotropin secretion exists in the fetal mouse as early as 17 days of gestation. Corbier *et al.* (1978) have studied changes in testicular weight and serum gonadotrophin and testosterone levels before, during, and after birth in the perinatal rat. A sudden and transient increase of serum and pituitary gonadotrophins occurs at birth, which is followed by rapid increase of absolute and relative testicular weights between 2 and 12 hours ($P < 0.001$) and by a transient increase of serum testosterone between 0 hours *in utero* and 2 hours. Similarly, premature newborn rats obtained by cesarian delivery on Day 20 of gestation also show an increase in testicular weight between 0 and 6 hours and an increase in serum testosterone levels between 0 and 2 hours, with only a slight increase in serum LH. These data have suggested that the hypophyseotesticular axis of the rat is stimulated at the moment of birth. This transient testicular crisis occurring at birth has been suggested to affect the process of masculinization of the central nervous system of the rat.

Piacsek and Goodspeed (1978) have studied FSH, LH, prolactin, and testosterone concentrations during the maturation of the pituitary-gonadal system in the male rat. It has been suggested that spermatogenesis is not complete until FSH and testosterone reach maximum levels, whereas prolactin appears to be involved in the stimulation of accessory sex organ growth. The pronounced variation in serum LH concentrations during the maturation period seems to reflect a progressive change in the sensitivity of the hypothalamic-pituitary axis to the negative feedback effect of gonadal steroids. Moger (1977) has observed that hemiorchidectomy causes a disturbance of the hypothalamic-pituitary-testicular axis in the prepubertal male rat, resulting primarily in elevated serum FSH concentrations. But this disturbance has little effect on the endocrine changes associated with puberty in the male rat. Swanson and McCarthy (1978) studying estradiol treatment and LH response have concluded that the prepubertal heifer, similar to the mature cow, has a functional hypothalamic-pituitary-gonadal feedback system.

Hall-Martin *et al.* (1978) while studying the development and maturation of male reproductive organs in the giraffe have observed that in the fetal testis the main hormone is androstenedione but in the adult testis testosterone is predominant.

Robinson and Bridson (1978) have observed that testosterone concentrations are elevated in the sera during the first 14 postnatal weeks in male rhesus monkey and pigtail macaques. Castration of male rhesus monkey eliminated the postnatal elevation, indicating that the primary source of male neonatal androgens is the testis. This is further supported by the observation that serum testosterone concentrations in neonatal females were the same as levels observed in castrated males. According to Preslock and Steinberger (1977) the mature testis of the baboon *Papio anubic* formed more testosterone, androstenedione, and progesterone from pregnenolone than did the immature testis. The immature testis formed more 17α-hydroxyprogesterone and 20α-dihydroprogesterone from pregnenolone than did the mature testis. These results have indicated that mature baboon testis has greater C_{17}–C_{20} lyase, 17β-hydroxysteroid dehydrogenase, and 5α-reductase activities than immature testis, whereas immature testis possesses greater 20α-reductase activity.

Colenbrander and Van Straaten (1977) have studied serum LH levels in relation to Leydig cell development in the pig. Colenbrander *et al.* (1978) have studied changes in serum testosterone concentrations in the male pig during development. Testosterone concentrations are elevated between 40 and 60 days p.c. but become low between 60 and 100 days p.c. when the testis descends. Elevated concentrations occur in the perinatal period and from eighteenth week after birth. Between 60 days p.c. and 16 weeks after birth the alterations in serum testosterone concentrations parallel those of testicular development, as judged by morphology and steroid histochemistry. Kraulis *et al.* (1978) have shown the priming effect of androgens on gonadotrophin surges in the immature rat. Ketelslexgers *et al.* (1978) have observed that the early development of testicular FSH receptors is followed by a conspicuous rise in plasma FSH, with a concomitant increase in testicular growth and LH receptor concentration. The resulting increase in gonadal sensitivity to LH appears to be responsible for the notable increase in secretion of testosterone which occurs during puberty in the presence of a relatively small alteration in the circulating LH levels. The sequence of changes observed in gonadotrophins and their testicular receptors is consistent with the view that FSH-induced testicular sensitivity to LH is an important factor in sexual maturation in the male rat. Robert and Delost (1978) have concluded that the increase in testosterone concentrations during puberty in the male guinea pig is due to a rise in the production rate and not to a decrease of the clearance rate.

Acknowledgments

It gives me great pleasure to acknowledge the essential contribution of Dr. M. Courot, Station De Physiologie De La Reproduction, Nouzilly, France. The manuscript was reviewed critically and perceptively by him. I owe to him essential corrections, additions, and suggestions. Some literature supplied by WHO under the Small Supplies Program is also acknowledged.

References

Abramovich, D. R., and Rowe, P. (1973). *J. Endocrinol.* **56,** 621.

Abramovich, D. R., Baker, T. G., and Neal, P. (1974). *J. Endocrinol.* **60,** 179.

Acevedo, H. F., Axelrod, L. R., Ishikava, E., and Takaki, M. D. (1963). *J. Clin. Endocrinol. Metab.* **23,** 885.

Ahluwalia, B., Williams, J., and Verma, P. (1974). *Endocrinology* **95,** 1411.

Ahluwalia, B., Williams, J., and Verma, P. (1975). *J. Reprod. Fertil.* **44,** 131.

Allen, B. M. (1904). *Am. J. Anat.* **3,** 89.

Ancel, P., and Bouin, P. (1926). *C.R. Assoc. Anat.* **21,** 1.

Aoki, A. (1970). *Protoplasma* **71,** 209.

Arai, Y., and Serisawa, K. (1973). *Proc. Soc. Exp. Biol. Med.* **143,** 656.

Arai, Y., Serisawa, K., and Kigawa, A. (1975). *Acta Endocrinol.* **79,** 387.

Armstrong, D. T. (1968). *Recent Pro. Horm. Res.* **24,** 255.

Armstrong, D. T., and Dorrington, J. H. (1977). *In* "Regulatory Mechanisms Affecting Gonadal Hormone Action. Advances in Sex Hormone Research" (J. A. Thomas and R. L. Singhal, eds.), Vol. 3, pp. 217–258. Univ. Park Press, Baltimore, Maryland.

Armstrong, D. T., Miller, L. S., and Knudsen, K. A. (1969). *Endocrinology* **85,** 393.

Armstrong, D. T., Moon, Y. S., Fritz, I. B., and Dorrington, J. H. (1975). *In* "Hormonal Regulation of Spermatogenesis" (F. S. French, V. Hansson, E. M. Ritzen, and S. N. Nayfeh, eds.), pp. 85–96. Plenum, New York.

Attal, J. (1969). *Endocrinology* **85,** 280.

Attal, J. (1970). Thése Doct. Sci., University of Paris.

Attal, J., and Courot, M. (1963). *Ann. Biol. Anim. Biochim. Biophys.* **3,** 219.

Attal, J., Andre, D., and Engels, J. A. (1972). *J. Reprod. Fertil.* **28,** 207.

August, G. P., Grumbach, M. M., and Kaplan, S. L. (1972). *J. Clin. Endocrinol. Metab.* **34,** 319.

Baillie, A. H. (1960). *Q. J. Microsc. Sci.* **101,** 475.

Baillie, A. H. (1964). *J. Anat.* **98,** 641.

Baillie, A. H. (1965). *J. Anat.* **99,** 507.

Baillie, A. H., and Griffiths, K. (1964). *J. Endocrinol.* **29,** 9.

Baillie, A. H., Niemi, M., and Ikonen, M. (1965). *Acta Endocrinol.* **48,** 429.

Baillie, A. H., Ferguson, M. M., and Hart. D.McK. (1966a). "Developments in Steroid Histochemistry." Academic Press, New York.

Baillie, A. H., Ferguson, M. M., and Hart, D.McK. (1966b). *J. Clin. Endocrinol. Metab.* **26,** 738.

Bartke, A. (1971). *J. Reprod. Fertil.* **25,** 153.

Bartke, A., Weir, J. A., Mathison, P., Roberson, C., and Dalterio, S. (1974). *J. Hered.* **65,** 204.

Bearn, J. G. (1968). *Acta Paediatr. Acad. Sci. Hung.* **9,** 159.

Beatty, R. A. (1960). *Mem. Soc. Endocrinol.* **7,** 45.

Beatty, R. A. (1964). *In* "Intersexuality" (C. N. Armstrong and A. J. Armstrong, eds.), pp. 17–143. Academic Press, New York.

Beaumont, H. M. (1960). *Int. J. Radiat. Biol.* **2,** 247.

Beaumont, H. M., and Mandl, A. M. (1963). *J. Embryol. Exp. Morphol.* **11,** 715.

Bennett, W. I., Dufau, M. L., Catt, K. J., and Tullner, W. W. (1973). *Endocrinology* 92, 813.

Bergada, C., and Mancini, N. E. (1973). *J. Clin. Endocrinol. Metab.* **37,** 935.

Berger, M., Jean, Ch., De Turckham, M., Veyssiere, G., and Jean, Cl. (1974). *C.R. Soc. Biol. (Paris)* **168,** 130.

Berger, M., Jean-Faucher, Ch., De Turckheim, M., Veyssiere, G., and Jean, Cl. (1975). *Arch. Int. Physiol. Biochem.* **83,** 239.

Berger, M., Chazaud, J., Jean-Faucher, Ch., De Turckheim, M., Veyssiere, G., and Jean, Cl. (1976a). *Biol. Reprod.* **15,** 561.

Berger, M., Chazaud, J., Jean-Faucher, Ch., De Turckheim, M., Veyssiere, G., and Jean, Cl. (1976b). *J. Steroid Biochem.* **7,** 649.
Bishop, D. W. (1969). *In* "Reproduction and Sexual Behaviour" (M. Diamond, ed.), pp. 261–286. Indiana Univ. Press, Bloomington.
Bjerregaard, P., Bro-Rasmussen, F., and Reumert, T. (1974). *Z. Zellforsch.* **147,** 401.
Black, J. L., and Erickson, B. H. (1965). *Anat. Rec.* **161,** 45.
Black, V. H. (1971). *Am. J. Anat.* **131,** 415.
Black, V. H., and Christensen, A. K. (1969). *Am. J. Anat.* **124,** 211.
Blackburn, W. R., Chung, K. W., Bullock, L., and Bardin, C. W. (1973). *Biol. Reprod.* **9,** 9.
Blackshaw, A. W. (1970). *In* "The Testis" (A. D. Johnson, W. R. Gomes, and N. L. Vandemark, eds.), Vol. II, pp. 73–123. Academic Press, New York.
Blackshaw, A. W. (1973). *J. Reprod. Fertil. Suppl.* **18,** 55.
Blackshaw, A. W., and Elkington, J. S. H. (1970a). *J. Reprod. Fertil.* **22,** 69.
Blackshaw, A. W., and Elkington, J. S. H. (1970b). *Biol. Reprod.* **2,** 268.
Blandau, R. J. (1969). *Am. J. Obstet. Gynecol.* **104,** 310.
Blandau, R. J., White, B. J., and Rumery, R. E. (1963). *Fertil. Steril.* **14,** 482.
Blizzard, R. M., and Alberts, M. (1956). *J. Pediatr.* **48,** 782.
Bloch, E. (1964). *Endocrinology* **74,** 833.
Bloch, E. (1967). *Steroids* **9,** 415.
Bloch, E., Tissenbaum, B., Rubin, L., and Deane, H. W. (1962). *Endocrinology* **71,** 629.
Bloch, E., Lew, M., and Klein, M. (1971a). *Endocrinology* **89,** 16.
Bloch, E., Lew, M., and Klein, M. (1971b). *Endocrinology* **88,** 41.
Bloch, E., Gupta, C., Felman, S., and Vandamme, O. (1974). *In* "Sexual Endocrinology of the Perinatal Period" (M. G. Forest and J. Bertrand, eds.), p. 177. INSERM, Paris.
Boczkowski, K. (1971). *Clin. Genet.* **2,** 379.
Boczkowski, K. (1973). *Clin. Genet.* **4,** 213.
Boon, D. A., Keenan, R. E., Slaunwhite, W. R., and Aceto, T. (1972). *Pediat. Res.* **6,** 111.
Booth, W. D. (1975). *J. Reprod. Fertil.* **42,** 459.
Bouin, P., and Ancel, P. (1903). *C.R. Soc. Biol.* **55,** 1682.
Brambell, F. W. R. (1930). "The Development of Sex in Vertebrates." Sidgwick & Jackson, London.
Braun, T., and Sepsenwol, S. (1974). *Endocrinology* **94,** 1028.
Bressler, R. S. (1975). *Anat. Rec.* **181,** 524.
Bressler, R. S. (1976). *Am. J. Anat.* **147,** 447.
Bressler, R. S., and Lustbader, I. J. (1978). *Andrologia* **10,** 291.
Bressler, R. S., and Ross, M. H. (1972). *Biol. Reprod.* **6,** 148.
Brokelmann, J. (1963). *Z. Zellforsch.* **59,** 820.
Brown-Grant, K., Fink, G., Greig, R., and Murray, M. A. F. (1975). *J. Reprod. Fertil.* **44,** 25.
Burgos, M. H., and Fawcett, D. W. (1955). *J. Biophys. Biochem. Cytol.* **2,** 233.
Burgos, M. H., Vitale-Calpe, E., and Aoki, A. (1970). *In* "Testis" (A. D. Johnson, W. R. Gomes, and N. L. Vandermark, eds.), pp. 531–649. Academic Press, New York.
Bustos-Obregon, E., and Courot, M. (1974). *Cell Tissue Res.* **150,** 481.
Bustos-Obregon, E., and Holstein, A. F. (1973). *Z. Zellforsch.* **141,** 413.
Bustos-Obregon, E., Courot, M., Fléchon, J. E., Hochereau-de Reviers, M. T., and Holstein, A. F. (1975). *Andrologia* **7,** 141.
Byskov, A. G., and Saxen, L. (1976). *Dev. Biol.* **52,** 193.
Cameron, D. F., and Markwald, R. R. (1975). *In* "Hormonal Regulation of Spermatogenesis" (F. S. French, V. Hansson, E. M. Ritzen, and S. N. Nayfeh, eds.), pp. 479–494. Plenum, New York.
Carithers, J. R., and Green, J. A. (1972a). *J. Ultrastruct. Res.* **39,** 239.
Carithers, J. R., and Green, J. A. (1972b). *J. Ultrastruct. Res.* **39,** 251.

Catt, K. J., Dufau, M. L., Neaves, W. B., Walsh, P. C., and Wilson, J. D. (1975). *Endocrinology* **97,** 1157.
Cavallero, C., Magrini, V., Dellepiane, M., and Cizely, T. (1965). *Ann. Endocrinol. (Paris)* **26,** 409.
Challis, J. R. G., Kim, C. K., Naftalin, F., Judd, H. L., Yen, S. S. C., and Benirschke, K. (1974). *J. Endocrinol.* **60,** 107.
Charny, C. W., Conston, A. S., and Meranze, D. R. (1952). *Ann. N.Y. Acad. Sci.* **55,** 597.
Chemes, H. E., Rivarola, M. A., and Bergada, C. (1976a). *J. Reprod. Fertil.* **46,** 279.
Chemes, H. E., Rivarola, M. A., and Bergada, C. (1976b). *J. Reprod. Fertil.* **46,** 283.
Chieffi, G., Materazzi, G., and Botte, V. (1964). *Atti Sco. Peloritana Sci. Fis. Mat. Nat.* **10,** 515.
Chiquoine, A. D. (1954). *Anat. Rec.* **118,** 135.
Chowdhury, A. K., and Steinberger, E. (1975). *Biol. Reprod.* **12,** 609.
Christensen, A. K. (1970). *In* ''The Human Tesits'' (E. Rosenaberg, ed.), pp. 75–93. Plenum, New York.
Christensen, A. K. (1975). *In* ''Handbook of Physiology'' (R. O. Greep and E. B. Astwood, eds.), Vol. V, pp. 57–94. American Physiological Society, Washington, D.C.
Christensen, A. K., and Gillim, S. W. (1969). *In* ''The Gonads'' (K. W. McKerns, ed.), pp. 415–488. Appleton, New York.
Christensen, A. K., and Mason, N. R. (1965). *Endocrinology* **76,** 646.
Christiansen, R. O., and Desautel, M. (1973). *Endocrinology* **94,** A-100.
Chubb, C., Ewing, L., and Irby, D. (1975). *Proc. Ann. Meet. Soc. Study Reprod., 8th,* p. 44 (Abstract).
Clark, J. M., and Eddy, E. M. (1975). *Dev. Biol.* **47,** 136.
Clements, J. A., Reyes, F. I., Winter, J. S. D., and Faiman, C. (1976). *J. Clin. Endocrinol. Metab.* **42,** 9.
Clermont, Y. (1969). *Am. J. Anat.* **126,** 57.
Clermont, Y. (1972). *Physiol. Rev.* **52,** 198.
Clermont, Y., and Bustos-Obregon, E. (1968). *Am. J. Anat.* **122,** 237.
Clermont, Y., and Perey, B. (1957). *Am. J. Anat.* **100,** 241.
Coffey, J. C., French, F. S., and Nayfeh, S. N. (1971). *Endocrinology* **89,** 865.
Coniglio, J. G., Grogan, W. M., and Rhamy, R. K. (1975). *Biol. Reprod.* **12,** 255.
Connell, C. J. (1972). *Z. Zellforsch. Mikrosk. Anat.* **128,** 139.
Connel, C. J. (1977). *Am. J. Anat.* **148,** 149.
Cooke, A. A., deJong, F. H., van der Molen, H. J., and Rommerts, F. F. G. (1972). *Nature (London)* **237,** 255.
Cotta, Y., Terqui, M., Pelletier, J., and Courot, M. (1975). *C.R. Acad. Sci. (Paris)* **280** (Ser D), 1473.
Courot, M. (1962). *Ann. Biol. Anim. Biochim. Biophys.* **2,** 25.
Courot, M. (1963). *In* ''Proceedings of an International Symposium on the Effects of Ionizing Radiation on the Reproductive System'' (Colorado, U.S.A.), pp. 279–286. Pergamon, New York.
Courot, M. (1970). *In* ''The Human Testis'' (E. Rosenberg and C. A. Paulsen, eds.), pp. 355–364. Plenum, New York.
Courot, M. (1971). Thése Doct. Sci., University of Paris.
Courot, M. (1976). *Andrologia* **8,** 187.
Courot, M., Hochereau-de Reviers, M. T., and Ortavant, R. (1970). *In* ''The Testis'' (A. D. Johnson, W. R. Gomes, and N. L. Vandermark, eds.), Vol. 1, pp. 339–432. Academic Press, New York.
Courot, M., Ortavant, R., and de Reviers, M. (1971). *Exp. Anim.* **4,** 201.
Courot, M., Hochereau-de Reviers, M. T., Monet-Kuntz, C., Locatelli, A., Pisselet, C., Blanc, M. R., and Dacheux, J. L. (1978). *J. Reprod. Fertil.* (in press).

Crim, L. W., and Geschwind, I. (1972). *Biol. Reprod.* **7,** 42.
Cutuly, E., and Cutuly, E. C. (1940). *Endocrinology* **26,** 503.
Davidson, J. M. (1974). *In* ''Endocrinologie Sexuelle de la Période Périnatale'' (M. G. Forest and J. Bertrand, eds.), pp. 377–394. INSERM, Paris.
Dalcq, A. M. (1967). *C.R. Acad. Sci.* **264,** 2386.
Davies, A. G. (1971). *J. Reprod. Fertil.* **25,** 21.
Davies, A. G., Davies, W. E., and Summer, C. (1975). *J. Reprod. Fertil.* **42,** 415.
Debeljuk, L., Arimura, A., and Schally, A. U. (1972). *Endocrinology* **90,** 585.
De Krester, D. M., Burger, H. G., and Husdon, B. (1973). *J. Clin. Endocrinol. Metab.* **38,** 787.
De la Balze, F. A., Mancini, R. A., Arrillaga, F., Andrada, J. A., Vilar, O., Gurtman, A. I., and Davidson, O. W. (1960). *J. Clin. Endocrinol. Metab.* **20,** 266.
De Moor, P., Verhoeven, G., and Heyns, W. (1973). *Differentiation* **1,** 241.
De Rooij, D. G. (1970). *In* ''Morphological Aspects of Andrology'' (A. F. Holstein and E. Horstmann, eds.), Vol. 1, pp. 13–16. Grosse Verlag, Berlin.
Dierichs, R., and Wrobel, K. H. (1973). *Z. Anat. Entwickl. Gesch.* **143,** 49.
Dierichs, R., Wrobel, K. H., and Schilling, E. (1973). *Z. Zellforsch.* **143,** 207.
Diez D'Aux, R. C., and Murphy, B. E. P. (1974). *J. Steroid Biochem.* **5,** 207.
Donovan, B. T., Ter Haar, M. B., and Peddie, M. J. (1974). *In* ''Endocrinologie Sexuelle de la Période Périnatale'' (M. G. Forest and J. Bertrand, eds.), pp. 161–176. INSERM, Paris.
Donovan, B. T., Ter Haar, M. B., Lockhart, A. N., MacKinnon, P. C. B., Mattock, J. M., and Peddie, M. J. (1975). *J. Endocrinol.* **64,** 511.
Dorrington, J. H., and Fritz, I. B. (1973). *Biochem. Biophys. Res. Commun.* **54,** 1425.
Dorrington, J. H., and Fritz, I. B. (1974). *In* ''Gonadotropins and Gonadal Function'' (N. R. Moudgal, ed.), pp. 500–511. Academic Press, New York.
Dorrington, J. H., and Fritz, I. B. (1975). *Endocrinology* **96,** 879.
Dorrington, J. H., Roller, N. F., and Fritz, I. B. (1974). *In* ''Hormone Binding and Target Cell Activation in the Testis'' (M. L. Dufau and A. R. Means, eds.), pp. 237–242. Plenum, New York.
Dubois, R. (1964). *C.R. Acad. Sci.* **258,** 3904.
Dubois, R. (1965). *C.R. Acad. Sci.* **260,** 5885.
Dubois, R. (1966). *C.R. Acad. Sci.* **262,** 2625.
Dufau, M. L., and Means, A. R. (Eds.) (1974). ''Hormone Binding and Target Cell Activation in the Testis.'' Plenum, New York.
Dufour, J. J. (1967). M.Sc. Thesis, University of Guelph.
Dyche, W. J. (1975). *Anat. Rec.* **181,** 349.
Dym, M., and Cavicchia, J. (1977). *Biol. Reprod.* **17,** 390.
Dym, M., and Clermont, Y. (1969). *Anat. Rec.* **163,** 181 (Abstract).
Dym, M., and Fawcett, D. W. (1970). *Biol. Reprod.* **3,** 308.
Dym, M., and Fawcett, D. W. (1971). *Biol. Reprod.* **4,** 195.
Eddy, E. M. (1974). *Anat. Rec.* **178,** 731.
Eddy, E. M., and Clark, J. M. (1955). *In* ''Electron Microscopic Concepts of Secretion. The Ultrastructure of Endocrine and Reproductive Organs'' (M. Hess, ed.), pp. 151–168. Wiley, New York.
Eddy, E. M., and Kahri, A. I. (1976). *Anat. Rec.* **185,** 333.
Eguchi, Y., Sakamoto, Y., Arishima, K., Morikawa, Y., and Hashimoto, Y. (1975). *Endocrinology* **96,** 504.
Eguchi, Y., Arishima, K., Nasu, T., Toda, M., Morikawa, Y., and Hashimoto, Y. (1978). *Anat. Rec.* **190,** 679.
Eiger, W., Neuman, F., and von Berswordt-Wallrabe, R. (1971). ''Hormones in Development'' (M. Hamburgh and E. J. W. Barrington, eds.), pp. 651–667. Appleton, New York.
Elias, H. (1971). *Anat. Rec.* **169,** 310.

Elias, W. (1974). *Verh. Anat. Ges.* **68,** 123.
Elkington, J. S. H., Sanborn, B. M., and Steinberger, E. (1975). *Mol. Cell. Endocrinol.* **2,** 157.
Elsaesser, F., König, A., and Smidt, D. (1972). *Acta Endocrinol.* **69,** 553.
Engel, W., and Frowein, J. (1974). *Cell* **2,** 75.
Erickson, B. H., and Martin, P. G. (1972). *Int. J. Radiat. Biol.* **22,** 517.
Erickson, B. H., Reynolds, R. A., and Brooks, F. T. (1972). *Radiat. Res.* **50,** 388.
Essenberg, J. M., Horowitz, M., Davidson, S., and Ryder, V. L. (1955). *Anat. Rec.* **121,** 393.
Everett, N. B. (1943). *J. Exp. Zool.* **92,** 48.
Everett, N. B. (1945). *Biol. Rev.* **20,** 45.
Ewing, L., and Brown, B. L. (1977). *In* "The Testis" (A. D. Johnson and W. R. Gomes, eds.), Vol. 4, pp. 239–287. Academic Press, New York.
Fakundig, J. L., Tindall, D. J., Dedman, J. R., Mena, C. R., and Means, A. R. (1976). *Endocrinology* **98,** 392.
Falin, L. J. (1969). *Acta Anat.* **72,** 195.
Fawcett, D. W. (1975). *In* "Handbook of Physiology" (R. O. Greep and E. B. Astwood eds.), Vol. V, pp. 21–55. American Physiological Society, Washington, D.C.
Fawcett, D. W., Long, J. A., and Jones, A. L. (1969). *Recent Prog. Horm. Res.* **25,** 315.
Feldman, S. C., and Bloch, E. (1974). *Proc. Ann. Soc. Study Reprod. 7th,* p. 138.
Feldman, S. C., and Bloch, E. (1978). *Endocrinology* **102,** 999.
Ficher, M., and Steinberger, E. (1968). *Steroids* **12,** 491.
Ficher, M., and Steinberger, E. (1971). *Acta Endocrinol. (Kbh)* **68,** 285.
Flickinger, C. J. (1967). *Z. Zellforsch.* **78,** 92.
Flint, A. P. F., and Armstrong, D. T. (1972). *In* "Gonadotrophins" (B. B. Saxena, L. G. Beling, and H. M. Gandy, eds.), pp. 269–286. Wiley, New York.
Folman, Y., Ahmad, N., Sowell, J. G., and Eik-Nes, K. B. (1973). *Endocrinology* **92,** 41.
Forest, M. G. (1978). Abstracts. Symposium Lectures. Vth, Inter. Cong. Horm. Steroids, New Delhi, p. 70.
Forest, M. G., Cathiard, A. M., and Bertrand, J. A. (1973). *J. Clin. Endocrinol. Metab.* **36,** 1132.
Forest, M. G., Sizonenko, P., Cathiard, A. M., and Bertrand, J. (1974). *J. Clin. Invest.* **53,** 819.
Foster, D. L. (1974). *In* "Endocrinologie Sexuelle de la Periode Perinatale" (M. G. Forest and J. Bertrand, eds.), pp. 143–155. INSERM, Paris.
Foster, D. L., Cook, B., and Balbandov, A. V. (1972). *Biol. Reprod.* **6,** 253.
Franchi, L. L., and Mandl, A. M. (1964). *J. Embryol. Exp. Morphol.* **12,** 289.
Frasier, S. D., Gafford, F., and Horton, R. (1969). *J. Clin. Endocrinol. Metab.* **29,** 1404.
French, F. S., Hansson, V., Ritzen, E. M., and Nayfeh, S. N. (Eds.) (1975). "Hormonal Regulation of Spermatogenesis." Plenum, New York.
Fritz, I. B. (1973). *Curr. Top. Cell. Regul.* **7,** 129.
Fritz, I. B., Kopec, B., Lam, K., and Vernon, R. G. (1974). *In* "Hormone Binding and Target Cell Activation in the Testis" (M. L. Dufau and A. R. Means, eds.), pp. 311–327. Plenum, New York.
Fritz, I. B., Rommerts, F. F. G., Louis, B. G., and Dorrington, J. H. (1976). *J. Reprod. Fertil.* **46,** 17.
Frowein, J. (1973). *J. Endocrinol.* **57,** 437.
Frowein, J., and Engel, W. (1974). *Nature (London)* **249,** 377.
Frowein, J., and Engel, W. (1975). *J. Endocrinol.* **64,** 59.
Frowein, J., Engel, W., and Weise, H. C. (1973). *Nature (London)* **246,** 148.
Galena, H. J., Pillai, A. K., and Turner, C. (1974). *J. Endocrinol.* **63,** 223.
Garren, L. D., Gill, G. N., Masui, H., and Walton, G. M. (1971). *Recent Prog. Horm. Res.* **27,** 433.
George, F. W., Catt, K. J., Neaves, W. B., and Wilson, J. D. (1978). *Endocrinology* **102,** 665.

Gier, H. T., and Marion, G. B. (1970). *In* "The Testis" (A. D. Johnson, W. R. Gomes, and N. L. Vandemark, eds.), Vol. 1, pp. 1–45. Academic Press, New York.
Gillman, J. (1948). *Contr. Embryol. Carnegie Inst.* **32,** 81.
Goldman, B. D., and Gorski, R. A. (1971). *Endocrinology* **89,** 112.
Goldman, A. S., and Klingele, D. A. (1974). *Endocrinology* **94,** 1.
Goldman, A. S., Bongiovanni, A. M., and Yakovak, N. C. (1966). *Proc. Soc. Exp. Biol. Med.* **121,** 757.
Goldman, A. S., Baker, M. K., and Stanek, A. E. (1972). *Proc. Soc. Exp. Biol. Med.* **140,** 1486.
Goldstein, J. L., and Wilson, J. D. (1972). *J. Clin. Invest.* **51,** 1647.
Gondos, B. (1970). *In* "Gonadotrophins and Ovarian Development" (W. R. Butt, A. C. Crooke, and M. Ryle, eds.), pp. 239–248. Livingstone, Edinburgh.
Gondos, B. (1973). *Differentiation* **1,** 177.
Gondos, B. (1975). *Ann. Clin. Lab. Sci.* **5,** 4.
Gondos, B. (1977). *In* "The Testis" (A. D. Johnson and W. R. Gomes, eds.), Vol. 4, pp. 1–38. Academic Press, New York.
Gondos, B., and Conner, L. A. (1972). *Biol. Reprod.* **7,** 118 (abstract).
Gondos, B., and Conner, L. A. (1973). *Am. J. Anat.* **136,** 23.
Gondos, B., and Hobel, C. J. (1971). *Z. Zellforsch.* **119,** 1.
Gondos, B., and Hobel, C. J. (1973). *Endocrinology* **93,** 1973.
Gondos, B., and Zemjanis, R. (1970). *J. Morphol.* **131,** 431.
Gondos, B., Renston, R. N., and Conner, L. A. (1973). *Am. J. Anat.* **136,** 427.
Gondos, B., Paup, D. C., Ross, J., and Gorski, R. A. (1974). *Anat. Rec.* **178,** 551.
Gondos, B., Renston, R. H., and Goldstein, D. A. (1976). *Am. J. Anat.* **145,** 167.
Gondos, B., Morrison, K. P., and Renston, R. H. (1977). *Biol. Reprod.* **17,** 745.
Gonzalez-Angulo, A., Hernandez-Jaurequi, P., and Marquez-Monter, H. (1971). *Am. J. Vet. Res.* **32,** 1665.
Goomer, N., Saxena, R. N., and Sheth, A. R. (1978). *J. Steroid Biochem.* **9,** 879 (Abstr. 324).
Greep, R. O., Fevold, H. L., and Hisaw, F. L. (1936). *Anat. Rec.* **65,** 261.
Griswold, M. D., Mably, E. R., and Fritz, I. B. (1976). *Mol. Cell. Endocrinol.* **4,** 139.
Griswold, M. D., Solari, A., Tung, P. S., and Fritz, I. B. (1977). *Mol. Cell. Endocrinol.* **7,** 151.
Groom, G. V., Groom, M. A., Cooke, I. D., and Bayns, A. R. (1971). *J. Endocrinol.* **49,** 335.
Gropp, A., and Ohno, S. (1966). *Z. Zellforsch.* **74,** 505.
Grota, L. J. (1971). *Neuroendocrinology* **8,** 136.
Grumbach, M. M., and Kaplan, S. L. (1973). *In* "Foetal and Neonatal Physiology, Proc. Sir Joseph Barcroft Cent. Symp," p. 463, Cambridge Univ. Press, London and New York.
Grumbach, M. M., Roth, J. C., Kaplan, S. L., and Kelch, R. P. (1974). *In* "Control of the Onset of Puberty" (M. M. Grumbach, G. D. Grave, and F. E. Mayer, eds.), pp. 115–181. Wiley, New York.
Grünwald, P. (1934). *Z. Anat. Entwicki. Gesch.* **103,** 1.
Grünwald, P. (1942). *Am. J. Anat.* **70,** 359.
Gupta, D. (1978). Abstracts. Symposium Lectures Vth, Inter. Cong. Horm. Steroids, New Delhi, p. 75.
Guraya, S. S. (1968). *Acta Morphol. Neerl. Scand.* **7,** 51.
Guraya, S. S. (1971). *Physiol. Rev.* **51,** 785.
Guraya, S. S. (1973a). *Acta Endocrinol.* (Suppl. 161) **72,** 1.
Guraya, S. S. (1973b). *Proc. Acad. Sci. India* **39B,** 311.
Guraya, S. S. (1973c). *Ann. Biol. Anim. Biochim. Biophys.* **13,** 229.
Guraya, S. S. (1974a). *Cell Tissue Res.* **150,** 497.
Guraya, S. S. (1974b). *Acta Morphol. Acad. Sci. Hung.* **22,** 283.

Guraya, S. S. (1974c). *In* "Gonadotropins and Gonadal Function" (N. R. Moudgal, ed.), pp. 220–236. Academic Press, New York.
Guraya, S. S. (1974d). *Acta Anat.* **90,** 250.
Guraya, S. S. (1975a). *J. Reprod. Fertil.* **42,** 59.
Guraya, S. S. (1975b). *J. Reprod. Fertil.* **43,** 67.
Guraya, S. S. (1975c). *J. Reprod. Fertil.* **45,** 141.
Guraya, S. S. (1976a). *Int. Rev. Cytol.* **44,** 365.
Guraya, S. S. (1976b). *Int. Rev. Cytol.* **47,** 99.
Guraya, S. S. (1977a). *Int. Rev. Cytol.* **51,** 49.
Guraya, S. S. (1977b). *Arch. It. Anat. Embryol.* **82,** 21.
Guraya, S. S. (1978). *Int. Rev. Cytol.* **55,** 171.
Guraya, S. S., and Bilaspuri, G. S. (1976a). *Ann. Biol. Anim. Biochim. Biophys.* **16,** 137.
Guraya, S. S., and Bilaspuri, G. S. (1976b). *Gegenbaurs Morphol. Jahrb., Leipzig* **122,** 147.
Guraya, S. S., and Bilaspuri, G. S. (1976c). *Indian J. Anim. Sci.* **46,** 388.
Guraya, S. S., and Uppal, J. (1977). *Andrologia* **9,** 371.
Guraya, S. S., and Uppal, J. (1978). *Acta Morphol. Neerl. Scand.* **16,** 287.
Guraya, S. S., Stegner, H. E., and Pape, C. (1974). *Cytobiologie* **9,** 100.
Haffen, K. (1970). *Adv. Morphog.* **8,** 285.
Hagenas, L., Ritzen, E. M., Ploen, L., Hansson, V., French, F. S., and Nayfeh, S. N. (1975). *Mol. Cell. Endocrinol.* **2,** 39.
Hall, P. F. (1970). *In* "The Testis" (A. D. Johnson, W. R. Gomes, and N. L. Vandemark, eds.), Vol. II, pp. 1–71. Academic Press, New York.
Hall, P. F., Irby, D. C., and de Kretser, D. M. (1969). *Endocrinology* **84,** 488.
Hamerton, J. L. (1968). *Nature (London)* **219,** 910.
Hansson, V., Reusch, E., Trygstad, O., Torgerson, O., Ritzen, E. M., and French, F. S. (1973). *Nature (London) New Biol.* **246,** 56.
Hansson, V., Weddington, S. C., Naess, O., Attramadal, A., French, F. S., Kotite, N., Nayfeh, S. N., Ritzen, E. M., and Hagenas, L. (1975). *In* "Hormonal Regulation of Spermatogenesis" (F. S. French, V. Hansson, E. M. Ritzen, and S. N. Nayfeh, eds.), pp. 323–336. Plenum, New York.
Hart, D. McK., Baillie, A. H., Calman, K. C., and Ferguson, M. M. (1966a). *J. Anat.* **100,** 801.
Hart, D. McK., Claman, K. C., Ferguson, M. M., Niemi, M., and Baillie, A. H. (1966b). Hart, McK. Ph.D. Thesis, Glasgow University.
Hashimoto, I., and Suzuki, Y. (1966). *Endocrinol. Jpn.* **13,** 326.
Hasler, M. J., Falvo, R. E., and Nalbandov, A. V. (1975). *Gen. Comp. Endocrinol.* **25,** 36.
Hatakeyama, S. (1965). *Acta Pathol. Jpn.* **15,** 155.
Hayashi, H., and Harrison, R. G. (1971). *Fertil. Steril.* **22,** 351.
Hayashi, H., and Sandoval, G. J. (1976). *Rev. Bras. Pesqui Med. Biol.* **9,** 55.
Heys, F. (1931). *Q. Rev. Biol.* **6,** 1.
Hilliard, J., Spies, H. G., Lucas, L., and Sawyer, C. H. (1968). *Endocrinology* **82,** 122.
Hilscher, W. (1970). *Andrology* **1,** 17.
Hilscher, B., Hilscher, W., Delbruck, G., and Lerouge-Benard, B. (1972). *Z. Zellforsch.* **125,** 229.
Hilscher, B., Hilscher, W., Bülthoff-Ohnolz, B., Kamer, U., Birke, A., Pelzer, H., and Gauss, G. (1974). *Cell Tissue Res.* **154,** 443.
Hitzeman, J. W. (1962). *Anat. Rec.* **143,** 351.
Hochereau-de Reviers, M. T. (1970). Thése, Doct. Sci., University of Paris.
Hochereu-de Reviers, M. T., and Courot, M. (1976). *Proc. Int. Congr. Anim. Reprod. Artif. Insem.,* 8th, Cracow. Vol. 3, pp. 85–87.
Hochereu-de Reviers, M. T., and Courot, M. (1978). *Ann. Biol. Anim. Biochim. Biophys.* **18,** 573.
Hodgen, G. D. (1977). *In* "The Testis" (A. D. Johnson and W. R. Gomes, eds.), Vol. 4, pp. 401–423. Academic Press, New York.

Hodgen, G. D., and Sherins, R. J. (1973). *Endocrinology* **93,** 985.
Hollinger, M. A. (1970). *Life Sci.* **9,** 533.
Holstein, A. F., Wartenberg, H., and Vossmeyer, J. (1971). *Z. Anat. Entwickl. Gesch.* **135,** 43.
Hooker, C. W. (1944). *Am. J. Anat.* **74,** 1.
Hooker, C. W. (1948). *Rec. Prog. Horm. Res.* **3,** 173.
Hovatta, O. (1972). *Z. Zellforsch. Mikroskop. Anat.* **131,** 299.
Huckins, C., and Clermont, Y. (1968). *Arch. Anat. Hist. Embryol.* **51,** 343.
Huckins, C., Mills, N., Besch, P., and Means, A. R. (1973). *Endocrinology* (Suppl. 92), A-94.
Hughes, G. (1972). *Int. J. Radiat. Biol.* **4,** 511.
Huhtaniemi, I., and Vikho, R. (1970). *Steroids* **16,** 197.
Huhtaniemi, I., Ikonen, M., and Vikho, R. (1970). *Biochem. Biophys. Res. Commun.* **38,** 715.
Huhtaniemi, I. T., Korenbrot, C. C., Serón-Ferré, M., Foster, D. B., Paper, J. T., and Jaffe, R. B. (1977). *Endocrinology* **101,** 839.
Ichihara, I. (1970). *Z. Zellforsch. Mikroskop. Anat.* **108,** 475.
Ihrig, T. J., Renston, R. H., Renston, J. P., and Gondos, B. (1974). *J. Reprod. Fertil.* **39,** 105.
Ikonen, M. (1965). Ph.D. Thesis. University of Helsinki, Finland.
Ikonen, M., and Niemi, M. (1966). *Nature (London)* **212,** 716.
Inano, H., Hori, Y., and Tamaoki, B. (1967). *Ciba. Found. Colloq. Endocrinol.* **16,** 105.
Jaffe, R. B., and Payne, A. H. (1971). *J. Clin. Endocrinol. Metab.* **33,** 592.
Jean, Cl., André, M., Jean, Ch., Berger, M., De Turckheim, M., and Veyssiere, G. (1975). *J. Reprod. Fertil.* **44,** 235.
Jeon, K. W., and Kennedy, J. R. (1973). *Dev. Biol.* **31,** 275.
Jirasek, J. E. (1967). *In* "Endocrinology of the Testis" (G. E. W. Wolstenholme and M. O. Connor, eds.), pp. 3–30. Little, Brown, Boston, Massachusetts.
Jirasek, J. E. (1971). "Development of the Genital system and Male Pseudohermaphroditism." The Johns Hopkins Press, Baltimore, Maryland.
Johnston, P. M. (1951). *J. Morphol.* **88,** 471.
Josso, N. (1970). *C.R. Acad. Sci. (Paris)* **271,** 2149.
Josso, N. (1971). *J. Clin. Endocrinol. Metab.* **32,** 404.
Josso, N. (1972). *J. Clin. Endocrinol. Metab.* **34,** 265.
Josso, N. (1973). *Endocrinology* **93,** 829.
Josso, N. (1974). *Pediatr. Res.* **8,** 755.
Jost, A. (1947). *Arch. Anat. Microsc. Morphol.* **36,** 151.
Jost, A. (1953). *Recent Prog. Horm. Res.* **8,** 379.
Jost, A. (1960). *Harvey Lect.* **55,** 201.
Jost, A. (1965). *In* "Organogensis" (R. L. DeHaan and H. Ursprung, eds.), pp. 611–628. Holt, New York.
Jost, A. (1966). *In* "The Pituitary Gland" (G. W. Harris and B. T. Donovan, eds.), Vol. 2, p. 299. Butterworths, London.
Jost, A. (1970). *Philos. Trans. R. Soc. London (Biol. Sci.)* **259,** 119.
Jost, A. (1971). *In* "Hermaphroditism, Genital Anomalies and Related Endocrine Disorders" (H. Jones and W. Wallace, eds.), pp. 16–64. Williams & Wilkins, Baltimore, Maryland.
Jost, A. (1972a). *Arch. Anat. Microsc. Morphol. Exp.* **61,** 415.
Jost, A. (1972b). *Johns Hopkins Med. J.* **130,** 38.
Jost, A., Vigier, B., and Prepin, J. (1972). *J. Reprod. Fertil.* **29,** 349.
Jost, A., Vigier, B., Prepin, J., and Perchellet, J. P. (1973). *Recent Prog. Horm. Res.* **29,** 1.
Jost, A., Magre, S., and Cressent, M. (1974). *In* "Male Fertility and Sterility" (R. E. Mancini and L. Martini, eds.), pp. 1–11. Academic Press, New York.
Judd, H. L., Parker, D. C., Siler, T. M., and Yen, S. S. C. (1974). *J. Clin. Endocrinol. Metab.* **38,** 710.

Knorr, D. W., Vanha-Pertula, T., and Lipsett, M. B. (1970). *Endocrinology* **86,** 1289.
Kormano, M. (1967). *Histochemie* **9,** 327.
Kormano, M., and Havatta, O. (1972). *Z. Anat. Entwickl.* **137,** 239.
Krueger, P. M., Hodgen, G. D., and Sherins, R. J. (1974). *Endocrinology* **95,** 955.
Kuehl, F. A., Patanelli, D. J., Tarnoff, J., and Humes, J. L. (1970). *Biol. Reprod.* **2,** 154.
Lacroix, A., Garnier, D. H., and Pelletier, J. (1977). *Ann. Biol. Anim. Biochim. Biophys.* **17,** 1013.
Lacy, D. (1962). *Br. Med. Bull.* **18,** 205.
Lacy, D. (1974). *In* "Male Fertility and Sterility" (R. E. Mancini and L. Martini, eds.), pp. 175–205. Academic Press, New York.
Lamont, K. G., Perez-Palacios, G., Perez, A. E., and Jaffe, R. B. (1970). *Steroids* **16,** 127.
Lee, V. W. K., de Kretser, D. M., Hudson, B., and Wang, C. (1975). *J. Reprod. Fertil.* **42,** 121.
Leeson, C. R., and Leeson, T. S. (1963). *Anat. Rec.* **147,** 243.
Leeson, T. S., and Leeson, C. R. (1970). *J. Urol.* **8,** 127.
Levina, S. E. (1968). *Gen. Comp. Endocrinol.* **11,** 151.
Linder, H. R. (1961). *J. Endocrinol.* **23,** 161.
Lipsett, M. B., and Tullner, W. W. (1965). *Endocrinology* **77,** 273.
Lording, D. W., and de Kretser, D. M. (1970). *J. Anat.* **106,** 191.
Lording, D. W., and de Kretser, D. M. (1972). *J. Reprod. Fertil.* **29,** 261.
Lostroh, A. J. (1969). *Endocrinology* **85,** 438.
Lu, C., and Steinberger, A. (1977). *Biol. Reprod.* **17,** 84.
MacArthur, E., Short, R. V., and O'Donnell, V. J. (1967). *J. Endocrinol.* **38,** 331.
MacIndoe, J. H., and Turkington, R. W. (1973). *Endocrinology* **92,** 595.
McCann, S. M., Ojeda, S., and Negro-Vilar, A. (1974). *In* "The Control of the Onset of Puberty" (M. M. Grumbach, G. D. Grave, and F. E. Mayer, eds.), pp. 1–31. Wiley, New York.
McKay, D. G., Hertig, A. T., Adams, E. C., and Danzinger, S. (1953). *Anat. Rec.* **117,** 201.
McKay, D. G., Hertig, A. T., Adams, E. C., and Danzinger, S. (1955). *Anat. Rec.* **122,** 125.
McKinney, T. D., and Desjardins, C. (1973). *Biol. Reprod.* **9,** 279.
Majumder, G. C., Lessin, S., and Turkington, R. W. (1975). *Endocrinology* **96,** 890.
Males, J. L., and Turkington, R. W. (1970). *J. Biol. Chem.* **245,** 6329.
Males, J. L., and Turkington, R. W. (1971). *Endocrinology* **88,** 579.
Mancini, R. E. (1968). *In* "Testiculo Humano" (R. E. Mancini, ed.), pp. 11–45. Panamericana, Buenos Aires.
Mancini, R. E., Narbaitz, R., and Lavieri, J. C. (1960). *Anat. Rec.* **136,** 477.
Mancini, R. E., Vilar, O., Lavieri, J. C., Andrada, J. A., and Heinrich, J. J. (1963). *Am. J. Anat.* **112,** 263.
Mathur, R. S., Wiqvist, N., and Diczfalusy, E. (1972). *Acta Endocrinol.* **71,** 792.
Matsumoto, A. (1971). *J. Fac. Sci. U. Tokyo* **12,** 345.
Matsumoto, K., and Yamada, M. (1973). *Endocrinology* **93,** 253.
Mauger, A., and Clermont, Y. (1974). *Arch. Anat. Microsc. Morphol. Exp.* **63,** 133.
Mauléon, P. (1961). *Proc. Int. Congr. Anim. Reprod., 4th, The Hague* **2,** 348.
Mauleon, P. (1969). *In* "Reproduction in Domestic Animals" (H. H. Cole and P. T. Cupps, eds.), pp. 187–215. Academic Press, New York.
Mayer, D. B. (1964). *Dev. Biol.* **10,** 154.
Means, A. R. (1973). *Adv. Exp. Biol. Med.* **36,** 431.
Means, A. R. (1974a). *Life Sci.* **15,** 371.
Means, A. R. (1974b). *In* "Gonadotropins and Gonadal Function" (N. R. Moudgal, ed.), pp. 485–499. Academic Press, New York.
Means, A. R. (1977). *In* "The Testis" (A. D. Johnson and W. R. Gomes, eds.), Vol. 4, pp. 163–188. Academic Press, New York.
Means, A. R., and Hall, P. F. (1971). *Cytobios* **3,** 17.

Means, A. R., and Huckins, C. (1974). *In* "Hormone Binding and Target Cell Activation in the Testis" (M. L. Dufau and A. R. Means, eds.), pp. 145–166. Plenum, New York.
Means, A. R., and Vaitukaitis, J. (1971). *Endocrinology* **90,** 39.
Meistrich, M. L., and Trostle, P. K. (1975). *Exp. Cell. Res.* **92,** 231.
Merchant, H. (1975). *Dev. Biol.* **44,** 1.
Merchant, H. (1976). *Am. J. Anat.* **145,** 319.
Merchant, H., and Zamboni, L. (1973). *In* "The Development and Maturation of Ovary and Its Functions" (H. Peters, ed.), pp. 95–100. Excerpta Medica, Amsterdam.
Merkow, L., Acevedo, H., Slifkin, M., and Caito, B. J. (1968). *Am. J. Pathol.* **53,** 47.
Meusy-Dessolle, N. (1974). *C.R. Acad. Sci. (Paris)* **278,** 1257.
Milewich, L., George, F. W., and Wilson, J. D. (1977). *Endocrinology* **100,** 187.
Mills, N. C., and Means, A. R. (1972). *Endocrinology* **91,** 147.
Mills, C. N., Mills, T. M., and Means, A. R. (1977). *Biol. Reprod.* **17,** 124.
Mintz, B. (1957). *Anat. Rec.* **127,** 333.
Mintz, B. (1959). *Arch. Anat. Microsc. Morphol. Exp.* **48,** 155.
Mintz, B., and Russel, E. S. (1957). *J. Exp. Zool.* **134,** 207.
Miyachi, Y., Nieschlag, E., and Lipsett, M. B. (1973). *Endocronology* **92,** 1.
Mizuno, M., Lobotsky, J., Lloyd, C. W., Kobayashi, T., and Murusawa, Y. (1968). *J. Clin. Endocrinol. Metab.* **28,** 1133.
Moens, P. B., and Go, V. L. W. (1972). *Z. Zellforsch.* **127,** 201.
Moger, W. H., and Armstrong, D. T. (1974). *Can. J. Biochem.* **52,** 744.
Monesi, V. (1965). *Exp. Cell Res.* **39,** 197.
Monn, E., Desautel, M., and Christiansen, R. O. (1972). *Endocrinology* **91,** 716.
Monet-Kuntz, C., Terqui, M., Locatelli, A., Hochereau-de Reviers, M. T., and Courot, M. (1977). *C.R. Acad. Sci. (Paris)* **283** (Ser. D), 1763.
Montamat, E. E., and Blanco, A. (1976). *Exp. Cell Res.* **103,** 241.
Moon, Y. S., and Hardy, M. H. (1973). *Am. J. Anat.* **138,** 253.
Moon, Y. S., and Raeside, J. I. (1972). *Biol. Reprod.* **7,** 278.
Moon, Y. S., Hardy, M. H., and Raeside, J. I. (1973). *Biol. Reprod.* **9,** 330.
Moyle, W. R., and Ramachandran, J. (1973). *Endocrinology* **93,** 127.
Musto, N., Hafiez, A. A., and Bartke, A. (1972). *Endocrinology* **91,** 1106.
Nagy, F. (1972). *J. Reprod. Fertil.* **28,** 389.
Nakai, T., Kigawa, T., and Sakamoto, S. (1971). *Endocrinol. Jpn.* **18,** 353.
Narbaitz, R., and Adler, R. (1967). *Acta Physiol. Lat. Am.* **17,** 286.
Nayfeh, S. N., Barefoot, S. W., Jr., and Baggett, B. (1966). *Endocrinology* **78,** 1041.
Nayfeh, S. N., Coffey, J. C., Kotite, N. J., and French, F. S. (1975a). *In* "Hormonal Regulation of Spermatogenesis" (F. S. French, V. Hansson, E. M. Ritzen, and S. N. Nayfeh, eds.), pp. 53–64. Plenum, New York.
Nayfeh, S. N., Coffey, J. C., Hansson, V., and French, F. S. (1975b). *J. Steroid Biochem.* **6,** 329.
Neaves, W. B. (1975). *Contraception* **11,** 571.
Nebel, B. R., Amarose, A. P., and Hackett, E. M. (1961). *Science* **134,** 832.
Neill, J. D. (1973). *Endocrinology* **90,** 1154.
Nicander, L., and Ploen, L. (1969). *Z. Zellforsch.* **99,** 221.
Nicander, L., Abdel-Raouf, M., and Crabo, B. (1961). *Acta Morphol. Neerl. Scand.* **4,** 127.
Niemi, M., and Ikonen, M. (1961). *Nature (London)* **189,** 592.
Niemi, M., and Ikonen, M. (1962). *Endocrinology* **70,** 167.
Niemi, M., and Ikonen, M. (1963). *Endocrinology* **72,** 443.
Niemi, M., Ikonen, M., and Hervonen, A. (1967). "Ciba Foundation Colloquia on Endocrinology, Endocrinology of the Testis." Vol. 16, pp. 31–55. Churchill, London.
Nieschlag, E., and Bieniek, H. (1975). *Acta Endocrinol.* **79,** 375.

Nieuwkoop, P. D. (1949). *Experientia* **5,** 308.
Noumura, T., Weisz, J., and Lloyd, C. W. (1966). *Endocrinology* **78,** 245.
Novi, A. M., and Saba, P. (1968). *Z. Zellforsch.* **99,** 221.
O, W. S., and Baker, T. G. (1976). *J. Reprod. Fertil.* **48,** 399.
Odell, W. D., Swerdloff, R. S., Jacobs, H. S., and Hescox, M. H. (1973). *Endocrinology* **92,** 160.
Odell, W. D., Swerdloff, R. S., Bain, J., Wollesen, F., and Grover, P. K. (1974). *Endocrinology* **95,** 1380.
Ohno, Y. S. (1967). "Sex Chromosomes and Sex-linked Genes." Springer-Verlag, Berlin and New York.
Ortavant, R., Courot, M., and Hochereau-de Reviers, M. T. (1969). *In* "Reproduction in Domestic Animals" (H. H. Cole and P. T. Cupps, eds.), pp. 251–276. Academic Press, New York.
Ortavant, R., Courot, M., and Hochereau-de Reviers, M. T. (1972). *J. Reprod. Fertil.* **31,**, 451.
Ortavant, R., Courot, M., and Hochereau-de Reviers, M. T. (1977). *In* "Reproduction in Domestic Animals" (H. H. Cole and P. T. Cupps, eds.), 3rd ed., pp. 203–227. Academic Press, New York.
Ortiz, E., Price, D., and Zaaijer, J. (1966). *K. Ned. Akad. Wet. Proc. Ser. C.* **69,** 400.
Ortiz, E., Zaaijer, J. J., and Price, D. (1967). *K. Ned. Akad. Wet. Proc. Ser. C***3,** 389.
Ozdzenski, W. (1972). *Arch. Anat. Microsc. Morphol. Exp.* **61,** 267.
Pahnke, V. G., Leidenberger, F. A., and Kunzig, H. J. (1975). *Acta Endocrinol.* **79,** 610.
Parlow, A. F., Coyotupa, J., and Kovacic, N. (1973). *J. Reprod. Fertil.* **32,** 163.
Pasteels, J. L., Gausset, P., Danguy, A., and Ectors, F. (1974). *In* "Endocrinologie Sexuelle de la Periode Perinatale" (M. G. Forest and J. Bertrand, eds.), pp. 13–35. INSERM, Paris.
Payne, A. H., and Jaffe, R. B. (1975). *J. Clin. Endocrinol. Metab.* **40,** 102.
Payne, A. H., Kelch, R. P., Murono, E. P., and Kerlan, J. T. (1977). *J. Endocrinol.* **72,** 17.
Pearse, A. G. E. (1968). "Histochemistry." Churchill, London.
Pelliniemi, L. J. (1970). *In* "Morphological Aspects of Andrology" (A. F. Holstein and E. Horstmann, eds.), pp. 5–8. Grosse Verlag, Berlin.
Pelliniemi, L. J. (1975a). *Anat. Embryol.* **147,** 19.
Pelliniemi, L. J. (1975b). *Am. J. Anat.* **144,** 89.
Pelliniemi, L. J. (1975c). Academic Dissertation Medical Faculty of the University of Turku. Turku, Finland.
Pelliniemi, L. J. (1976a). *In* "International Congress on Cell Biology, 1st, Sept. 5–10." Boston, Mass. (Abstract).
Pelliniemi, L. J. (1976b). *Tissue Cell* **8,** 163.
Pelliniemi, L. J., and Niemi, M. (1969). *Z. Zellforsch.* **99,** 507.
Picon, R. (1970). *C.R. Acad. Sci. (Paris)* **271,** 2370.
Picon, R. (1971). *C.R. Acad. Sci. (Paris)* **272,** 98.
Picon, R. (1976). *J. Endocrinol.* **71,** 231.
Picon, R., and Ottowicz, J. (1967). *Arch. Anat. Microsc. Morphol. Exp.* **56,** 1.
Pinkerton, J. H. M., McKay, D. C., Adams, E. C., and Hertig, A. T. (1961). *Obstet. Gynecol.* **18,** 152.
Podesta, E. J., and Rivarola, M. A. (1974). *Endocrinology* **95,** 455.
Pointis, G., Mahoudeau, J. A., and Cedard, L. (1978). *J. Steroid Biochem.* **9,** 874 (Abstr. 300).
Politzer, G. (1933). *Z. Anat. Entwickl. Gesch.* **100,** 331.
Prasad, M. R. N. (1974). *In* "Gonadotropins and Gonadal Function" (N. R. Moudgal, ed.), pp. 199–204. Academic Press, New York.
Price, D., and Ortiz, E. (1965). *In* "Organogenesis" (R. L. De Hann and H. Ursprung, eds.), pp. 629–652. Holt, New York.
Price, D., Ortiz, E., and Deane, H. W. (1964). *Am. Zool.* **4,** 327 (Abstract).
Price, D., Ortiz, E., and Zaaijer, J. J. P. (1967). *Anat. Rec.* **157,** 27.
Price, D., Zaaijer, J. J. P., and Ortiz, E. (1969). *K. Ned. Akad. Wet. Proc. Ser. C* **72,** 370.

Price, D., Zaaijer, J. J. P., Ortiz, E., and Brinkmann, A. O. (1975). *Am. Zool.* **15,** 173.
Raeside, J. I. (1976). *J. Reprod. Fertil.* **46,** 423.
Rajaniemi, J. H., and Niemi, M. (1974). *Steroids Lipids Res.* **5,** 297.
Ramirez, V. D., and McCann, S. M. (1965). *Endocrinology* **76,** 412.
Rawlings, N. C., Hafs, H. D., and Swanson, L. V. (1972). *J. Anim. Sci.* **34,** 435.
Reiter, E. O., Fuldauer, V. G., and Root, A. W. (1976). *Steroids* **28,** 829.
Renston, R. H., Ihrig, T. J., Renston, J. P., and Gondos, B. (1975). *J. Reprod. Fertil.* **43,** 91.
Resko, J. A. (1970a). *Endocrinology* **87,** 680.
Resko, J. A. (1970b). *Endocrinology* **86,** 1444.
Resko, J. A. (1977). *In* "Regulatory Mechanisms Affecting Gonadal Hormone Action. Advances in Sex Hormone Research" (J. A. Thomas and R. L. Singhal, eds.), Vol. 3, pp. 139–168. Univ. Park Press, Baltimore, Maryland.
Resko, J. A., Feder, H. H., and Goy, R. W. (1968). *J. Endocrinol.* **40,** 485.
Resko, J. A., Malley, A., Begley, D., and Hess, D. L. (1973). *Endocrinology* **93,** 156.
Reyes, F. L., Winter, J. S. D., and Faiman, C. (1973). *J. Clin. Endocrinol. Metab.* **37,** 74.
Reyes, F. I., Boroditsky, R. S., Winter, J. S. D., and Faiman, C. (1974). *J. Clin. Endocrinol. Metab.* **38,** 612.
Reynaud, A. (1950). *Arch. Anat. Microsc. Morphol. Exp.* **39,** 518.
Reynaud, A., and Frilley, M. (1947). *Ann. Endocrinol. (Paris)* **8,** 400.
Rice, B. F., Johanson, C. A., and Sternberg, W. H. (1966). *Steroids* **1,** 79.
Rigaudiere, N. (1973). *C.R. Soc. Biol. (Paris)* **167,** 968.
Ritzen, E. M., Hagenas, L., Hansson, V., and French, F. S. (1975). *In* "Hormonal Regulation of Spermatogenesis" (F. S. French, V. Hansson, E. M. Ritzen, and S. N. Nayfeh, eds.), pp. 353–366. Plenum, New York.
Rivarola, M. A., Podesta, E. J., and Chemes, H. E. E. (1972). *Endocrinology* **91,** 537.
Roosen-Runge, E. C. (1973). *J. Reprod. Fertil.* **35,** 339.
Roosen-Runge, E. C., and Anderson, D. (1959). *Acta Anat.* **37,** 125.
Ross, M. H. (1967). *Am. J. Anat.* **121,** 525.
Ross, M. H. (1976). *Anat. Rec.* **186,** 79.
Russel, L. (1977a). *Am. J. Anat.* **148,** 301.
Russell, L. (1977b). *Am. J. Anat.* **148,** 313.
Russell, L. (1978). *Anat. Rec.* **190,** 99.
Russo, J. (1970). *Z. Zellforsch.* **104,** 14.
Russo, J., and de Rosas, J. C. (1971). *Am. J. Anat.* **130,** 461.
Sanborn, B. M., Elkington, J. S. H., Chowdhury, M., Tcholakian, R. K., and Steinberger, E. (1975a). *Endocrinology* **96,** 304.
Sanborn, B. M., Elkington, J. S. H., Steinberger, A., Steinberger, E., and Meistrich, M. L. (1975b). *In* "Hormonal Regulation of Spermatogenesis" (F. S. French, V. Hansson, E. M. Ritzen, and S. N. Nayfeh, eds.), pp. 293–309. Plenum, New York.
Sanborn, B. M., Elkington, J. S. H., Steinberger, A., Steinberger, E., and Meistrich, M. L. (1976). *In* "Regulatory Mechanisms of Male Reproductive Physiology" (C. H. Spilman, T. J., Lobel, and K. T. Kirton, eds.), pp. 45–58. Elsevier, Amsterdam.
Sanyal, M. K., and Villee, C. A. (1977). *Biol. Reprod.* **16,** 174.
Sapsford, C. S. (1962). *Austr. J. Zool.* **10,** 178.
Sapsford, C. S. (1964). *Austr. J. Zool.* **12,** 127.
Satyaswaroop, P. G., and Gurpide, E. (1975). *In* "Hormonal Regulation of Spermatogenesis" (F. S. French, V. Hansson, E. M. Ritzen, and S. N. Nayfeh, eds.), pp. 165–180. Plenum, New York.
Secchiari, P., Martorana, F., Pellegrini, S., and Luisi, M. (1976). *J. Anim. Sci.* **42,** 405.
Schlegel, R. J., Farias, E., Russo, N. C., Moore, J. R., and Gardner, L. I. (1967). *Endocrinology* **81,** 565.

Seguchi, H., and Hadziselimovic, F. (1974). *Verh. Anat. Ges.* **68,** 133.
Serra, G. B., Pérez-Palacios, G., and Jaffe, R. B. (1970). *J. Clin. Endocrinol. Metab.* **30,** 128.
Setchell, R. V., Davies, R. V., and Main, S. J. (1977). "The Testis" (A. D. Johnson and W. R. Gomes, eds.), Vol. 4, pp. 190–238. Academic Press, New York.
Sholl, S. A. (1974). *Steroids* **24,** 703.
Short, R. V. (1972). *In* "Proc. of International Symposium on the Genetics of the Spermatozoon" (R. A. Beatty and S. Gluecksohn-Waelsch, eds.), pp. 325–345. Dept. of Genetics of the University of Edinburgh and of the Albert Einstein College of Medicine, New York.
Siiteri, P. K., and Wilson, J. D. (1974). *J. Clin. Endocrinol. Metab.* **38,** 113.
Simpson, M. E., and Evans, H. M. (1946). *Endocrinology* **39,** 281.
Solomon, S. (1966). *J. Clin. Endocrinol. Metab.* **20,** 762.
Sowell, J. G., Folman, Y., and Eik-Nes, K. B. (1974). *Endocrinology* **94,** 346.
Spiegelman, M., and Bennet, D. (1973). *J. Embryol. Exp. Morphol.* **30,** 97.
Steinberger, E. (1971). *Physiol. Rev.* **51,** 1.
Steinberger, E., and Duckett, G. E. (1965). *Endocrinology* **76,** 1184.
Steinberger, E., and Ficher, M. (1968). *Steroids* **11,** 351.
Steinberger, A., and Steinberger, E. (1967). *J. Reprod. Fertil., Suppl.* **2,** 117.
Steinberger, A., and Steinberger, E. (1971). *Biol. Reprod.* **4,** 84.
Steinberger, A., and Steinberger, E. (1977). *In* "The Testis" (A. D. Johnson and W. R. Gomes, eds.), Vol. 4, pp. 371–399. Academic Press, New York.
Steinberger, E., and Steinberger, A. (1969). *In* "The Gonads" (K. W. McKerns, ed.), pp. 715–737. Appleton, New York.
Steinberger, E., and Steinberger, A. (1972). *In* "Regulation of Organ and Tissue Growth" (R. J. Gess, ed.), pp. 299–314. Academic Press, New York.
Steinberger, E., and Steinberger, A. (1975). *In* "Handbook of Physiology" (R. O. Greep and E. B. Astwood, eds.), Section 7, Vol. V, pp. 1–20. American Physiological Society, Washington, D.C.
Steinberger, E., Ficher, M., and Steinberger, A. (1970). *In* "The Human Testis" (E. Rosenberg and C. A. Paulsen, eds.), pp. 333–352, Plenum, New York.
Steinberger, E., Root, A., Ficher, M., and Smith, K. D. (1973). *J. Clin. Endocrinol. Metab.* **37,** 746.
Steinberger, A., Elkington, J. S. H., Sanborn, B. M., Steinberger, E., Heindel, J. J., and Lindsey, J. N. (1975a). *In* "Hormonal Regulation of Spermatogenesis" (F. S. French, V. Hansson, E. M. Ritzen, and S. N. Nayfeh eds.), pp. 399–411. Plenum, New York.
Steinberger, A., Lindsey, J. N., Heindel, J. J., Sanborn, B., and Elkington, J. S. H. (1975b). *Endocr. Res. Commun.* **2,** 261.
Stoll, R., Laffiton, L., and Moraud, R. (1973). *C.R. Soc. Biol.* **167,** 1092.
Strott, C. A., Sundel, H., and Stahlman, M. T. (1974). *Endocrinology* **65,** 1327.
Swerdloff, R. S., Walsh, P. C., Jacobs, H. S., and Odell, W. D. (1971). *Endocrinology* **88,** 120.
Swerdloff, R. S., Jacobs, H. S., and Odell, W. D. (1972). *In* "Gonadotropins" (B. B. Saxena, C. G. Beling, and H. M. Gandy, eds.), pp. 546–566. Wiley, New York.
Swierstra, E. E., and Foote, R. H. (1965). *Am. J. Anat.* **116,** 401.
Tamaoki, B., Inano, H., and Nakano, H. (1969). *In* "The Gonads" (K. W. McKerns, ed.), pp. 547–614. Appleton, New York.
Taylor, T., Coutts, J. R. T., and MacNaughton, M. C. (1974). *J. Endocrinol.* **60,** 321.
Tarkowski, A. K. (1970). *Philos. Trans. B* **259,** 107.
Tindall, D. J., Shrader, W. T., and Means, A. R. (1974). *In* "Hormone Binding and Target Cell Activation in Testis" (M. L. Dufau and A. R. Means, eds.), pp. 167–175. Plenum, New York.
Tindall, D. J., Vitale, R., and Means, A. R. (1975). *Endocrinology* **97,** 636.
Tindall, D. J., Cunningham, G. R., and Means, A. R. (1978). *J. Biol. Chem.* **253,** 166.
Tonutti, E. (1954). *Sem. Hop. (Paris)* **30,** 2135.

Tramontana, S., Botte, V., and Chieffi, G. (1967). *Arch. Ostetr. Ginecol.* **72,** 434.
Tsang, W. N., Collins, P. M., and Lacy, D. (1973). *J. Reprod. Fertil.* **34,** 513.
Tseng, M. T., Alexander, N. J., and Kittinger, G. W. (1975). *Am. J. Anat.* **143,** 349.
Tsujimura, T., and Matsumoto, K. (1974). *Endocrinology* **94,** 288.
Tung, P. S., and Fritz, I. B. (1975). *In* "Hormonal Regulation of Spermatogenesis" (F. S. French, V. Hansson, E. M. Ritzen, and S. N. Nayfeh, eds.), pp. 495–508. Plenum, New York.
Tung, P. S., Dorrington, J. H., and Fritz, I. B. (1975). *Proc. Natl. Acad. Sci.* **75,** 1838.
Turolla, E., Magrini, V., and Arcari, G. (1965). *Folia Endocrinol.* **4,** 1.
Van Beurden, W. M. O., Roodnat, B., De Jong, F. H., Mulder, E., and van der Molen, H. J. (1976). *Steroids* **28,** 8470.
Van Der Molen, H. J., Grootegoed, J. A., De Greef-Bijleveld, M. J., Rommerts, F. F. G., and Van der Vusse, G. J. (1975). *In* "Hormonal Regulation of Spermatogenesis" (F. S. French, V. Hansson, E. M. Ritzen, and S. N. Nayfeh, eds.), pp. 3–24. Plenum, New York.
Van Straeten, H. W. M., and Wensing, C. J. G. (1978). *Biol. Reprod.* **18,** 86.
Van Wagenen, G., and Simpson, M. E. (1965). "Embryology of the Ovary and Testis. *Homo sapiens* and *Macaca mulatta.*" Yale Univ. Press, New Haven, Connecticut.
Vanha-Perttula, T. (1973). *J. Reprod. Fertil.* **32,** 45.
Vanha-Perttula, T., and Nikkanen, V. (1973). *Acta Endocrinol.* **72,** 376.
Vermeulen, A. (1973). *In* "The Endocrine Function of the Human Testis" (V. H. T. James, M. Serio, and L. Martini, eds.), pp. 145–160. Academic Press, New York.
Vernon, R. G., Go, V. L. V., and Fritz, I. B. (1971). *Can. J. Biochem.* **49,** 761.
Vernon, R. G., Kopec, B., and Fritz, I. B. (1974). *Mol. Cell Endocrinol.* **1,** 167.
Veyssiere, G., Berger, M., Jean-Faucher, Ch., De Turckheim, M., and Jean, Cl. (1975). *Arch. Int. Physiol. Biochim.* **83,** 667.
Veyssiere, G., Berger, M., Jean-Faucher, Ch., De Turckheim, M., and Jean, Cl. (1976). *Endocrinology* **99,** 1263.
Vilar, O. (1970). *In* "The Human Testis" (E. Rosenberg and C. A. Paulsen, eds.), pp. 95–111. Plenum, New York.
Vilar, O. (1973). *In* "The Endocrine Function of the Human Testis" (V. H. T. James, M. Serio, and L. Martini, eds.), pp. 257–268. Academic Press, New York.
Villee, D. (1974). *In* "Endocrinologie Sexuelle de la Periode Perinatale" (M. G. Forest and J. Bertrand, eds.), pp. 247–254. INSERM, Paris.
Vitale, R. (1975). *Anat. Rec.* **181,** 501.
Vitale, R., Fawcett, D. W., and Dym, M. (1973). *Anat. Rec.* **176,** 333.
Vossmeyer, J. (1971). *Z. Anat. Entwickl. Gesch.* **134,** 146.
Warren, D. W., Haltmeyer, G. C., and Eik-Nes, K. B. (1972). *Biol. Reprod.* **7,** 94.
Warren, D. W., Haltmeyer, G. C., and Eik-Nes, K. B. (1973). *Biol. Reprod.* **8,** 560.
Warren, D. W., Haltmeyer, G. C., and Eik-Nes, K. B. (1975). *Endocrinology* **96,** 1226.
Wartenberg, H. (1974). *Verh. Anat. Ges.* **68,** 63.
Wartenberg, H., Holstein, A. F., and Vossmeyer, J. (1971). *Z. Anat. Entwickl. Gesch.* **134,** 165.
Wells, L. J. (1950). *Arch. Anat. Microsc. Morphol. Exp.* **39,** 499.
Wells, L. J. (1962). *In* "The Ovary" (S. Zuckerman, ed.), Vol. 2, pp. 131–154. Academic Press, New York.
Weniger, J. P., and Zeis, A. (1972). *C.R. Acad. Sci.* **275,** 1431.
Weniger, J. P., and Zeis, A. (1975). *Arch. Anat. Microsc. Morphol. Exp.* **64,** 61.
Weniger, J. P., Ehrhardt, J. D., and Fritig, B. (1967). *C.R. Acad. Sci. (Paris)* **262,** 1911.
Whitehead, R. H. (1904). *Am. J. Anat.* **3,** 167.
Wiebe, J. P. (1976). *Endocrinology* **98,** 505.
Wiebe, J. P. (1978). *Endocrinology* **102,** 775.
Wilson, J. D., and Siiteri, P. K. (1972). *Exc. Med. Int. Congr. Ser.* **273,** 1051.

Wilson, J. D., and Siiteri, P. K. (1973). *Endocrinology* **92,** 1182.
Winter, J. S. D., and Faiman, C. (1972). *Pediat. Res.* **6,** 126.
Witschi, E. (1948). *Contrib. Embryol. Carnegie Inst.* **32,** 67.
Witschi, E. (1951). *Recent Prog. Horm. Res.* **6,** 1.
Witschi, E. (1962). *In* "The Ovary" (Int. Acad. Pathol. Monograph), No. 3, pp. 1–23. Williams & Wilkins, Baltimore, Maryland.
Wolfe, H. J., and Cohen, R. B. (1964). *J. Clin. Endocrinol. Metab.* **24,** 616.
Wolff, E. (1962). *In* "The Ovary" (S. Zuckerman, ed.), Vol. 2, pp. 81–130. Academic Press, New York.
Woods, M. C., and Simpson, M. E. (1961). *Endocrinology* **69,** 91.
Xama, M., and Turkington, R. W. (1972). *Endocrinology,* **91,** 415.
Yamada, M., and Matsumoto, K. (1974). *Endocrinology* **94,** 77.
Yamada, M., Yasue, S., and Matsumoto, K. (1972). *Acta Endocrinol.* **71,** 393.
Yamada, M., Yasue, S., and Matsumoto, K. (1973). *Endocrinology* **93,** 81.
Yoshizaki, K., Matsumoto, K., and Samuels, L. T. (1978). *Endocrinology* **102,** 978.
Zamboni, L., and Merchant, H. (1973). *Am. J. Anat.* **137,** 299.
Zarrow, M. X., and Clark, J. H. (1969). *Endocrinology* **84,** 340.

References to Addendum

Arshan, M., Zaidi, A. A., and Qazi, M. H. (1978). *Int. J. Androl.* Suppl. 2, 319.
Ayala, S. S., Brenner, R. R., and Dumm, C. G. (1977). *Lipids* **12,** 1017.
Bartke, A. (1976). *In* "Sperm Action, Progress in Reproductive Biology" (P. P. Hubinont, M. L'Hermite, and J. Schwers, eds.), Vol. I, pp. 136–152. S. Karger, Basel.
Bressler, R. S. (1978). *Ann. Biol. Anim. Biochem. Biophys.* **18**(2B), 535.
Bousquet, J., Vanhems, E., and Dubuisson, L. (1978). *C. R. Seanc. Soc. Biol.* **172,** 94.
Byskov, A. G. (1974). *Nature (London)* **252,** 396.
Byskov, A. G. (1975). *J. Reprod. Fertil.* **45,** 201.
Byskov, A. G. (1978a). *Ann. Biol. Anim. Biochem. Biophys.* **18**(2B), 327.
Byskov, A. G. (1978b). *Int. J. Androl.* Suppl. 2, 29.
Catt, K. J., and Dufau, M. L. (1976). *Biol. Reprod.* **14,** 1.
Chakravarti, R. N., Sinha, M. K., and Dash, R. J. (1978). *J. Steroid. Biochem.* **9,** 819, Abstract 31.
Charreau, E. H., Attramadal, A., Torjensen, P. A., Purvis, K., Calandra, R., and Hansson, V. (1977). *Molec. Cell. Endocrinol.* **6,** 303.
Childs, G. W., Hon, C., Russell, L. R., and Gardner, P. J. (1978). *J. Histochem. Cytochem.* **261,** 545.
Clausen, O. P. F., Purvis, K., and Hansson, V. (1978). *Ann. Biol. Anim. Biochem. Biophys.* **18**(2B), 541.
Colenbrander, B., and Van Straaten, H. W. (1977). *Acta Morphol. Neerl. Scand.* **15,** 102 (Abstract).
Colenbrander, B., de Jong, F. H., and Wensing, C. J. G. (1978). *J. Reprod. Fertil.* **53,** 377.
Corbier, P., Kerdelhue, B., Picon, R., and Roffi, J. (1978). *Endocrinology* **103,** 1985.
Dorrington, J. H., and Armstrong, D. T. (1975). *Proc. Natl. Acad. Sci. U.S.A.* **72,** 2677.
Dorrington, J. H., and Fritz, I. B. (1974). *Endocrinology* **94,** 395.
Dorrington, J. H., Roller, N. F., and Fritz, I. B. (1975). *Mol. Cell. Endocrinol.* **3,** 57.
Dorrington, J. H., Fritz, I. B., and Armstrong, D. T. (1976a). *Molec. Cell. Endocrinol.* **6,** 117.
Dorrington, J. H., Fritz, I. B., and Armstrong, D. T. (1976b). *Progr. Int. Congr. Endocrinol. 5th,* Abstract 768, 316.

Elkington, J. S. H., Sanborn, B. M., and Steinberger, E. (1975). *Molec. Cell. Endocrinol.* **2,** 157.
Evain, D., Morera, A. M., and Saez, J. M. (1976). *Int. Congr. Endocrinol. 5th, Hamburg,* Abstract, 524.
Fritz, I. B. (1977). *In* "Biochemical Actions of Hormones" (G. Litwack, ed.), Vol. 5. Academic Press, New York.
Fritz, I. B., Louis, B. G., Tung, P. S., Griswold, M., Rommerts, F. G., and Dorrington, J. H. (1975). *In* "Hormonal Regulation of Spermatogenesis" (F. S. French, V. Hansson, E. M. Ritzén, and S. N. Nayfeh, eds.), pp. 413–420. Plenum Press, New York.
Fritz, I. B., Griswold, M. D., Louis, B. G., and Dorrington, J. H. (1976a). *Molec. Cell. Endocrinol.* **5,** 289.
Fritz, I. B., Rommerts, F. G., Louis, B. G., and Dorrington, J. H. (1976b). *J. Reprod. Fertil.* **46,** 17.
Fritz, I. B., Louis, B. G., Tung, P. S., and Dorrington, J. (1978). *Ann. Biol. Anim. Biochem. Biophys.* **18**(2B), 555.
Gilula, N. B., Fawcett, D. F., and Aoki, A. (1976). *Devel. Biol.* **50,** 142.
Grootegoed, J. A., Peters, M. J., Mulder, E., Rommerts, F. G., and Van Der Molen, H. J. (1977). *Molec. Cell. Endocrinol.* **9,** 159.
Gross, D. S., and Baker, B. L. (1979). *Am. J. Anat.* **154,** 1.
Gulyas, B. J., Tullner, W. W., and Hodgen, G. D. (1977). *Biol. Reprod.* **17,** 650.
Hall-Martin, A. J., Skinner, J. D., and Hopkins, B. J. (1978). *J. Reprod. Fertil.* **52,** 1.
Hansson, V., French, F. S., Weddington, S. C., Nayfeh, S. N., and Ritizén, E. M. (1974). *In* "Hormone Binding and Target Cell Activation of the Testis" (M. L. Dufau and A. R. Means, eds.), pp. 287–290. Plenum Press, New York.
Hansson, V., Ritzén, E. M., French, F. S., and Nayfeh, S. N. (1975). *In* "Handbook of Physiology" (D. W. Hamilton and R. O. Greep, eds.), Vol. 5, Sect. 7, pp. 173–201. American Physiological Society, Washington, D.C.
Hansson, V., Purvis, K., Calandra, R., French, F. S., Kotite, N., Nayfeh, S. N., Ritzen, E. M., and Hagenas, L. (1976a). *Excerpta Med. Int. Congr. Ser. No. 402,* pp. 410–416. Amsterdam.
Hansson, V., Clandra, R., Purvis, K., Ritzén, E. M., and French, F. S. (1976b). *Vitam. Horm.* **34,** 187.
Hansson, V., Purvis, K., Ritzén, E. M., and French, F. S. (1978). *Ann. Biol. Anim. Biochem. Biophys.* **18**(2B), 565.
Hsueh, A. J. W., Dufau, M. L., and Catt, K. J. (1976). *Biochem. Biophys. Res. Commun.* **72,** 1145.
Huhtaniemi, I. T., Korenbrot, C. C., and Jaffe, R. B. (1977). *J. Clin. Endocrinol. Metab.* **44,** 963.
Josso, N., Picard, J. Y., and Tran, D. (1977). *Recent Progr. Hormone Res.* **33,** 117.
Kellokumpu-Lehtinen, P., Santti, R., and Pelliniemi, L. J. (1979). *Anat. Rec.,* in press.
Ketelslexgers, J.-M., Hetzel, W. D., Sherins, R. J., and Catt, K. J. (1978). *Endocrinology* **103,** 212.
Kotite, N. J., Nayfeh, S. N., and French, F. S. (1978). *Biol. Reprod.* **18,** 65.
Kraulis, I., Traikov, H., Sharpe, M., Ruf, K. B., and Naftolin, F. (1978). *Endocrinology* **103,** 1822.
Kula, K. (1977). *Folia Morphol. (Warsz)* **26,** 167.
Louis, B. G., and Fritz, I. B. (1977). *Molec. Cell. Endocrinol.* **7,** 9.
Luciani, J. M., Devictor, M., and Stahl, A. (1977). *J. Embryol. Exp. Morphol.* **38,** 175.
Madhwa Raj, H. G., and Dym, M. (1976). *Biol. Reprod.* **14,** 489.
Mauleon, P. (1978). *In* "Control of Ovulation" (D. B. Crighton, N. B. Haynes, G. R. Foxcroft, and G. E. Lamming, eds.), pp. 141–158. Butterworth, London.
Means, A. R., and Vaitukaitis, J. (1972). *Endocrinology* **90,** 39.
Means, A. R., Fakunding, J. L., Huckins, C., Tinall, D. J., and Vitale, R. (1976). *Recent Progr. Hormone Res.* **32,** 477.

Merchant-Larios, H. (1978). Workshop on the Development and Maturation of the Reproductive Organs and Functions 4th, Tours, France, October 18–20, 1978.
Meyer, R., Posalaky, Z., and McGinley, D. (1977). *J. Ultrastruct. Res.* **61,** 271.
Mills, N. C., and Means, A. R. (1977). *Biol. Reprod.* **17,** 769.
Mills, N. C., Van, N. T., and Means, A. R. (1977). *Biol. Reprod.* **17,** 760.
Moger, W. H. (1977). *Biol. Reprod.* **17,** 661.
Molen, J. (1975). *Molec. Cell. Endocrinol.* **2,** 171.
Mulder, E., Van Beurden-Lamers, W. M. O., De Boer, W., Brinkmann, A. O., and Van Der Molen, H. J. (1974). *In* "Hormone Binding and Target Cell Activation in the Testis" (M. L. Dufau and A. R. Means, eds.), pp. 343–355. Plenum Press, New York.
Mulder, E., Peters, M. J., De Vries, J., and Van Der Molen, J. (1975). *Molec. Cell. Endocrinol.* **2,** 171.
Musto, N., Gunsalus, G., and Bardin, C. W. (1978). Abstracts. Symposium Lectures. Int. Congr. Horm. Steroids, 5th, New Delhi, India, 30 October–4 November 1978, p. 88.
O, W. S., and Baker, T. G. (1978). *Ann. Biol. Anim. Biochem. Biophys.* **18**(2B), 351.
Ohno, S. (1976). *Cell* **7,** 315.
Ohno, S. (1978). "Major Sex-Determining Genes." Springer-Verlag, Berlin.
Piacsek, B. E., and Goodspeed, M. P. (1978). *J. Reprod. Fertil.* **52,** 29.
Preslock, J. P., and Steinberger, E. (1977). *Gen. Comp. Endocrinol.* **33,** 547.
Pelliniemi, L. J., Lauteala, L., and Salonius, A.-L. (1977). *J. Anat.* **124,** 244 (Abstract).
Pelliniemi, L. J., Kellokumpn-Lehtinen, P.-L., and Lauteala, L. (1979). *Ann. Biol. Anim. Biochem. Biophys.,* in press.
Purvis, K., Calandra, R., Attramadala, A., Haug, E., Ritzén, E. M., and Hansson, V. (1976a). Int. Symp. on Androgens and Antiandrogens, Milan, April 1976.
Purvis, K., Calandra, R., Naess, O., Attramadal, A., Torjesen, P. A., and Hansson, V. (1976b). *Nature (London)* **265,** 169.
Purvis, K., Calandra, R., Haug, E., and Hansson, V. (1977a). *Molec. Cell. Endocrinol.* **7,** 203.
Purvis, K., Haug, E., Clausen, O. P. F., Naess, O., and Hansson, V. (1977b). *Molec. Cell. Endocrinol.* **8,** 317.
Purvis, K., Torjesen, P. A., Haug, E., and Hansson, V. (1977c). *Molec. Cell. Endocrinol.* **8,** 73.
Purvis, K., Clausen, O. P. F., and Hansson, V. (1978a). *J. Reprod. Fert.* **52,** 379.
Purvis, K., Clausen, O. P. F., and Hansson, V. (1978b). *Endocrinology* **102,** 1053.
Purvis, K., Clausen, O. P. F., and Hansson, V. (1978c). *Ann. Biol. Anim. Biochem. Biophys.* **18**(2B), 595.
Ramaley, J. A. (1979). *Biol. Reprod.* **20,** 1.
Rivelis, C., Prepin, J., Vigier, B., and Jost, A. (1976). *C.R. Acad. Sci. Paris Ser. D.* **282,** 1429.
Robert, A., and Delost, P. (1978). *J. Reprod. Fertil.* **53,** 273.
Robinson, J. A., and Bridson, W. E. (1978). *Biol. Reprod.* **19,** 773.
Rohloff, D., and Mulling, M. (1977). *Zuchthyg* **12,** 30.
Rommerts, F. F. G., Grootegoed, J. A., and Van Der Molen, J. (1978). *Ann. Biol. Anim. Biochem. Biophys.* **18**(2B), 607.
Sanborn, B. M., Tsai, Y. H., Steinberger, A., and Steinberger, E. (1978). *Ann. Biol. Anim. Biochem. Biophys.* **18**(2B), 615.
Sar, M., Stumpf, W. E., McLean, W. S., Smith, A. A., Hansson, V., Nayfeh, S. N., and French, F. S. (1975). *In* "Hormonal Regulation of Spermatogenesis" (F. S. French, V. Hansson, E. M. Ritzén, and S. N. Nayfeh, eds.), pp. 311–319. Plenum Press, New York.
Sarkar, S., Dubey, A. K., Banerji, A. P., and Shah, P. N. (1978). *J. Reprod. Fertil.* **53,** 285.
Sharpe, R. M. (1976). *Nature (London)* **264,** 644.
Sholl, S. A., and Goy, R. W. (1978). *Biol. Reprod.* **18,** 160.
Sinha, M. K., Dash, R. J., and Chakravarti, R. N. (1978). *J. Steroid Biochem.* **9,** 823, Abstract 52.

Sivashankar, S., Prasad, M. R. N., Thampan, T. N. R. V., Sheela Rani, C. S., and Moudgal, N. R. (1977). *Indian J. Exp. Biol.* **15,** 345.

Solari, A., and Fritz, I. B. (1978). *Biol. Reprod.* **18,** 329.

Steinberger, A., and Steinberger, E. (1976). *In* "Sperm Action, Progress in Reproductive Biology" (P. P. Hubinot, M. L. L'Hermite, and J. Schwers, eds.), Vol. I, pp. 42–54. S. Karger, Basel.

Steinberger, A., Tanki, K. J., and Siegal, B. (1974). *In* "Hormone Binding and Target Cell Activation in the Testis" (M. L. Dufau and A. R. Means, eds.), pp. 177–191. Plenum Press, New York.

Steinberger, E. (1978). Abstracts. Symposium Lectures. *Int. Congr. Horm. Steroids. 5th,* New Delhi, India, 30 October–4 November 1978, p. 21.

Steinberger, E., Steinberger, A., and Sanborn, B. M. (1977). *In* "Recent Progress in Andrology, Sereno Symp. (A. Fabbrini and E. Steinberger, ed.). Academic Press, New York.

Swanson, L. V., and McCarthy, S. K. (1978). *Biol. Reprod.* **18,** 475.

Tcholakian, R. K., and Steinberger, A. (1978). *J. Steroid Biochem.* **9,** 827, Abstract 68.

Tung, P. S., Dorrington, J. H., and Fritz, I. B. (1975). *Proc. Natl. Acad. Sci. (Wash.)* **72,** 1838.

Upadhyay, S., Luciani, J. M., and Zamboni, L. (1978). *Workshop on the Development and Maturation of the Reproductive Organs and Functions 4th,* Tours, France, October 18–20, 1978.

Van Beurden, W. M. O., Roodnat, B., and Van der Molen, H. J. (1978). *Int. J. Androl.* Suppl. 2, 374.

Vanha-Perttula, T. (1978). *Ann. Biol. Anim. Biochem. Biophys.* **18**(2B), 633.

Wiebe, J. P., and Tilbe, K. S. (1978). *J. Steroid Biochem.* **9,** 822, Abstract 46.

Wilson, E. M., and Smith, A. A. (1975). *In* "Hormonal Regulation of Spermatogenesis" (F. S. French, V. Hansson, E. M. Ritzén, and S. N. Nayfeh, eds.), pp. 281–286. Plenum Press, New York.

Zamboni, L., Mauleon, P., and Bezard, J. (1978). *Workshop on the Development and Maturation of the Reproductive Organs and Functions 4th,* Tours, France, October 18–20, 1978.

INTERNATIONAL REVIEW OF CYTOLOGY, VOL. 62

Transitional Cells of Hemopoietic Tissues: Origin, Structure, and Development Potential

JOSEPH M. YOFFEY

Department of Anatomy and Embryology, Hebrew University–Hadassah Medical School, Jerusalem, Israel

ISBN 0-12-364462-3

I. Introduction

Transitional cells are a heterogeneous group of cells whose primary habitat is the bone marrow, where most of them are to be found. In the smaller laboratory animals such as the mouse, an appreciable number of transitional cells are present in the spleen, though it is possible that these may be of secondary importance in relation to similar cells in the marrow (Kretchmar and Conover, 1970; Silini *et al.*, 1976). In recent years evidence has been steadily accumulating that transitional cells are the stem cells of the hemopoietic system, and probably of the lymphomyeloid complex as a whole.

Since the early days of hematology different members of the transitional cell group have been described and given a variety of names (Section III,A). But an understanding of the group as a whole, albeit as yet incomplete, has only become possible because the introduction of modern techniques has enabled us to analyze with considerable accuracy their kinetic properties and their capacity for differentiation. The recent extensive review by Rosse (1976) on transitional cells and lymphocytes in bone marrow has dealt with many fundamental aspects of the problem. The present review contributes additional data and shifts the emphasis in the interpretation of some of the data already available.

The term ''transitional'' was first employed under the mistaken impression that the majority of the body's lymphocytes were produced in the scattered lymphoid tissues and, for the most part, entered the blood through the thoracic duct (Yoffey, 1932–1933). From the blood they were then believed to enter the marrow, to serve as hemopoietic stem cells. In the marrow it was thought that they underwent enlargement, first of the pachychromatic nucleus, whose chromatin broke up to become leptochromatic, and then of the cytoplasm which gradually developed increasing degrees of basophilia, so that the cells would finally become typical blast cells committed to one or other of several possible lines of differentiation. On the basis of this interpretation, the enlarging cells were at first termed ''transitional lymphocytes'' (Yoffey and Courtice, 1956). Further support for this interpretation seemed to be forthcoming when PHA and other mitogens were introduced, and it became possible to observe the pachychromatic small lymphocytes of the blood undergoing precisely what appeared to be such a sequence of changes: enlarging, becoming leptochromatic, and finally developing basophilic cytoplasm to become the typical PHA blast cell. Although the PHA blast cell does not undergo hemopoietic differentiation, the morphological resemblance to the blast cells in the bone marrow seemed to be very striking, and at first sight appeared fully to confirm the interpretation originally placed upon the stem cell role of the small lymphocyte.

However, subsequent investigations (Osmond and Everett, 1964; Harris and Kugler, 1965) showed that, far from small lymphocytes in the marrow enlarging to form transitional cells, precisely the reverse was usually the case. It was the division of transitional cells which gave rise to small lymphocytes, as was later observed in cultures by Rosse (1972a) and Osmond *et al.* (1973). Nevertheless, for a number of reasons the term "transitional" has been retained (Section III.A), although the designation "lymphocyte" obviously had to be discarded. The fundamental importance of the transitional cell compartment lies in the fact that, in the bone marrow of mammals, it contains in varying proportions the uncommitted pluripotential stem cells of the entire lymphomyeloid complex, as well as cells already committed to specific lines of development. But in making this statement it is important to clarify one point at the outset. There may not be a single stem cell throughout life.

In the fetus, for example, the changing sites of hemopoiesis seem to be associated with different stem cells in yolk sac and liver, although once myeloid hemopoiesis has become established, the transitional cells appear to be an even more prominent constituent of the bone marrow than in postnatal life (Yoffey *et al.*, 1961b). Maximow (1910–1911, 1927) maintained that, in the earliest stages of development, the first stem cells in the marrow are those of the primitive mesenchyme. It may also be the case that, under certain experimental conditions, the localized regeneration of bone marrow is initially associated with a type of stem cell different from that which is present once myeloid hemopoiesis is fully established (Tavassoli and Crosby, 1968; Patt and Maloney, 1970; Knospe *et al.*, 1972). Indeed, it is readily conceivable that any primitive undifferentiated and uncommitted cell can, if appropriately stimulated, undergo hemopoietic differentiation. But all these considerations do not alter the fact that, once the definitive hemopoietic pattern has become established and the bone marrow fully developed, it is in the transitional compartment that the hemopoietic stem cells are to be found.

II. Morphology of Transitional Cells

A. Size

Transitional cells show a spectrum of sizes, ranging from the small transitional cell about 7 μm in diameter to the large transitional cell 12 μm+ in diameter. Measurements of cell (or nuclear) size in smears can only be approximately correct, since the size can vary with such factors as speed of spread and the medium in which the cells are suspended. Nevertheless, such measurements have proved of considerable use in radioautographic studies with the light microscope (Moffatt *et al.*, 1967; Osmond, 1967; Yoffey *et al.*, 1965a; Everett and Caffrey, 1967; E. D. Thomas *et al.*, 1965; Keiser *et al.*, 1967; Rosse, 1973, and many others). An additional disadvantage of the measurement of cell size in smears is that it is laborious and time consuming. More recently, large numbers of transitional cells have been obtained in suspension and measured in a Coulter counter (Yoffey *et al.*, 1978b; Patinkin *et al.*, 1979). From these measurements it appears that in guinea pig marrow there are two or three size groups in the transitional cell compartment. The possible significance of these findings is discussed in Section V,A,C. In addition to the three main size groups, there are a number of cells of intermediate size, so that between the smallest and the largest transitional cells there exists a spectrum of sizes.

The size range of transitional cells has been illustrated repeatedly. Light microscope illustrations of the range of transitional cells in guinea pig marrow have been published by Yoffey (1957) and Yoffey *et al.* (1965a), the last-named authors illustrating a size spectrum both of unlabeled cells (Figs. 1–13, loc. cit.) and of cells labeled with thymidine (Fig. 25 ibid.). Illustrations in color after staining with MacNeal's stain have been published by Harris and Kugler (1963), and at higher magnifications by Rosse and Yoffey (1967) and Rosse (1970b). Everett and Caffrey (1967) illustrate transitional cells in rat marrow and Keiser *et al.* (1967), in the dog. Sharp *et al.* (1976) and Riches *et al.* (1976b) illustrate transitional cells in murine bone marrow. Ultramicroscopic photographs are also available (e.g., rat: Ben-Ishay and Yoffey, 1972; mouse: Yoffey and Weinberg, 1976; Dicke *et al.*, 1973; guinea pig: Yoffey and Courtice, 1970). Inspection of these illustrations brings out very clearly the fact that there are a number of characteristic features in transitional cell morphology.

B. High N:C Ratio

One of the characteristic features is the high N:C ratio. In air-dried Romanowsky-stained smears this high ratio tends to be exaggerated, and the cytoplasmic shrinkage may cause the cytoplasm to appear as a tuft at one pole of the cell. In ultramicroscopic preparations with much less shrinkage there is still a high N:C ratio, but there is usually a complete cytoplasmic rim surrounding the nucleus (Fig. 1).

C. Nuclear Structure

In the typical transitional cell the nucleus is leptochromatic, in sharp contrast to the pachychromatic nucleus of the small lymphocyte (Fig. 1). The nuclear membrane is thin, contrasting with the juxtamembranous condensation of

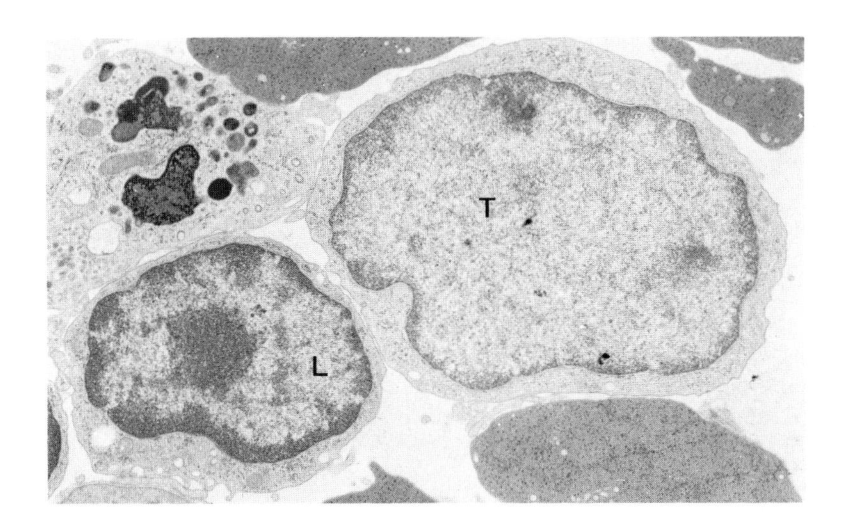

FIG. 1. A pachychromatic small lymphocyte (L) and a leptochromatic transitional cell (T), which contains mainly monoribosomes and corresponds to the pale transitional cell of light microscopy. Some transitionals contain mainly polyribosomes, and these are the basophilic cells. The small lymphocyte has a characteristic condensation of chromatin at the nuclear membrane. (Adapted from Yoffey and Weinberg, 1976.) ×7500.

chromatin in the lymphocyte. The interior of the nucleus is less dense than that of the lymphocyte, since in the transitional cells there is a much higher ratio of parachromatin to chromatin. The one or two nucleoli present are usually pale, without the thick coating of nucleolar-associated DNA found in the small lymphocyte (Pathak *et al.*, 1956). The transitional cell nucleus often has indentations, sometimes quite deep. Rosse (1971), in cultures of bone marrow, observed that the transitional cell nucleus could be very active, even in stationary cells, showing a marked streaming of nuclear material and the repeated formation and disappearance of deep invaginations of the nuclear membrane. The precise significance of these changes is not known, but it seems likely that, insofar as they increase the surface of the nuclear membrane, they may perhaps be an indication of enhanced nuclear and cytoplasmic activity, possibly in relation to DNA and RNA synthesis (Sections VI,B and VII,A).

One proviso must be made in regard to the leptochromasia of transitional cells. This is most marked in the large transitional cells, which are the most active members of the compartment in relation to DNA synthesis. In the smaller transitional cells, many of which are not in DNA synthesis (Section VII,B) there is some thickening of the nuclear chromatin, although it is never as pachychromatic as in the small lymphocyte. The smaller transitional cells appear to be the resting phase of the compartment, as judged by the absence of DNA synthesis. Furthermore, as Rosse (1971, 1972a) has shown, when the smaller transitionals divide to give rise to small lymphocytes, 5 to 6 hours elapse before the daughter cells acquire the morphology characteristic of the small lymphocyte. During this time, one sees cells with varying degrees of pachychromasia, and such cells are sometimes referred to as L-T cells.

D. Cytoplasm: Degree of Basophilia

The cytoplasm of transitional cells may be pale or deeply basophilic, or show varying degrees of basophilia on light microscopy after the customary hematological strains. The relationship between pale and basophilic transitionals is as yet imperfectly understood. Apart from the degree of basophilia, there tends to be rather more cytoplasm in the pale than in the basophilic transitionals (Yoffey, 1966), so that even in air-dried smears the pale transitionals may have a complete rim of cytoplasm. In studies which involve drawing a distinction between pale and basophilic transitionals, it is important to record the numbers of the intermediate group which one cannot place with certainty (Jones *et al.*, 1967; Moffatt *et al.*, 1967; Rosse and Yoffey, 1967). This is essential when considering the relationship between the pale and the basophilic cells.

In an analysis of the complex kinetic data (discussed at length by Rosse, 1976) there would seem to be two fundamentally different possibilities. The first is that the pale, intermediate, and basophilic transitionals are all completely unrelated cell groups. This could in fact imply not only the existence of three distinct cell groups, but possibly also a subdivision of the intermediate group into further subgroups depending on their degree of basophilia.

The second possibility which seems more likely is that the pale and the basophilic transitionals are interrelated, either because pale transitionals can develop increasing amounts of basophilia and finally become fully basophilic, or because the basophilic cells can undergo a number of divisions in the course of which they gradually become less basophilic. This latter type of change occurs in the course of the long production pathway in lymph nodes, when the highly basophilic lymphoblast undergoes a number of divisions ending in the production of much less basophilic small lymphocytes. While the process of formation of pale transitionals by progressive loss of basophilia cannot be excluded, it certainly is not the only way in which pale transitionals can arise, for they are also capable of DNA synthesis and self-replication. In fact, Moffatt *et al.* (1967) reported a high mitotic index in the larger pale transitionals (cf. Miller and Osmond, 1973). So, unless one postulates a dual mechanism for the origin of pale transitional cells, the evidence seems to favor replication rather than an origin from basophilic cells through loss of basophilia. There remains always, of course, the possibility of recruitment from an outside source. If the pale transitionals are self-replicating, then the intermediate transitionals would be stages between the pale and basophilic cells. This of course does not rule out the possibility of self-replication in the basophilic group.

The large basophilic transitionals merge through a continuous series of intermediate stages into the various blast cells, the most numerous of which are the proerythroblasts and myeloblasts. While some transitionals have only a small tuft of basophilic cytoplasm, one also finds a whole series of cells with progressively increasing amounts of cytoplasm, and at the stage when there is enough basophilic cytoplasm to surround the nucleus completely, the cell is regarded as a fully formed blast cell. But the classification of the stages between the basophilic transitional and the fully formed blast cell presents a problem, since it is difficult in such a cell series to decide where the transitional cells end and the blast cells begin. For purposes of classification, cells are arbitrarily recorded as transitional cells until the cytoplasm extends halfway round the cell, and as blast cells when the cytoplasm extends more than halfway. This type of problem does not arise with the pale transitionals.

The development of basophilia is presumably due to the increased formation of cytoplasmic RNA, and ultrastructural studies seem to confirm this in terms of the increasing number of polyribosomes (Ben-Ishay and Yoffey, 1972; Yoffey and Weinberg, 1976). But even on ultrastructural examination it may be difficult to distinguish between basophilic transitional cells and proerythroblasts, if the latter do not yet contain some distinctive constituent such as ferritin. Transitional cells which contain only monoribosomes are presumably the counterpart of the pale transitionals of light microscopy.

The more intensely basophilic transitionals tend to be more numerous among the larger members of the group. Thus in the 10 μm+ size group in normal marrow (guinea pig) over 50% were recorded as basophilic, whereas about 25% were pale and 20% intermediate (Moffat *et al.*, 1967; Jones *et al.*, 1967). The difference in labeling index is much greater than the percentage distribution of the three groups would imply, so that in the case of the pale transitional cells the radioautographic data of Moffatt *et al.* (1967) agree with the finding of Rosse (1970a) that a considerable number of pale transitional cells are out of cycle. It should, however, be noted that the data of Moffatt *et al.* (1967) were obtained in normal guinea pigs. It remains to be seen whether the labeling index of the pale transitionals undergoes alteration in different functional conditions of the marrow.

III. The Transitional Cell Compartment as a Whole: A Brief General Survey

Before discussing in greater detail the role of the transitional cells, it will be useful to give a brief general account of the transitional cell compartment as a whole and of some of the ways in which it is now believed to function. Some of the points now presented briefly will be amplified in subsequent sections. There are two opposing processes at work in the compartment (Fig. 2). One process is the constant proliferation of its cells without any change, a process of self-replication which in itself would merely result in their continual increase. The other is the differentiation of its cells into the various blast cells, a process which would have the opposite effect and diminish the number of cells in the compartment. The size of the compartment at any given time reflects the dynamic equilibrium between these two processes. Some of the ways in which this equilibrium can be disturbed are discussed in Section VIII (cf. Boggs *et al.*, 1972).

Self-replication involves an apparently simple process in which large transitionals divide to give rise to small ones, and these in turn enlarge. But there are several variables in this apparently simple process. The speed with which the small transitionals enlarge, and the large transitionals divide, determines the rate at which new stem cells are formed, and it is possible that this rate can undergo considerable variation.

Differentiation occurs in a number of directions, but in quantitative terms the three main lines of differentiation are into proerythroblasts, myeloblasts, and lymphocytes. The quantitative technique applied to the study of these three cell groups has not as yet been applied to a similar study of monocyte or megakaryocyte production. In the normal steady state it seems to be the larger transitionals which follow the erythropoietic or granulopoietic line of develop-

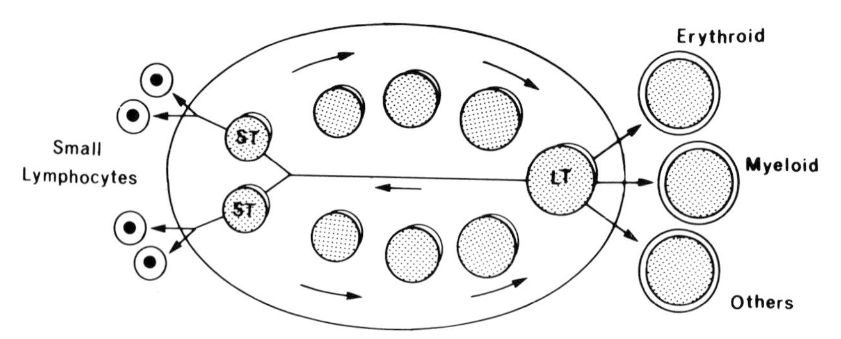

FIG. 2. A scheme of the transitional cell compartment in bone marrow. The compartment is capable (a) of self-maintenance, when large transitional cells (LT) divide into small ones (ST) which can then enlarge; (b) of differentiation and giving rise to the various groups of blood cells, mainly small lymphocytes, erythrocytes, and granulocytes. There is a spectrum of cell sizes in the compartment, and also of cells with varying degrees of basophilia. The transitional cell compartment contains both the committed and the uncommitted stem cells. Small lymphocytes, for the most part immunologically committed, are formed by division of the smaller transitional cells. But there may be a minor subgroup of uncommitted lymphocytes which are capable of enlarging and reentering the transitional cell compartment. (From Yoffey, 1977.)

ment, via the stage of the basophilic blast cell. But in some experimental situations the transitionals may apparently undergo an earlier differentiation (Sections V,B and VIII,F), without going through the blast cell stage. The small lymphocytes result from the division of the smaller transitional cells and appear to be predominantly B lymphocytes (Reviewed by Yoffey, 1975; and Rosse, 1976). These are just as much differentiated cells as are the erythrocytes and granulocytes, but the final stages of their development, when they participate in immune reactions, may occur long after they have left the marrow. There is also evidence that some T-cell precursors are present in bone marrow (Komuro and Boyse, 1973; Cohen and Patterson, 1975; Press *et al.*, 1977).

One of many as yet unsolved problems is whether, in addition to the large numbers of B lymphocytes, there may be present in the marrow a special subgroup of small lymphocytes which are not immunologically committed, and which are capable of enlarging and reentering the transitional cell compartment.

Terminology

In view of the fundamental importance now attached to transitional cells, it has become increasingly desirable to resolve the growing confusion in terminology. Although as noted the term "transitional" was first introduced on the basis of a mistaken hypothesis, and was intended initially as a morphological description for the continuous spectrum of transitions between the pachychromatic small lymphocyte at one end of the spectrum and the large basophilic blast cell at the other, it still appears to have much in its favor. In the stem cell scheme now adopted (Fig. 2), the range of cells, namely, from the small to the large transitional and then the blast cell, is almost as extensive as was originally thought to be the case. It is true there is still a gap between the small lymphocyte and the small transitional cell, but this constitutes only a narrow band in a much wider spectrum of cell sizes, over the greater part of which the term "transitional" is as fully applicable as when it was first introduced. Furthermore, if it should finally turn out to be the case that there are some small lymphocytes, even if few in number, which remain unspecialized and are capable of enlarging and reentering the transitional cell compartment, the term "transitional" would be as valid in such cases as when it was first introduced. One further advantage of the term is that it conveys some idea of the dynamic nature of the compartment, in which one might otherwise tend to think that one was dealing with a relatively static collection of cells.

There is not doubt, when one surveys the hematological literature, that from the early days of hematology, transitional cells have been repeatedly observed and given a variety of names. But attention was usually focused on one or another type of cell in the compartment which happened to be particularly conspicuous in the material examined, whether experimental or clinical. In consequence the compartment was never clearly recognized as a whole—although some observers came very near to doing so—and indeed could not have been so recognized until techniques became available for the more sophisticated types of kinetic study now employed. Naegeli's (1900, 1931) preoccupation with the myeloblast is a case in point. He could not find an origin for the myeloblast from mesenchymal cells, even in the embryo, where the myeloblast appeared all of a sudden, without any obvious precursor, in both blood and tissues. His "micromyeloblast" (1931, Abb. 67, cells 2 and 3) is a typicall small transitional cell, which he would almost certainly have identified as a small lymphocyte precursor had tritiated thymidine been at his disposal. Instead, following Ehrlich, he concluded that lymphocytes had nothing to do with the marrow tissue proper, but were "extraparenchymatous," a view which was adopted by the majority of hematologists for many years. Pappenheim (1907), in his description of large and small "haematogones," seems to have been dealing mainly with transitional cells. At a later date (Pappenheim and Ferrata, 1910, and subsequent publications) the terminology was changed to "lymphoid" cells and "lymphoidocytes." Here too the concern was obviously with transitional cells, since this 1910 paper was based to a large extent on the bone marrow of guinea pigs, in which transitional cells figure very conspicuously. The "primitive" cells of Cunningham *et al.* (1925) are also typical transitional cells, but like so many hematologists they too followed Naegeli in refusing to recognize lymphocytes as an integral part of the marrow. They derived the primitive cells from a somewhat ill-defined reticulum cell. Rhoads and Miller (1938) recorded an increase in primitive cells in the marrow in some cases of aplastic anemia of the hyperplastic type, while similar cells have been described in recovery from agranulocytosis (Blackburn, 1947–1948; Pisciotta *et al.*, 1964). Davidson *et al.* (1943), in the bone marrow of patients with refractory anemia, observed numerous cells which they described as "Q" cells, which appear to have possessed transitional cell morphology, and which disappeared as the anemia improved. Tillmann *et al.* (1976) noted increased numbers of transitional cells in the marrow in cases of transient red cell aplasia. Of the many other names given to transitional cells one may note the following: "lymphocyte-like" (Burke and Harris, 1959; Harris, 1961; Cudkowicz *et al.* 1964b; Bennett and Cudkowicz, 1967), "lymphoid" (E. D. Thomas *et al.*, 1965; Osmond *et al.*, 1973; Kempgens *et al.*, 1973, 1976), "hemopoietic stem cell" (Dicke *et al.*, 1973). "leptochromatic lymphocyte" (De Gowin and Gibson, 1976), "leptochromatic mononuclear cell" (Sharp *et al.*, 1976; Riches *et al.*, 1976a,b), hematopoietic precursor cell (Murphy *et al.*, 1971), "presumptive hematopoietic stem cell" (Rubinstein and Trobaugh, 1973), "X" cell (Simar *et al.*, 1968).

Of the various names which have been suggested, the term "transitional" seems least open to objection, since it refers to the distinctive morphological attributes of a cell group which also has very distinctive kinetic properties. Terms such as "lymphoid" are unsatisfactory, since cells which may be described as lymphoid are to be found throughout the lymphomyeloid complex, so that the term covers a multitude of cells which are certainly not members of the transitional group. The use of the term "lymphoid" is sometimes justified on the grounds that transitional cells give rise to lymphocytes. But by the same token they could also be designated "erythroid" or "granuloid," since they also give rise to cells of the erythrocytic and granulocytic series.

If the stem cell scheme presented in Fig. 2 is correct, it follows that stem cells must exist in different sizes, and there is now growing evidence (Section VIIC,D) that this is in fact the case. Of all the various cells which have at different times been suggested as stem cells, the transitionals are the only group which can be shown to exist in different sizes. Moreover, they are the only group which possesses the migratory and proliferative properties shown by stem cells in a number of experimental situations.

IV. Transitional Cells in the Fetus

A. Yolk Sac

A full account of the histology of yolk sac hemopoiesis as studied by the light microscope has been given by Bloom (1938). More recently, ultrastructural studies have been performed by Sorenson (1961), Fukuda (1973a), and others. Most observers seem to have followed Maximow (1927) in stating that the stem cells in the yolk sac are of mesenchymal origin, though a derivation from entoderm has also been suggested (Gladstone and Hamilton, 1941). The ultrastructural studies of Fukuda (1973a) could readily be interpreted in accordance with this view (see his Fig.1, p. 199). Whatever the origin of the cells, they are characteristically large and basophilic and bear no resemblance to transitional cells; but the question which arises is whether they may be the progenitors of transitional cells. This is a point to be emphasized, since Moore and Metcalf (1970), and others, have maintained that the yolk sac in the mouse—and presumably in other mammals—is the primary source of hemopoietic stem cells, which are thought to migrate from the yolk sac to the liver and the rest of the lymphomyeloid complex. This concept appears to have its origin in studies of the avian yolk sac (Moore and Owen, 1967), a much more substantial source of cells than the vestigial yolk sac of mammals. If the cells of the mammalian yolk sac are indeed the ultimate source of all the transitional cells in the organism, they must be capable of a remarkable degree of proliferation. They must also undergo a striking change in morphology as they migrate, for neither in liver nor bone marrow do they occur in the form in which they are found in the yolk sac.

B. Liver

Hepatic hemopoiesis in mammals is much more extensive than that in the yolk sac, and has been carefully studied by a number of observers, by both light and electron microscopy. As in the case of the yolk sac, here too most observers seem to have accepted Maximow's (1927) view that the hemocytoblasts which appear in large numbers in the liver are mesenchymal derivatives (Bloom, 1938; Zamboni, 1965; Sorenson, 1963; Rifkind *et al.*, 1969; Jones, 1970). Thomas and Yoffey (1964) thought that the hemocytoblasts were derived from the undifferentiated cells of the liver parenchyma, and would therefore be entodermal in origin, in agreement with the view first put forward over a century ago by Toldt and Zuckerkandel (1875). The histogenetic problems have been more fully discussed elsewhere (Thomas and Yoffey, 1964; Yoffey, 1971). As compared with bone marrow, which is the third and definitive site of fetal hemopoiesis, fetal liver contains very few lymphocytes or granulocytes (Thomas *et al.*, 1960; Rosenberg, 1969).

Some transitional cells are present in the liver, and have been illustrated by Thomas (1973), and Fukuda (1973b, 1974). Fukuda gives illustrations both of undifferentiated mononuclear cells and of transitional cells. Hoyes *et al.* (1973) illustrate a small population of pale cells, similar to transitional cells in relation to the periportal bile ducts, and they regard this as evidence of their origin from entoderm. They interpret their findings as indicating that these cells become transformed into lymphoid elements which resemble transitional cells. However, although some transitional cells are present in fetal liver, it is difficult to ascertain how many there are and how important an element they may be in hepatic hemopoiesis.

One puzzling feature about the nature of the stem cells in fetal liver is the fact that transfusions of fetal liver cell suspensions have frequently proved to be far less effective in conferring protection than suspensions of marrow cells. Micklem and Loutit (1966) have reviewed this aspect of the problem, and have drawn attention to the frequent occurrence of lymphoid tissue hypoplasia in syngeneic fetal liver chimeras. The fact that this hypoplasia can be overcome if lymph node cells are transfused at the same time serves only to highlight the difference between liver and marrow suspensions. Thomas (1971) found that after transfusion of fetal liver cells to irradiated animals, the regenerating bone marrow had a normal composition. This makes it all the more difficult to understand why lymphoid tissue hypoplasia should develop. A more recent review by Lowenberg (1975) suggests that stem cells from transplanted fetal liver either differentiate less than marrow stem cells or die in larger numbers. Whether or not this is the case, his findings confirm the general impression that transplanted fetal liver cells are not as effective as bone marrow cells in conferring protection.

C. Bone Marrow

The bone marrow during fetal life contains large numbers of transitional cells, as first reported by Yoffey *et al.* (1961b) and Yoffey and Thomas (1964). They studied the marrow in 50 normal human fetuses delivered by hysterotomy and ranging in age from 8 to 28 weeks. Marrow was obtained from femur, humerus, clavicle, ribs, vertebrae, and sternum, and was studied both in smears and by sectioning *in situ* after decalcification. The study of sections confirmed the finding of Hammar (1901) that the primary marrow is loose connective tissue in which there are dilated thin-walled vessels. This is followed by infiltration with lymphoid cells, some of which are small lymphocytes, others transitional cells. Transitional cells can usually be distinguished from small lymphocytes in sections as well as in smear preparations because they are larger and do not stain so densely. The dense staining of the small lymphocyte is due to the fact that there is a thick layer of chromatin—the juxtamembranous condensation—next to the nuclear membrane, and a relatively large nucleolus, or more precisely, as already noted, a nucleolus surrounded by nucleolar-associated DNA (Pathak *et al.*, 1956). There is very little parachromatin. The transitional cell on the other hand has only a thin nuclear membrane, a good deal more light-staining parachromatin, and a nucleolus with very little DNA.

No morphological evidence was seen of the transformation of mesenchymal cells into lymphoid cells, as so consistently maintained by Maximow (1927). Maximow's studies were performed for the most part on sectioned material, and his "hemocytoblasts" appear to have been a mixture of basophilic transitional cells and fully formed blast cells. He described the formation of hemocytoblasts in close association with the periosteum, and their subsequent entry into the marrow, but Yoffey and Thomas (1964) could not find any instances of this in the fetal marrows which they examined, 50 in all, ranging from 8 to 28 weeks.

If the transitional cells and lymphocytes in the primary marrow are not formed locally, they must migrate to the marrow from some extramyeloid source, but we have no information what this source may be. For reasons already given, the yolk sac does not seem a very likely source. The thymus has begun to develop by the time the marrow appears, but the thymus has generally been regarded as an importer rather than an exporter of cells. A possible source about which one might speculate is the intestinal lymphoid tissue, for at the time the bone marrow is beginning to develop, there are quite well-marked lymphocytic accumulations in the terminal ileum, and these are associated with large lymphatic vessels containing what appear to be numerous transitional cells (Yoffey, 1968). If these lymphatics do in fact contain migrating cells, we still do not know the direction of the migration, whether from or to the intestinal lymphoid tissue. But for phylogenetic reasons the speculation is at least an intriguing one, since at the beginning of vertebrate evolution the intestine seems to be a primary hemopoietic area, as shown for example by the spiral valve in cyclostomes (Jordan and Speidel, 1930), even before the spleen is fully developed and long before the bone marrow puts in an appearance.

D. Heavy Fetal Demand for Stem Cells

Transitional cells appear to be present in the bone marrow from the outset, and have been illustrated in photomicrographs (Yoffey *et al.* 1961b, Fig. 1) from which it is evident that the size spectrum is already present. Lymphocytes and transitional cells average around 25% of the cells in fetal marrow (range 10–45%), and of these the transitionals constitute about 15%. These large numbers of transitional cells are found during a period of several months when there is a peak demand for stem cells. One must attribute this primarily to the needs of red cell production. Between the twelfth and twenty-fifth weeks of gestation the erythrocyte content per unit volume of blood is more than doubled, while the body weight increases 17-fold (Thomas and Yoffey, 1962). If the increase in blood volume during this period of development is of the same order, then the circulating erythrocyte population will increase more than 30-fold.

This must be a conservative estimate, since it does not take into account the replacement of effete red cells, while in addition there is suggestive evidence that the mean red cell life is shorter during the early stages of development than in the adult (Mollison, 1948). In the third trimester there is a further rapid increase in the number of erythrocytes per cubic millimeter of blood (Thomas and Yoffey, 1962). From the data of Hudson (1965) it is clear that the available volume of bone marrow in relation to body weight is a good deal less in the fetus than in the adult, and this must be a limiting factor in the capacity of the bone marrow to meet the great demand for red cells. The erythropoietic role of the liver is therefore essential during the fetal period to compensate for the temporary inadequacy of the marrow. But even with this hepatic assistance, the maximal level of red cell production is still required from the bone marrow, especially during the later stages of pregnancy, when hepatic hemopoiesis is being phased out in preparation for its postnatal metabolic activities. It is doubtless the high rate of erythropoiesis in the marrow which necessitates a large number of transitional cells to function as stem cells. This may apply to some extent in the early postnatal period, when transitional cells constitute about 2% of the total nucleated cells for the first few months (Rosse *et al.,* 1977). At the same time the lymphocytes increase to about 50% of the nucleated cells of the marrow. These results for human marrow (cf. Gairdner *et al.,* 1952) recall the similar findings of Harris and Burke (1957) in the rat, where the transitional cells are described as "lymphocytelike," and of Sabin *et al.* (1936) in the rabbit, where they are referred to as "primitive" cells. Miller and Osmond (1974) followed the changes in C3H mice from birth to 16 weeks and found an appreciably higher level of transitional cells ("medium lymphoid" and "large lymphoid") than Rosse *et al.* (1977) in man. Miller and Osmond (1975) further extended these observations by quantitative kinetic studies on young, pubertal, and adult C3H mice.

E. Comparison of Fetal and Adult Marrow

Once the adult stage is reached, when in the normal steady state red cells are formed only to replace normal wear and tear, the need for stem cells becomes minimal, and because of their small numbers their identity has frequently been a matter of controversy. Dicke *et al.* (1973), using an enrichment technique (discontinuous albumin density gradient), illustrate hemopoietic stem cells in mouse, monkey, and man, and the structure common to all is that of a typical transitional cell. Kempgens *et al.* (1973) made a careful study of transitional ("lymphoid") cells in the marrow of eight healthy adults, where they found them, though few, to resemble the transitional cells in animal marrow. In man they constituted 0.85% of the total marrow cells, and their overall labeling index with thymidine was 32.5%. They also noted that whereas the pale transitionals possessed only the diploid number of chromosomes, the basophilic transitionals had a range from diploid to tetraploid (see their Abb. 2, p. 438). In a later paper (Kempgens *et al.*, 1976) they reported that in seven cases of renal anemia the transitional cell content of the marrow increased, ranging from 1.4 to 6.4% of the nucleated cells. Though the transitional cells were proliferating more slowly than in normal subjects, their numbers nevertheless increased, presumably because they were not differentiating into red cells. This type of condition, in which there is an element of maturation arrest, is one in which transitionals tend particularly to accumulate. In drug-induced agranulocytosis (Pisciotta *et al.*, 1964) there is a marked increase in transitional cells in the period during which they are not differentiating into granulocytes, but their numbers fall rapidly during the recovery stage, when they begin to differentiate into granulocytes once more. Some cases of aplastic anemia, especially those of the hyperplastic type, may also have an element of maturation arrest. Rhoads and Miller (1938) compared the marrow in those cases of aplastic anemia which they considered to be due to maturation arrest with the marrow in cases of agranulocytosis and concluded that the two conditions had much in common. Boggs and Boggs (1976) interpret the pathogenesis of aplastic anemia in terms of a defective pluripotential stem cell and a disturbance of the normal dynamic equilibrium between self-replication and differentiation. Transitional cells (Q cells) also accumulate in the marrow in some types of refractory anemia (Davidson *et al.*, 1943) (cf. Tillmann *et al.*, 1976). Kim *et al.* (1976) used an ingenious technique to remove the more mature erythroid cells from thalassemic and nonthalassemic marrow, leaving inter alia "early erythroid precursors" which, as far as one can judge from their illustrations, appear to have transitional cell morphology (see their Fig. 1B, p. 770).

In general, it appears that the transitional cell content of human marrow, high in the fetus, is at its lowest in the healthy adult in the normal steady state, when it is also appreciably lower than the figures usually obtained in the small laboratory animals.

F. Species Variation in Stem Cell Requirement

There seem to be two main factors responsible for species variations in stem cell requirement: (1) the life of the circulating red cell, and (2) the number of mitoses in the course of the red cell production pathway. Whereas in man the life of the circulating red cell is of the order of 120 days, in the mouse it is about 40 days (Rodnan *et al.*, 1957), in the rat 60 to 70 days (Berlin and Lotz, 1951), and in the guinea pig about 80 days (Everett and Yoffey, 1959). Other things being equal, e.g., the ratio between the volume of hemopoietic tissue and that of circulating red blood cells, as also the maturation time and number of mitoses between the stem cell and the mature erythrocyte, then in the normal steady state the mouse should require three times as many stem cells as in man. But other things may not be equal. In man there is a maturation period to 5 to 6 days, during which four to five mitoses may occur (Badenoch and Callender, 1954; Lajtha and Oliver, 1960), whereas in the rat the period is more like 2 to 3 days (Belcher *et al.*, 1954; Tarbutt, 1969). From the sedimentation velocity data of McCool *et al.* (1970) there would appear to be only two mitoses in the course of erythrocyte maturation in the rat, a result somewhat at variance with the findings of other workers. In the guinea pig Starling and Rosse (1976) estimate that there are between two and four mitoses between the proerythroblast and the orthochromatic erythroblast. Though there are conflicting views on the number of mitoses between stem cell and mature erythrocyte, it seems not improbable that in a maturation period of 2 to 3 days there may well be two mitoses less than in a maturation period of 4 to 5 days. Even one mitosis less will double the number of stem cells required, while two mitoses less would quadruple the number. Considerations of this nature may go some way toward explaining the difference in transitional cell content between the bone marrow of man and that of smaller laboratory animals.

G. Transitional Cells in Fetal Blood

It now appears to be generally accepted that some hemopoietic stem cells are present in blood throughout life, though in greater numbers before than after birth. Since the first demonstrations of stem cells in the blood of the mouse

(Popp, 1960; Goodman and Hodgson, 1962), dog (Cavins *et al.*, 1964), guinea pig (Malinin *et al.*, 1965), rat (Lord, 1967), and rabbit (Grigoriu *et al.*, 1971), they have also been found in man (Barr *et al.*, 1975). There is now evidence that the number of stem cells in the fetal circulation may be considerably greater than after birth.

In the case of the mouse, the stem cell content of the blood can be measured either in terms of CFU-S or the capacity to confer protection. Barnes *et al.* (1964) noted that when lethally irradiated mice were transfused with blood leukocytes to confer protection, 10^7 cells were required if the leukocytes were from the blood of adult mice, whereas if from the blood of fetal mice only 10^4 to 10^5 were needed. In other words, fetal blood contained more than 100 times as many stem cells as adult blood.

1. *Transitional Cells in Human Fetal Blood*

In man, throughout the greater part of fetal life, the blood leukocytes consist largely of what were at first termed lymphocytes (Thomas and Yoffey, 1962). Subsequent examination brought out the fact that in midfetal life a high percentage of the cells which had first been described as lymphocytes were transitional cells (Winter *et al.*, 1965). A representative selection of these has been illustrated in color (Yoffey, 1971). Although the percentage of transitional cells in fetal blood diminishes during the later stages of pregnancy, they are still present in appreciable numbers in the blood of the fetus at full term (Faulk *et al.*, 1973). About 2% of these cells show spontaneous DNA synthesis, as opposed to 0.2% in the adult (cf. Prindull, 1974; Prindull *et al.*, 1975). In general, the fetal cells have a much higher N:C ratio than those of the adult.

Faulk *et al.* (1973), in addition to confirming the much greater incidence of cells in spontaneous DNA synthesis in cord blood, performed ultramicroscopic studies. They found that, in addition to typical transitional cells, there was an unexpectedly large number of cells which were neither typical pachychromatic small lymphocytes, nor fully leptochromatic transitional cells, but possessed an intermediate type of nuclear structure. Morris *et al.* (1975) illustrate a "blast-like" cell from cord blood which is a characteristic transitional cell (cf. Fig 8.4, p. 648, Yoffey and Courtice, 1970).

There are significant differences in RNA as well as DNA synthesis. Winter *et al.* (1965) noted that in cord blood twice as many cells labelled spontaneously with tritiated uridine as in the blood of adults. Prindull *et al.* (1976) reported that cytoplasmic DNA synthesis, in all probability mitochondrial, occurred more frequently in cells of cord than adult blood. Some of the problems of RNA and DNA synthesis have been reviewed by Prindull *et al.* (1977b).

Knudtzon (1974) investigated the colony-forming capacity of cord blood cells and concluded that there were about 40 times more CFU-C in cord than in adult blood, the colonies being predominantly granulocytic. Prindull *et al.* (1978)

obtained essentially similar results when comparing the number of CFU-C in fetal and adult blood. From the work of Moore *et al.* (1973) it would appear that the CFU-C are transitional cells.

2. *Response to Mitogens*

Prindull (1971) compared the growth properties of cord blood and adult lymphocytes and concluded that the lymphocytes of the newborn proliferated more actively than those of the adult. Weber *et al.* (1973) compared the speed of the response to PHA of neonatal and adult lymphocytes and noted that the neonatal lymphocytes responded more rapidly. They attributed this to two factors. For most of the neonatal lymphocytes the response was of the usual type, except that they calculated the G_1 period was 4 hours less than in similar cultures of blood from adults. But they also postulated that a small number of lymphocytes had been "prestimulated" *in vivo,* and that this prestimulation accounted for their entry into S phase soon after exposure to PHA.

Yoffey *et al.* (1978a) studied the response of cord blood lymphocytes to PHA and to Con A and confirmed the main results of Weber *et al.* (1973) (cf. Stites *et al.,* 1972), but explained the early response of cord blood lymphocytes on the basis of nuclear structure. In mitogen-stimulated cultures of adult blood lymphocytes, the essential first stage is the loosening of chromatin as the pachychromatic nucleus becomes leptochromatic. Only when a sufficient degree of leptochromasia has been attained can DNA synthesis begin. Since, as already noted, a large number of cord blood lymphocytes are already partly leptochromatic, they can attain the necessary degree of leptochromasia sooner. Whatever the explanation, these findings provide further evidence that, in addition to the undoubted transitional cells present in cord blood, neonatal lymphocytes and perhaps what are modified transitional cells are a distinctive cell group in the fetus and neonate.

V. The Size Spectrum in the Transitional Cell Compartment

Reference has already been made to the existence of a size spectrum in the transitional cell compartment. Because the size data are of fundamental importance in an analysis of the stem cell problem, it was decided to obtain more detailed information. A full account of the technical procedures has been published elsewhere (Patinkin *et al.,* 1979; Yoffey *et al.,* 1978b) and only the essential features will be mentioned here. For the accurate measurement of large numbers of transitional cells and lymphocytes in a Coulter counter, it was necessary to obtain populations of these cells in as pure a state as possible, free from contamination by other cells. As the starting point, guinea pigs were used between 7 and 10 days of rebound (Section VIII) after 7 days of hypoxia at 0.5 atm. The studies of Jones *et al.* (1967) showed that the marrow in these animals contains not only greatly increased numbers of transitional cells and lymphocytes, but also markedly diminished numbers of nucleated erythroid cells and a trend toward diminished production of myeloid cells. Passage of rebound marrow through albumin density and Ficoll gradients reduces to two or three the number of components to be measured in each fraction in which the size distribution can be easily analyzed.

Four peaks were obtained in all. Peak 1 was a population of small cells (volume 53–59 μm^3) composed mainly of pachychromatic small lymphocytes with a few larger cells, which may be the L-T cells already mentioned (Section II).

Peak 2 was a population of somewhat larger cells (volume 154–160 μm^3), consisting of small transitional cells with little or no cytoplasm.

Peak 3 was a population of still larger cells (volume 200–218 μm^3) entirely of medium to large transitional cells with a large leptochromatic nucleus and clearly visible pale to basophilic cytoplasm. These cells are found exclusively in the lighter density fractions of 17 and 19% albumin.

Peak 4 was a population of very large transitional cells with a large leptochromatic nucleus (volume 350–400 μm^3) with relatively more—and usually much more basophilic—cytoplasm. These cells were somewhat inconstant, and their numbers when present were smaller than those in the other peaks (Fig. 3).

A. Significance of Size Spectrum for Lymphocyte Production

The existence of at the most three size groups in the transitional cell compartment affords additional evidence that marrow lymphocytes—for the most part presumably B lymphocytes—are formed via a *short production pathway* involving only two or three mitoses. The shortness of the pathway explains the observation of Everett and Caffrey (1967) that in radioautographs, after the administration of tritiated thymidine *in vivo,* there is markedly denser labeling in marrow lymphocytes than in those of the thymus, in which there is a *long production*

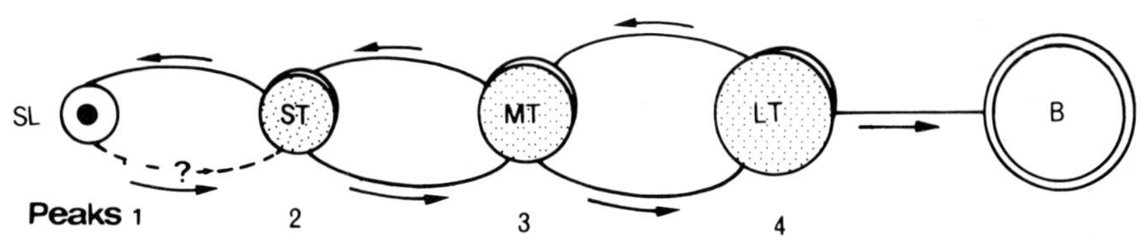

FIG. 3. There are three basic size groups in the transitional cell compartment, namely, large (LT), medium (MT), and small (ST) transitionals. In self-maintenance there can presumably be cycling between the small and medium, as well as between the medium and large transitionals. (– – –?– – –) Possible reentry of subgroup of uncommitted small lymphocytes (SL) into transitional cell compartment. Lymphocyte (B) production can involve at the most only three mitoses, constituting the short production pathway. (Modified from Yoffey *et al.*, 1978b.)

pathway estimated to consist of eight mitoses (Leblond and Sainte-Marie, 1960; Yoffey and Courtice, 1970). As a result of the greater number of mitoses in the case of the thymus, those small lymphocytes emerging from cells which label at the beginning of the pathway, and therefore undergo eight mitoses, show extreme label dilution (Yoffey *et al.*, 1961a). The finding by Robinson *et al.* (1965), after continuous infusion of tritiated thymidine *in vivo,* of two groups of lymphocytes in the bloodstream—one heavily and one lightly labeled—could be explained on the basis of the intermingling in the bloodstream of the lymphocytes emerging from the long and short production pathways (cf. Craddock, 1965; Rosse 1971, 1978).

B. B-Lymphocyte Formation from Transitional Cells

The formation of B lymphocytes in the marrow has been discussed at length by Rosse (1976). The present review therefore will deal only briefly with some problems of their origin from transitional cells. One of the main lines of approach has been to separate the B lymphocytes from their precursor cells. Basten *et al.* (1972), working with murine bone marrow, effected this separation by passing the cell suspension through a specially designed column. The precursor cells did not bind anti-immunoglobin serum, whereas the small lymphocytes did. The precursor cells were colony-forming units, i.e., transitional cells, as discussed in Section VI.

Lafleur *et al.* (1972) removed B cells from a marrow suspension by velocity sedimentation. They then injected the remainder of the suspension, containing the precursor cells, into irradiated mice and found that they gave rise to B lymphocytes. Detectable numbers of B lymphocytes were found within 3 days after transplantation, following which B-cell activity increased with a doubling time of 24 hours. In fact (Section VII,B) this is quite slow growth for the larger transitional cells and suggests rather that the proliferation may be taking place at the smaller end of the transitional compartment, through cycling between the medium and small transitionals (Fig. 3).

Lafleur *et al.* (1973) found an occasional precursor cell which was already taking up anti-immunoglobin serum. Osmond and Nossal (1974a,b) postulated that transitional (''lymphoid'') cells which were going to form B cells developed subthreshold concentrations of surface immunoglobulin immediately before reaching the stage of the mature B lymphocyte, whose surface immunoglobulin was readily detectable (cf. Ryser and Vassali, 1974) The fact that here and there, as noted by Lafleur *et al.* (1973), one could find an occasional transitional cell with detectable surface immunoglobulin is perhaps comparable in the process of B-cell development with the asynchronous erythroblast which (Section VIII,F) sometimes appears in erythroid differentiation. In both instances a specific cytoplasmic constituent becomes apparent earlier than is usually the case, while the cell still has transitional cell morphology.

The formation of B lymphocytes from transitional cells raises some interesting problems of cell differentiation. If all the receptors on a given B cell are specific for the particular antigen which they bind (Raff *et al.*, 1973), can one pluripotential cell give rise to large numbers of B lymphocytes each of which differs from all the others in possessing its own specific antigen receptor? Yung *et al.* (1973), who protected lethally irradiated mice with fetal liver and investigated the subsequent development of their immune responses, concluded that this was actually the case, and that from one stem cell, in the course of about 20 days, there could arise a panel of lymphocytes capable of recognizing something like 10,000 antigens.

In the case of fetal liver, the pluripotential stem cell may or may not have been a transitional cell. But in a similar experiment performed by Trentin *et al.* (1967) with adult marrow, the transitional cells would certainly be the cells primarily involved. Yung *et al.* (1973) thought that their results supported the clonal selection theory, whereas Trentin *et al.* (1967) felt that the requisite mutation rate was far too rapid for the development of so many antigen-recognition specificities. From what we know now, however, about the proliferative capacity of transitional cells, they could certainly be able to multiply at the speed required for the postulated mutation rate. They are in fact the only primitive cells in the marrow capable of so multiplying (Section VII).

Whatever the correct immunological interpretation, experiments such as these raise a fundamental problem in connection with the transitional cell compartment. How, on one hand, do transitional cells give rise to comparatively uniform populations of blood cells, such as erythrocytes and granulocytes, and, on the other hand, to such diverse populations as the B lymphocytes with their manifold antigen-recognition properties? In considering the development potential of transitional cells, this is a question to which we have as yet no answer.

C. The Size Spectrum and Hemopoietic Stem Cells

Since the transitional cell compartment contains stem cells not only for lymphocytes but also for all the other blood cells, the size spectrum of the compartment carries with it the necessary corollary that hemopoietic stem cells should also exist in different sizes. There is now a growing body of evidence that such indeed is the case (Section VII,C).

VI. Colony-Forming Units and Transitional Cells

Evidence from a number of investigators indicates that the colonyforming units which proceed from the marrow to the spleen (CFU-S) are transitional cells. The colony-forming units arising in fetal liver have not been as thoroughly investigated as those from the bone marrow, and their precise identity has not been so convincingly established. To the extent that there are transitional cells in fetal liver (Section IV,B) they would presumably be the hepatic counterpart of the marrow CFU-S.

A. Direct Injection of Transitional Cells into Spleen

Following the work of Turner *et al.* (1967) and Hurst *et al.* (1969), Murphy *et al.* (1971) further investigated the changes in murine bone marrow during rebound (Section VI,A). After 3 days of rebound, marrows were treated by density gradient centrifugation. They found that 80 to 95% of a light-density fraction was composed of mononuclear cells which had typical transitional cell structure, namely, a leptochromatic nucleus, one or more nucleoli, sparse endoplasmic reticulum, a conspicuous golgi zone, and a centrosome in the region of the nuclear indentation. A suspension of these cells was injected directly into the shielded exteriorized spleen of mice which had received a lethal dose of irradiation 48 hours previously. The animals were killed at times ranging from 15 minutes to 8 days later. The injected cells developed into typical spleen colonies, with cells of the erythrocyte, granulocyte, and megakaryocyte series.

B. Endocolonization from Shielded Marrow

De Gowin *et al.* (1972) and De Gowin and Gibson (1976) made ingenious use of the endocolonization technique. With one leg shielded, mice were irradiated with 850 rad, and 3 hours later the shielded leg was also irradiated with 850 rad. Over the next 3 to 4 days, repletion of the stem cell compartment took place and at various times tritiated thymidine was injected intravenously and the spleen examined in two stages. One hour after thymidine injection a portion of the spleen was removed and was found to contain heavily labeled mononuclear cells resembling "medium to large leptochromatic lymphocytes." The remainder of the spleen was removed from the same mouse 24 to 48 hours later and was found to contain lightly labeled erythroblasts, myeloid cells, and lymphoid cells. Grain counts suggested that erythroblasts and their precursors had undergone about four divisions, myeloid cells and their precursors two to three divisions, and lymphoid cells and their precursors two to three divisions during the 48-hour period. They conclude that the migrating endogenous cells are pluripotential hemopoietic stem cells. They resemble "medium to large leptochromatic lymphocytes and replicate during stem cell compartment repletion." Figure 1 on p. 319 of De Gowin and Gibson (1976) depicts a typical labeled large transitional cell, which is described as a "labelled mononuclear cell resembling a large leptochromatic lymphocyte." However, by the authors' own definition, the cell is clearly not a lymphocyte, but a pluripotential stem cell which gives rise to all the other blood cells in addition to lymphocytes.

The experiments of Orlic *et al.* (1968), though somewhat different in their experimental design, have some features in common with those of De Gowin and Gibson (1976). They made an ultrastructural study of erythropoiesis—resulting from the administration of erythropoietin—in the spleens of mice during rebound. In these spleens there was, at the start of the experiment, the usual depression of erythropoiesis which occurs during rebound. They used alternate thick and thin sections, and by autoradiography in the thick section they could see labeled cells which they interpreted as newly activated erythropoietin-sensitive cells. These cells could then be examined in greater detail in the alternate thin sections in which their ultrastructure could be observed. Within 1 hour after administration of erythropoietin, they observed cells which became labeled with both [^{3}H-]thymidine and [^{3}H-]uridine. The labeled cells appear to have been erythropoietin-sensitive transitional cells, as shown in their Figs. 1,2, and 3. The early increase in RNA synthesis needs to be evaluated in relation to the initial RNA content of the cells. The position in this respect is not altogether clear. But if the initial RNA content is low, then the erythropoietin-responsive cells would be pale transitionals, and the uridine uptake would be the counterpart of the change from paleness to basophilia as seen with the light microscope. Rosse (1973) has shown that in stimulated marrows (bleeding + hypoxia) the basophilic transitionals label with uridine as heavily as proerythroblasts and basophilic erythroblasts.

C. Transitional Cell Enrichment

Techniques to obtain greatly enriched populations of transitional cells have made it possible to investigate their stem cell role with greater ease and certainty than when they are only a small fraction of a large cell population. An interesting example of this is the study of the *in vitro* colony-forming cell (CFU-C) by Moore *et al.* (1972; cf. Dicke *et al.*, 1973) who worked with the bone marrow of Rhesus monkeys. By means of buoyant density gradient separation they obtained a light-density distribution profile of cells which gave rise to hemopoietic colonies (macrophage and granulocyte) in agar culture. With high-resolution density gradient separation applied to a light-density fraction of bone marrow, they succeeded in obtaining a 100-fold enrichment of *in vitro* CFU-C. The most highly enriched fractions contained the majority of the CFU-C present in the original marrow. Fractions were regularly obtained in which practically one-fourth of the cells were colony formers.

D. Active Proliferation of CFU-C

That the CFU-C were actively proliferating cells could readily be shown by the thymidine suicide technique. Metcalf (1972) found that CFU-C from murine bone marrow had an *in vitro* suicide rate of 45%, whereas Harris and Hoelzer (1978) found a CFU-C suicide rate of 41% in normal rat marrow, and about 25% in leukemic marrow. Iscove *et al.* (1970) reported a suicide rate of 35% for CFU-C in murine marrow. In terms of the percentage of cells involved, the thymidine suicide rate is presumably the same as the labeling index would be if thymidine of lower specific activity had been used. Hydroxyurea, killing cells in S, gives more or less comparable information. Rickard *et al.* (1970) found that approximately 50% of CFU-C from murine marrow were killed by hydroxyurea.

Moore *et al.* (1972) showed not only that CFU-C possessed the high labeling index and morphology of transitional cells, but also that single cells could give rise to colonies containing both macrophages and granulocytes, thus demonstrating a common origin for these two groups. Dicke *et al.* (1973) were able to obtain a 50-fold enrichment of CFU-C from human bone marrow, and here too, on both light and electron microscopy, the cells obtained were typical transitional cells.

E. CFU-S in Marrow after Nitrogen Mustard

Another form of what might be termed enrichment *in vivo* was employed by Sharp *et al.* (1976) and Riches *et al.* (1976a,b). They administered mustine hydrochloride to mice, following which the nucleated cell content of the marrow falls to 20% of control values, whereas the CFU-S, at first depressed to 10% of control values, rose sharply to 60% on Day 4, while from the fifth day onward there was evidence of varying degrees of marrow regeneration. The marrow at Day 4, greatly enriched in transitional cells, also contained a correspondingly greater number of CFU-S. The transitional cells are occasionally referred to as such, but more often they are described as "leptochromatic mononuclear cells." The illustrations, including radioautographs, are of typical transitional cells. It is difficult to compare these results in the mouse with those of Thomas *et al.* (1965) in the dog, but it could well be that the "lymphoid" cells observed by those authors are the canine equivalent of the murine CFU-S, more especially since the demonstration by Keiser *et al.* (1967) of the presence of transitional cells in canine bone marrow.

VII. Kinetics of Transitional Cell Compartment

A. Labeling with Tritiated Thymidine

Since transitional cells are an actively proliferating group, it is not surprising that so many of them label with tritiated thymidine. The first observation on the DNA labeling of transitional cells was made by Everett *et al.* (1959; cited by

Yoffey, 1960, Table II, p. 31). In one guinea pig it was noted that 2 hours after the administration of tritiated thymidine, 20 out of 35 transitional cells were labeled, at a time when there were no labeled small lymphocytes. The significance of this and similar findings was not adequately appreciated at the time. Five years later Osmond and Everett (1964) reported that in guinea pig marrow 35% of the transitionals labeled with tritiated thymidine, whereas in the rat, Everett and Caffrey (1967) noted that about 50% of the transitionals labeled after the intravenous injection of thymidine. Studies now in progress on the transitional cells in rat bone marrow give a labeling percentage of the same order or even higher (Yoffey and Yaffe, 1978).

B. The Proliferation Gradient

The kinetics of the transitional cell compartment are very complex, and as yet imperfectly understood. One of the more obvious features is the proliferation gradient, by which is meant the gradual fall in the labeling index from the large to the small transitionals (Moffatt *et al.*, 1967) (Fig. 4). The gradient is a funda-

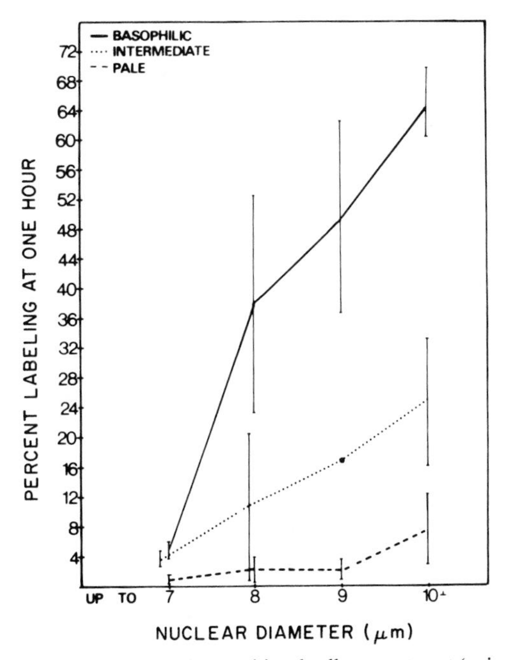

Fig. 4. The proliferation gradient in the transitional cell compartment (guinea pig) 1 hour after *in vivo* labeling of bone marrow cells with tritiated thymidine. The highest labeling index is to be found in the basophilic cells, the lowest in the pale transitionals. The intermediate cells are between these two groups. In all three groups, the larger cells have a considerably higher labeling index than the small ones. (Drawn from the data of Moffatt *et al.*, 1967.)

mental property of the compartment. It seems to involve at least three known factors, which may somehow be interrelated, and probably several factors which are still the subject of conjecture. The known factors are (a) cytoplasmic basophilia, (b) nuclear diameter, and (c) length of the S period. As to nuclear diameter, in general, the larger the cell, the higher the labeling index. The difference is more marked in the basophilic group, in which the figures for labeling percentage are: ≤7 μm, 4.9%; 8 μm, 38.2%; 9 μm, 49.5%; 10 μm+, 65.4%. In the pale group the index was very much lower: ≤7 μm, 1.4%; 8 μm, 2.4%; 9 μm, 2.2%; 10 μm+, 7.7% (Moffat *et al.*, 1967). For the group with intermediate degrees of basophilia the figures were: ≤7 μm, 3.6%; 8 μm, 10.9%; 9 μm, 17.5%; 10 μm+, 25.3%. Another noteworthy feature in the transitional cell compartment is the variation in the duration of the S phase, which according to Miller and Osmond (1973) is 3.5 hours in the largest transitionals, with a cell cycle time of 6.4 hours, whereas in the smaller transitionals S is 10.9 hours. The high labeling index and low mitotic index in the basophilic transitionals led Moffat *et al.* (1967) to suggest that they might be cells which were already committed to a given line of differentiation, and that while DNA synthesis began at the transitional stage, mitosis would not occur until the cell had entered its specific blast cell compartment. The pale transitionals could be regarded as a resting phase of the stem cell population, and this view seemed to be supported by the work of Rosse (1970a), who noted that many pale transitionals could be out of cycle for several days. This interpretation seemed to fit in well with the view emerging from CFU-S studies that an appreciable number of pluripotential stem cells were in a resting nonproliferative state. But it must be borne in mind that the same considerations could apply to the smaller basophilic transitionals, which are also proliferating at a low rate. Whichever resting cells are concerned, whether pale or basophilic, the entry of a small number of nonproliferating stem cells into a highly proliferative state could provide an obvious amplifying mechanism when more stem cells are required. If this interpretation is correct, the cells with various intermediate degrees of basophilia could be those which were passing from the low to the high proliferative stage.

C. Stem Cells of Varying Size

In considering the size spectrum of transitional cells, it was suggested that if they are in fact stem cells, then these latter must also exist in different sizes. Reference has already been made to the observation by De Gowin and Gibson (1976) that the labeling stem cells which they found in the spleen were "medium to large leptochromatic lymphocytes." More accurate assessment of the distribution of large and small stem cells has come from the use of the velocity sedimentation technique developed by Peterson and Evans (1967) and Miller and Phillips (1969). This depends basically on the fact that large cells sediment more rapidly than small ones. According to Anderson *et al.* (1969), when mammalian cells divide and their volume is approximately halved, the sedimentation velocity of the larger cell will be approximately 1.59 times that of the daughter cells. In a self-maintaining compartment it would therefore be reasonable to expect a difference in sedimentation velocity of this order between the large and small cells, and in general, differences of this order of magnitude are in fact found.

In practice, the line of demarcation between the largest and the smallest stem cells is never quite so clear cut as this analysis might lead one to expect, since if small cells are continually growing into larger ones there are bound to be intermediate sizes, as indicated schematically in Fig 2. Furthermore, with three size peaks in the transitional cell compartment, two possibilities of self-replication call for consideration, namely, cells cycling between large and medium or between medium and small transitionals.

One of the unknown factors in the control of the stem cell compartment is the way in which the proportion can be varied both of small cells undergoing enlargement and of large transitionals which either differentiate or divide to replenish the smaller cells. In this connection it cannot be too strongly emphasized that in all the available data on varying stem cell sizes, it is the transitional cell compartment alone which has the requisite size spectrum to account for evidence of size heterogeneity in the stem cell compartment. Likewise, only the transitional cells possess the labeling properties which would agree with the thymidine suicide data on stem cells.

D. Velocity Sedimentation Data

The results obtained by velocity sedimentation show on the whole a fair measure of agreement on the smaller cells, which sediment at around 4 mm/hour, but at the larger end there is a considerable degree of spread when one compares the results of different observers. Metcalf and Macdonald (1975) separated murine marrow cells by sedimentation velocity, and then cultured them in agar with three different types of colony-stimulating factor. They obtained two types of growth, colonies and clusters. Colony-forming cells had a single peak (S 4.4 mm/hour), whereas cluster-forming cells had two peaks, with a sedimentation velocity of 5.7 mm/hour for the larger main peak.

Metcalf *et al.* (1977) irradiated mice sublethally (250 rad), and 2 days later found that the sedimentation velocity had risen from 4.3 to 7.8 mm/hour, after which it slowly returned to normal over a period of 3 weeks. The great increase in sedimentation velocity correlates well with the work of Harris (1956, 1960), who noted that in the first 2 weeks or so after sublethal irradiation (150r) in guinea pigs there was a great increase in proliferating transitional cells which are presumably the larger members of the transitional compartment, with a high labeling index (Fig. 4) and a short cycle time. This interpretation is further supported by the finding that, 2 days after 250 rad, the thymidine suicide rate had gone up from 35 to 70% (Metcalf *et al.*, 1977). It is pertinent at this point to compare these sublethal irradiation data with the studies by Sutherland *et al.* (1971) on marrow regenerating 6 days after transplantation, where they found a high value for CFU-C at 7.3 mm/hour, as against the usual peak value of 5 mm/hour. Here too, regenerating marrow has actively proliferating transitionals, the "lymphocyte-like" cells of Cudkowicz *et al.* (1964b). As is evident from the proliferation gradient (Fig. 4), the most active proliferation takes place in the largest cells.

Heath *et al.* (1976) used the velocity sedimentation technique in an analysis of the action of erythropoietin on stem cells, which they found to sediment in two peaks differing in their response to erythropoietin. One, sedimenting at 5.7 to 7.0 mm/hour responded rapidly to a low dose of erythropoietin and formed colonies of hemoglobin-synthesizing cells which reached their peak numbers in 2 days, ending up as small colonies (up to 65 cells) of nonnucleated cells. These stem cells responding rapidly to erythropoietin they termed CFU-E. By way of contrast, in cultures which were begun with a high dose of erythropoietin, with which they were continually fed, crops of new colonies appeared after about 6 days, and they were termed "bursts," the stem cells being known as BFU-E. CFU-E sedimented at 5.7 to 7.0 mm/hour, whereas BFU-E sedimented at 3.7 to 4.1 mm/hour. The CFU-E would presumably be the large transitional group immediately preceding the proerythroblast stage, and the findings tempt one to infer that it is the larger transitionals which are the ERC. The BFU-E, with a sedimentation velocity of 3.7 to 4.1 mm/hour, would be at the smallest possible end of the transitional cell spectrum, which Heath *et al.* (1976) interpret as the "EPO-responsive progenitor, the committed erythroid stem cell." This interpretation is further discussed in Section VIII, D,E,F, but for the moment one can only reemphasize that, with the erythropoietin-responsive cells as with the granulocyte–macrophage progenitors, the CFU-C, there are no other stem cells beside the transitionals which possess the size spectrum which fits in with the observed facts.

Monette *et al.* (1974) fractionated mouse bone marrow cells by velocity sedimentation, finding the usual size spread. However, treatment with hydroxyurea, which acts selectively on cells in DNA synthesis, destroyed a high proportion of large transitionals and left undamaged mainly small CFU-S, i.e., small transitionals, the vast majority of which are not in DNA synthesis (Fig. 4) and which sedimented at a velocity of 4.2 mm/hour. If one correlates these data with those of Miller and Osmond (1973), which give a cell cycle time for the larger transitionals of around 6 hours, it would appear that the supply of stem cells can be increased very rapidly by activating the small transitionals and switching them from the relatively quiescent small end to the highly proliferative large end of the compartment. A cell cycle time of around 6 hours could on this basis mean a very considerable increase in the number of stem cells in the course of 24 hours. It is pertinent here to cite the data of Boggs *et al.* (1972) who reported a generation time of 6.4 hours for CFU-S.

It is possible that the process can be accelerated still more. According to Monette *et al.* (1968), the cell cycle time for the less mature erythroid cells in the rat is 4 hours. These early erythroid cells are very difficult to distinguish from the transitional cells which are their precursors (Section VII,F) (cf. Rosse and Beaufait, 1978). If the large transitionals can have a cell cycle time of this brevity, then the amplifying potential of the transitional compartment could be very considerable indeed when greatly increased numbers of stem cells are required for augmented erythropoiesis, provided that at the same time the number of small transitionals which enlarge is increased, and the rate of their growth is speeded up. In other words, there has to be an appropriate balance of differentiation and self-replication, as maintained by Boggs and Boggs (1976).

E. Quantitative Data on Transitional Cell Proliferation

Osmond *et al.* (1973) have contributed some important quantitative data to our understanding of the transitional ("lymphoid" in their terminology) cell population of guinea pig bone marrow. They analyzed the transitional cells in terms of nuclear size, nuclear structure, cytoplasmic basophilia, and proliferative activity. They observed varying degrees of pachychromasia in the transitional cells, up to 40% in the cells ranging from 8.0 to 8.9 μm, and 7.0% in the 9.0 to 9.9 μm group. The 10-μm+ cells were almost entirely leptochromatic. Cells with the more pachychromatic nuclei do not label and may be regarded as the resting stage of the transitional compartment.

They confirmed the data of Moffatt *et al.* (1967) in finding that of the transitionals which showed DNA synthesis, the majority (86%) were basophilic. The mean labeling index for the three groups was: basophilic, 69.8%; intermediate, 8.9%; and pale, 3.3%. They also made important observations on the varying duration of the S period and calculated that this ranged from 3.5 ± 0.3 hours for the largest transitionals, to 10.9 ± 0.9 hours for the smallest. The overall production by the large transitionals was equivalent to 2.2% of the entire transitional cell population per hour, a rate of production which "considerably exceeded the requirements for renewal of small lymphocytes." Since in the normal steady state the transitional compartment remains unchanged, the production in excess of that required for small lymphocytes must be needed for other purposes, namely, to act as precursors for other cells, mainly erythrocytic and granulocytic. Double centrifugation in a sucrose–serum gradient yielded a virtually pure population (96.0 ± 1.9%) of transitional cells, which had a very high protective capacity, confirming once again the view that transitional cells give rise to other cells besides lymphocytes.

F. The Normal Steady State

In the normal steady state the transitional compartment is stable, despite the continued differentiation into the various types of cells which are leaving the marrow. But at the same time as these cells are differentiating, some of the large transitionals divide to maintain the supply of small transitionals from which they are derived. The generally accepted view is that in the normal steady state only a small number of stem cells are in cycle (e.g., Lajtha and Oliver, 1960; Becker *et al.*, 1965). Using the thymidine suicide technique, Becker *et al.* (1965) concluded that in adult mice the number of stem cells in DNA synthesis was "almost imperceptible," whereas in situations where the hemopoietic system is expanding, 40 to 65% of CFU-S could be in DNA synthesis. According to Blackett (1968), in the rat a higher proportion of stem cells is in cycle (30%) than the 10% usually accepted for the mouse. But in either case there is a substantial reserve of stem cells whose proliferation can be increased by various experimental procedures. Some of the more important of these are discussed in the following section.

VIII. Modulation of the Transitional Cell Compartment

A number of experimental procedures have been directed toward a study of the bone marrow in order to throw light on the stem cell problem. These procedures fall into two main groups. In one group, efforts were made to induce an increased demand for stem cells, which might then be produced in greater numbers and so be more readily identified. In another group, a sustained demand for stem cells placed too great a strain on the stem cell compartment, which became markedly depleted.

A. Sublethal Irradiation

One of the early contributions was made by Harris (1956), who subjected guinea pigs to sublethal irradiation (150 rad whole body) and then followed the changes in the bone marrow by a quantitative technique. The results were striking. An initial abrupt fall in all the radiosensitive nucleated cells, most marked in the transitionals (Harris, 1960), was succeeded by a series of changes which Harris followed for 36 days. Regeneration began with the transitional cells and lymphocytes, which multiplied very actively from the fourth day onward until by Day 16 they were about twice their control value. Recovery of the nucleated erythroid cells was at first very slow, but became more marked from Days 14 to 20, with a slight overshoot above control values. Myeloid regeneration started about Day 16 and rose to 50% above control value. From Days 16 to 20, as the erythroid and myeloid regeneration were reaching the peak of their activity, transitional cells and lymphocytes fell abruptly to less than half their control levels. The striking increase in transitional cells before the great spurt in erythropoietic and granulopoietic regeneration, and their marked fall when this regeneration began, seemed to fit in very well with the view that at least some of the transitionals might be stem cells, the precursors of blast cells and also of small lymphocytes (Harris and Kugler, 1965; Osmond and Everett, 1964). Further support for the stem cell role was obtained by the finding (Harris *et al.*, 1963), confirming earlier observations, that transitional cells labeled actively with thymidine and underwent frequent mitoses.

Other studies on the effects of irradiation were performed by Simar *et al.* (1968) in the mouse and Blackett *et al.* (1964) in the rat. The last-named authors also found a great increase in ''lymphoid'' cells after irradiation. Chervenick and Boggs (1971) studied the effects of irradiation by means of the endogenous spleen colony technique, which is of special interest in view of the fact that the colony-forming units are members of the transitional cell compartment (Section VI).

B. Radiomimetic Substances

Experiments with radiomimetic substances gave results very like those of irradiation, with the possible difference, as noted by Haas *et al.* (1968), that recovery changes after nitrogen mustard seem to be more rapid than after irradiation. Thomas *et al.* (1965) administered a nearly toxic dose of HN_2 to dogs and after 3 to 6 days examined the stem cell content of the regenerating marrow by transfusing a suspension of its cells into an irradiated dog, whose marrow then began to regenerate. The first cells to appear in the course of the regeneration became evident after 36 hours. They were ''heavily labelled small 'lymphoid' cells. . . . At 60 hours, most of the labelled cells were of the 'haemocytoblast' type. At 120 hours, though the marrow was still hypocellular, young myeloid and erythroid cells predominated.'' From the work of Keiser *et al.* (1967) it would appear that the ''lymphoid'' cells in the dog are transitional cells.

C. Marrow Transplantation

Following the pioneer studies of Jacobson *et al.* (1949) on the effect of spleen shielding in conferring radiation protection in mice, numerous experiments were performed on the transfusion of suspensions of marrow cells, which were found to recolonize the entire lymphomyeloid complex and not only the bone marrow (for literature see Micklem and Loutit, 1966; and Yoffey, 1974). Jacobson *et al.* (1954) observed that marrow cells protected in accordance with the numbers given, with a maximum upper limit, as might be expected. Virtually none of the earlier workers tried to identify the cells concerned in conferring protection.

Cudkowicz *et al.* (1964a) transfused isologous marrow into irradiated mice and used the spleen colony technique to count the stem cells in the regenerating marrow of the recipients. When the recipient bone marrow was examined within 10 days after transplantation, it contained a fair number of myeloblasts and erythroblasts, but its ability to give rise to spleen colonies was greatly reduced. They concluded from this that myeloblasts and erythroblasts were not stem cells. The marrow did not regain its normal stem cell content until after 30 days or more, and this they were able to correlate with the reappearance in the marrow of normal numbers of "lymphocyte-like" cells which must have been transitional cells since they flash-labeled with thymidine, with which the pachychromatic small lymphocytes do not. Cudkowicz *et al.* (1964b) also studied the effects of serial transplantation. Thirty days after the first transplantation, the regenerating marrow had a normal stem cell content, while after the second and third transplantations, although protection was conferred, the marrow content of "lymphocyte-like" cells was beginning to diminish. The fourth transplantation did not confer protection, and the recipient marrow was then found to contain very few "lymphocyte-like" i.e., transitional, cells.

Experiments of this type would seem to suggest that if the demands on the stem cell compartment become too heavy, it finally is unable to cope with them. The dual role of the transitional cell compartment, self-maintenance on one hand and differentiation on the other, implies that the size of the compartment at any given time is the result of a very dynamic equilibrium between these two processes. In the regeneration of irradiated marrow by transfused stem cells, there would appear to be a very strong demand for differentiation to give rise to the various blood cells required, and this could conceivably result in all or most of the transfused stem cells differentiating, with few or none left to maintain the compartment subsequently. Since after the primary transplantation this does not occur, it seems reasonable to accept an interpretation such as that of Chervenick and Boggs (1971), that if the stem cell compartment is small, there may be "a phase of cell growth refractory to differentiating stimuli" (cf. van Bekkum and Weyzen, 1961; De Gowin and Johnson, 1967; Lajtha *et al.*, 1964). But presumably after prolonged and heavy demand for new blood cells, this essentially defensive mechanism in the stem cell compartment, to prevent it disappearing completely, finally breaks down. (cf. Vos and Dolmans, 1972).

Whatever the mechanism involved, these multiple transplantation experiments give rise to a condition which is the reverse of that which follows sublethal irradiation. After sublethal irradiation the marrow has increased numbers of transitional cells, whereas in the transplantation experiments the transitional cells disappear. Marrows rich in transitional cells have an enhanced protective value (Harris and Kugler, 1963; Osmond *et al.*, 1973), whereas those poor in these cells have diminished protective capacity.

D. Transitional Cell Enrichment and Enhanced Protection

If transitional cells are indeed stem cells, then an obvious line of attack is to find out whether enriched suspensions of transitional cells are more effective than normal marrow in conferring protection on lethally irradiated animals. A number of observers have obtained results which indicate that this is in face the case. Harris and Kugler (1963, 1967) transfused guinea pig marrow 13 days after sublethal irradiation, at which time the marrow contained about 75% of lymphocytes and transitionals, of which 13% were transitional. In rats, Morrison and Toepffer (1967) obtained a concentrated suspension of lymphocytes and transitional cells from normal marrows by dextran gradient separation, and found here also that the cells gave more effective protection than normal marrow against lethal irradiation. In these enriched suspensions there were always a fair number of small lymphocytes, which made the answer not as clear-cut as it would otherwise have been. This difficulty was overcome by Osmond *et al.* (1973), who employed double centrifugation in a sucrose–serum gradient of guinea pig marrow, following the earlier work of Osmond and Yoshida (1970), and thereby obtained suspensions containing more than 98% transitional (''lymphoid'' cells (See Fig. 1, loc. cit., p. 132).

E. Transitional Cells in Erythroblastic Islands

The erythroblastic islands have provided an interesting demonstration of transitional cell differentiation along erythroid lines. It has long been recognized that increased erythropoiesis involves the differentiation of additional stem cells (Kindred, 1942; Yoffey, 1957; Erslev, 1959, 1964; Alpen and Cranmore, 1959; and many others). It seemed the obvious thing to look for these stem cells in the erythroblastic islands, which consist essentially of a central macrophage, the *central reticular cell,* surrounded by erythroblasts in various stages of differentiation. However, during rebound (Section VIII,G) the period of posthypoxic polycythemia, most of the nucleated erythroid cells disappear, and the majority of the erythroblastic islands consist of a central reticular cell, surrounded by nonerythroid cells, mainly small lymphocytes and an occasional transitional cell (Ben-Ishay and Yoffey, 1971, 1972). When erythropoiesis is resumed sooner or later, one can then see quite clearly, on ultramicroscopic examination, that in close contact with the central reticular cell there is either a transitional cell with polyribosomes or a proerythroblast. This seems to be the earliest stage in the formation of a new erythroblastic island.

The transitional cell and the proerythroblast are so alike morphologically that it may be difficult to distinguish them (see Section VIII,F) without the use of special biochemical and biophysical techniques. Thus Rosse (1973) and Rosse and Trotter (1974) followed up the earlier work of Rosse *et al.* (1970), starting with guinea pigs in rebound (Section VIII,G) in which erythropoiesis has almost disappeared. These animals were bled and reexposed to hypoxia, so that erythropoiesis was very strongly stimulated in a marrow in which there were very few hemoglobin-containing cells. Hemoglobin synthesis was recognized on radioautography of cells which had incorporated ^{55}Fe. It was also detected on light microscopy by means of the benzidene reaction and the absorption of light in the hemoglobin absorption waveband (4046Å). In electron micrographs, cytoplasmic hemoglobin was detected by the increase in electron density after treating the tissue with diaminobenzidene (DAB) and OsO_4. ''In normal marrow, proerythroblasts were the earliest cells in which haemoglobin could be detected, but during the early phase of erythropoietic stimulation haemoglobin was demonstrated in transitional cells with all the methods employed.''

F. Asynchrony in Erythroblast Formation

The studies of Rosse (1973) and Rosse and Trotter (1974) provided confirmation of the earlier findings of Yoffey *et al.* (1965b) in studies of animals subjected to secondary hypoxia. In normal marrow proerythroblasts were the earliest members of the erythroid series in which hemoglobin could be detected, whereas in secondary hypoxia hemoglobin occasionally became evident in cells which still had the leptochromatic nucleus of the transitional cell, corresponding from the point of view of hemoglobin content to what would normally be the stage of the polychromatic erythroblast. The presence of hemoglobin was confirmed by the Lepehne (1919) reaction and by the use of monochromatic light of the appropriate waveband.

In all these experiments use was made of rebound marrow, which contains an increased number of transitional cells. The appearance of hemoglobin in transitional cells seems to short circuit the usual erythrocyte production pathway, in which hemoglobin does not become evident until after the proerythroblast stage has been reached (see Alpen and Cranmore, 1959), following which two or more mitoses may occur before the red cell becomes mature. It has been known since the work of Suit *et al.* (1957; cf. Alpen and Cranmore 1959) that some red cells may mature from the stem cell without any divisions taking place, in which case there is no dilution of grain count between the stem cell and the mature erythrocyte in animals given ^{59}Fe. It may well be that it is the premature development of hemoglobin in the transitional cell, before the proerythroblast stage is reached, which inhibits the usual sequence of mitoses and is responsible for the phenomenon of ''skipped mitoses.'' Whether this is the case or not, the appearances indicate that erythroid cells can develop directly from transitional cells, in which hemoglobin is synthesized earlier than usual in relation to the nuclear changes. Because of the failure of cytoplasmic and nuclear changes to follow the usual time sequence, the hemoglobin-containing transitional cells were described as "asynchronous erythroblasts" (Yoffey *et al.*, 1965b).

In this connection the experiments of Zucali *et al.* (1974) are of interest. Using a discontinuous bovine serum albumin density gradient, they obtained from rebound rat marrow fractions which responded to erythropoietin (by heme synthesis) six to seven times more strongly than unfractionated rat marrow. The strongly responding fraction was rich in transitional cells. More recently Rosse and Beaufait (1978) investigated the identity of the cells responsible for heme synthesis by radiautography after the uptake of ^{55}Fe. They fractionated the marrow by the density gradient technique of Yoshida and Osmond (1971) and obtained two fractions, F1 and F2. F1 was a light cell fraction rich in transitional cells, but containing no blast cells or heme-synthesizing cells. The fractions were cultured for 48 hours with and without the addition of ESF, and the uptake of ^{59}Fe was measured. Radioautography with ^{55}Fe revealed that a number of transitional cells incorporate the isotope in response to ESF, and must be regarded as erythropoietin-sensitive cells. This of course does not mean that all transitional cells are ERC, since many are committed to other lines of development, e.g. lymphocytes, macrophages, or granulocytes. But there is as yet no way of distinguishing the different groups of cells by their morphology.

The results just presented, together with those given previously (Section VI), all indicate that the various colony-forming cells are members of the transitional cell compartment. For the macrophage–granulocyte precursors (CFU-C) there is the evidence of Moore *et al.* (1972) and Dicke *et al.* (1973). For the erythropoietin-responsive cells (ERC) there is the evidence of Yoffey *et al.* (1965b), Zucali *et al.* (1974), and Rosse and Beaufait (1978). For CFU-S there is the evidence of Murphy *et al.* (1971), Dicke *et al.* (1973), De Gowin and Gibson (1976), and Riches *et al.* (1976a,b). Many other investigations referred to in this review all point in the same direction. The cumulative effect of these observations admits of only one conclusion.

G. Hypoxia and Rebound

One of the most useful of all tools in effecting and analyzing changes in the transitional cell compartment has been hypoxia and its sequelae, notably the subsequent rebound. Hypoxia presumably acts by stimulating the endogenous secretion of erythropoietin (reviewed by Krantz and Jacobson, 1970), and possibly in other ways as yet unknown. In considering the effects of hypoxia it is important to recognize its limitations. Beyond a certain point hypoxia gives rise to serious stress and metabolic changes. Campbell (1934) noted that in the case of mice and rabbits growth occurred normally until an altitude of 18,000 ft

(12% O_2) was reached. At simulated altitudes of 20,000 ft animals could not be acclimatized to live in health. For this reason, experiments at a simulated altitude of 20,000 ft or more are of doubtful value if continued for more than a few days.

1. *Primary Hypoxia, Rebound, and Secondary Hypoxia*

The hypoxia to which a normal animal is subjected is termed *primary hypoxia* (Moffatt *et al.*, 1964a). If primary hypoxia continues for a period of several days, long enough for polycythemia to develop, and the animal is then kept in ambient air, the bone marrow enters the state of *rebound* (Moffatt *et al.*, 1964b). Rebound animals can be subjected to hypoxia, and erythropoiesis thereby stimulated a second time, in some cases accentuated by bleeding (Rosse *et al.*, 1970; Rosse and Trotter, 1974). Hypoxia applied to an animal in rebound is *secondary hypoxia* (Yoffey *et al.* 1965b). It was in fact the differential response to primary and secondary hypoxia which gave one of the early clues to the role of the transitional cell compartment.

The polycythemia which follows erythropoietic stimulation is described as *active polycythemia*. *Passive polycythemia* results from red cell transfusion (Jacobson *et al.*, 1960; Weitz-Hamburger *et al.*, 1971). In the course of the production of active polycythemia, severe hypoxia can raise the plasma erythropoietin 20 to 50 times above normal (Krantz and Jacobson, 1970), and presumably the effects on the erythropoietin level of diminishing intensities of hypoxia are on a progressively falling scale.

2. *The Effect of Increasing Demand for Stem Cells*

For the accurate assessment of the experimentally induced changes in the various compartments of the marrow, a quantitative technique is desirable. Such a technique was developed for guinea pig marrow initially (Yoffey, 1966), and it is in this animal that some of the major results have been obtained. In analyzing these results it must be borne in mind that, in quantitative terms, the main demand for stem cells is shared between the largest three groups in the marrow, namely, erythroid, myeloid, and lymphoid (Fig. 2). As has already been noted, in the normal steady state the stem cell compartment contains a considerable number of resting cells, which constitute an ample reserve for additional requirements, so that up to a point stem cell production can be increased to keep pace with the increased demand. But beyond this point, when there are not enough stem cells for all three groups, deficiencies become manifest. This can readily be seen if one compares the effects of moderate and severe erythropoietic stimulation. At both 10,000 and 20,000 ft erythropoiesis is stimulated, more strongly in the latter than in the former, as might be expected (Yoffey *et al.*, 1966). At 10,000 ft, while the erythroid cells in the marrow are increased, the lymphocytes and transitional cells are unchanged, whereas the myeloid cells fall (Yoffey *et al.*, 1967). At 20,000 ft, when the erythropoietic stimulation is much stronger, the lymphocytes and transitional cells also fall sharply. These findings suggest that at 20,000 ft, when the transitional cells are needed mainly to meet the increased demand for stem cells necessary for erythropoiesis, the demand is approaching close to the limit of their capacity. There is a significant fall in the number of transitional cells, and fewer than usual are now available for granulocyte and lymphocyte production, both of which fall sharply, though they do not disappear completely. This type of *correlated* response, i.e., increased production of one group of cells, the erythroid, and diminished production of the two other groups, namely, the myeloid and lymphoid, suggests that so many transitional cells are being directed into the erythropoietic line of differentiation that not enough are left for the other two groups.

Harris *et al.* (1966) came to a basically similar conclusion when studying the reaction of guinea pig marrow to a combination of hypoxia and bleeding. They too interpreted their results in terms of a common stem cell reacting to the increased demand for erythropoietic differentiation. They noted also increased proliferation of the basophilic transitionals (described as "mononuclear cells with high N:C ratio"), and observed "a complete morphological sequence between this type of cell and the proerythroblast." Rickard *et al.* (1971) working with mice obtained results which they also thought suggested stem cell competition.

Although the transitional cells in the marrow fall markedly when there is strong erythropoietic stimulation, as shown by the changes mentioned in guinea pig marrow at a simulated altitude of 20,000 ft, they never seem to disappear completely. Prindull *et al.* (1977a) investigated the effect on transitional cells of successive stimulation of granulopoiesis and erythropoiesis, but here too, though the transitional cells fell markedly, they never completely disappeared. The failure to make transitional cells disappear on severe stimulation of erythropoiesis by hypoxia recalls the experiments of Kubanek *et al.* (1968), who gave mice erythropoietin at 8-hour intervals for 23 days without any sign of stem cell depletion.

It is pertinent at this point to cite the work of Rosse (1973), who performed an interesting experiment to demonstrate the capacity of transitional cells to switch, according to the stimulus given, into erythropoiesis or lymphocyte production. Guinea pigs in rebound were given an injection of tritiated thymidine *in vivo,* and this labeled all transitional cells in DNA synthesis. In the polychythaemic animal, in which red cell production is at a very low level, the majority of the labeled transitionals give rise to small lymphocytes.

If however, after the initial labeling of the transitionals, the animals were then given an erythropoietic stimulus, large numbers of labeled erythroid cells were formed and not labeled small lymphocytes. These results suggest that one and the same cell, when given the appropriate stimulus, can differentiate into either erythrocytes or lymphocytes. This interpretation raises an interesting problem. Since the early experiments of Jacobson *et al.* (1960), a good deal of evidence has accumulated to show that even in the prolonged depression of erythropoiesis associated with transfusion polycythemia, the ERC turn over actively for many weeks without differentiation (Krantz and Jacobson, 1970; Kubanek *et al.*, 1973). The question then arises why the ERC do not undergo progressive increase during polycythemic depression? Rosse (1976) has suggested that this does not occur because many of the proliferating ERC give rise to lymphocytes, which are constantly leaving the marrow. These lymphocytes constitute what Rosse has termed a "sink" mechanism. Whether any or all of these are typical B lymphocytes has yet to be established.

3. *Rebound*

Though hypoxia at 20,000 ft yielded very interesting results, in view of Campbell's (1943) conclusions the bulk of the observations were subsequently made at a simulated altitude of 17,000 ft, approximately 0.5 atm. During 7 days of primary hypoxia the marrow erythroid cells rise sharply, reaching a peak at 4 to 5 days and then fall somewhat, though still above their normal level. (Yoffey *et al.*, 1968). During this period lymphocytes and myeloid cells fall sharply, while transitional cells show a slight increase, not quite reaching the significance level. When the guinea pigs are taken out of the decompression chamber and kept in ambient air, the *rebound* period commences. In the first group of experiments, when rebound was maintained for 7 days, erythropoiesis was strongly depressed, though it never completely disappeared. The depression of erythropoiesis continues as long as the animals are polycythemic. By contrast, at the beginning of rebound the myeloid cells return to their control level, and lymphocytes and transitional cells to well above this level; hence the name rebound is applied to this phase.

At the end of seven days of rebound the transitional cells were found to be three to four times the control level. This stage, namely 7 days of rebound after 7 days of primary hypoxia at 17,000 ft, is an excellent one at which to obtain a marrow with greatly increased numbers of transitional cells.

In the interpretation of the transitional changes in hypoxia, it was assumed that they were proliferating more actively than usual to keep pace with the increased requirement of stem cells for erythropoiesis. Studies now in progress on rat bone marrow seem to be in accord with this assumption. In a number of rats, a suspension of marrow cells was labeled *in vitro* with tritiated thymidine, and the percentage of labeled transitionals was then counted, during each of 7 days of hypoxia (17,000 ft) and 7 days of rebound (Yoffey and Yaffe, 1978). The results (Fig. 5) do in fact show a significant increase in the percentage of labeled transitionals during hypoxia. Surprisingly, however, there is a sharp fall in the labeling index during the first 24 hours. This recalls the marked fall in transitionals 24 hours after blood withdrawal (Rosse *et al.*, 1970) (Fig. 3, p. 83), though these latter experiments were performed on guinea pigs and the data were

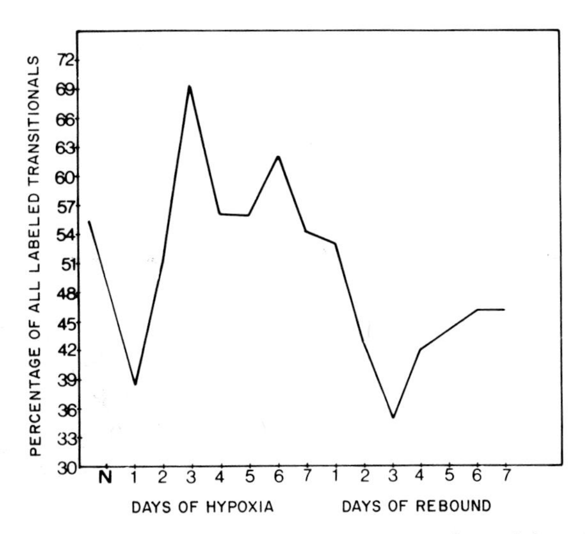

FIG. 5. Labeling index of transitional cells in rat bone marrow of normal, hypoxic, and rebound animals, after incubation for 1 hour with tritiated thymidine. N = Normal. After 24 hours of hypoxia there is a marked fall in the percentage of labeling cells, followed by a sharp rise to a peak on Day 3 of hypoxia. A further significant fall during the first 3 days of rebound is followed by a gradual rise. (Unpublished data of Yoffey and Yaffe, 1978.)

in absolute counts. Furthermore, they were performed on animals at a stage of rebound when the level of transitionals in the marrow was significantly above normal. The fall in marrow transitionals at the commencement of erythropoietic stimulation may be due to their migration from the marrow, as suggested by Rencricca *et al.* (1970) and Boyum *et al.* (1972), investigating the migration of CFU-S.

4. *The Biphasic Transitional Cell Response in Rebound*

The enhanced proliferation of transitionals during hypoxia does not result in their significant increase, presumably because they enter the erythroid compartment as quickly as they are formed. At the commencement of rebound, when erythropoiesis falls abruptly to a very low level, the transitionals seem for a time to be unable to slow down their rate of growth, and their proliferation therefore continues at the accelerated pace of the hypoxic period. If this interpretation is correct, it would provide an explanation of the striking increase in transitional cells during the early part of rebound, an increase which is difficult to explain on any other basis.

The increased proliferation of transitional cells in rebound raises a further problem, for if continued indefinitely they would come to occupy the greater part of the marrow and leave little or no room for other cells. Jones *et al.* (1967) investigated what happened when rebound was continued for 12 days, i.e., 5 days more than the 7-day rebound period in which the transitionals increased so strikingly. The results were surprising. From the seventh to the twelfth days of rebound the transitionals fell sharply, to their control level or less, while at the same time the small lymphocytes rose to twice their control level on the eighth and ninth days of rebound and then fell again. These results suggested the coming into play of an escape mechanism when the transitional cells reached too high a level. But we have as yet no idea of the way in which, following the increased proliferation of transitional cells in hypoxia, there is brought about their characteristic biphasic response in rebound, viz. their great increase in the earlier and their subsequent fall in the later stages of rebound.

Although most of the rebound studies have been performed in guinea pigs, Turner *et al.* (1967) observed rebound changes in the mouse, while Griffiths (1969) described rebound phenomena in the rat. Recent studies in the rat (see Fig. 5) have shown a fall in the percentage labeling of transitional cells at the beginning of rebound, with a subsequent rise toward the end of rebound. It is difficult to compare these percentage data with the absolute counts made in guinea pigs. But the absolute counts obtained by Griffiths (1969) with data from Days 5 to 12 of rebound show a low figure for labeled transitionals on Day 5 and a sharp rise on Day 6. In the mouse, Hurst *et al.* (1969) concluded that the increased number of CFU-S in rebound marrow was in all probability related to the increased number of transitional cells. Okunewick *et al.* (1969), Okunewick and Fulton (1970), and Tribukait and Forssberg (1964), though they did not actually identify the transitional cells in murine bone marrow, all found evidence of the increased stem cell content of the marrow in rebound.

The marked rise in marrow lymphocytes during rebound is possibly explicable on the basis of the ''sink'' mechanism suggested by Rosse (1976). What is not easy to explain is the secondary fall in myeloid cells after the eighth day of rebound (Jones *et al.*, 1967). At this time erythropoiesis is minimal, so stem cell competition seems very unlikely. The secondary fall in myeloid cells occurs at the commencement of the transitional cell decline. The two falls may possibly be related to one another.

Rosse *et al.* (1970) made use of rebound guinea pigs to confirm the role of transitional cells in erythropoiesis. When animals were bled on the ninth day of rebound, at which time there was very little erythopoiesis, one could see, after the initial fall already mentioned, a significant increase in the number of transitional cells, contrasting with the continued decline in unbled animals and immediately preceding the appearance of a new crop of erythroid cells.

5. *Passive (Transfusion) Polycythemia*

Weitz-Hamburger *et al.* (1971) followed up the earlier studies of Jacobson *et al.* (1960) by performing differential counts of bone marrow in mice made polycythemic by transfusion. They found a marked accumulation in the marrow of "lymphoid" elements, in the same way as these were found in rebound animals, guinea pigs, rats, and mice. The basic mechanism appears to be the same. Transitional cells which were proliferating at their normal rate to enter the erythroid compartment no longer do so once the animal is polycythemic, and they therefore accumulate in the marrow. There are, however, two differences between active and passive polycythemia. Passive polycythemia can be maintained for long periods by giving repeated transfusions, whereas active polycythemia is of relatively short duration. In the guinea pig, for example, following hypoxia for 7 days at 17,000 ft, the polycythemia has fallen sufficiently after 14 to 16 days of rebound for erythropoiesis to be resumed. The second difference is that during hypoxia, to meet the increased demand for erythroid cells, the transitional cells proliferate much more actively than in normal animals. Consequently in rebound these actively proliferating transitionals accumulate more rapidly and in larger numbers than in the transitionals proliferating at their normal rate in transfusion polycythemia.

Passive polycythemia can be maintained for 3 months by repeated transfusions (Jacobson *et al.*, 1960), during which time erythropoiesis is at a very low level. Nevertheless the ERC are still capable of responding, for when the transfusions ceased, then as soon as the hematocrit fell to normal, erythropoiesis quickly reappeared. This raises over a longer period the same problem as arises in the shorter period of rebound. If the ERC are turning over all the time, they should continually increase, which does not seem to happen. In addition to the "sink" mechanism suggested by Rosse (1973), there may be considerable lengthening of the generation time (Kretchmar *et al.*, 1970; Morse *et al.*, 1970).

IX. Some Concluding Observations

In view of all the evidence which has accumulated, there can now be little doubt that the transitional cell compartment contains the essential immunohemopoietic stem cells on whose correct functioning the integrity of the entire lymphomyeloid complex ultimately depends. It is a distinctive cell population in its morphology, size spectrum, kinetic properties, and migratory capacity. In general, the proliferative rate is high in the larger cells, low in the smaller ones, and intermediate in between these two extremes. The small transitionals are in all probability the resting stage of the stem cell compartment.

Some of the properties of stem cells which have emerged from a large number of experimental studies may be summarized in a series of stem cell postulates:

1. Stem cells must be present in appreciable numbers, or be capable of rapid proliferation, so that when there are heavy demands for them they will not be exhausted.
2. Stem cells must exist in two forms, proliferating and nonproliferating.

3. Within the stem cell compartment the proportion of proliferating to non-proliferating cells can vary in accordance with stem cell requirements.

4. Stem cells must be capable of varying degrees of mobilization and migration into and out of the hemopoietic tissues.

In considering the stem cell role of transitional cells, it is important to bear in mind that there is no other cell group in the marrow which has the postulated stem cell properties. The autoradiographic studies of Caffrey *et al.* (1966) and the continuous labeling studies of Fliedner *et al.* (1968), Haas *et al.* (1969), and Haas *et al.* (1973) have effectively ruled out all the classicial contenders for the role of stem cell, notably, reticulum cells or endothelium.

Transitional cells still present many problems. They are a heterogeneous population. In morphology, there is the striking difference between the pale and basophilic transitionals. Functionally, there is the difference between the pluripotential and the committed cells. The present position with regard to the heterogeneous transitional cell compartment resembles in some ways that which prevailed in relation to the small lymphocytes before new techniques made it possible to sort out the lymphocytes into their different groups.

We are not yet able to distinguish morphologically between the various committed cells unless the commitment begins at an early stage, while the cells still have transitional cell morphology. Examples of this are the asynchronous erythroblasts, or the occasional transitional cells which have a surface immunoglobulin layer.

The resting stage of the stem cell, the small transitional, may be confused with the small lymphocyte. The kinetics of this cell have been studied in stimulated marrow by Haas *et al.* (1973). The possibility still remains open that there may be a specialized subgroup of small lymphocytes which are not immunologically conditioned, and which may perhaps be capable of enlarging and reentering the active transitional cell compartment.

Finally, for such a fundamentally important group as the transitional cells, the question of control requires much more investigation. In view of the great variations in the size of the compartment, it seems possible that there is a humoral control apart from the feedback mechanisms usually postulated. Boyum *et al.* (1972), for example, found that when marrow cells were cultured in diffusion chambers, CFU-S proliferated more rapidly in irradiated than in normal hosts. Stohlmann *et al.* (1973) made a similar observation on Millipore chambers implanted into neutropenic (cyclophosphamide-treated) animals. Observations of this kind suggest the existence of some diffusible matter obtaining access to the interior of the chamber and influencing stem cell growth. For such a fundamentally important group as the transitional cells, it would provide an invaluable new approach to their study if a substance could be isolated which controlled their proliferative activity.

ACKNOWLEDGMENTS

The studies in progress referred to in this review have been supported by grants from the Joint Research Fund of the Hebrew University and Hadassah, and from the Ann Langer Cancer Research Foundation.

REFERENCES

Alpen, E. L., and Cranmore, D. (1959). *In* "The Kinetics of Cellular Proliferation" (F. Stohlman, Jr., ed.), p. 290, Grune & Stratton, New York.
Anderson, E. C., Bell, G. I., and Peterson, D. F. (1969). *Biophys. J.* **9,** 246.
Badenoch, J., and Callender, S. T. (1954). *Br. J. Radiol.* **27,** 381.
Barnes, D. W. H., Ford, C. E., and Loutit, J. F. (1964). *Lancet* **i,** 1395.
Barr, R. D., Whang-Peng, J., and Perry, S. (1975). *Science* **190,** 284.
Basten, A., Warner, N. L., and Mandel, T. (1972). *J. Exp. Med.* **135,** 627.
Becker, A. J., McCulloch, E. A., Siminovitch, L., and Till, J. E. (1965). *Blood* **26,** 296.
Bekkum, D. W., van, and Weyzen, W. W. H. (1961). *Pathol. Biol. (Paris),* **9,** 888.
Belcher, E. H., Gilbert, I. G. F., and Lamerton, L. F. (1954). *Br. J. Radiol.* **27,** 387.
Ben-Ishay, Z., and Yoffey, J. M. (1971). *J. Reticuloendothel. Soc.* **10,** 482.
Ben-Ishay, Z., and Yoffey, J. M. (1972). *Lab. Invest.* **26,** 637.
Bennett, M., and Cudkowicz, G. (1967). *In* "The Lymphocyte in Immunology and Hemopoiesis" J. M. Yoffey, ed.), p. 182. Arnold, London.
Berlin, N. I., and Lotz, C. (1951). *Proc. Soc. Exp. Biol. (N.Y.)* **78,** 788.
Blackburn, E. K. (1947–48). *J. Clin. Pathol.* **1,** 295.
Blackett, N. M. (1968). *J. Natl. Cancer Inst.* **41,** 908.
Blackett, N. M., Roylance, P. J., and Adams, K. (1964). *Br. J. Haematol.* **10,** 453.
Bloom, W. (1938). *In* "Handbook of Hematology" (H. Downey, ed.), Vol 2, p. 865. Hoeber, New York.
Boggs, D. R., and Boggs, S. S. (1976). *Blood* **48,** 71.
Boggs, S. S., Chervenick, P. A., and Boggs, D. R. (1972). *Blood* **40,** 375.
Boyum, A., Carsten, A. L., Laerum, O. D., and Cronkite, E. P. (1972). *Blood* **40,** 174.
Burke, W. T., and Harris, C. (1959). *Blood* **14,** 409.
Caffrey, R. W., Everett, N. B., and Rieke, W. O. (1966). *Anat. Rec.* **155,** 41.
Campbell, J. A. (1934). *Br. J. Exp. Pathol.* **16,** 39.
Cavins, J. A., Scheer, S. C., Thomas, E. D., and Ferrebee, J. W. (1964). *Blood* **23,** 38.
Chervenick, P. A., and Boggs, D. R. (1971). *Blood* **27,** 568.
Cohen, J. J., and Patterson, C. K. (1975). *J. Immunol.* **114,** 374.
Craddock, C. G. (1965). *Acta Haematol.* **33,** 19.
Cudkowicz, G., Upton, A. C., Shearer, G. M., and Hughes W. L. (1964a). *Nature (London)* **201,** 165.
Cudkowicz, G., Upton, A. C., Smith, L. H., Gosslee, D. G., and Hughes, W. L. (1964b). *Ann. N.Y. Acad. Sci.* **114,** 571.
Cunningham, R. S., Sabin, F. R., and Doan C. A. (1925). *Contrib. Embryol. Carnegie Inst.* **16,** 227.
Davidson, L. S. P., Davis, L. J., and Innes, J. (1943). *Edinburgh Med. J.* **50,** 226.
De Gowin, R. L., and Gibson, D. P. (1976). *Blood* **47,** 315.
De Gowin, R. L., and Johnson, S. (1967). *Proc. Soc. Exp. Biol. Med.* **126,** 442.

De Gowin, R. L., Hoak, J. C., and Miller, S. H. (1972). *Blood* **40,** 881.

Dicke, K. A., van Noord, M. J., Maat, B. Schaefer, U. W., and van Bekkum, D. W. (1973). *In* "Ciba Symposium: Haemopoietic Stem Cells" (G. E. W. Wolstenholme and M. O'Connor, eds.), p. 47. Associated Scientific Publ., Amsterdam.

Erslev, A. J. (1959). *Blood* **14,** 386.

Erslev, A. J. (1964). *Blood* **24,** 331.

Everett, N. B., and Caffrey, R. W. (1967). *In* "The Lymphocyte in Immunology and Haemopoiesis" (J. M. Yoffey, ed.), p. 108, Arnold, London.

Everett, N. B., and Yoffey, J. M. (1959). *Proc. Soc. Exp. Biol. Med.* **101,** 318.

Everett, N. B., Reinhardt, W. O., and Yoffey, J. M. (1959). Cited by Yoffey, 1960, in Ciba Foundation Symposium on Haemopoiesis, p. 31.

Faulk, W. P., Goodman, J. R., Maloney, M. A., Fudenberg, H. H., and Yoffey, J. M. (1973). *Cell. Immunol.* **8,** 166.

Fliedner, T. M., Haas, R. J., Stehle, H., and Adams, A. (1968). *Lab. Invest.* **18,** 249.

Fukuda, T. (1973a). *Virchows. Arch. Zellpathol. B* **14,** 197.

Fukuda, T. (1973b). *Virchows. Arch. Zellpathol. B* **14,** 31.

Fukuda, T. (1974). *Virchows. Arch. Zellpathol. B* **16,** 249.

Gairdner, D., Marks, I., and Roscoe, J. D. (1952). *Arch. Dis. Child.* **27,** 128.

Gladstone, R. J., and Hamilton, W. J. (1941). *J. Anat. (London),* **76,** 9.

Goodman, J. W., and Hodgson, G. S. (1962). *Blood* **19,** 702.

Griffiths, D. A. (1969). *Blood* **34,** 696.

Grigoriu, G. Antonescu, M., and Iercan E. (1971). *Blood* **37,** 187.

Haas, R. J., Bohne, F., and Fliedner, T. M. (1968). *Proc. Monaco Symp. IAEA Vienna,* p. 205.

Haas, R. J., Bohne, F. E., and Fliedner, T. M. (1969). *Blood* **34,** 791.

Haas, R. J., Hans-Dieter, F., Fliedner, T. M., and Fache, I. (1973). *Blood* **42,** 209.

Hammar, J. A. (1901). *Anat. Anz.* **19,** 567.

Harris, C. (1961). *Blood* **18,** 691.

Harris, C., and Burke, W. T. (1957). *Am. J. Pathol.* **33,** 931.

Harris, P. F. (1956). *Br. Med. J.* **ii,** 1032.

Harris, P. F. (1960). *Acta Haematol. (Basel)* **23,** 293.

Harris, P. F., and Kugler, J. H. (1963). *Acta Haematol. (Basel)* **32,** 146.

Harris, P. F., and Kugler, J. H. (1965). *Acta Haematol. (Basel)* **33,** 351.

Harris, P. F., and Kugler, J. H. (1967). *In* "The Lymphocyte in Immunology and Haemopoiesis" (J. M. Yoffey, ed.), p. 135. Arnold, London.

Harris, P. F., Haigh, G., and Kugler, J. H. (1963). *Acta Haematol. (Basel)* **29,** 166.

Harris, P. F., Harris, R. S., and Kugler, J. H. (1966). *Br. J. Haematol.* **12,** 419.

Harriss, E. B., and Hoelzer, D. (1978). *Blood* **51,** 221.

Heath, D. S., Axelrad, A. A., McLeod, D. L., and Streeve, M. M. (1976). *Blood* **47,** 777.

Hoyes, A. D., Riches, D. J., and Martin, B. G. H. (1973). *J. Anat. (London)* **115,** 99.

Hudson, G. (1965). *Br. J. Haematol.* **11,** 446.

Hurst, J. M., Turner, M. S., Yoffey, J. M., and Lajtha, L. G. (1969). *Blood* **33,** 859.

Iscove, N. N., Till, J. E., and McCulloch, E. A. (1970). *Proc. Soc. Exp. Biol. Med.* **134,** 33.

Jacobson, L. O., Marks, E. K., Gaston, E. O., Robson, M., and Zirkle, R. E. (1949). *Proc. Soc. Exp. Biol. Med.* **70,** 740.

Jacobson, L. O., Marks, E. K., and Gaston, E. O. (1954). *In* "Radiobiology Symposium" (M. Bacq and P. Alexander, eds.), p. 122. Butterworth, London.

Jacobson, L. O., Goldwasser, E., and Gurney, C. W. (1960). *In* "Ciba Foundation Symposium on Haemopoiesis" (G. E. W. Wolstenholme and M. O'Connor, eds.), p. 423. Churchill, London.

Jones, H. B., Jones, J. J., and Yoffey, J. M. (1967). *Br. J. Haematol.* **13,** 934.

Jones, R. O. (1970). *J. Anat. (London)* **107,** 301.

Jordan, H. E., and Speidel, C. C. (1930). *Am. J. Anat.* **46,** 355.
Keiser, G., Cottier, H., Bryant, B. J., and Bond, V. P. (1967). *In* "The Lymphocyte in Immunology and Haemopoiesis" (J. M. Yoffey, ed.), p. 149. Arnold, London.
Kempgens, U., Mayer, M., Muller, U., and Queisler, W. (1973). *Verh. Dtsch. Gesellsch. Inn. Med.* **79,** 437.
Kempgens, U., Mayer, M., Schlossardt, S., Muller, U., and Queiser, W. (1976). *Acta Haematol.* **56,** 193.
Kim, H. C., Marks, P. A., Rifkind, R. A., Maniatis, G. M., and Bank, A. (1976). *Blood* **47,** 767.
Kindred, J. E. (1942). *Am. J. Anat.* **71,** 207.
Knospe, W. H., Gregory, S. A., Husseini, S. G., Fried, W., and Trobaugh, F. E. (1972). *Blood* **39,** 331.
Knudtzon, S. (1974). *Blood* **43,** 357.
Komuro, K., and Boyse, E. A. (1973). *Lancet* **i,** 740.
Krantz, B., and Jacobson, L. O. (1970. "Erythropoietin and the Regulation of Erythropoiesis." Univ. of Chicago Press, Chicago, Illinois.
Kretchmar, A. L., and Conover, W. R. (1970). *Blood* **36,** 772.
Kretchmar, A. L., McDonald, T. P., and Lange, R. D. (1970). *In* "Hemopoietic Cellular Proliferation" (F. Stohlman, Jr., ed.), p. 150. Grune & Stratton, New York.
Kubanek, B., Tyler, W. S., Ferrari, L., Porcellini, A., Howard, D., and Stohlman F., Jr. (1968). *Proc. Soc. Exp. Biol. Med.* **127,** 770.
Kubanek, B., Bock, O., Heit, W., Bock, E., and Harriss, E. B. (1973). *In* "Haemopoietic Stem Cells," Ciba Foundation Symposium (G. E. W. Wolstenholme and M. O'Connor, eds.), p. 243. Excerpta Medica, Amsterdam.
Lafleur, L., Miller, R. G., and Phillips, R. A. (1972). *J. Exp. Med.* **135,** 1363.
Lafleur, L., Miller, R. G., and Phillips, R. A. (1973). *J. Exp. Med.* **137,** 954.
Lajtha, L. G., and Oliver, R. (1960). *In* "Haemopoiesis: Cell Production and Its Regulation" (G. E. W. Wolstenholme and M. O'Connor, eds.), p. 289. Churchill, London.
Lajtha, L. G., Gilbert, C. W., Porteous, D. D., and Alexanian R. (1964). *Ann. N.Y. Acad. Sci.* **113,** 742.
Leblond, C. P., and Sainte-Marie, G. (1960). *In* "Haemopoiesis: Cell Production and Its Regulation" (G. E. W. Wolstenholme and M. O'Connor, eds.), p. 152. Churchill, London.
Lepehne, G. (1919). *Beitr. Pathol. Anat. Allg. Pathol.* **62,** 165.
Lord, B. I. (1967). Nature (*London*) **214,** 924.
Lowenberg, B. (1975). "Fetal Liver Cell Transplantation." Radiobiological Institute, Rijswijk, The Netherlands.
McCool, D., Miller, R. J., Painter, R. H., and Bruce W. R. (1970). *Cell Tissue Kinet.* **3,** 55.
Malinin, Th. I., Perry, V. P., Kerby, C. C., and Dolan, M. F. (1965). *Blood* **25,** 693.
Maximow, A. (1910–11). *Arch. Mikrosc. Anat.* **76,** 1.
Maximow, A. (1927). *In* "Handbuch der Mikroskopischen Anatomie des Menschen" (W. von Mollendorff, ed.), Vol. 2, p. 232, Springer, Berlin.
Metcalf, D. (1972). *Proc. Soc. Exp. Biol. Med.* **139,** 511.
Metcalf, D., and MacDonald, H. R. (1975). *J. Cell. Physiol.* **85,** 643.
Metcalf, D., Johnson, G. R., and Wilson, G. (1977). *Exp. Hematol.* **5,** 299.
Micklem, H. S., and Loutit, J. F. (1966) "Tissue Grafting and Radiation," pp. 145–147. Academic Press, New York.
Miller, R. G. A., and Phillips, R. A. (1969). *J. Cell. Physiol.* **73,** 191.
Miller, S. C., and Osmond, D. G. (1973). *Cell Tissue Kinet.* **6,** 259.
Miller, S. C., and Osmond, D. G. (1974). *Exp. Hematol.* **2,** 227.
Miller, S. C., and Osmond, D. G. (1975). *Cell Tissue Kinet.* **8,** 97.
Moffatt, D. J., Rosse, C., Sutherland, I. H., and Yoffey, J. M. (1964a) *Acta Anat.* **68,** 26.

Moffatt, D. J., Rosse, C., Sutherland, I. H., and Yoffey, J. M. (1964b). *Acta Anat.* **59,** 188.
Moffatt, D. J., Rosse, C., and Yoffey, J. M. (1967). *Lancet* **ii,** 547.
Mollison, P. L. (1948). *Lancet* **i,** 513.
Monette, F. C., Lobue, J., Gordon, A. S., and Alexander, P. J. (1968). *Science* **162,** 1132.
Monette, F. C., Gilio, J., and Chalifoux, P. (1974). *Cell Tissue Kinet.* **7,** 443.
Moore, M. A. S., and Metcalf, D. (1970). *Br. J. Haematol.* **18,** 279.
Moore, M. A. S., and Owen, J. J. T. (1967). *Lancet* **ii,** 658.
Moore, M. A. S., Williams, N., and Metcalf, D. (1972). *J. Cell. Physiol.* **79,** 283.
Moore, M. A. S., Williams, N; and Metcalf, D. (1973). *J. Natl. Cancer Inst.* **50,** 603.
Morris, R. B., Nichols, B. A., and Bainton, D. F. (1975). *Dev. Biol.* **44,** 223.
Morrison, J. H., and Toepffer, J. F. (1967). *Am. J. Physiol.* **213,** 923.
Morse, B. S., Rencricca, N. J., and Stohlman, F., Jr. (1970). *In* ''Hemopoietic Cellular Proliferation'' (F. Stohlman, ed.), p. 160. Grune & Stratton, New York.
Murphy, M. J., Bertles, J. F., and Gordon, A. S. (1971). *J. Cell Sci.* **9,** 23.
Naegeli, O. (1900). *Dtsch. Med. Wochenschr.* **26,** 287.
Naegeli, O. (1931). ''Blutkrankheiten und Blutdiagnostik,'' 5th ed. Springer, Berlin.
Okunewick, J. P., and Fulton, D. (1970). *Blood* **36,** 239.
Okunewick, J. P., Hartley, K. M., and Dorden, J. (1969). *Radiat. Res.* **38,** 530.
Orlic, D., Gordon, A. S., and Rhodin, J. A. G. (1968). *Ann. N.Y. Acad. Sci.* **149,** 198.
Osmond, D. G. (1967). *In* ''The Lymphocyte in Immunology and Haemopoiesis'' (J. M. Yoffey, ed.), p. 120. Arnold, London.
Osmond, D. G., and Everett, N. B. (1964). *Blood* **23,** 1.
Osmond, D. G., and Nossal, G. J. V. (1974a). *Cell. Immunol.* **13,** 117.
Osmond, D. G., and Nossal, G. J. V. (1974b). *Cell. Immunol.* **13,** 132.
Osmond, D. G., and Yoshida, Y. (1970). *Proc. Can. Fed. Biol. Soc.* **13,** 152.
Osmond, D. G., and Yoshida, Y. (1971). *Proc. Annu. Leucocyte Culture Conf., 4th,* p. 97.
Osmond, D. G., Miller, S. C., and Yoshida, Y. (1973). *In* ''Haemopoietic Stem Cells,'' Ciba Foundation Symposium (G. E. W. Wolstenholme and M. O'Connor, eds.), p. 131. Associated Scientific Publ., Amsterdam.
Pappenheim, A. (1907). *Folia Haematol. (Leipzig.)* **4,** 1, 142.
Pappenheim, A., and Ferrata, A. (1910). *Folia Haematol. Pt. 1 (Arc.)* **10,** 78.
Pathak, V. B., Reinhardt, W. O., and Yoffey, J. M. (1956). *J. Anat. (London)* **90,** 568.
Patinkin, D., Grover, N. B., and Yoffey, J. M. (1979). *Br. J. Haemato.* **41,** 309.
Patt, H. M., and Maloney, M. A. (1970). In ''Haemopoietic Cellular Proliferation'' (F. J. Stohlam, ed.), p. 56. Grune & Stratton, New York.
Peterson, E. A., and Evans, W. H. (1967). Nature (*London*) **214,** 824.
Pisciotta, A. V., Santos, A. S., and Keller, C. (1964). *J. Lab. Clin. Med.* **63,** 445.
Popp, R. A. (1960). *Proc. Soc. Exp. Biol. Med. (N.Y.)* **104,** 722.
Press, O. W., Rosse, C., and Clagett, J. (1977). *J. Exp. Med.* **146,** 735.
Prindull, G. (1971). *Blut* **23,** 7.
Prindull, G. (1974). *Z. Kinderheilk.* **118,** 197.
Prindull, G., Prindull, B., Ron, A., and Yoffey, J. M. (1975). *J. Pediat.* **86,** 773.
Prindull, G., Ron, A., and Yoffey, J. M. (1976). *Klin. Wochenschr.* **54,** 637.
Prindull, G., Prindull, B., and Yoffey, J. M. (1977a). *Acta Anat.* **100,** 95.
Prindull, G., Prindull, B., Schroter, W., and Yoffey, J. M. (1977b). *Eur. J. Pediat.* **126,** 243.
Prindull, G., Prindull, B., and v. d. Meulen, N. (1978). *Acta Paediat. Scand.* **67,** 413.
Raff, M. C., Feldmann, M., and de Petris, S. (1973). *J. Exp. Med.* **137,** 1024.
Rencricca, N. J., Rizzoli, V., Howard, D., Duffy, P., and Stohlman F., Jr. (1970) *Blood* **38,** 764.
Rhoads, C. P., and Miller, D. K. (1938). *Arch. Pathol.* **26,** 648.

Riches, A. C., Sharp, J. G., Littlewood, V., Briscoe, C. V., and Thomas, D. B. (1976a) *J. Anat. (London)* **122,** 717.
Riches, A. C., Brynmor, T. D., Briscoe, C. V., Littlewood, V., and Sharp, J. G. (1976b). *Wadsley Med. Bull.* **6,** 35.
Rickard, K. A., Shadduck, R. K., Howard, D. E., and Stohlman, F., Jr. (1970). *Proc. Soc. Exp. Biol. Med.* **134,** 152.
Rickard, K. A., Rencricca, N. J., Shadduck, R. K., Monette, F. C., Howard, D. E., Garrity, M., and Stohlman, F., Jr. (1971). *Br. J. Haematol.* **21,** 537.
Rifkind, R. A., Chui, D., and Epler, H. (1969). *J. Cell Biol.* **40,** 343.
Robinson, S. H., Brecher, G., Lourie, I. S., and Haley, J. E. (1965). *Blood* **26,** 281.
Rodnan, G. P., Ebaugh, F. G., Jr., and Fox, M. R. S. (1957). *Blood* **12,** 355.
Rosenberg, M. (1969). *Blood* **33,** 66.
Rosse, C. (1970a). *Nature (London)* **227,** 73.
Rosse, C. (1970b). *Zeiss Inf.* **73,** 90.
Rosse, C. (1971). *Blood* **38,** 372.
Rosse, C. (1972a). *Proc. Leucocyte Culture Conf., 6th* p. 55. Academic Press, New York.
Rosse, C. (1973). In "Haemopoietic Stem Cells," Ciba Foundation Symposium (G. E. W. Wolstenholme and M. O'Connor, eds.), p. 105. Associated Scientific Publ., Amsterdam.
Rosse, C. (1976). *Int. Rev. Cytolo.* **45,** 155.
Rosse, C., and Beaufait, W. (1978). *Anat. Rec.* **190,** 525.
Rosse, C., and Trotter, J. A. (1974). *Blood* **43,** 885.
Rosse, C., and Yoffey, J. M. (1967). *J. Anat. (London)* **102,** 113.
Rosse, C., Griffiths, D. A., Edwards, A. E., Gaches, C. G. C., Long, A. L. H., Wright, J. L. W., and Yoffey, J. M. (1970). *Acta Haematol. (Basel)* **43,** 80.
Rosse, C., Kraemer, M. J., Dillon, T. L., McFarland, R., and Smith, N. J. (1977). *J. Lab. Clin. Med.* **89,** 1225.
Rubinstein, A. S., and Trobaugh, F. E., Jr. (1973). *Blood* **42,** 61.
Ryser, J., and Vassali, P. (1974). *J. Immunol.* **113,** 740.
Sabin, F. R., Miller, F. R., Smithburn, K. L., Thomas, R. M., and Hummel, L. E. (1936). *J. Exp. Med.* **64,** 97.
Sharp, J. G., Brynmor, T. D., Briscoe, C. V., Littlewood, V., and Riches, A. C. (1976). *Wadsley Med. Bull.* **6,** 23.
Silini, G., Andreozzi, U., and Pozzi, L. V. (1976). *Cell Tissue Kinet.* **9,** 341.
Simar, I. J., Haot, J., and Betz, E. H. (1968). *Eur. J. Cancer* **4,** 529.
Sorenson, G. D. (1961). *Lab. Invest.* **10,** 178.
Sorenson, G. D. (1963). *Ann. N.Y. Acad. Sci.* **111,** 45.
Starling, M. R., and Rosse, C. (1976). *Cell Tissue Kinet.* **9,** 191.
Stites, D. P., Wybran, J., Carr, M. C., and Fudenberg, H. H. (1972). *In* "Ontogeny of Acquired Immunity," Ciba Foundation Symposium (G. E. W. Wolstenholme and M. O'Connor, eds.), p. 113. Associated Scientific Publ., Amsterdam.
Stohlman, F., Jr. (1961). *Proc. Soc. Exp. Biol. Med.* **107,** 751.
Stohlman, F., Jr., Quesenberry, P., Niskanen, E., Morley, A., Tyler, W., Rickard, K., Symann, M., Monette, F., and Howard, D. (1973). *In* "Haemopoietic Stem Cells." Ciba Foundation Symposium (G. E. W. Wolstenholme and M. O'Connor eds.), p. 205. Associated Scientific Publ., Amsterdam.
Suit, H. D., Lajtha, L. G., Oliver, R., and Ellis, F. (1957). *Br. J. Haematol.* **3,** 165.
Sutherland, D. J. A., Till, J. E., and McCulloch, E. A. (1971). *Cell Tissue Kinet.* **4,** 479.
Tarbutt, R. G. (1969). *Br. J. Haematol.* **16,** 9.
Tavassoli, M., and Crosby, W. H. (1968). *Science* **161,** 54.

Thomas, D. B. (1971). *J. Anat.* **110,** 297.

Thomas, D. B. (1973). *In* "Haemopoietic Stem Cells." Ciba Foundation Symposium (G. E. W. Wolstenholme and M. O'Connor, eds.), p. 71. Associated Scientific Publ., Amsterdam.

Thomas, D. B., and Yoffey, J. M. (1962). *Br. J. Haematol.* **8,** 290.

Thomas, D. B., and Yoffey, J. M. (1964). *Br. J. Haematol.* **10,** 193.

Thomas, D. B., Russell, P. M., and Yoffey, J. M. (1960). *Nature (London)* **187,** 876.

Thomas, E. D., Fliedner, T. M., Thomas, D., and Cronkite, E. P. (1965). *J. Lab. Clin. Med.* **65,** 794.

Tillmann, W., Prindull, G., and Schroter, W. (1976). *Eur. J. Pediat.* **123,** 51.

Toldt, C., and Zuckerkandl, E. (1875). *SB Akad. Wiss. Wien. Math-Nat* **K172,** 241. 241.

Trentin, J., Wolf, N., Cheng, V., Faulberg, W., Weiss, D., and Bonhag, R. (1967). *J. Immunol.* **98,** 1326.

Tribukait, B., and Forssberg A. (1964). *Naturwissenschaften* **51,** 12.

Turner, M. S., Hurst, J. M., and Yoffey, J. M. (1967). *Br. J. Haematol.* **13,** 942.

Vos, O., and Dolmans, M. J. A. S. (1972). *Cell Tissue Kinet.* **5,** 371.

Weber, T. H., Santesson, B., and Skoog, V. T. (1973). *Scand. J. Haematol.* **11,** 177.

Weitz-Hamburger, A., Lobue, J., Sharkis, S. J., Gordon, A. S., and Alexander, P., Jr. (1971). *J. Anat. (London)* **109,** 549.

Winter, G. C. B., Byles, A. B., and Yoffey, J. M. (1965). *Lancet* **ii,** 932.

Yoffey, J. M. (1932–33). *J. Anat. (London)* **67,** 250.

Yoffey, J. M. (1957). Brookhaven Symp. Biol. **10,** 1.

Yoffey, J. M. (1960). *In* "Ciba Foundation Symposium on Haemopoiesis" (G. E. W. Wolstenholme and M. O'Connor, eds.), pp. 1–36. Churchill, London.

Yoffey, J. M. (1966) "Bone Marrow Reactions." Arnold, London.

Yoffey, J. M. (1968). *J. Anat. (London)* **102,** 583.

Yoffey, J. M. (1971). *Isr. J. Med. Sci.* **7,** 825.

Yoffey, J. M. (1974). "Bone Marrow in Hypoxia and Rebound." Thomas, Springfield, Illinois.

Yoffey, J. M. (1975). *Isr. J. Med. Sci.* **11,** 1230.

Yoffey, J. M. (1977). *Ad. Microcircul.* **7,** 49.

Yoffey, J. M., and Courtice, F. C. (1956). "Lymphatics, Lymph and Lymphoid Tissue." Edward Arnold, London.

Yoffey, J. M., and Courtice, F. C. (1970). "Lymphatics, Lymph and the Lymphomyeloid Complex." Academic Press, New York.

Yoffey, J. M., and Thomas, D. B. (1964). *J. Anat. (London)* **95,** 613.

Yoffey, J. M., and Weinberg, A. (1976). *Proc. Eur. Congr. Electron Microsc. 6th,* p. 590.

Yoffey, J. M., and Yaffe, P. (1978). Unpublished data.

Yoffey, J. M., Reinhardt, W. O., and Everett, N. B. (1961a). *J. Anat. (London)* **95,** 293.

Yoffey, J. M., Thomas, D. B., Moffatt, D. J., Sutherland, I. H., and Rosse, C. (1961b). Ciba Foundation Study Group No. 10 (G. E. W. Wolstenholme and M. O'Connor, eds.), p. 45, Churchill, London.

Yoffey, J. M., Hudson, G., and Osmond, D. G. (1965a). *J. Anat. (London)* **99,** 841.

Yoffey, J. M., Rosse, C., Moffatt, D. J., and Sutherland, I. H. (1965b). *Acta Anat.* **62,** 476.

Yoffey, J. M., Smith, N. C. W., and Wilson, R. S. (1966). *Scand. J. Haematol.* **3,** 186.

Yoffey, J. M., Smith, N. C. W., and Wilson, R. S. (1967). *Scand. J. Haematol.* **4,** 145.

Yoffey, J. M., Jeffreys, R. V., Osmond, D. G., Turner, M. S., Tahsin, S. C., and Niven, P. A. R. (1968). *Ann. N.Y. Acad. Sci.* **149,** 179.

Yoffey, J. M., Ron, A., Prindull, G., and Yaffe, P. (1978a). *Clin. Immunol.* Immunopathol. **9,** 491.

Yoffey, J. M., Patinkin, D., and Grover, N. B. (1978b). *Isr. J. Med. Sci.* **14,** 1247.

Yoshida, Y., and Osmond, D. G. (1971). *Blood* **37,** 73.
Yung, L., Wyn-Evans, T. C., and Diener, E. (1973). *Eur. J. Immunol.* **3,** 224.
Zamboni, L. (1965). *J. Ultrastruct. Res.* **12,** 525.
Zucali, J. R., Van Zant, G., Rakowitz, F., and Gordon, A. S. (1974). *Exp. Hematol.* **2,** 250.

Human Chromosomal Heteromorphisms: Nature and Clinical Significance

RAM S. VERMA AND HARVEY DOSIK

Division of Hematology and Cytogenetics, The Jewish Hospital and Medical Center, and Department of Medicine, State University of New York, Downstate Medical Center, Brooklyn, New York

I. Introduction

A genetic variant which occurs within a breeding population at a higher frequency than would be maintained by recurrent mutation is defined as a polymorphism (Ford, 1940). In the Paris Conference (1971), the term ''variant'' is recommended for use in describing the situation where deviations from the norm of chromosome morphology are observed. Recently, in the Supplement to the Paris Conference (1975) the term ''heteromorphism'' was used. Because this term is the most common one, we intend to use it throughout the text.

Even before the advent of banding techniques, it was recognized that certain homologs of human chromosomes exhibit consistent morphologic differences. Due to recent advances in banding techniques it has become possible to recognize the heteromorphic regions with great precision. An increasing number of heteromorphic sites are being recognized in the human genome. Several questions regarding human chromosomal heteromorphisms have now been answered, e.g.:

(1) the types of heteromorphisms are consistent within a person, but variable from person to person;

ISBN 0-12-364462-3

(2) the extent of heteromorphisms is continuous, not discrete;
(3) the most common sites are the short arms of the acrocentric chromosomes (13–15, 21, and 22), the long arm of the Y chromosome, the centromere (cen) of chromosomes 3 and 4, and the secondary constriction (h) regions of chromosomes 1, 9, and 16.
(4) their mode of inheritance is Mendelian;
(5) often there is no direct relationship between a heteromorphism identified by one technique and that identified by another;
(6) all contain varying amounts of different classes of highly redundant DNA;
(7) perhaps they may have important clinical significance (Lubs, 1977).

Generally, heteromorphisms of human chromosomes have been detected by QFQ (Q-bands by fluorescence using quinacrine; nomenclature suggested by Paris Conference, 1971, Supplement, 1975), CBG (C-bands by barium hydroxide using Giemsa), and RFA (R-bands by fluorescence using acridine organe) (Craig-Holmes *et al.*, 1973; Geraedts and Pearson, 1974; Bobrow and Madan, 1974; McKenzie and Lubs, 1975; Muller *et al.*, 1975; Lin *et al.*, 1976; Robinson *et al.*, 1976; Van Dyke, 1977; Verma and Lubs, 1974, 1975a,b, 1976; Verma *et al.*, 1977a,b,c, 1978, and Verma and Dosik, 1978). Furthermore, heteromorphisms of the nucleolar organizer regions of acrocentric chromosomes have been noted by a silver staining technique (Hayata *et al.*, 1977). The frequencies of heteromorphisms presented here will provide important baseline data for future clinical, population, and somatic cell hybridization cytogenetic studies.

II. Types, Classification, and Frequencies

The heteromorphic chromosomes of the human genome can be classified by size, position, staining intensity, and/or any combination of these types using various banding techniques. The classification of different types of heteromorphisms has been based on estimation rather than actual measurements. Furthermore, variation is continuous rather than discrete. However, several attempts have been made to classify them on the basis of arbitrary scales, and different codes (levels) have been assigned. Since each banding technique provides a unique type of heteromophism, they will be described separately.

A. QFQ Heteromorphisms

The QFQ technique identifies heteromorphic bands which can be classified according to their fluorescent intensity. Different intensities are classified into five levels established at the Paris Conference (1971) with the brightest level being assigned a code of 5 and the least fluorescent a code of 1 (Table I). The bands in the centromeric regions of chromosomes 3 and 4, the short arm (bands

TABLE I
CRITERIA FOR QFQ INTENSITY LEVEL[a]

Code	Description	Comparison
1	Negative	No fluorescence
2	Pale	As in distal lp
3	Medium	As in major bands 9q
4	Intense	As in major distal 13q bands
5	Brilliant	As in distal Y

[a]Paris Conference (1971) and McKenzie and Lubs (1975).

p11 and p13) of the acrocentric chromosomes (13, 14, 15, 21, and 22), and the long arm of Y are found heteromorphic by the QFQ technique (Fig. 1).

1. *Acrocentric Chromosomes*

There are two bands (p11 and p13) in the short arm of human acrocentric chromosomes which show intensity heteromorphisms. Comparing the results with frequencies reported by several investigators, it is seen that, although there are coincidental frequencies on some chromosomes, there are substantial differences on others. Chromosome 13 was found to be the most heteromorphic (Table II), whereas chromosome 21 was the least heteromorphic.

2. *Y Chromosome*

Even before the advent of banding techniques, the length of the human Y chromosome was known to vary from person to person and from one ethnic group to another (Cohen *et al.*, 1966). The distal one- to two-thirds of the long arm of the human Y chromosome is brightly fluorescent when the QFQ technique is employed. It has been suggested that the genetically active material is located in the nonfluorescent segment and this segment has been considered invariable in size (Bobrow *et al.*, 1971; Laberge and Gagne, 1971; Knuutila and Grippenberg, 1972), whereas the genetically inactive brilliant fluorescent segment is variable in size (Paris Conference, 1971). Now, there is evidence that the length of the nonfluorescent segment is also variable (Schnedl, 1971b; Soudek *et al.*, 1973, and Verma *et al.*, 1978). Since it has been demonstrated that there is variation in the nonfluorescent segment as well as in the fluorescent segment of the Y chromosome, future investigations of Y chromosomal abnormalities should attempt to determine if variation in the nonfluorescent segment plays a role in the abnormalities.

3. *Chromosomes 3 and 4*

It can be seen from Table II that intensity heteromorphisms are quite common in the centromeric region (3q11) of chromosome 3 (range, 41–68.4%) in a normal population. Nevertheless, there is wide variation among various studies in

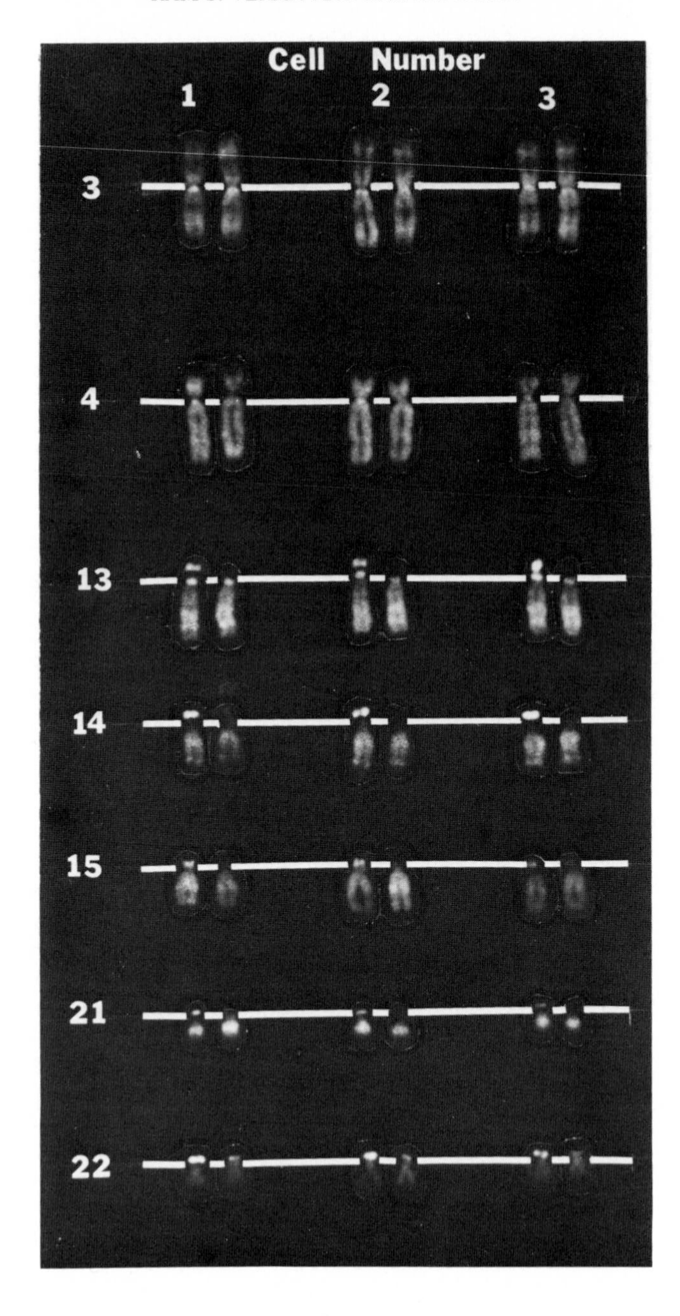

FIG. 1. Variable bright fluorescent bands in chromosomes 3, 4, 13–15, 21, and 22 by QFQ technique.

TABLE II

QFQ Heteromorphisms in Normal Individuals

Reference	Population	Sample size	Chromosome number						
			3	4	13	14	15	21	22
Verma *et al.* (1978); Verma and Dosik (1979)	Adult	100	62	14	56.5	10	10	15.5	10
Lin *et al.* (1976)	Newborn	930	55.5	14.1	33.0	1	1	1.2	0.6
Buckton *et al.* (1976)	14-year-old	109	68.4	33.5	37.6	13.3	10.5	16.5	15.1
	Newborn	482	64.9	48.3	46.8	10.3	12.5	10.3	18
McKenzie and Lubs (1975)	Newborn	77	41.0	41.0	46.7	4.5	1.3	2.6	9.0
Mikelsaar *et al.* (1975)	Adult	208	65.0	27.8	88.5	9.8	6.2	8	44.0
Muller *et al.* (1975)	Newborn	376	55.3	13.1	81.8	15.9	13.6	19.6	62.3
Geraedts and Pearson (1974)	Adult	221	48.4	2.7	50.0	14.3	21.5	24.4	21.9

the frequency of QFQ heteromorphisms (Mikelsaar *et al.*, 1974). The differences seen may be due to population differences, but they may also be due to the use of different classification systems for heteromorphisms. Using the intensity of the centromeric region of chromosome 3, it has become possible to identify inversions of the pericentromeric heterochromatin. Allderdice (1973) reported a pericentric inversion in chromosome 3 (inv (3) (p15 q12)) in two phenotypically normal persons. This observation was confirmed by Soudek *et al.* (1974). The frequency of inversion of fluorescent heterochromatin in chromosome 3 was the same in a sample of 370 retarded persons as in a sample of 222 normal individuals (4%). It can be concluded that inversion heteromorphisms of chromosome 3 are not associated with mental retardation (Soudek and Sroka 1978; Mikelsaar *et al.*, 1978).

The incidence of bright variable bands in the centromeric region of chromosome 4 is different from that in chromosome 3. Furthermore the range of variation is wider for this chromosome as compared to chromosome 3. No inversion heteromorphism is recorded for chromosome 4.

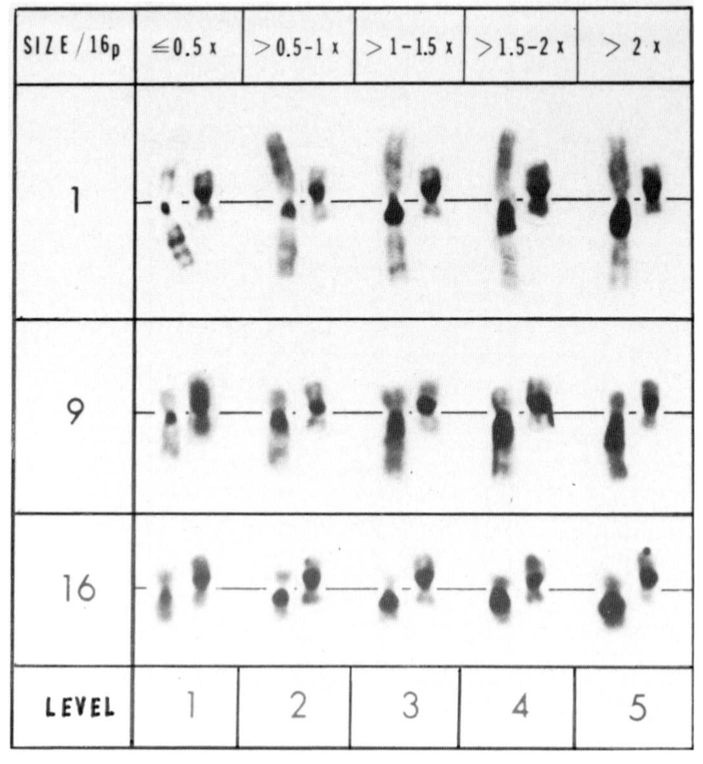

FIG. 2. Classification of secondary constriction regions (qh) of chromosomes 1, 9, and 16 by CBG technique. (Courtesy of Dr. S. R. Patil.)

B. CBG Heteromorphisms

A staining procedure by which the constitutive heterochromatin can be identified is described by Arrighi and Hsu (1971) and commonly known as C-banding. It stains the secondary constriction (h) regions of chromosomes 1, 9, and 16, the distal one- to two-thirds of chromosome Y, the centromeres of all chromosomes, and the satellites of the acrocentric chromosomes. It is well recognized that variation in the size of C-bands represents a continuous distribution, not a discrete one. These regions are also heteromorphic in their position (de la Chapelle, 1974; Hansmann, 1976; Van Dyke *et al.*, 1977, 1979, and Soudek and Sroka, 1977). Craig-Holmes *et al.* (1975) and McKenzie and Lubs (1975) classified constitutive heterochromatin regions into three size categories: normal (n), large (+), and small (−). Muller *et al.* (1975), Lubs *et al.* (1977a), and Verma *et al.* (1978) used five categories (Fig. 2); very small, small, intermediate, large, and very large (Table III). Recently we classified the inversion heteromorphisms into five categories: no inversion (NI), partial inversion-minor (MIN), half inversion (NI), partial inversion-major (MAJ), and complete inversion (CI) (Fig. 3, Table IV). Because of the different criteria used for classification of C-band heteromorphisms, a comparison of published data would be meaningless. Recently, we studied 80 normal Caucasians using the CBG technique. The frequency of size and inversion heteromorphisms is presented in Tables V and VI. The most heteromorphic chromosome was 9. Furthermore, we have suggested that there is a possible relationship between size and inversion. Employing the CBG technique, McKenzie and Lubs (1975) have shown that the centromeric region of each chromosome of the human genome is heteromorphic. Heteromorphisms of chromosome 19 (Crossen, 1975), 6, and 12 (Sofuni *et al.*, 1974) have been described. It has been presumed or expressively stated that the sites of dark C-bands and strongly fluorescent Q-bands on the long arm of the Y

TABLE III

CRITERIA FOR "SIZE" POLYMORPHISMS OF SECONDARY CONSTRICTION (h) REGIONS IN CHROMOSOMES 1, 9, AND 16

Level	Description[a]	Basis[b]
1	Very small	$\leq 0.5 \times 16p$
2	Small	$>0.5 - 1 \times 16p$
3	Intermediate	$>1\text{-}1.5 \times 16p$
4	Large	$>1.5 - 2 \times 16p$
5	Very large	$>2 \times 16p$

[a]See Paris Conference (1971); Supplement to Paris Conference (1975), and Patil and Lubs (1977).

[b]For pictorial examples with criteria descriptions, see Fig. 2.

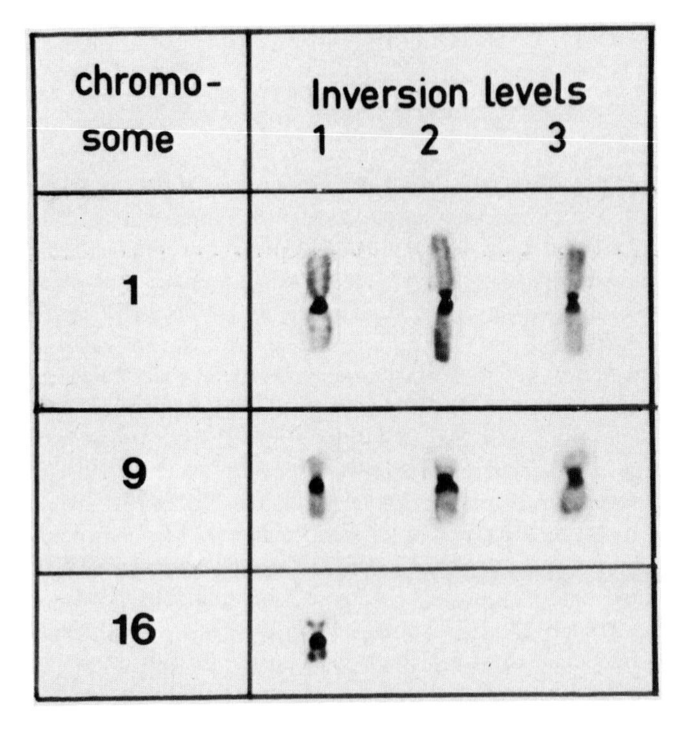

FIG. 3. Pictorial examples of different levels of inversions (see text); also see Verma *et al.* (1979).

chromosome are identical (Cervenka, 1971; Pearson *et al.*, 1973; Jalal *et al.*, 1974). This concept has been challenged by Soudek and Laraya (1976). They showed that a small Y chromosome with no Yq fluorescence displayed constitutive heterochromatin on the end of Yq, and concluded that C- and Q-bands on Yq therefore need not be necessarily identical.

TABLE IV

CRITERIA FOR "INVERSION" POLYMORPHISMS OF SECONDARY CONSTRICTION (h) REGIONS IN CHROMOSOMES 1, 9, AND 16

Level	Description	Definition[a]
1	No inversion (NI)	h region is confined to the long arm
2	Partial inversion—minor (MIN)	Less than half of h region present on the short arm
3	Half inversion (HI)	Half of h region present on the short arm and the other half on the long arm
4	Partial inversion—major (MAJ)	More than half of the h region present on the short arm
5	Complete inversion (CI)	Complete shift of h region from the long to the short arm

[a]From Verma *et al.* (1979).

TABLE V

FREQUENCIES OF DIFFERENT-SIZE POLYMORPHISMS BY SEX[a]

	Size level																
	Very small (1)				Small (2)				Intermediate (3)				Large (4)				
Sex	1	9	16	T	1	9	16	T	1	9	16	T	1	9	16	T	Total
Male	8	14	40	62	34	44	35	113	38	20	5	63		2		2	240
Female	8	25	37	70	36	40	36	112	34	14	7	55	2	1		3	240
Total (%)				132 (27.5)				225 (46.87)				118 (24.58)				5 (1.04)	480

[a]From Verma *et al*. (1979).

TABLE VI
FREQUENCIES OF DIFFERENT TYPES OF INVERSION POLYMORPHISMS BY SEX IN CHROMOSOMES 1, 9, AND 16[a]

		Inversion levels[b]		
Sex	Chromosome	1 (NI)	2 (MIN)	3 (HI)
Male	1	70	10	
	9	72	6	2
	16	80		
Female	1	74	5	1
	9	70	8	2
	16	80		
Total		446	29	5
Percentage		92.92	6.04	1.04

[a]Inversion levels which are not seen are not included here.
[b]NI, No inversion; MIN, partial inversion—minor; HI, half inversion (see Verma *et al.*, 1979).

C. RFA HETEROMORPHISMS

Heteromorphisms of the short arm of human acrocentric chromosomes by acridine orange reverse banding (RFA) have been well documented (Verma and Lubs, 1974). Color and size heteromorphisms were noted using RFA (Fig. 4,5). Color variations were classified into six different colors: red, red-orange, orange-yellow, pale-yellow, bright-yellow, and pale green. Size heteromorphisms were classified into five sizes: very small, small, average, large, and very large (Verma and Lubs, 1975b) (Tables VII, VIII). In order to establish an interrelationship between QFQ and RFA heteromorphisms, 100 normal Caucasians were studied by sequential QFQ and RFA techniques. It was concluded that there was no consistent relationship between negative or brilliant QFQ variants and the various colors observed with RFA (Verma *et al.*, 1977a,b,c). RFA color heteromorphisms for chromosomes 13, 14, 15, 21, and 22 were 33.0, 38.0, 28.0, 50.0, and 24.5%, whereas QFQ frequencies were 56.6, 10.0, 10.0, 15.5, and 10.0%, respectively (Table IX). The frequencies of size heteromorphisms using RFA for chromosomes 13, 14, 15, 21, and 22 were 22.5, 19.5, 14.5, 19, and 17% respectively (Table X, Figs. 6, 7) Furthermore, the RFA technique detected more variation in the human acrocentric chromosomes than any other method. Previous reports have suggested a very low frequency (5%) of size variation in human acrocentric chromosomes when they are stained with Giemsa (Zankl and Zang, 1971; Lubs and Ruddle, 1970; Mikelsaar *et al.*, 1973; Buckov *et al.*, 1974; Hamerton *et al.*, 1972).

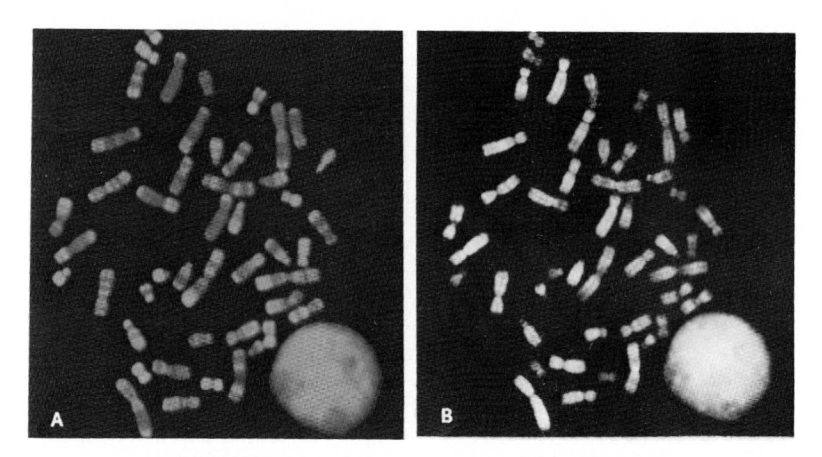

FIG. 4. (A) RFA-Banded human chromosomes. Every acrocentric chromosome is heteromorphic. This figure has been reproduced from a color slide. See Verma *et al.* (1977a). (B) QFQ-Banded chromosomes of the same metaphase shown in (A).

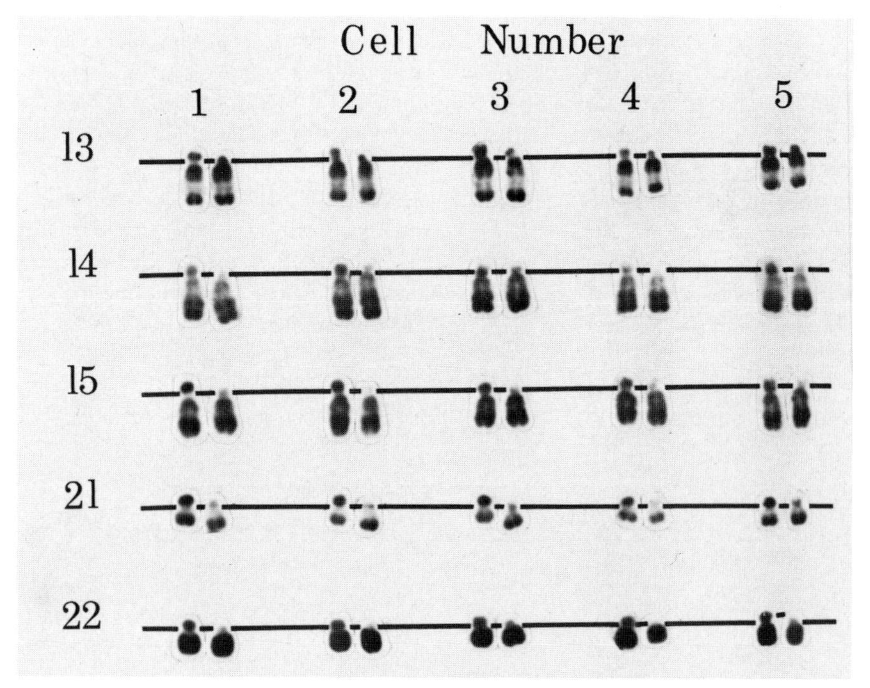

FIG. 5. Consistent homolog differences in the short arms of acrocentric chromosome by RFA technique. Chromosomes are printed in black and white from color transparencies (see Verma and Lubs, 1976).

TABLE VII
CRITERIA FOR CLASSIFICATION OF RFA HETEROMORPHISMS (COLOR)[a]

Code	Color	Like
1	Red	3c
2	Red-orange	Mid 2q(q21–q24)
3	Orange-yellow	18p
4	Pale yellow	Distal 1q(q32 and 42) (also 20p)
5	Bright yellow	Distal 1p (p3) (also 22q and 19)
6	Pale green	Present only in secondary constriction regions of acrocentric chromosomes

[a]From Verma and Lubs (1975b).

D. NOR HETEROMORPHISMS

Different techniques have recently been developed to stain the nucleolar organizer regions (NOR) of human acrocentric chromosomes. (Eiberg, 1974a,b; Denton *et al.*, 1977; Bloom and Goodpasture, 1975; Ved Brat *et al.*, 1979). The NOR technique revealed darkly stained segments (N-bands) at the regions corresponding to the satellite stalks. It has been demonstrated that the size of NOR regions in each person is constant from cell to cell, but variable between individuals (Matsui and Sasaki, 1973; Hayata *et al.*, 1977). Employing the RFA technique, it was demonstrated that the color of NOR is pale green. No color variation was noted in these regions (Verma and Lubs, 1975b); nevertheless, size variation was noted even using the RFA technique.

TABLE VIII
CRITERIA FOR CLASSIFICATION OF RFA HETEROMORPHISMS (SIZE)[a]

Code	Size	13–15	21–22
1	Very small	Virtually absent	Virtually absent
2	Small	$<0.5 \times$ 18p	$\leq 0.25 \times$ 18p
3	Average	≥ 0.5 to $1.5 \times$ 18p	$>0.25 \times$ 18p
4	Large	>1.5 to $2.0 \times$ 18p	= 18p
5	Very large	$>2.0 \times$ 18p	> 18p

[a]From Verma and Lubs (1975b).

TABLE IX

COLOR DISTRIBUTION BETWEEN MALES AND FEMALES OF RFA HETEROMORPHISMS[a]

	Acrocentric chromosome																			
	13				14				15				21				22			
Color (code)	M	F	T	%	M	F	T	%	M	F	T	%	M	F	T	%	M	F	T	%
Red (1)	2	1	3	1.5	0	2	2	1.0	1	1	2	1.0	1	0	1	0.5	2	3	5	2.5
Red-orange (2)	1	3	4	2.0	1	2	3	1.5	2	3	5	2.5	8	10	18	9.0	4	2	6	3.0
Orange-yellow (3)	66	68	134	67.0	56	68	124	62	70	74	144	72.0	22	20	42	21.0	8	12	20	10.0
Pale yellow (4)	8	7	15	7.5	11	14	25	12.5	12	8	20	10.0	47	53	100	50.0	76	75	151	75.5
Bright yellow (5)	4	5	9	4.5	5	5	10	5.0	3	0	3	1.5	8	3	11	5.5	2	3	5	2.5
Pale green (6)	20	15	35	17.5	26	9	36	18.0	12	14	26	13.0	14	14	28	14.0	8	5	13	6.5

[a]From Verma *et al.* (1978).

TABLE X

SIZE DISTRIBUTION BETWEEN MALES AND FEMALES BY RFA[a]

Size (code)		Acrocentric chromosome																			
		13				14				15				21				22			
		M	F	T	%	M	F	T	%	M	F	T	%	M	F	T	%	M	F	T	%
Very small	(1)	0	0	0	0	0	0	0	0	0	0	0	0	0	0	0	0	0	0	0	0
Small	(2)	19	23	42	21	16	16	32	16	10	5	15	7.5	8	10	18	9.0	4	7	11	5.5
Average	(3)	80	75	155	77.5	79	82	161	80.5	83	88	171	85.5	80	82	162	81.0	83	83	166	83
Large	(4)	1	2	3	1.5	5	2	7	3.5	6	7	13	6.5	12	8	20	10.0	11	9	20	10
Very large	(5)	0	0	0		0	0	0	0	1	0	1	0.5	0	0	0	0	2	1	3	1.5

[a]From Verma *et al.* (1978).

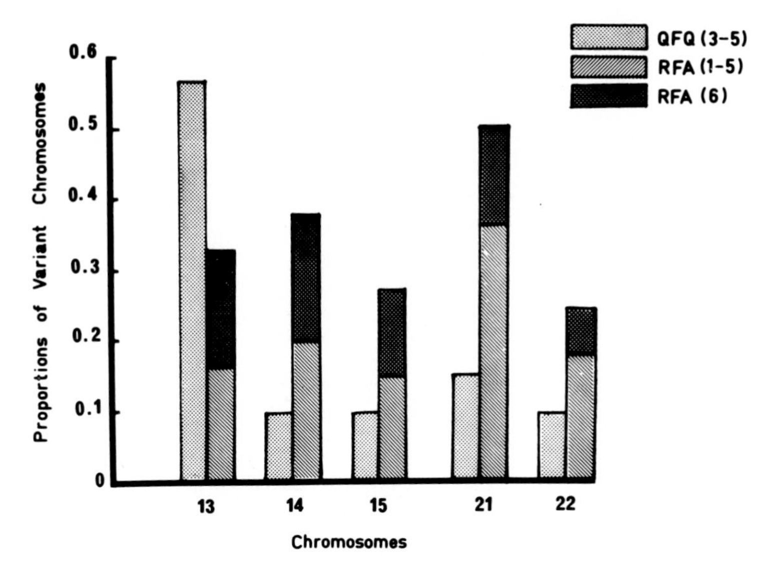

FIG. 6. Comparison of frequencies of QFQ and RFA color heteromorphisms. Color level 3 is average for D group and level 4 for G group chromosomes by RFA. Pale green (6) was seen in secondary constriction regions only (see Verma *et al.*, 1978).

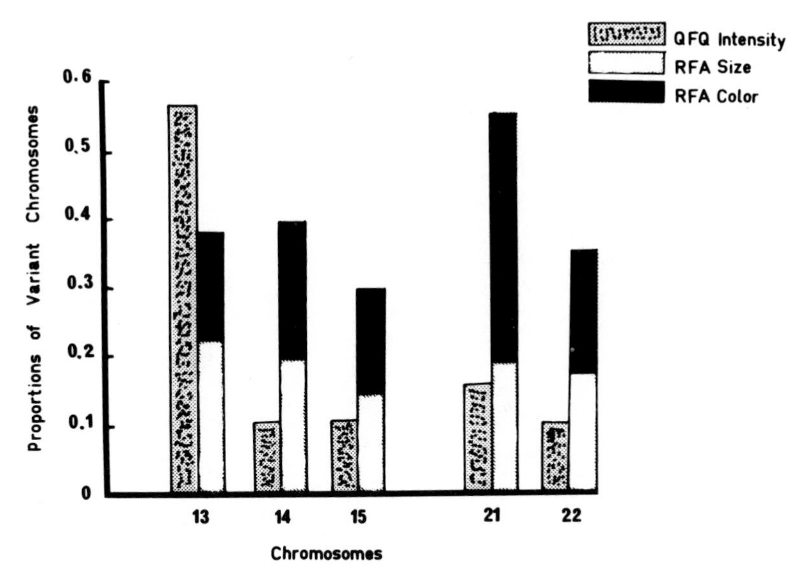

FIG. 7. The frequencies of RFA size and color heteromorphisms are combined and are compared with QFQ heteromorphisms.

III. Heritability, Segregation, Twin Zygosity, and Pitfalls

The existence of heteromorphisms can be tested by family studies. This has been demonstrated by Van Dyke *et al.* (1977) by their studies on twins. A large number of family studies indicate that heteromorphisms are inherited in a Mendelian fashion (Verma and Lubs, 1976; Magenis *et al.*, 1976; Phillips, 1977); nevertheless, a few studies have implied that apparent noninherited variation between parent and offspring, somatic mosaicism, and preferential segregation occur (Craig-Holmes *et al.*, 1975; Sekhon and Sly, 1975; Fitzgerald, 1973; and Palmer and Schroder, 1971). Nakagome *et al.* (1977) observed a huge satellite (22S+) in a mentally retarded and malformed girl. It consisted of constitutive heterochromatin on the basis of the CBG technique. The variant was not observed in either parent. Genetic studies revealed that the father was, indeed, the biological father of the proband. The noninherited variation may have arisen from a mismatching of the repetitive DNA sequences with subsequent unequal crossing over. Further, the observed mosaic patterns provide suggestive evidence that such an event occurs in somatic cells as well as during meiosis.

IV. Photometric Method for Quantification

The heteromorphic regions are evaluated by subjective criteria. Since the heteromorphic regions show continuous variation, it is usually difficult to classify them precisely. Heteromorphisms are only useful if they produce consistent results. Therefore, reproducible measurements are required. In recent years, several investigators have made an attempt to classify the heteromorphic regions with automated systems using computers (Casperson *et al.*, 1970; Van der Ploeg *et al.*, 1974; Disteche and Bontemps, 1976; Schnedl *et al.*, 1977). Preliminary results suggest that quantification of the heteromorphisms by computer has excellent prospects (Lubs *et al.*, 1976) (Fig. 8).

FIG. 8. Computerized curves of both homologs of chromosome 22 superimposed (note the enlarged area of the short arm of chromosome 22; see Lubs *et al.* (1976).

V. Structural Variation and Lateral Asymmetry

When the first banding patterns in human and other chromosomes were visualized using the QFQ technique, they were immediately attributed to major differences in the base composition of the DNA along the chromatids. Several investigators have proved that the major factor in the formation of Q-banding is the increased AT content in the brightly fluorescent bands and the relatively high GC content in the dull areas inbetween (Weisblum and Haseth, 1972). Several staining techniques have been adopted for localizing highly repetitious DNA in chromosomes (CBG: Arrighi and Hsu, 1971; G-11: Bobrow *et al.*, 1972; QFQ: Caspersson *et al.*, 1970; RFA: Verma and Lubs, 1975b). Based on staining properties heterochromatin is different from euchromatin. Consitutive heterochromatin is composed of highly repetitive DNA located in specific sites on the chromosomes, and it is associated with visible chromosome heteromorphisms. It has been well documented that every human chromosome contains one or more regions of constitutive heterochromatin which is associated with the centromere or the short arm (usually acrocentric chromosomes). The heterochromatin is variable in size, position, and color depending upon the staining technique used. Recently, Lin *et al.* (1974) and Lin and Alfi (1976, 1978) have demonstrated the lateral asymmetry in the secondary constriction regions of chromosome 1 using

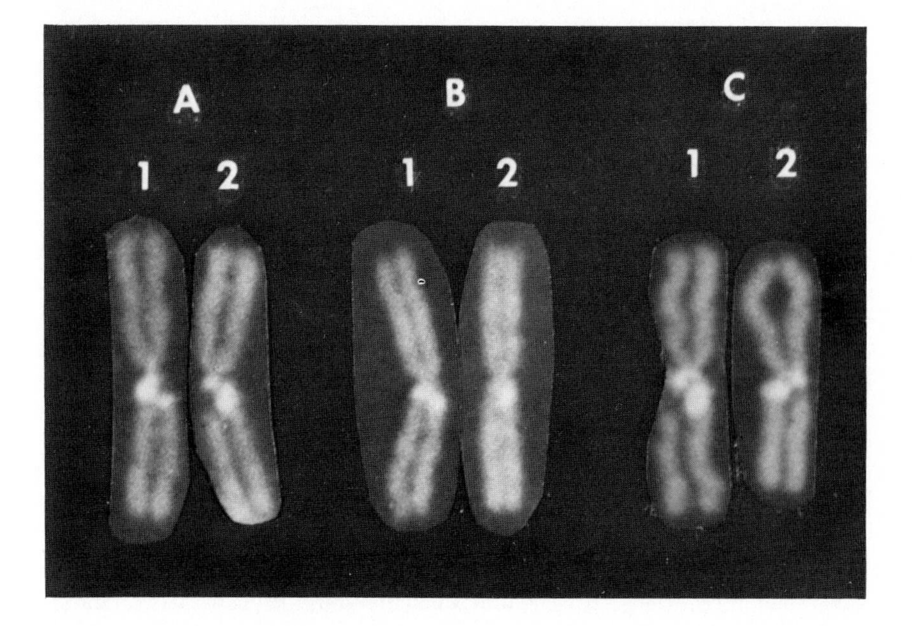

FIG. 9. Lateral asymmetry of chromosome 1, from three individuals. (Courtesy of Dr. M. S. Lin.)

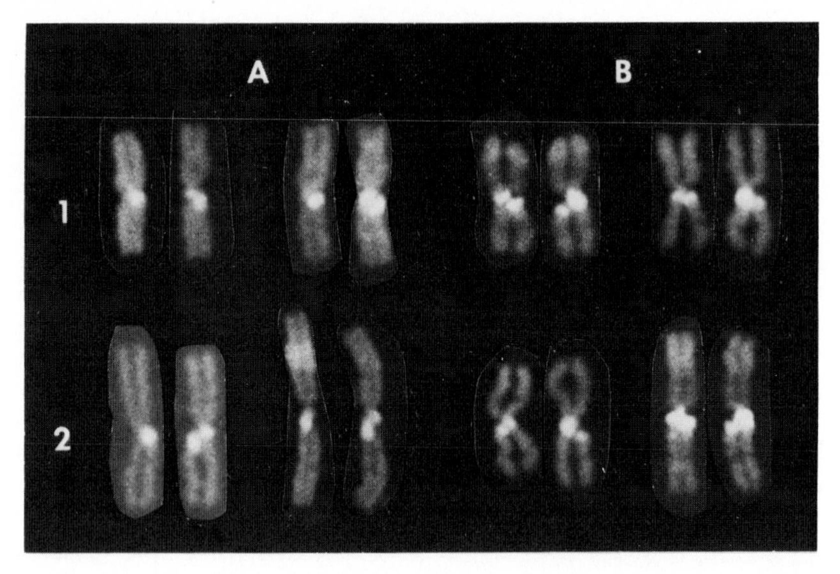

FIG. 10. Consistency of homolog differences in lateral asymmetry heteromorphisms of chromosome 1 in replication studies from two subjects. (Courtesy of Dr. M. S. Lin.)

the BrdU-DAP1 technique. Their results suggest a high frequency of asymmetry of heterochromatin in the population (Figs. 9,10). Furthermore, their results support the unequal distribution of thymine residues in the two chains of satellite DNA in the 1qh regions of human chromosomes. Lateral asymmetry in C-band heterochromatin regions of chromosomes 1, 15, 16, and the Y have also been demonstrated by Angell and Jacobs (1975). The short arms of human acrocentric chromosomes have been shown to be the sites of gene coding for 18 S and 28 S ribosomal RNA (rRNA) (Henderson *et al.*, 1972; Evans, 1973; and Evans *et al.*, 1974). The heteromorphic nature of ribosomal cistrons in human acrocentric chromosomes is not known. Nevertheless, it has been estimated that approximately 440 copies of the rRNA genes are distributed among the human acrocentric chromosomes (Bross and Krone, 1972; Bross *et al.*, 1973; Schmid and Krone 1974; Dittes *et al.*, 1975).

VI. Application of Heteromorphisms

Chromosomal heteromorphisms found in human chromosomes provide a useful tool for several studies, since they are inherited in a Mendelian fashion, are stable, and are presumed to have a low mutation rate. By using heteromorphic markers of chromosome 21, the origin of the extra chromosome in Down syndrome can be determined (Licznerski and Lindsten, 1972; Robinson, 1973; Wagenbichler *et al.*, 1976; Mikelsen *et al.*, 1976; and Verma and Dosik, 1978). The mechanism producing triploidy in human abortuses has been studied (Jonasson *et al.*, 1972; Lauritsen, 1976). The origin of additional chromosomes by the use of heteromorphisms has recently been discussed by Langenbeck *et al.* (1976), Jacob and Morton (1978), and Kajii and Nikawa (1977). Maternal cell contamination of amniotic fluid culture can be detected (Hauge *et al.*, 1975). Heteromorphisms have been used to establish paternity (Schnedl, 1974; Schwinger, 1974; de la Chapelle *et al.*, 1967; and Weleber *et al.*, 1976). This has also been done prenatally (Jonasson *et al.*, 1972).

Other important applications of heteromorphisms as markers will be in elucidating the chromosomal mechanisms involved in the production of mosaics, in studying chimeras, and in following the fate of transfused or transplanted cells. Their use in zygosity testing in twins and other multiple births has great application. Furthermore, heteromorphic variants could be used much more extensively in gene mapping in somatic cell hybridization (Donahue *et al.*, 1968; and Magenis *et al.*, 1970).

VII. Clinical Significance

The biological and clinical significance of chromosomal heteromorphisms is poorly understood. Nevertheless, the literature suggests that there is an association between one or more particular heteromorphisms and clinical abnormalities (Boue et al., 1975). Lubs and Ruddle (1970) found a significant increase in giant satellites in a G group chromosome in children who had a major congenital anomaly detected at birth in comparison with those who did not. Furthermore, Lubs *et al.* (1976) have concluded that QFQ and CBG heteromorphisms have no clinical significance with respect to the IQ of children. In a number of studies, it has been suggested that carriers of extreme heteromorphisms may produce children with chromosomal abnormalities (Halbrecht and Shabtay, 1976). Nielsen *et al.* (1974) found a higher frequency of 9qh+ heteromorphisms in the parents of children with major chromosome abnormalities than in a random sample of newborn babies. Jacob *et al.* (1975) have suggested a lowered fertility in individuals who are carriers of extreme heteromorphisms. Furthermore, some investigators have found a longer Y chromosome in criminals (Soudek and Laraya, 1974), whereas others have not found any length differences between criminals and noncriminal controls (Benezech *et al.*, 1976). Patil and Lubs (1977) have suggested that the long Y chromosome is an important cause of fetal loss. In summary, the clinical significance of heteromorphisms requires further testing in carefully controlled studies.

VIII. Conclusion

Several banding techniques have not only helped in the precise identification of human chromosomal abnormalities, but have led to a great understanding of heteromorphic areas in the human genome. Many reported cases were thought to be abnormal using conventional techniques, but have proved to be heteromorphic normal variants. There are several problems in quantifying these heteromorphisms because of the technical variables associated (Verma and Dosik, 1979). The mechanism of the origin of heteromorphisms is poorly understood and the data on mutation rates are controversial. In general, they are inherited from generation to generation and have valuable applications in clinical medicine.

ACKNOWLEDGMENT

Supported by National Cancer Institute Contract NO1-CP-43251.

REFERENCES

Allderdic, P. W. (1973). *Am. J. Hum. Genet.* **25,** 11a.
Angell, R. R., and Jacobs, P. A. (1975). *Chromosoma* **51,** 301.
Arrighi, F. E., and Hsu, T. C. (1971). *Cytogenetics* **10,** 81.
Benezech, M., Nöel, E., Travers, E., and Mottet, J. (1976). *Hum. Genet.* **32,** 77.
Bloom, S. E., and Goodpasture, C. (1976). *Hum. Genet.* **34,** 199.
Bobrow, M., and Madan, K. (1974). *Ann. Genet.* **17,** 81.
Bobrow, M., Pearson, P. L. Pike, M. C., and El Alfi, O. S. (1971). *Cytogenetics* **10,** 190.
Bobrow, M., Madan, K., and Pearson, P. L. (1972). *Nature (London) New Biol* **238,** 122.
Bochkov, N. P., Kuleshov, N. P., Chebotaren, A. N., Alekhin, V. I., and Michian, S. A. (1976). *Humangenetik* **22,** 139.
Boue, J., Taillemite, J. L., Hazael-Massieux, P., Leonard, C., and Boue A. (1975). *Humangenetik* **30,** 217.
Bross, K., and Krone, W. (1972). *Humangenetik* **14,** 137.
Bross, K., Dittes H., Krone, W., Schmid, M., and Vogel W. (1973). *Humangenetik* **20,** 223.
Buckton, K. E., O'Riordan, M. L., Jacobs, P. A., Robinson, J. A., Hill R., and Evans, H. J. (1976). *Ann. Hum. Genet.* **40,** 99.
Casperson, T., Zech, L., and Johansson, C. (1970). *Exp. Cell Res.* **62,** 490.
Cervenka, J. (1971). *Clin. Genet.* **2,** 275.
Cohen, M. M., Shaw, M. W. And McCluer, J. W. (1966). *Cytogenetics* **5,** 34.
Craig-Holmes, A. P., Moor, F. B., and Shaw, M. W. (1973). *Am. J. Hum. Genet.* **25,** 181.
Craig-Holmes, A. P., Moore, F. B., and Shaw, M. W. (1975). *Am. J. Hum. Genet.* **27,** 178.
Crossen, P. E. (1975). *Clin. Genet.* **8,** 218.
de la Chapelle, A., Fellman, J., and Unnerus, V. (1967). *Ann. Genet.* **10,** 60.
de la Chapelle, A., Schroder, J., Stenstrand, K., Fellman, J., Herva, R., Saarni, M., Anttolainen, P. J. L, Tallila, I., Tervila, L., Husa, L., Tallqvist, G., Robson, E. B., Cook, P. J. L., and Sanger, R. (1974). *Am. J. Hum. Genet.* **26,** 746.

Denton, T. E., Brooke, W. R., and Howell, W. M. (1977). *Stain Technol.* **52,** 311.
Disteche, C., and Bontemps, J. (1976). *Chromosoma* **54,** 39.
Dittes, H., Krone, W., Bross, K., Schmidt, M., and Vogel, W. (1975). *Humangenetik* **26,** 47.
Donahue, R. P., Bias, W. B., Renwick, J. H., and McKusick, V. A. (1968). *Proc. Natl. Acad. Sci U.S.A.* **61,** 494.
Eiberg, H. (1974a) *Lancet* **II,** 836.
Eiberg, H. (1974b). *Nature (London),* **248,** 55.
Evans, H. J. (1973). *Br. Med. Bull.* **29,** 196.
Evans, H. J., Buckland, R. A., and Pardue, M. L. (1974). *Chromosoma* **48,** 405.
Fitzgerald, P. H. (1973). *Cytogenet. Cell Genet.* **12,** 404.
Ford, E. B. (1940). *In* "Polymorphism and Taxanomy. The New Systematics" (J. Huxley, ed.), p. 493-513. Oxford Univ. Press (Clarendon), London and New York.
Geraedts, J. P. M., and Pearson, P. L. (1974). *Clin. Genet.* **6,** 247.
Goodpasture, C., and Bloom, S. E. (1975). *Chromosoma* **53,** 37.
Halbrecht, I., and Shabtay, F. (1976). *Clin. Genet.* **10,** 113.
Hamerton, J. L., Ray, M., Abbot, A., Williamson, C. H., and Ducasse, G. C. (1972). *Can. Med. Assoc. J.* **106,** 776.
Hansmann, I. (1976). *Hum. Genet.* **31,** 247.
Hauge, M., Poulsen, H., Halberg, A., and Mikkelsen, M. (1975). *Humangenetik* **26,** 187.
Hayata, I., Oshimura, M., and Sandberg, A. A. (1977). *Hum. Genet.* **36,** 55.
Henderson, A. S., Warburton, D., and Atwood, K. C. (1972). *Proc. Natl. Acad. Sci. U.S.A.* **69,** 3394.
Jacob, P. A. (1977). *Prog. Med. Genet.* **2,** 251.
Jacob, P. A., Frackiewicz, A., Law, P., Hilditch, E. J., and Morton, N. E. (1975). *Clin. Genet.* **8,** 169.
Jacob, P. A., and Morton, N. E. (1978). *Hum. Hered.* (in press).
Jalal, S. M., Pfeiffer, R. A., Pathak, S., and Hsu, T. C. (1974). *Humangenetic* **24,** 59.
Jonasson, J., Therkelsen, A. J., Lauritsen, J. G., and Lindsten, J. (1972). *Hereditas* **71,** 168.
Kajii, T., and Niikawa, N. (1977). *Cytogenet. Cell Genet.* **18,** 109.
Knuutila, S., and Grippenberg, V. (1972). *Hereditas* **70,** 307.
Laberge, C., and Gagne, R. (1971). *Johns Hopkins Med. J.* **128,** 79.
Langenbeck, V., Hansmann, I., Hinney, B., and Honig, V. (1976). *Hum. Genet.* **33,** 89.
Lauritsen, J. G. (1976). *Acta Obstet Gynecol. Scand.* **52,** 1.
Licznerski, G., and Lindsten, J. (1972). *Hereditas* **70,** 153.
Lin, M. S., and Alfi, O. S. (1976). *Chromosoma* **57,** 219.
Lin, M. S., and Alfi, O. S. (1978). *Cytogenet. Cell Genet.* **21,** 243.
Lin, M. S., Latt, S. A., and Davidson, R. L. (1974). *Exp. Cell Res.* **86,** 392.
Lin, C. C., Gedeon, M. M., Griffith, P., Smink, W. K., Newton, D. R., Wilkie, L., and Sewell, L. M. (1976). *Hum. Genet.* **31,** 315.
Lubs, H. A. (1977). *In* "Occurence and Significance of Chromosome Variants. Molecular Human Cytogenetics" (R. S. Sparks, D. E. Comings, and C. F. Fox, eds.), pp. 443-455. Academic Press, New York.
Lubs, H. A., and Ruddle, F. H. (1970). *Pfizer Med. Monogr.* **119,** 142.
Lubs, H. A., Verma, R. S., and Ledley, R. S. (1976). *Chromosomes Today* **5,** 429.
Lubs, H. A., Patil, S. R., Kimberling, W. J., Brown, J., Cohen, M., Gerald, P., *et al.* (1977a). *In* "Population Cytogenetics" (E. B. Hook, and I. H. Porter, eds.), pp. 133-159. Academic Press, New York.
Lubs, H. A., Kimberling, W. J., Hecht, F., Patil, S. R., Brown, J., Gerald, P., and Summitt, R. L. (1977b). *Nature (London)* **268,** 631.
Magenis, R. E., Hecht, F., and Lovrien, E. W. (1970). *Science* **170,** 85.

Magenis, E., Palmer, C. G., Wang, L., Brown, M., Chamberlin, J., Parks, M., Merritt, A. D., Rivas, M., and Yu, P. L. (1976). *In* "Population Cytogenetics" (E. B. Hook and I. H. Porter, eds.), pp. 179–188. Academic Press, New York.

Matsui, S., and Sasaki, M. (1974). *Nature (London)* **246,** 148.

Mikelsaar, A-V. N., Tuur, S. J., and Kaosaar, M. E. (1973). *Humangenetik* **20,** 89.

Mikelsaar, A-V. N., Viikmaa, M. H., Tuur, S. J., and Kaosaar, M. E. (1974). *Humangenetik* **23,** 59.

Mikelsaar, A-V. N., Kaosaar, M. E., Tuur, S. J., Vikmaa, M. H., Talvik, T. A., and Laats, J. (1975). *Humangenetik* **26,** 1.

Mikelsaar, A-V. N., Illus, T., and Kivi, S. (1978). *Hum. Genet.* **41,** 109.

Mikkelsen, M., Hallberg, A., Poulsen, H. (1976). *Hum. Genet.* **32,** 17.

2Muller, H. J., Klinger, H. P., and Glasser, M. (1975). *Cytogenet. Cell Genet.* **15,** 239.

Nakagome, Y., Kitagawa, T., Iinuma, K., Matsunaga, E., Shinoda, T., and Ando, T. (1977). *Hum. Genet.* **27,** 255.

Nielsen, J., Friedrich, U., Hreidarsson, A. B., and Zeuthen, E. (1974). *Clin. Genet.* **5,** 316.

Palmer, C. G., and Schroder, J. (1971). *J. Med. Genet.* **202.**

Paris Conference (1971). Standardization in Human Cytogenetics: Birth Defects. Original Article Ser. VIII. The National Foundation, New York.

Paris, Conference (1975). (Supplement) Standardization in Human Cytogenetics: Birth Defects. Original Article Ser. XI, p. 9. The National Foundation, New York.

Patil, S. R., and Lubs, H. A. (1977). *Hum. Genet.* **38,** 35.

Pearson, P. L., Geraedts, J. P. M., and Vander-Linden, A. G. J. M. (1973). *In* "Modern Aspects of Cytogenetics: Constitutive Heterochromatin in Man, pp. 201–213. Stuttgart-New York.

Phillips, R. B. (1977). *Can. J. Genet. Cytol.* **19,** 405.

Ploeg, M., Vander Duijn, P., van Ploem, J. S. (1974). *Histochemistry* **42,** 9.

Robinson, J. (1973). *Lancet* **I,** 131.

Robinson, J. A., Buckton, K. E., Spowart, G., Newton, M., Jacobs, P.A., Evans, H. J., and Hill, R. (1976). *Ann. Hum. Genet.* **40,** 113.

Schmid, M., and Drone, W. (1974). *Humangenetik* **23,** 267.

Schnedl, W. (1971a). *Hum. Genet.* **12,** 188.

Schnedl, W. (1971b). *Hum. Genet.* **12,** 59.

Schnedl, W. (1974). *Int. Rev. Cytol. Suppl.* **4,** 237.

Schnedl, W., Roscher, V., and Czaker, R. (1977). *Hum. Genet.* **35,** 185.

Schwinger, E. (1974). *Beitr. Gerichtl. Med.* **32,** 163.

Sekhon, G. S., and Sly, W. S. (1975). *Am. J. Hum. Genet.* **27,** 79a.

Sofuni, T., Tanabek, K., Ohtaki, K., Shimba, H., and Awa, A. A. (1974). *Jpn. J. Hum. Genet.* **19,** 251.

Soudek, D., and Laraya, P. (1974). *Clin. Genet.* **6,** 225.

Soudek, D., and Laraya, P. (1976). *Hum. Genet.* **32,** 339.

Soudek, D., and Sroka, H. (1977). *Clin. Genet.* **12,** 285.

Soudek, D., and Sroka, H. (1978). *Hum. Genet.* **44,** 109.

Soudek, D., and Sroka, H. (1979). *Hum. Genet.* (in press).

Soudek, D., Langmuir, V., and Stewart, D. J. (1973). *Hum. Genet.* **18,** 285.

Soudek, D., O'Shaughnessy, S., Laraya, P., and McCreary, B. D. (1974). *Hum. Genet.* **22,** 343.

Van Dyke, D. (1977). *Mamm. Chrom. Newslett.* **18,** 85.

Van Dyke, D. L., Palmer, C. G., Nance, W. E., and Yu, P. L. (1977). *Am. J. Hum. Genet.* **29,** 431.

Van Dyke, D. L., Palmer, C. G., Nance, W. E., and Yu, P. L. (1979). *Am. J. Hum. Genet.* **29,** 431.

Ved Brat, S., Verma, R. S., and Dosik, H. (1979). *Stain Tech.* **54,** 107.

Verma, R. S., and Dosik, H. (1978). *Jpn. J. Hum. Genet.* **23,** 17.

Verma, R. S., and Dosik, H. (1979). *Can. J. Genet. Cytol.* **21,** 109.

Verma, R. S., and Dosik, H. (1979). *Clin. Genet.* **15,** 450.
Verma, R. S., and Lubs, H. A. (1974). *Mamm. Chrom. Newslett.* **16,** 35.
Verma, R. S., and Lubs, H. A. (1975a). *Am. J. Hum. Genet.* **27,** 110.
Verma, R. S., and Lubs, H. A. (1975b). *Hum. Genet.* **30,** 225.
Verma, R. S., and Lubs, H. A. (1976). *Hum. Hered.* **26,** 315.
Verma, R. S., Dosik, H., and Lubs, H. A. (1977a). *J. Hered.* **64,** 262.
Verma, R. S., Dosik, H., and Lubs, H. A. (1977b). *Ann. Hum. Genet. (London)* **41,** 257.
Verma, R. S., Dosik, H., and Lubs, H. A. (1977c). *Hum. Genet.* **38,** 231.
Verma, R. S., Dosik, H., Scharf, T., and Lubs, H. A. (1978). *J. Med. Genet.* **15,** 227.
Verma, R. S., Dosik, H., and Lubs, H. A. (1979). *Am. J. Med. Genet.* **2,** 331.
Wagenbichler, P., Killian, W., Rett, A., and Schnedl, W. (1976). *Hum. Genet.* **32,** 13.
Weisblum, B., and Haseth, P. L. de (1972). *Proc. Natl. Acad. Sci. U.S.A.* **69,** 629.
Weleber, R. G., Verma, R. S., Kimberling, W. J., Fieger, H. G., and Lubs, H. A. (1976). *Ann. Genet.* **19,** 241.
Zankl, H., and Zang, D. (1971). *Humangenetik* **13,** 160.

Subject Index

I

M

N

P

S

T

Contents of Previous Volumes

Volume 4

Volume 5

Volume 6

Volume 7

Volume 8

Volume 9

Volume 10

Volume 11

Volume 12

Volume 13

Volume 14

Volume 15

Volume 16

Volume 17

Volume 18

Volume 19

Volume 20

Volume 21

Volume 22

Volume 23

Volume 24

Volume 25

Volume 26

Volume 27

Volume 28

Volume 29

Volume 30

Volume 31

Volume 32

Volume 33

Volume 34

Volume 35

Volume 36

Volume 37

Volume 38

Volume 39

Volume 40

Volume 41

Volume 42

Volume 43

Volume 44

Volume 45

Volume 46

Volume 47

Volume 48

Volume 49

Volume 50

Volume 51

Volume 52

Volume 53

Volume 54

Volume 55

Volume 56

Volume 57

Volume 58

Volume 59

Volume 60

Volume 61